Junkers Ju 388

Junkers Ju 388

Development, Testing, and Production of the Last Junkers High-Altitude Aircraft

Christoph Vernaleken and Martin Handig

Schiffer Military History
Atglen, PA

Translated from the German by David Johnston
Book Design by Ian Robertson.

Library of Congress Catalog Number: 2006923637

Printed in China.
ISBN: 0-7643-2429-2

This book was originally printed under the title,
Junkers Ju 388 by Aviatic Verlag

We are interested in hearing from authors with book ideas on related topics.

Published by Schiffer Publishing Ltd.
4880 Lower Valley Road
Atglen, PA 19310
Phone: (610) 593-1777
FAX: (610) 593-2002
E-mail: Info@schifferbooks.com.
Visit our web site at: www.schifferbooks.com
Please write for a free catalog.
This book may be purchased from the publisher.
Please include $3.95 postage.
Try your bookstore first.

In Europe, Schiffer books are distributed by:
Bushwood Books
6 Marksbury Avenue
Kew Gardens
Surrey TW9 4JF, England
Phone: 44 (0) 20 8392-8585
FAX: 44 (0) 20 8392-9876
E-mail: Info@bushwoodbooks.co.uk.
Visit our website at: www.bushwoodbooks.co.uk
Free postage in the UK. Europe: air mail at cost.
Try your bookstore first.

Contents

Foreword and Acknowledgments

To our knowledge, this book is the first published attempt at a comprehensive monograph covering the development, testing, and production of the Junkers Ju 388. It is also dedicated to the fate of the machines captured by the Allies, and to the story of the last surviving machine. In view of the, in some respects, extensive surviving materials concerning production and testing of the Ju 388, it is actually amazing that no previous attempt has been made to document its history.

Almost 60 years have passed since the Ju 388's first flight. After such a long time it is impossible to expect to be able to compile a history of this aircraft that is complete in all respects. During the search for fragments of this history we have visited numerous archives in Germany and abroad. The documents used for this book come from the holdings of the special collections of the following: Deutsches Museum in Munich (DMM); the Archives Division of the National Air and Space Museum (NASM) in Washington, DC; the Bundesarchiv-Militärarchiv in Freiburg (BA-MA); the National Archives (NARA), Washington, DC; the archive of the Imperial War Museum (IWM), London; the Public Records Office (PRO), Kew; the EADS Unternehmensarchiv Ottobrunn; the historical archive of BMW AG; the factory archive of MTU AeroEngines AG; and the Landeshauptarchiv Sachsen-Anhalt, Oranienbaum. We are extremely grateful for the support of these archives, without which this book would not have been possible. For their generous assistance, especially with photographic material, we wish to thank Hans-Ullrich Willbold and Wolfgang M¸hlbauer of the EADS Unternehmensarchiv.

There are also considerable stores of documents in private archives, gathered partly from estates, partly in years-long research from the named and other archives. Here our special thanks go to Manfred Griehl, who always had an open ear for our questions, and supported us with numerous photos and documents from his archive. Invaluable assistance was provided by Peter Achs, whose extracts from the log books of Junkers company pilots provided an outstanding insight into the testing activities at Dessau and Merseburg. Oliver Thiele not only assisted us with important documents, but also gave us valuable information about the whereabouts of monthly and weekly reports from the former testing stations. Art Medcalf did not hesitate to loan us several microfilms from across the big pond; Peter Petrick, Barry Rosch, G¸nther Ott, and James V. Crow supplied us with photos. We are very grateful to Stephen Ransom for his help in dating Junkers company photographs. We would also like to thank Ryan Bettag, Ulf Balke, Heiner R. Baltabol, Reinhold Brasack, Dr. Lutz Budrafl, Christy Butler, Sven Carlsen, Carl E. Charles, Tom Davidson, Tom Dietz, Udo Hafner, Thomas Hampel, Rainer Hˆnle, Matthias Jens, Steffen Kunze, Russ Lee, Alfred Misna, Christoph Mˆller, Dan O'Connell, Christoph Regel, Friedemann M. Schell, Paul L. Tremblay, and Dave Williams for their kind support.

Since many of those who took part in the development of the Ju 388 have already died, the book includes few accounts by eyewitnesses. Nevertheless, we have Dietmar Kuropka and Wolfgang Sommerfeld of the "Arbeitskreis Junkerswerke und Fliegerhorst Bernburg" to thank for contributions by several former Junkers employees.

For the color side-views we thank Simon Schatz of Vienna who, basing his work on original Junkers drawings and photographs, undertook the attempt to accurately depict in color aircraft previously seen only in contemporary black and white photographs.

Finally, we would like to thank our families and friends who encouraged us to continue our work, and who showed understanding when we often had little time for them. Deserving of special mention are Barbara, Andreas, and Norbert Vernaleken, as well as Hans Handig and Susanne Skol for their proofreading of the manuscript.

The high reproduction costs and publication fees that most public archives charge for photographs represented a major problem. It was not our intention to produce a picture book; however, a monograph about a *Luftwaffe* aircraft without photographs to illustrate the text is unthinkable. In any case, photos of a type of which only about 75 to 100 examples were built, and which saw virtually no action with regular units are not as abundant as in the case of the Messerschmitt Bf 109, for example. Nevertheless, the resources that are generally available in public archives, in some cases after decades, are astounding. For example, they contain original Junkers factory albums with photographs of outstanding quality. Unfortunately, for reasons of cost we have only been able to use a small selection of these photos—many detail photos must remain unpublished. All those interested are therefore urged to visit the archives themselves, to see the relevant material, and appeal to those responsible for its publication in order to bring about a change in the situation.

Many of the photos presented here have been published before; however, we do claim to have correctly identified the majority of the photos in this book for the first time. In previous publications a Ju 388 L-0 with the manufacturer's code DW+YY has usually been wrongly identified as the "Ju 388 V1" and, for lack of careful research, many "experts" identify the Ju 388 V8, manufacturer's code PG+YB, as the V3.

In writing this book we faced the task of presenting the most complete picture possible, based on surviving documents, and fragments of the Ju 388's history. The required detective work was difficult, but it also brought us much joy. In order to make it clear to the reader where our information came from, we have listed our sources in the footnotes. While the first chapters include many references to other printed sources, later chapters are based almost exclusively on original documents from the named archives and collections. For documents bearing a U.S. stamp for microfilming, we have provided the reel (R) and frame (F) for the appropriate microfilm (MF), even if our copy came from the Deutsches Museum, for example, because the identification of the microfilm allows the interested reader a quick and efficient method of accessing the sources used.

Not all of the chapters of the Ju 388 story are equally well documented. Because of gaps in the available documentation, in some cases we were forced to use conjecture. We hope that we have clearly identified these as such to the reader. The focal points of our account, resulting from the materials available, are the prototypes and series production, plus the question of power plants and the associated extensive testing. The multifaceted story of the captured aircraft concludes the story. It also seemed worthwhile to us to provide comparisons of selected contemporaries of the Ju 388, and to provide in brief some details that are little known.

We hope that our first work provides a complete picture of the last Junkers high-altitude aircraft, even if it cannot be excluded that previously unknown material on the Ju 388 may surface in the future. If we have succeeded in this, we will have achieved the main goal of our joint effort.

In all our enthusiasm for historical aircraft, it must not, however, be forgotten that the aircraft described here was developed and produced as a weapon of war under the National-Socialist regime, and that forced labor, prisoners of war, and concentration camp inmates took part in its production.

On the internet we continue to maintain our website (http//www.Ju388.de), from which this book came. On our website we will continue to publish new and supplementary information about the Ju 388, as well as corrections to our book. The easiest way to submit feedback (praise, constructive criticism, etc) on this book is to send an email to webmaster@Ju388.de. We welcome all correspondence.

Vienna and Eiterfeld, September 2003

Chapter 1: Background

The *Luftwaffe* had great plans for the Ju 388. In the new aircraft plan adopted on 8 July 1944 (Delivery Plan 226), the Ju 388 was one of the five remaining types (with the Me 262, Ar 234, Do 335, and Ta 152), which were to fill all roles. Production of all other aircraft types was canceled.[1] Approximately 550 Ju 388s and 525 Do 335s were to be produced each month. The bulk of these would be bombers, and together with the 500 Ar 234s to be built each month, they were expected to result in a qualitative and quantitative leap forward for the bomber arm.[2]

The Ju 388 was a development of the Junkers Ju 188, which in turn was based on the famous Ju 88. The Ju 388's most significant innovation compared to its predecessors was that it was conceived as a fast, high-flying aircraft with a pressurized cockpit and high-altitude engines. The Ju 388 K high-altitude bomber and the Ju 388 L reconnaissance aircraft were to avoid interception by flying at altitudes that the enemy's air defenses could not reach. The Ju 388 J night fighter and heavy fighter variant was intended to intercept enemy aircraft operating at high-altitude.

The basic idea behind this development was to initiate mass production of a high-speed, high-altitude aircraft, using major components from a type already in production (Ju 88/188) to save time and minimize production preparation and start-up costs. The Ju 388 was therefore designed to achieve its best performance at the power plants' maximum boost altitude. In contrast to other high-altitude aircraft, which were designed to fly as high as possible, the Ju 388 was to fly as fast as possible at its operational altitude. Consequently, the extended wings seen on most high-altitude types were dispensed with.[3]

Factory photograph of a presentation model of the Ju 388 J- night fighter and high-altitude *Zerstörer*, taken at about the end of August 1943. Note the FHL 131 Z remotely controlled tail turret with two MG 131s, antennas for the FuG 220, and the splinter camouflage scheme used in 1943. The rudder and propellers are different than the later production versions. Griehl Collection

As early as mid-1942, the *Luftwaffe* had begun pressing for the soonest possible development of a high-speed, high-altitude bomber with which to attack "vital targets in the British motherland." Attacks on such targets with existing equipment were given little chance of success. This information is contained in the *Luftwaffe* Chief of Staff's response to Program Study 1011 of July 1942.[4] According to this, Hitler had been demanding high-speed, high-altitude bombers since 1940, but in the summer of 1942 the general staff had neither "technical proposals" (meaning concrete aircraft design proposals), nor tactical-technical development guidelines. Another important factor in the renewed pressure for high-altitude aircraft late in the war was the fear on the part of the *Luftwaffe* command and State Aviation Ministry (RLM) that Germany would be unable to cope with attacks by American high-altitude bombers like the Boeing B-29. As early as 3 August 1943, at a conference held by the Chief of Aircraft Procurement and Supply, it was speculated that American aircraft would soon be able to attack from heights as great as 10000 meters, and that neither the Me 209 nor Me 410 would be able to do anything about it. A version of the Ju 188 equipped with BMW 801 J engines was seen as the only type capable of intercepting high-altitude bombers, and at that time it was already clear to those responsible that "in light of this, the entire planned production program was ill-conceived." The anticipated arrival in Europe of the B-29 not only provided a clear impulse for development of the Ju 188 into the Ju 388 high-altitude aircraft, it also served as "justification" for the type almost until the end of the war. The following revealing statement is found in a report by the Heinkel-Hirth-Motoren GmbH, dated 13 November 1944:

"Once the B-29 enters service, which is expected to happen between the end of this year and the beginning of next year, the Ju 388 will be the principal interceptor used against this aircraft, because the type's expected operating altitude is 12 km, and at present the Ju 388 is the only aircraft that can operate at that altitude with a sufficiently powerful armament and suitable radius of action.[5]"

In fact, as early as 1941 the USAAF was planning to deploy extremely long range, high-altitude bombers. The Consolidated-Vultee B-36 was conceived as a much larger and more powerful successor to the Boeing B-29 and Consolidated B-32. These plans for an intercontinental strategic bomber were probably in response to the threat to Great Britain posed by the Third Reich, and the growing tension in the Pacific. A mock-up of the B-36 was inspected that same year.

The Boeing B-29, Lockheed B-30[7], Douglas B-31, Consolidated B-32,[8] and Martin XB-33A[9] "Super Marauder," all four-engine designs, grew out of a specification for a "very long range aircraft" issued by the USAAF in September 1940. In principle, all of these types were supposed to replace the Boeing B-17 and Consolidated B-24 in 1944; however, only the B-29 and B-32 reached the prototype stage. It was planned to first equip units of the 15th Air Force in the Mediterranean Theater with the B-32, followed by those of the 8th Air Force, or "Mighty Eighth."

View from the cockpit of the Ju 388 that underwent extensive testing in the USA. Visible in the distance are B-29s. Rosch Collection

The discussions held in the State Ministry of Aviation suggest that the German intelligence service was able to provide the *Luftwaffe* command with amazingly detailed information on American air armaments. It is also noteworthy that the *Luftwaffe*'s top decision makers correctly recognized the potential of the B-29, even though they sometimes lacked sufficient knowledge of their own aircraft. They continued receiving information on production figures, and the anticipated formation and deployment of a large B-29 unit in England for operations against Germany[10] until the end of January-beginning of February 1945. The first prototype of the B-29 began flight trials in September 1942. The aircraft was plagued by engine problems, as the Ju 388 would later be. Thanks to its outstanding aerodynamic shape and four supercharged 2,200-hp Wright 3350 Cyclone engines, the B-29 was capable of striking targets at distances up to 3,250 km, carrying 10 tons of bombs at speeds of up to 357 mph (575 kph) at a maximum altitude of 33,600 feet (10241 m). In addition to a pressurized cabin,[11] the aircraft was equipped with an extensive defensive armament, which was mainly remotely-controlled, and radar for navigation and target finding. It is not surprising that the USSR and China produced copies of the B-29 in the postwar years, as the type's performance was far in advance of any comparable type. The type remained state of the art until the introduction of jet bombers, apart from the B-36 interlude. Together with the atom bomb and rocket technology imported from Germany, strategic bombers like the B-29 formed the basis of the USA's rise to superpower status. The advantage lay in the country's almost limitless industrial potential, and the near peacetime conditions that prevailed during the B-29's design and production start-up in 1942-43. These conditions favored the development of such a large bomber. The requirement for the long-range heavy bombers named earlier was probably based on American fears of having to fight a war across the Atlantic against National-Socialist Germany after the fall of Great Britain, which was still a distinct possibility in 1940. In fact, this requirement ceased to exist sometime in 1941. The Japanese attack on Pearl Harbor on 7 December 1941, however, changed the situation, and the B-29 was earmarked for service in the Pacific.[12]

Aircraft companies in the USA were also developing new twin-engined bombers, fighters, and reconnaissance aircraft. At the same time, England was pursuing the development of high-altitude fighters and bombers, including the Vickers-Armstrong Wellington Mk. V and Mk. VI. A few Wellingtons powered by Merlin high-altitude engines saw service as pathfinders[13] until they were replaced by versions of the DH 98 Mosquito.

The *Luftwaffe* needed more than just high-altitude and heavy fighters to intercept enemy bombers, however. The Battle of Britain in 1940 had forced the military leadership to admit that its bombers could not survive against British air defenses in daylight. The emphasis, therefore, shifted to night attacks. The situation worsened as the war went on, making offensive operations in the west almost impossible. In early 1944 a final attempt was made to respond in kind to the ceaseless attacks on German cities. Dubbed "Operation Steinbock" by the Germans and "Baby Blitz" by the British, the operation lacked sufficient forces, and was therefore doomed to failure. The *Luftwaffe* began the offensive with about 500 aircraft (Do 217, Ju 88/188, and He 177), but early in the campaign about 100 of these had to be transferred to Italy to counter the Allied landings at Anzio. Although the attacks were made exclusively by night, by May losses had risen to more than 300 aircraft. By that point at the latest, it became apparent that losses to enemy night fighters were unbearably high.

The RLM saw the solution to these problems in qualitative superiority, for at least some in the *Luftwaffe* command realized that once America entered the war at the latest, Germany could not hope to match the production output of the Allied nations. Also, the threat of Allied bombers appearing over Germany with fighter escort left only one way out: to fly higher or faster than the attackers, preferably both. This was technically demanding, and required more capable power plants. Superiority was just as vital for bombers and reconnaissance aircraft as it was for German fighters, which had been reduced to a defensive role. The *Luftwaffe* also lacked modern guidance systems. While it did have "EGON"—equivalent to the British "Oboe" blind-bombing system[15]—it lacked a modern ground-mapping radar like the USAAF's H2R Rotterdam. British signals intelligence soon became aware of the "EGON" guidance system, and jamming seriously reduced its effectiveness, further hampering the German night raids.

Until superior aircraft types became available, the *Luftwaffe* command turned to the so-called "Revenge Weapons," the V1/Fieseler Fi 103 and V2/Peenemünde A4. These innovative unmanned missiles did not place any demands on highly-trained aviation personnel.

In mid-April 1943 the Ju 188 was included among the designs contained in Aircraft Delivery Plan 233/1 for 1943-44. The plan called for production of bomber and reconnaissance variants powered by the BMW 801 E, or a high-altitude version of the BMW 801. Long-range plans envisaged variants powered by the Jumo 213 and Jumo 222 (beginning late 1944). At that time the Ju 288 was viewed as the solution for night raids against London, and production was expected to begin in late 1944. The design emphasized heavy armament, good armor protection, and a heavy bomb load.

In mid-May 1943 *Oberst* Rowehl of the High-Altitude Flight Experimental Station (VfH) presented his ideas on the future of high-altitude combat operations.[16] He described initial experiments with a Ju 88 powered by BMW 801 D engines, with nitrous-oxide injection[17] as a preliminary step toward the Ju 88 H. He also reported that tests with a Ju 188 equipped with BMW 801 TJ engines had gone surprisingly well. At combat power, maximum speeds of 585 kph had been reached at a height of 11500 meters. The advantages of the "non-displaced exhaust-driven turbosupercharger"[18] were confirmed. Junkers claimed that the engine modifications suggested by VfH and the aerodynamically-improved engine installation by Junkers Flugzeug und Motorenbau (JFM) would enable the Ju 188 to achieve 650 kph at 12500 meters at combat power. The first pre-production engines were to be expected in June 1943. Rowehl continued:

"Since heights above 11 000 m in unpressurized aircraft place a great strain on the crew, and as pressure suits are too bulky for the cockpit of the Ju 88 A-4, it is urgent that production of the Ju 188 with the three-man pressurized cockpit be accelerated. It is considered practical and desirable that two development facilities be made available for the `customizing' of engines and airframes for special missions."

He proposed the BMW factory in Spandau, and the Henschel facility in Schönefeld. Rowehl's proposals were a sign of things to come. Henschel continued work on the Hs 130 for experimental and special missions, while Junkers decided to develop a production high-altitude machine with a pressurized cockpit based on the Ju 188.

An intensive high-level discussion took place in the RLM on 10 August 1943. In response to plans to build very small numbers of the Hs 130 E in 1945-46, Major Hoffmann, representing the Technical Office GL/C E2, demanded an examination of the procurement possibilities.

"It will either be the Hs 130 or an alternative high-altitude aircraft with a improved level of performance with really promising target dates and numbers. The aircraft in question would be powered by four 801 TN engines and could be produced from the Hs 130 or the Ju 188 with the `3-man pressurized cockpit'. In both cases the result would be service ceilings above 14 000 meters and maximum speeds of 700 kph at about 13 000 m. The range of the four-engine Ju 188 would be somewhat greater than that of the Hs 130. One particular advantage of the Ju 188 solution is that it is based on a type already in mass production. In any case, the Henschel firm will have to be involved for the necessary development capacity."

High-altitude aircraft were again the theme just 10 days later, on 20 August 1943.[19] Pointing out that the Hs 130 E with Höhenzentrale had already been flown, Major Hoffmann requested a decision as to which high-altitude aircraft was to be pursued. Discussed as an alternative to the Hs 130 was the Ju 388, a high-altitude development of the Ju 188 in three versions: J (fighter/heavy fighter/night fighter); K and L. "The 488 is the four-engined successor type," continued Hoffmann. Performances were roughly similar (with only provisional data for the Ju 388). He also remarked:

"The Hs 130 offers more altitude, the 388 greater range. In view of increased enemy effort in the high-altitude sector, a decision must be made as soon as possible."

Generaldirektor Frydag, chairman of the Aircraft Steering Committee, proposed that the 488 be accepted if the goal was large numbers. But if the objective was the achievement of higher altitudes quickly, they would have to build the four-engined version of the Hs 130. Prof. Hertel, head of aircraft development at Junkers, added that the Ju 488 would have the wings of the Ju 388, the tail section of the Ju 288, a new fuselage, and a new wing center section. In his opinion, it was important to produce large numbers of aircraft as soon as possible, and "at the same time a high-altitude engine based on the DB 603 would be desirable." Generalfeldmarschall Erhard Milch,[20] together with the *General der Kampfflieger*, the representative of the RLM's technical office (GL/C), *Oberst* Petersen (KdE[21]), and *Oberst* Rowehl (VfH), decided that "initially, emphasis is to be placed exclusively on the Ju 388." This decision opened the way for the Ju 388. Milch also stated that *Generaldirektor* Frydag and Henschel were to lend Junkers all possible support. In particular, the Henschel design bureau was to assist in the production of jigs for the Ju 388. The question of a four-engined development was deferred for production capacity reasons. For the time being it would be scrutinized as a project and compared to a jet-powered development. At the same time, at the suggestion of the head of the GL/C and the VfH, the number of Hs 130 Es with Höhenzentrale to be built was reduced to 30.

By the end of 1943 it was becoming apparent to Junkers Flugzeug und Motorenwerke AG (JFM[22]), the largest producer of air armaments in Germany, that its Ju 288 program was about to be cancelled. The root cause was the suspension of work on the Jumo 222 engine, which had been designed specially for this aircraft. The Jumo 222 did not enter production until late in the war, and just 286[23] examples were completed. The lack of piston engines in the 2,500 to 3,000 h.p. class, combined with difficulties in developing effective superchargers brought about by raw material shortages, overshadowed almost every late war project for the *Luftwaffe*. The only possible solution was a switch to turbojet-powered aircraft, although at that stage jet engines still suffered from very high fuel consumption rates.

In Junkers' view, therefore, after the failure of the Ju 288 bomber program, the Ju 388 represented its last attempt[24] to maintain its position as the *Luftwaffe*'s principal supplier by producing a development of the successful Ju 88/188 series. As the *Luftwaffe*'s last piston-engined bomber, the Ju 388 could have filled the gap, pending the introduction of the advanced Ju 287 jet bomber.

Of course, a multitude of complex causes had brought about this situation, especially dwindling resources of raw materials, the systematic destruction of German manufacturing facilities by Allied bombing, and the resulting drop in industrial capacity. Additional factors included animosities between personalities at the decision-making level in the RLM and armaments producers, and the technical and organizational incompetence of Germany's political and military leadership.

From the beginning, development of the Ju 388 suffered because of the unsolved problems "inherited" from its Ju 88/188 predecessors. The most serious of these was the lack of a production-ready high-performance power plant, which meant that development of a new engine had to proceed in parallel with, or even behind, development of the aircraft.

The second serious disadvantage that limited the operational effectiveness of the Ju 88/188 design, was its inadequate defensive armament. Technical difficulties in overcoming this deficit had already had an adverse effect on the Ju 288, and the "direct" predecessors of the Ju 388 in the shape of the prototypes of the planned Ju 188 C, G, and H series with manned tail turrets.

Why had Junkers been selected to develop a multipurpose high-altitude aircraft? In designing the Ju 388, Junkers could draw upon considerable experience in building high-altitude aircraft. On 26 May 1929 Willy Neuenhofen had set an absolute altitude record of

The Ju 188 E, direct forerunner of the Ju 388. It remained in service with the French Navy until 1947. Vernaleken Collect on

The VfH and Versuchsverband d. ObdL operated the rare Ju 88 B, the starting point for development of the Ju 188, as a long range reconnaissance aircraft. Griehl Collection

The Hs 130 E was a large aircraft. Visible beneath the fuselage are elements of the Höhenzentrale. This consisted of a third engine driving a special mechanical high-altitude supercharger. Griehl Collection

12739 meters[25] while flying an almost production W 34 be/b3e with the registration D-1119, WNr. 2600. This record stood until 1934, when Italian Captain Donati set a new record of 14433 meters.[26] At the beginning of the 1930s the German Aviation Research Institute (DVL) began a systematic program of scientific research into high-altitude flight. Early on, the institute developed a close relationship with Junkers.[27] In 1928, in cooperation with the DVL, the company developed the Ju 49, an experimental high-altitude machine that first flew in 1931. In 1935 Junkers produced the EF 61 high-altitude bomber. The DVL's main role was the development of pressurized cockpits and high-altitude engines on behalf of the RLM.

In addition to these research aircraft, Junkers later developed the Ju 86 P and R, high-flying derivatives of the Ju 86 G, a military version of the Ju 86 commercial aircraft used by *Lufthansa* and other companies before the war. Design work was completed in November 1939, and flight testing of the Ju 86 P began in February 1940. Two more prototypes, the Ju 86 PV2 and PV3, joined the test program in March.[28] In 1942 Junkers produced the Ju 86 R by increasing the Ju 86 P's wingspan, and fitting more effective four-blade propellers. With the Ju 86 P and R series, Junkers demonstrated that it was possible to produce effective high-altitude reconnaissance aircraft, and even nuisance raiders by installing pressurized cockpits and high-altitude engines in conventional aircraft. For economical reasons (fuel consumption), Junkers selected an entirely new path: high-altitude diesel engines. Lower exhaust gas temperatures also simplified the task of designing exhaust-driven turbosuperchargers for the diesel engines. With its impressive service ceiling of 12 000 meters,[29] the Ju 86 P showed the way for the Ju 388 in two respects: it differed from the earlier version in the adoption of a completely new pressurized cockpit and high-altitude engines. Its Achilles Heel, however, lay in its lack of defensive armament. Junkers took this into account in designing the Ju 388. Following the short-term failure of the Jumo 222, Junkers was forced to look elsewhere for power plants for its Ju 388, and the most suitable engine available in late 1943 was the BMW 801 TJ. Delays in the development of the Jumo 222 meant that plans to introduce the engine as a standard power plant for production aircraft could not be realized.

All serious attempts by other companies to develop high-altitude aircraft were either unsuccessful, or abandoned at the pre-pro-

The Junkers Ju 288—here the V9—was to have been built in large numbers, and was therefore of great economic significance to the Junkers company, especially since the type was to be powered by the company's own Jumo 222. Griehl Collection

The Ju 49 high-altitude research aircraft was developed by Junkers in cooperation with the DVL beginning in 1928. The aircraft first flew in 1931. Griehl Collection

The EF 61 research aircraft, with its (interim) pressurized cockpit, was a milestone in high-altitude research by the JFM company in cooperation with the DVL.

The EF 61 first flew on 4 March 1937. It was the first aircraft to feature the annular radiators that would later become a Junkers trademark. Griehl Collection

The Ju 86 P differed from other versions of the aircraft in its pressurized cockpit and large Jumo 207 diesel engines.

Close-up of the Jumo 207 in a Ju 86 P. Griehl Collection

One of the few examples of the Do 217 P. Equipped with a pressurized cockpit and a third engine in the fuselage driving a high-altitude supercharger, the aircraft was developed into a high-altitude bomber. Griehl Collection

duction stage at the latest. None of the types designed specifically as high-altitude aircraft, such as the single-engine Blohm & Voss Bv 155, or the Henschel Hs 128 and 130, nor modified versions of existing types, such as the Do 217 H and P, proceeded beyond the testing phase.

The Focke-Wulf Ta 152 was the only high-altitude type to reach the production stage. Powered by a Jumo 213 E engine, this single-engine aircraft had a service ceiling of 13650 to 14800 meters.[30] So impressive was its performance that the Ta 152 remained in every delivery plan until the surrender.

The situation was very different in the field of "Turbine Air Jet Machines" (TL), or turbojet-powered aircraft. The Ar 234 had been under test since mid-1943, and in August Erich Sommer of the OKL Test Unit carried out the first jet reconnaissance missions over the heavily-defended invasion area. The Arado operated beyond the reach of Allied fighters, something no other operational type was capable of. This proved that jet aircraft were the way of the future for reconnaissance. Thus, to a certain degree, the Ju 388 was conceptually obsolete, even before its development was completed. The demand for conventional designs at that time can probably be explained by the *Luftwaffe* command's lack of trust in the new jet technology, and the interruptions in production associated with the lengthy conversion period. The most serious shortcoming of the new jets compared to conventional piston-engined aircraft was their limited range, which can be attributed to the high fuel consumption of the early jet engines. External fuel tanks provided some relief, however, they negated to some extent the jet's speed advantage. The jets were also extremely vulnerable during takeoff and landing, especially in the face of Allied air superiority. These problems could be partly offset through the use of takeoff-assist rockets and braking parachutes. The early jet aircraft also required paved runways. In most cases new concrete runways had to be constructed, and most of these were quickly destroyed by Allied bombs. In action, the high maintenance requirements of the jet engines had a negative impact on operational readiness. As it turned out, the jet engines would share the latter shortcoming with the BMW 801 TJ-0.

The basic concept of the Ju 88/188/388/488 series, to produce variants of one basic type for a variety of roles, was extremely forward-looking, pointing the way toward present-day multi-role aircraft like the Panavia Tornado, Sukhoi Su 27 Flanker, and the McDonnell-Douglas F/A-18 Hornet. Junkers succeeded in developing a "building block" principle, which made it possible to replace various modules—cockpit, fuselage, wings, engines—with more modern versions to keep pace with changing operational requirements. No other combat aircraft, not even the British Mosquito, which was comparable in many areas, covered as broad a spectrum of development work, variants, and operational roles as the Junkers Ju 88 and its successors. The cause for this was the importance of the Ju 88 program to Junkers which, together with

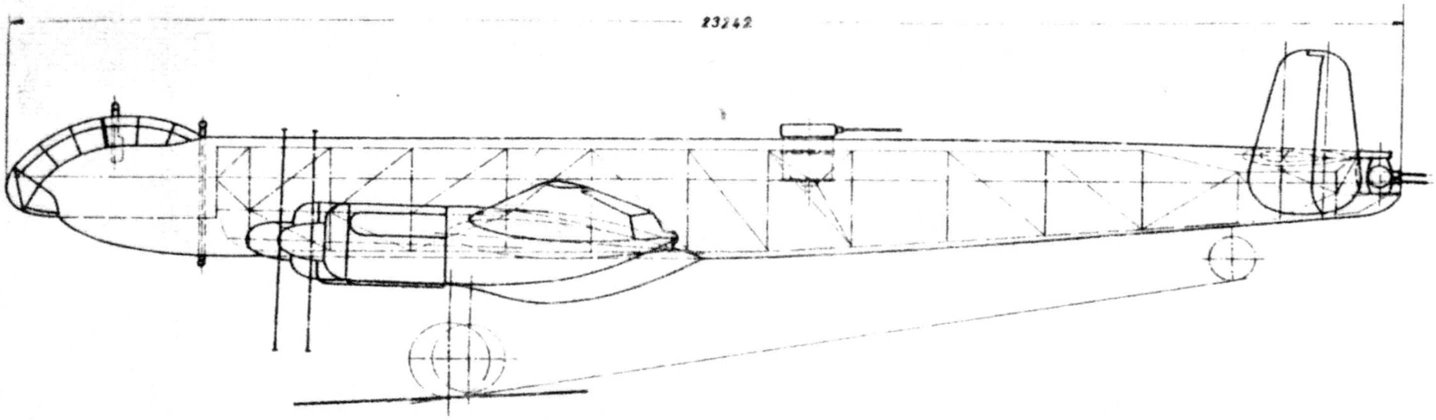

Side view of the Ju 488 V403 from the Junkers general arrangement drawing S 188 00-9947. It was created using the building block principle, with the pressurized cockpit and outer wings of the Ju 388 and the tail section of the Ju 288. Griehl Collection

Despite its poor quality, this photograph from an American report on German aircraft types in France documents the state of construction of the central fuselage section and wings of the Ju 488 V401. Imperial War Museum

Unfortunately of poor quality, this photograph shows the fuselage of the V401 "flying mockup" under construction. Haubner/Misna Collection

the responsible decision makers in the RLM, especially Erhard Milch,[32] was eager to avoid disturbing mass production. As early as September 1942, Milch regarded the Ju 188 as the universal combat aircraft for the *Luftwaffe*, and he envisaged that it would remain operational in various versions until 1947.[33]

When, late in the Second World War, consideration was given to the development and use of strategic bombers, an area in which the *Luftwaffe* had so far failed to achieve any success, Junkers understandably wanted to continue its building block concept, and base its design on the Ju 388. The new type was to incorporate as many components as possible that were already proven or in production. This would allow production to continue, and minimize the usually tedious testing required of a new design. Another advantage of the building block system not to be overlooked was that servicing was, for the most part, a known commodity with established supply lines. The design of the Ju 488 four-engined bomber was therefore based as much as possible on components of the Ju 388. The first design was essentially a "stretched" Ju 388, with a fuselage insert (similar to that of the Ju 88 G-10 and H), and a rectangular wing center-section to accommodate the additional engine nacelles and undercarriage. The intended roles of the Ju 488 were heavy bomber and long-range reconnaissance, and it was to replace the Ju 388 in those roles. It was anticipated, however, that the Ju 388 would be retained for the Zerstörer, or heavy fighter, role.[34] Components of two Ju 388 K-0s, namely WNr. 230 152 and WNr. 230 154, were to be used to construct "two flying mockups of the special version of the Ju 188 (V401 and V402)," the first two prototypes of the Ju 488.[35] Although it was in some respects created in parallel with the Ju 388, the Ju 488, in fact, represented the ultimate stage in Ju 388 development, and thus the entire Ju 88 family of aircraft.

Not even a single Ju 488 was ever completed, however. The V401 and V402 were to have been built under contract to Junkers by the Société Industrielle D'Aviation Latécoère in Toulouse. The receipt of design and development work from Junkers enabled Latécoère, S.N.C.A.S.E., and Breguet to retain their skilled workers in their own factories. The work began in early January 1944. As the central development office, Latécoère assumed responsibility for final assembly, while Breguet acted as subcontractor. Photographs confirm that fuselages, a new wing center section, and at least the horizontal stabilizer for the V401 existed. According to the CIOS[36] report on *Luftwaffe* production in France, 166 personnel worked on the Ju 488 project, for which wings came from Germany. As well, Ju 288 wings were allegedly sent from Holland for the project.[37] Work on the two prototypes was at an advanced stage when it was decided to transfer the project back to Germany. The Résistance struck first, however. On the night of 16-17 July 1944, just before the components were to be transported by rail to Germany for final assembly by Junkers, explosives set by members of the Résistance destroyed several of the almost complete components, especially the fuselages in the area of the wing fuselage joint. Rebuilding was out of the question. All work was halted as a result of the attack.

Components of the Ju 388 were also used in the last true bomber development by Junkers[38] to reach the test stage, the Ju 287. The first two prototypes of the Ju 287, which were also built as "flying mockups" to save time, used He 177 fuselages, components from various other types, and elements of the Ju 388's tail section. As a result, the Ju 287 was able to enter flight trials on 8 August 1944, almost parallel to the Ju 388.[39] Flight trials were conducted in September 1944 in spite of numerous bombing raids, although the site was Brandis, near Leipzig, not the main Junkers facility in Dessau.

This photograph of the Ju 287 V1 clearly shows that the horizontal and vertical tail surfaces of the Ju 388 were used in its construction. The notch at the bottom of the rudder indicates that this was a production standard tail. Handig Collection

Testing of the Ju 287 was officially brought to an end in late September by an order from the Armaments Staff—successor to the Fighter Staff.[40] The valuable engines were removed, after which the two prototypes, which were undoubtedly the most advanced bomber developments of their time, were simply parked at the edge of Brandis airfield—at a time when testing of the Ju 388 was proceeding apace. When American troops arrived in April 1945, they discovered the two wrecked machines at Brandis airfield. Both were entered in an Air Intelligence Section inventory of captured types dated 10 May 1945, together with the remains of a parallel development to the Ju 388, the Hs 130 A V6, as "Type Ju 187 (287???) – Condition: burnt out."

Chapter 2: Genesis of the Ju 388

Further Development of the Ju 88 and Ju 188

After Generalfeldmarschall Erhard Milch[42] took over the post of Generalluftzeugmeister (Chief of Aircraft Procurement and Supply) from Ernst Udet in 1941, the watchword became "Production" instead of "Development,"[43] and so increased production numbers generally took precedence over advances in development. It was understood that every new development or modification had a negative effect on production figures. Consequently, the chances of a completely new aircraft reaching quantity production were poor.

Throughout the war, therefore, Junkers—like all the other German aircraft manufacturers—kept an established product, in this case the Ju 88/188 series, under constant development, while continuing mass production. Admittedly, this was not completely at the company's own initiative; rather, it was the result of close interaction with the RLM and the *Luftwaffe*. At a time when the RLM had put the development of new types on the back burner, this approach enabled aircraft manufacturers to meet future demands on their aircraft by "responding to wishes and suggestions for improve-

Chief of Aircraft Procurement and Supply Erhard Milch, successor to Ernst Udet. Handig Collection

Servicing a Ju 88 D reconnaissance aircraft of the weather reconnaissance unit Wetterkette Nord. Vernaleken Collection

ment from the front." In spite of the war economy, armaments concerns were still run as businesses, and they were well aware of the need to retain their influence through new armaments contracts. It was necessary, therefore, that they not only make themselves subject to the unpredictable demands of the RLM, but to meet, or even better, exceed, these demands with marketing-oriented developments of their own products (catchword "Model Grooming"). In general, subsequent developments of the Ju 188 focused on improved speed and altitude performance, combined with reduced range and payload. By achieving the maximum possible service ceiling, it was anticipated that bombers and reconnaissance aircraft would be able to avoid interception by enemy fighters. It was hoped that the installation of more powerful engines and strengthening of defensive armament would result in significant improvements. The planned Ju 188 C, H, N, and R series were all the result of these concepts.

The Ju 88 underwent a process of steady improvement. Reconnaissance machines were derived from the bomber variants. The Ju 88/188 reconnaissance aircraft that saw widespread service with the long-range reconnaissance units was based on the Ju 88 A series,[44] with the dive brakes removed and two cameras installed behind Frame 15. In addition, the number of external stores racks beneath the wings was reduced to two. The first interim reconnaissance machines, the Ju 88 D-1 to D-7, began appearing in 1941. These in turn gave rise to the long-range Ju 88 H variant with a lengthened fuselage. The D-Series also formed the basis for fast reconnaissance and high-altitude variants. Specially modified late D-series aircraft with the ventral gondola removed served with Kommando Rowehl. These machines were powered by BMW 801 D engines, and most had GM 1 power boosting; however, the engines proved unreliable, and the machines were not popular. In autumn 1942 in Dessau, a Ju 88 D-6 with BMW 801s was supposedly converted to take BMW P8035 turbo-supercharged engines (based on the BMW 801 E).[45] The type was supposed to go into production as the Ju 88 D-7, and ended the career of the Do 217 P-1, which was then under test, but then it, too, was cancelled by the RLM.

Final developments in the Ju 88 series of bomber and reconnaissance aircraft were the Ju 88 S and T. These machines were fitted with a streamlined plexiglass nose cap reminiscent of the first prototypes. As well, the ventral gondola was deleted, and more powerful engines (BMW 801 and Jumo 213) were installed.

The concept of continuous development obviously bore fruit, for in early September 1942 Junkers received a contract from the RLM's Technical Office for the "further development of night fighters and high-speed bombers from the Ju 188." The contract specified development of a new three-man cockpit and construction of a mock-up.[46] In March 1943 the airframe construction department began preparations for the building of two prototypes each of the Ju 188 R (night fighter), Ju 188 S (high-speed bomber), and Ju 188 T (high-altitude reconnaissance aircraft). Plans were also made for the construction of seven more Ju 88 H-1s, plus development and construction of 10 prototype Ju 88 H-2s. There were already firm

Photo of an early Ju 88 S. Its most noticeable features are its BMW 801 engines, spherical nose cap, and absence of the ventral gondola. Haubner/Misna Collection

A Ju 88 night fighter with BMW 801 engines at a display of captured enemy aircraft in England. The R-series differed from the Ju 88 C-6 in having more powerful engines. Handig Collection

German markings were applied for display purposes. The British have obviously removed the aircraft's radar antennas. Handig Collection

orders for the conversion of a Ju 88 A-4 into a high-speed bomber equipped with BMW 801 engines with GM 1 power boosting (cost 161,000 RM), and development of the FA 15 remotely-controlled barbette later used in the Ju 388. Twenty prototypes were ordered at a total cost of 1,334 million RM.[47] At the same time the technical sales department was able to book new contracts for the development of the Ju 188 J-1 (high-speed night fighter), Ju 188 K-1 (high-speed bomber), Ju 88 L-1 (high-speed reconnaissance aircraft), and Ju 188 U-1 (personal transport for Generalfeldmarschall Milch). As well, Junkers was to develop and build prototypes for the Ju 188 G-3 series with a bulged bomb bay under the fuselage. In the second half of 1943 the Ju 188 J, K, and L series were renamed the Ju 388 J, K, and L. We have been unable to discover documents confirming the date, or giving an official reason for the change.

By early 1943 at the latest, the RLM obviously came to the realization that there was an urgent need for a new high-performance night fighter. The standard night fighter types (Me 110 G, Ju 88 C-6, Do 217 N) were becoming increasingly inferior to their Allied counterparts. It was already becoming apparent that the second-generation Bf 210/Me 410 was a failure. Apart from the Ju 188 J and R, the only other type under consideration was the He 219, originally conceived as a reconnaissance aircraft, but redesigned as a night fighter. Milch favored the Ju 188, for he had to take care not to threaten monthly production totals by introducing a completely new type. Thanks to mass production of the Ju 88, with which it shared many components, the Ju 188 had a broad production base. The large number of Junkers factories and licensed plants made production of the Ju 188 less vulnerable to air attack. Milch therefore saw a heavily armed Ju 188 R with modern airborne radar as the way of the future. General Kammhuber, the commander of the night fighter arm, favored the He 219. Comparative trials were conducted at Rechlin on 25-26 March 1943, and the He 219 V1 proved to be about 25 kph faster than the Ju 188, while in a mock combat it was able to out-turn the Junkers. The He 219's excellent performance resulted in a contract for production of 50 aircraft per month, the maximum that available production capacity would allow. Overall, the He 219 gave more of an impression of a heavy fighter, whereas the verdict on the Ju 188 was that it was a very maneuverable bomber.[48]

The starting point for night fighter and *Zerstörer* development: the Ju 188 R mockup with heavy fixed armament, additional armor, and antenna array. Deutsches Museum

A rare in flight photograph of a Ju 88 G-1 (2Z+HN) of the 5. *Staffel* of NJG 6. The aircraft is equipped with Lichtenstein SN2 airborne radar. Vernaleken Collection

Junkers is known to have conducted preliminary studies for a night fighter/heavy fighter based on the Ju 188 in July 1942. Factory photographs exist in which may be seen the mock-up[49] of a Ju 188 cockpit with three fixed, forward-firing MG 151/20 cannon in the forward fuselage under the cockpit floor. Two more MG 151s are mounted beneath the opening for the standard production dorsal turret, and in the spherical nose cap. The mock-up was also equipped with antennas for the FuG 212 "Lichtenstein C1" radar. Wooden panels behind the struts of the forward canopy framing simulated armor plate, increasing protection for the crew, as armor protection in the Ju 188 was inadequate for the heavy or night fighter roles. This measure eliminated several glazed panels, however, as it did in the Ju 88 G and Ju 388 J.

Evaluation of this configuration revealed the superiority of the technically simpler Ju 88 G, which had the fuselage of the Ju 188, including its larger fin and rudder, and a revised canopy with better aerodynamics and heavier armor than that of the earlier Ju 88 C and R series. The G series also differed from the Ju 88 C/R in having more powerful engines and a revised armament, in which the cannon in the aircraft's nose were replaced by two 20-mm MG 151s in a ventral tray.

Delivery of the Ju 88 V58—the prototype of the Ju 88 G-1—took place between July and September 1943.[50] Next to the Ju 88 A-4, the Ju 88 G night fighter and night close-support aircraft was the most successful variant of the entire series, with production continuing literally until the arrival of Allied troops. Both versions also formed the basis of Mistel conversions.

In this context it is also interesting that, apart from the previously mentioned Junkers monthly report of March 1943, there had been no reference to a light high-altitude night fighter as the logical answer to the feared DH 98 Mosquito. The only drawing of a Ju 188 night fighter project with what is obviously the production Ju 188 cockpit appears in a summary of Junkers designs for the USSR compiled by Prof. Baade's design team.[51] In addition to three fixed MG 151 cannon beneath the cockpit floor, whose arrangement was similar to that of the previously-described mock-up, the dorsal turret (B2-Stand) was equipped with two MG 151s, and converted into a rotating turret for "firing horizontally and obliquely upwards." The drawing also showed a fixed oblique armament, consisting of two MG 151/20 cannon behind the second bomb bay. The fact that this well known Junkers designer referred to it as a "project" may serve as proof that such a machine—whose power plants were not described in any detail—probably proceeded no further than the design stage.

All traces of the Ju 188 R, S, and T disappear in the Junkers company's monthly reports, however. These reports, which each month provided a list of the production aircraft delivered to the RLM and the state of remaining contracts, were supplemented by quarterly reports that gave a somewhat more detailed picture of new developments, orders, pre-orders, and delivered aircraft. By September 1944 all references to the Ju 188 R, S, and T had disappeared, leading one to conclude that the pending contracts had been withdrawn. There is also no reference to the Ju 188 S in the surviving delivery plans. These were used by the RLM's Technical Office (GL/C) to determine aircraft production for the next one to two years. The only variants of the Ju 188 contained in Delivery Plan 223 were the Ju 188 F-1 trop, Ju 188 A-2, and Ju 188 D-2 trop, while the only versions of the Ju 88 to be built were the C-6 trop and G-1.[52] In Delivery Plan 223/1 dated 15 April 1943, the Ju 388 still appears under the designations Ju 188 J, K, and L, production of which was to begin at ATG in the summer of 1944.[53]

Prototypes of the Ju 88 R, S, and T were definitely built, and a contract was issued for a small production series; however, these machines received little mention in the delivery plans.[54] The delivery of a Ju 88 R-1 converted from a Ju 88 C-6 is confirmed by the March 1943 monthly report. The same report makes reference to an advance order for the conversion of 50 aircraft into Ju 88 Ts, and the delivery of 50 spare engines, with a total cost of 55 million RM.

Ju 88 T-1 in service. Removal of the ventral gondola and the new spherical nose cap significantly improved the aircraft's aerodynamics. Haubner/Misna Collection

Ju 88 T-3 with Jumo 213 engines and flame dampers. An identifying feature of the reconnaissance version is the absence of the bombsight fairing beneath the nose. Griehl Collection

By June the airframe manufacturer had received an order worth 4,476,000 RM for the conversion of 50 Ju 88 A-4s into S-1s. At the same time, the estimated cost of converting 50 Ju 88 D-1s into T-1s was revised upwards to 4,343,000 RM. In the same quarter the Technical Operations Division not only delivered the Ju 88 V55 (WNr. 140 377, VL+KY, first flight on 28 Dec. 1942), prototype for the Ju 88 S-1 series, but 24 Ju 88 S-1 conversions and 17 Ju 88 T-1s, as well.[56] Eleven Ju 88 S-1s and 15 Ju 88 T-1s were delivered in the next quarter, and according to its own records, Junkers was behind schedule by six and eight machines, respectively.[56] At the end of December, after Merseburg had delivered 22 Ju 88 S-1 conversions and 18 Ju 88 T-1s, a contract was concluded for 50 Ju 88 S-1s and 50 Ju 88 T-1s. At the same time, a new order was received for the construction of 146 Ju 88 S-3s.[57,58] A few examples of the Ju 88 S and T are believed to have had BMW engines with high-altitude superchargers. Following the cancellation of the Ju 88 T-2 with GM 1 and MW 50 boost, of which just a single prototype was built—the last production reconnaissance version of the Ju 88, the T-3—was equipped with Jumo 213 A engines. It has been claimed that two of these were equipped with GM 1 power boosting, one with Jumo 213 E engines, and the other with DB 603 Es.[59]

This examination once again makes it clear that Junkers made numerous efforts to develop more capable night fighters, bombers, and reconnaissance aircraft from existing types and variants. It is highly likely that the Ju 188 R, S, and T proceeded no further than the drawing board and mock up construction. Not until the Ju 388 and Ju 88 G did the RLM issue Junkers another contract for quantity production.

Order for the First Ju 388s

As related earlier, in mid-May 1943 *Oberst* Rowehl of the Experimental Station for High-Altitude Flight submitted a report on initial experiments with a Ju 88 powered by BMW 801 D engines with nitrous-oxide injection,[60] as preliminary work on the Ju 88 H. Furthermore, testing of a Ju 188 equipped with BMW 801 TJ engines had gone surprisingly well, resulting in a maximum speed of 585 kph reached at combat power at an altitude of 11.5 km. The exhaust-driven turbosupercharger installed directly on the engine had proved a success. Aerodynamic improvements were to enable 650 kph to be achieved at 12500 meters at combat power. The first pre-production engines were to be expected in June 1943.[61] Rowehl continued:

"...as heights above 11000 m in unpressurized aircraft place a great strain on the crew and pressure suits are too bulky for the cockpit of the Ju 88 A-4, it is urgent that production of the Ju 188 with the three-man pressurized cockpit be accelerated."

This is in keeping with the fact that, on 24 May 1943, Junkers received an order by construction directive for "ten Ju 188 L-0 high-speed reconnaissance machines."[62] These aircraft, which later became the Ju 388 L pre-production series, included those with the *Werknummern* 300 001 to 300 010,[63] typical serial numbers for new designs. Conversion of the first machine was to take place at Junker's main plant in Dessau, using the fuselage, wings, and empennage of production Ju 188 E/F series aircraft. The remainder would be produced in Merseburg. The main difference between these aircraft and the production version was the newly-designed three-man pressurized capsule, which was bolted to the fuselage of a Ju 188 E by way of a new spacing framework. The aircraft were to be powered by two BMW 801 TJ-0 engines driving four-bladed propellers. This resulted in changes to the fuselage in front of Frame 9, the wings, and firewalls. Like that of the later Ju 388 L-1, the cockpit was to have fittings for heated flight suits and an oxygen system for three men. Unlike the production Ju 188, the required 15 spherical steel oxygen bottles were housed in the starboard wing instead of the fuselage, as in the later Ju 388.

Wing deicing was by means of a "Kärcher Stove" (a gasoline-powered heater), virtually the same system as in the Ju 88 G-1. Mounts for three cameras were to be installed in the fuselage behind Frame 15. Armament was to consist of a flexibly mounted MG 131 in the dorsal position with electrical firing and 500 rounds of ammunition, plus two fixed rearward-firing MG 131s in a tray beneath the rear bomb bay. The armament system was also to include an Rf 1 A or Rf 2 B periscope for the pilot. Provision was also to be made for *Rüstsätze* (equipment sets) M1 and M8 (bomb-carrying equipment used by the Ju 188), and installation of the Lotfe 7 D[64] bombsight. This was obviously to allow the aircraft to be converted into a bomber if required. For further details see the reproduction of the original production directive on the following pages. The Ju 188/388's simple defensive armament of fixed cannon in a ventral tray is understandable given previous operational experience. The appearance of the first true high-altitude versions of the Spitfire had spelled the end for the Ju 86 P/R high-altitude reconnaissance aircraft, which were incapable of defending themselves.

Ju 388 L-0 WNr. 3000 291 (DW+YY), seen here at Merseburg in early April 1944, was similar in appearance to the Ju 388 L-0 as described in the construction directive. Visible beneath the fuselage is the bulged fairing (Waffentropfen), which housed the aircraft's fixed rearwards firing armament. The associated mount for the rearview periscope may be seen atop the cockpit. A mechanic or pilot is sitting in the cockpit. EADS

Junkers Flugzeug und Motorenwerke **Dessau, 24 May 1943**
Limited Company
Aircraft Construction Master Plant
JFM-FTV-Au Scha./Rl.

Construction Directive
For
Production of 10 Ju 388 L-0 High-Speed Reconnaissance Aircraft
(Also serves as production directive)

Contract No: **7 188 26 10/001**
Completion Date: as per Production Bureau timetable

By order of the RLM GL/C-E 2, 10 Ju 188 aircraft are to be converted using major components. The aircraft will be given the serial numbers 300001 to 300010. The first aircraft will be converted in Dessau, the rest in Merseburg. The major components are to be delivered by FG.

I Fuselage
Ju 188 version without cockpit, electrical equipment, and radio wiring.

II Empennage
Ju 188-type vertical and horizontal tail surfaces, from reserve stocks.

III Wing
Complete wing with undercarriage and control surfaces, Ju 188 F version, from reserve stocks.

The conversion will take place as per the Ju 188 L-0 component plan. Essentially, the following modifications are to be carried out, plus installation of newly designed components, and the following designated systems:

1 Cockpit
Three man pressurized cockpit, attached to the fuselage by threaded ball socket connectors, and thus removable. Easily removable panel between fuselage and cockpit for access to pressure tight run-throughs, etc. Three part cockpit canopy with double glazing. Moisture traps to prevent fogging. Rear part jettisonable.

2 Fuselage
External fittings for attachment of cockpit, plus changes resulting from fitting out.

3 Control Surfaces
Inner area of landing flaps sheeted with steel for protection against exhaust gases.

4 Controls
New controls in cockpit.

5 Wing
Firewall and attachment points modified for BMW 801 TJ-0.
Removal of cockpit heat and wing deicing systems.
New nacelle extension.
Removal of Kuto cable cutters.

6 Power Plants
Two BMW 801 TJ-0 engines.
Engines will be delivered by BMW.

7 Propellers
Four-blade VDM metal propellers.

8 Fuel and Lubricant Systems:
Modifications in fuselage ahead of Frame 9, in the wings, and on firewall.

9 Tanks
Fuel, oil, deicing, and injection tanks Equipment State C. Hydraulic reservoir new.

10 Control System:
New control system.

11 Equipment
Folding seats for pilot, radio operator, and observer. Pilot's and observer's seats armored. All seats for use with seat type parachutes.

12 Electrical Systems:
New electrical equipment, which essentially differs from the usual equipment as follows:
a. Heated clothing connections in cockpit for crew of three.
b. Electrically operated air cooling ring, fire extinguishing system, cameras, guns.
c. Contents Indicators/Instruments: oil, fuel, hydraulic pressure, and oil temperature.

13 Radio Equipment
FuG 10 with TZG 10
Peil G-6 and APZA 6
Fu Bl 2 F
FuG 101
FuG 25a
FuG 216
NZG 16

14 Oxygen System:
Type 41 oxygen systems in cockpit for crew of three.
15 spherical steel bottles in starboard wing.

15 Climate Control System
Armatures and wiring in wings and cockpit.

16 Fire Protection System
Fire extinguishing system for engines and wing.

17 De-Icing System:
Kärcher heater in fuselage for deicing horizontal stabilizer.
Wing and propeller deicing as in Ju 188.

18 Camera Equipment:
Mounts for three cameras behind Frame 15, and wiring for heating and exposure interval controller.

19 Guns: movable
Experimental installation of MG 131 in dorsal position, left side feed, hand operated with electrical firing, as per special *E-Stelle* Tarnewitz switch.

20 Guns: fixed
2 MG 131 with ED and EA, right and left hand feeds in type 131/5 A rigid mount in weapons blister beneath rear fuselage bay. Electrically fired SVK 2/131/151 E weapons selector switch.
Rf 1 A or Rf 2 B rearview periscope for pilot.
Ammunition: 500 rounds per gun.

21 Dropped Weapons:
Permanent installations plus wiring for *Rüstsatz* M 1 and M 8.
Mount and wiring for installation of Lotfe 7 D bomb sight.

Aircraft Construction Technical Operations
(signature)

The Ju 388's oxygen bottles were housed in the starboard wing in order to make room for the tail turret's ammunition belt feed system. This was a change from the Ju 188, whose oxygen bottles were housed in the fuselage. Deutsches Museum

On 6 August 1943, the first addendum to the construction directive ordered 10 more Ju 388 L-0s, raising the total number to 20.[65] This change was reflected in the Technical Operations Division's monthly report for September 1943, which spoke of an increase in the conversion order to 20 examples each of the Ju 388 L-0 and K-0.[66]

The second block of 10 machines, built exclusively at Merseburg, were L-0 machines with the *Werknummern* 300 291 to 300 300. They were given the manufacturer's codes DW+YY to DW+ZH. This series included, for example, DW+YY, *Werknummer* 300 291, and DW+ZG, WNr. 300 299. Photographs exist of both aircraft, and a close examination of those depicting DW+YY reveal another characteristic of the Ju 388—an air intake and exhaust vent for the Kärcher heater on the left side of the fuselage.

In previous descriptions of the Ju 388, DW+YY was usually wrongly called the "Ju 388 V1," or "Ju 388 L V1." In original documents, however, it is the Ju 388 J-1 with the *Werknummer* 500 001 and manufacturer's code PE+IA that is always referred to as the Ju 388 V1.

The appearance of the Ju 388 pre-production series can probably be explained by the desire to test significant new components, namely the pressurized cockpit and the BMW 801 TJ engine, without spending a great deal of money, by simply mounting them on existing production machines. It was thus possible to produce 20 test beds in short order. This is confirmed by a letter from Junkers to Göring on 14 December 1943, in which the company justifies why the first two L-0 aircraft, the Ju 388 V7 (PG+YA), WNr. 300 001, and V8 (PG+YB), WNr. 300 002, should be retained for factory testing:

A side view of DW+YY, which is usually wrongly identified as the V1. The Ju 388's marked similarity to the Ju 188 is obvious. Visible on the fuselage behind the wing trailing edge are the air intake and exhaust vent for the Kärcher heater. The antennas for the FuG 101 and FuG 217 (vertical style) may be seen beneath the wing. Ott Collection

"The components installed in the various versions of the Ju 388, such as high-altitude capsule with climate-control system and BMW 801 TJ/0 high-altitude engines, must undergo thorough testing, as they have not been sufficiently tested to date. Only the aircraft of the Ju 388 L-0 series can be used for these extraordinarily important tests, as their simplified construction means that they can be completed much more quickly than the actual prototype aircraft, the V1 to V6. The latter are intended for further development tasks, namely of the remotely-controlled tail turret, the periscope, the offensive armament, and the bulged fuselage bomb bay, and, because of their greater design and construction costs, will not be available for testing until February 1944."

At the end of May 1943, when the 10 Ju 388 L-0s were ordered, the construction directive for the two Ju 388 J-1s mentioned in Junkers' monthly report for March 1943 was changed, or stated more precisely. Dated 16 May 1943, Addendum I to this construction directive stated that the two Ju 188s to be converted as night fighters were to receive the *Werknummern* 500 001/V4 and 500 002/V2.[68] It is interesting to note that these are the serial numbers assigned to the later Ju 388 V1 and V2. The same applies for the two prototypes of the Ju 188 K-1 bomber. Once again, the prototype numbers assigned in Addendum I to the construction directive dated 31 May 1943 were later changed: WNr. 500 003/V6 and WNr. 500 004/V7 were ultimately delivered as the V3 and V4.[69] The revised numbering was probably due to the change from a version of the Ju 188 to a completely new type, the Ju 388.

Amazingly, despite the fact that construction directives existed for the "successor" Ju 388, the Ju 188 S continued to be the subject of discussion at the highest levels. Either they were lagging hopelessly behind reality, or "Ju 188 S" was used as a synonym for a high-speed bomber developed from the Ju 188, and did not refer to a variant of the Ju 188. At a conference held in Rechlin at the end of June 1943, Milch once again demanded a Ju 188 for operations against England.[70] This type was to become available "as soon as possible." The "Ju 188 could not be expected to begin leaving the production lines until August 1944," Milch responded that "measures were to be taken to achieve an increase in the maximum speed of the 188 production aircraft while carrying a 1000 kg (desired 1800 kg) bomb load by the winter of 43-44." *Oberst* Petersen (KdE) declared that he expected that "the units should be able to achieve approximately 620 kph at 10 000 m with GM." This was to be accomplished by removing the dive brakes (which were still present on the first series of Ju 188 E aircraft), external bomb racks and dorsal turret, and by the "removal of the new bulged bomb bay." To this Prof. Hertel responded that work on this development was complete, and that Milch could inspect a prototype machine on 30 June 1943. Petersen was obviously unaware that no Ju 188s had yet been delivered with such a bomb bay, and that with the external racks removed, the largest bomb the aircraft could carry internally was 50 kg.

Junkers' monthly reports make it possible to follow precisely the development of the Ju 388. In the next quarterly report of June 1943 the company recorded a contract worth 442,000 RM for the development of a prototype tail turret for the Ju 188. Unfortunately, the report does not say if this was to be a manned or remotely-controlled turret. It was speculated that if the contract was received at the beginning of the quarter, the Ju 188 V1 might be converted for this purpose. According to Junkers records, Wendel test flew the aircraft on 12 June 1943.

As well, there were two pre-orders that might be seen as preliminary studies for the Ju 388. One expanded a previous advance

A cockpit mockup, which in the Junkers company album bears the caption "Ju 188 S." It is one of the few clues to the company's involvement with this series. Note the striking similarity to the Ju 388 in shape and layout. Griehl Collection

order for a Ju 88 with a bulged bomb bay to also include a Ju 88 A-15, the other called for a Ju 88 to be equipped with BMW 801 TJ engines. Furthermore, the spare parts operation was to deliver 100 examples of the FA 15 remote-control system later used in the Ju 388 tail turret for a half-million Marks.

The most important new advance order for the genesis of the Ju 388 envisaged the conversion of 10 Ju 188 airframes to Ju 188 L-0 standard at a cost of 3,275,000 RM. At the same time, the Technical Operations Division reported an order for 10 Ju 188 L-0 high-altitude, long-range reconnaissance aircraft.[71] Junkers had issued an internal order for these aircraft at the end of May (see above).

At a meeting of department heads in the RLM on 9 July 1943, it was decided to produce the Ju 188 in much larger numbers than currently planned, and to re-equip the *Luftwaffe*'s bomber units in three phases. Mass production was to begin with the Ju 188 A-2, E, and F versions. All three variants carried a moderately heavy armament without tail turret. According to the planners, these aircraft were to be followed by bombers with a remotely-controlled tail turret and larger bomb bay. In addition to the Ju 188 G-2 and H-2, purely high-altitude variants of the types Ju 188 J, K, and L (the later Ju 388 J, K, and L) were also planned.

Milch expressed doubts, however, and conjectured that the heavily-armed Ju 188 G and H would be too slow both for day and night operations. The efficient air defenses over the British Isles and the west required a fast night bomber with remotely-controlled tail turret and rear warning radar system.

As an interim solution he proposed a Ju 188 without dorsal turret. *Oberst* Peltz replied that the large bomb bay and tail turret reduced the Ju 188's range too drastically, that speed was further reduced, and that acceptable results could only be achieved by reducing the bomb load to just 1000 kg.[72] While the Ju 188 Schnellbomber was being discussed for future attacks on England, the Royal Air Force was systematically reducing German cities to soot and ashes. There was little the *Luftwaffe* could do to counter the night raids; the night fighter arm's situation was precarious. At a meeting held by the Chief of Aircraft Procurement and Supply on 16 July, therefore, the subject turned to aircraft for the General of Night fighters. The latter was supposed to be raising three new Gruppen, and on 1 April 1943 had already reported a shortfall of 170 aircraft. Milch therefore urged that all available Bf 110s be transferred to the night fighters. In addition, the general of Night fighters demanded a greater allocation of Ju 88 C-6s. In this con-

The Ju 388V8, WNr. 300 002 (PG+YB). With its bulged bomb bay and tail turret, the aircraft looks like a Ju 388 L-1, however, all company documents identify the machine as a Ju 388 L-0. EADS

text, it was also mentioned that the Ju 188 was to enter service as a night fighter in December 1944, with production of 20 aircraft per month beginning in January 1945. Stabsing. Friebel expressed the opinion that the Ju 88 R, a development of the Ju 88 C, would be "equally good as" the Ju 188 J; consequently, there was nothing to be gained by advancing the target date. Did Friebel know that the decisive difference between the Ju 88 R and the later Ju 388 J was that the latter was a high-altitude machine? His detail knowledge of the two types was apparently limited to the fact that "both have the BMW 801."

Later in the meeting, the General of Night fighters made an urgent request for 30-mm guns for all He 219, Ta 154, Ju 88 G, and Ju 188 night fighters to use against four-engined bombers. An ideal armament would consist of two MG 151, two MK 108, and two MK 103 cannon.[73]

Just why the Ju 188 S had been selected as a "Schnellbomber" for future operations against the British Isles became apparent at a conference held by the Chief of Aircraft Procurement and Supply on 19 July 1943. Future attacks on England were to be carried out with the Ju 188 S because, while the Me 410, Ta 154, and the jet bomber were fast enough, they lacked sufficient range.

Experience had shown that it took 4 1/2 to 5 years for a new aircraft to reach the production stage. Just developing a new canopy to the production stage took 1 1/5 years. Milch therefore concluded that the Ju 188 medium bomber would have to remain in production for three more years, because larger machines would not be available in sufficient numbers, and the fast Do 335 did not have sufficient range and payload. After three years under development, production of the Do 335 was expected to begin in April 1945 at an initial rate of two aircraft per month. The ultimate monthly production figure had yet to be decided upon; 150-200 aircraft per month was considered feasible in 1946. A "future jet bomber" with a range of 2400 km, with a 1000 kg bomb load was not expected until 1948-49.

On 30 July 1943 discussions were held as to which aircraft should be used by the night fighter arm to shadow enemy bombers using SN 2 radar. Kammhuber, the general of night fighters, believed that the Ju 188 was best suited for this task. In his opinion the Ju 88 was too small, a "lame crow" (inadequate performance), and lacked sufficient range. As a temporary solution he proposed the Ju 88 G and R with SN 2 and wide-angle search.

The question of whether the Ju 188 ("which is coming sooner or later anyway") should be "forced" to 150 examples per month instead of the He 219, was discussed on 3 August 1943. It was speculated that American aircraft would soon be able to operate at heights above 10000 meters, where both the Me 209 and Me 410 were at the limit of their performance. The only machine considered suitable for operations at that height was "the Ju 188 with BMW 801 I." In the same breath it was remarked that "from this aspect the entire planned construction program is wrong."

Servicing a Ju 88 C, the most important long range night fighter of the early war years. This example obviously lacks airborne radar. Vernaleken Collection

Which High-altitude Aircraft Is the Right One?

An intense discussion of high-altitude aircraft took place in the RLM on 10 August 1943. In view of the plans to build small numbers of the Hs 130 E with Höhenzentrale (a third engine in the fuselage driving a large supercharger) in 1945-46, the Technical Office's (Abt. GL/C-3 E-2) representative, Major Hoffmann, demanded an investigation of the procurement possibilities. In his view, what was needed was a four-engined aircraft with a service ceiling above 14000 m, and maximum speeds of more than 700 kph at a height of 13000 m. Such a machine could be produced from both the Hs 130 and the Ju 188 with the three-man cockpit. The four-engined Ju 188 had the advantages of greater range, and being based on a type already in mass production.

Just 10 days later, on 20 August, Major Hoffmann pressed for a decision as to which high-altitude aircraft should be pursued, pointing out that the Henschel Hs 130 with Höhenzentrale had already flown.[74] The alternative to the Hs 130 was the Ju 388, a high-altitude development of the Ju 188, with the Ju 488 as a four-engined follow-on type. Hoffmann pointed out the advantages of the two alternatives:

"The Hs 130 offers more height, the Ju 388 more range. In view of advances by the enemy in the high-altitude sector, a decision must be made as soon as possible."

Generaldirektor Frydag, chairman of the Aircraft Steering Committee, proposed adopting the 488 if the object was large numbers. If higher operating altitudes had to be achieved quickly, however, they would have to produce a four-engined version of the Hs 130. At this meeting Prof. Hertel raised the idea of a developed version of the DB 603 high-altitude engine to power the Ju 388. To this Milch replied:[75]

"I know that if we produce the 388 we will have to keep working on improvements, and I am also aware of how difficult it is to continue development of a type already in production. The single kilometer above 13 is more valuable than the 8 kilometers from 0 to 8 km. It also costs more. It proves to be very expensive. This work must go on, of course, and if we can make improvements in the engine field, then that absolutely must happen. An aircraft is not finished at the moment it enters production, rather, this marks

The Hs 130, here an E-series aircraft with *Höhenzentrale*, was developed as a high-altitude aircraft from the beginning, and flew much earlier than the Ju 388. The choice fell on the Ju 388, however, and the type progressed no further than a series of prototypes. Griehl Collection

A Bf 110 *Zerstörer* of ZG 2. The Messerschmitt twin-engined fighter was already inferior to its Allied opponents during the Battle of Britain. Vernaleken Collection

the beginning of further development. This is the fastest way to achieve success."

It is obvious that plans already existed to initiate production and testing of the Ju 388 simultaneously.

Milch decided that "for the time being, the main focus is to be placed exclusively on the Ju 388." This decision opened the way for the 388. The question of a four-engined follow on was set aside, and for the time being was to remain in the project stage. At the same time it was decided to build just 30 Hs 130 Es with the Höhenzentrale.

At this meeting Frydag expressed regret that they had placed the Me 410 into production, instead of staying with the Ju 88, for this would have made it easier to now switch to the Ju 388. His colleagues also spoke out against the Me 410. *Oberst* Rowehl declared that the 188 possessed far superior handling with comparable speed and range. Werner Baumbach, a highly decorated bomber pilot, expressed similar views. Milch subsequently contemplated replacing the Me 410s for the bomber arm with Ju 188s. According to von Lossberg, the Ju 188 was also superior to the Me 410 in the night fighter role, though he would rather have the He 219. But the switches for production of the Me 410 had already been thrown, and the production line was already running. It is not surprising, therefore, that four days later it was decided that a switch by Henschel from the Me 410 to the Ju 388 could not be justified, because of the potential production loss of 850 machines.

An important conference was held on 2 and 3 September 1943 in the "Reichsjägerhof," with Göring in attendance.[76] In response to the "sensitive blows" struck by American daylight bombing, the Reichsmarschall declared that the following measures must be taken:

"...continued uninterrupted work on the buildup of the air defense plus marshalling of offensive forces against targets in the south and west. Against this background, investigations were to be carried out as to which direction the enemy would turn if faced by strong defenses. The enemy could develop;

a. very large, heavily-armed bombers (in terms of armament, bomb load and load factors)
b. bombers capable of higher altitudes and
c. high speed aircraft similar to the Mosquito."

This assessment was remarkably accurate, for the Allies were in fact pursuing all three avenues. According to Göring, a "Super-Heavy Destroyer" armed with heavy-caliber "Special Weapons" would have to be placed into service to counter "Super Bombers." He considered the Me 262, which was capable of heights of 11500 to 12000 meters, to be suitable for the high-altitude fighter role and Mosquito interception. While General of Bombers Peltz viewed the Ju 188 as suitable for high-altitude night attacks, Göring wanted to see "offensive weapons projects"—high-altitude bombers, high speed bombers, large bombers, and jet bombers. He reasoned that the *Luftwaffe* was lagging 1 1/2 to 2 years behind the Allies in development and procurement, and that this lead could be overcome by "continuous pursuit." Despite all efforts, this would only be possible to a certain degree; therefore, it was necessary to limit types to the Ju 188 while concentrating on the development and planning of jet bombers. Two classes were to be created: one as small as possible, and the other able to carry 2 tons of bombs over a distance of 1000 km (probably meaning the Ju 287). He believed in this path, for he saw a decisive advantage in the jet bomber, because Germany could only hope to achieve qualitative superiority over the enemy. All efforts were therefore to be dedicated to the "fastest possible" research, development, and production of jet bombers. Milch was personally instructed to "consciously expedite" this planning proposal.

Junkers' monthly report for September 1943 records contracts for development of the three-man cockpit for the Ju 188, plus de-

With its extended fuselage, the Ju 88 H had increased range for missions over the North Atlantic. The aircraft seen here is fitted with FuG 200 search radar. Griehl Collection

The Ju 88 A-4 remained the backbone of the German bomber units until the end of the war. Vernaleken Collection

sign and construction of two Ju 388 K-0 high-speed bombers (2.1-million Reichsmark). A separate order called for the design and construction of two examples each of the Ju 388 J-1 night fighter (941,000 RM) and Ju 388 L-1 long-range reconnaissance aircraft (891,000 RM).[77] These must have been orders for the V1 to V6 prototypes. There were also fixed contracts for development of the Ju 88 H-1 and H-2 long-range reconnaissance machines. Ten prototypes of each version were to be constructed.[78] The costs of 1.74 and 1.45 million for the two series must be seen in relationship to the price of a single Ju 290, which clearly exceeded 900,000 RM. Advance orders were also received for the production of 20 Ju 388 K-0s (4.28-million RM). These were to be conversions, using major components of the Ju 188. Also noteworthy are orders for the prototype installation of the Jumo 222 in the Ju 188, and the development, design, and construction of the Ju 188 G-2. Efforts to develop high-altitude combat aircraft were obviously not limited to the Ju 388, for the same monthly report lists an advance order for the airframe construction division to carry out an experimental installation of the TK 11 high-altitude supercharger (Hirth exhaust-driven turbosupercharger 9-2281) in the Ju 88 A-4. This machine was delivered at the end of December, Flugkapitän Matthies having flown the Ju 88 V93, the flying test-bed for the TK 11 on 1 November 1943, and new orders were received to convert 75 Ju 88 A-4s to the TK 11.[79]

The first addendum to the September report, which is unfortunately not dated,[89] contained a great surprise for Junkers; for the first time, 586 Ju 388 Js appeared in the RLM's summary of production aircraft to be manufactured using major components from other types.[81] In the months that followed, however, this order was not pursued.[82]

The Ju 388 K-0 and L-0 were chosen as the central theme of a meeting held by the Technical Office on 12 October 1943, the discussion topic being the "Revenge Program, Bombers." As a result, Director Thiedemann of Junkers would have to construct 500 examples of the Ju 388, instead of the 350 (plus 100 Ju 188 S-3s) specified in the first Delivery Plan 225. Raw materials and labor forces were not even assured for the then current Delivery Plan 223, however, and this "Revenge Program" was wishful thinking with unrealistic production figures. The measures proposed by Milch to produce an effective increase in bomber production in 1944 included halting production of the He 111, falling back on "transport capacity," and increased use of "concentration camp operations." Later in the discussions, Thiedemann reported that work was proceeding on 20 Ju 388 K-0s "to bridge the gap until the Ju 388 enters production." With the proviso that 600 workers be provided as soon as possible, he proposed increasing the K-0 series by another 20 aircraft from major components of the Ju 188. The idea of fitting some of the K-0s with a pressurized cockpit was rejected, as it would offer no short term benefits. Thiedemann also stated that the 20 planned Ju 388 L-0 reconnaissance machines could still be completed as bombers. *Oberst* von Barsewitsch (General of Reconnaissance, 1943-45) considered the idea unsupportable, citing "the serious shortages and extremely heavy losses in reconnaissance aircraft." Generalmajor Kleinrath (head of Department 6, Quartermaster-General's Office) agreed with him. Milch therefore decided to initially allow work to proceed on 40 Ju 388 K-0s and 20 Ju 388 L-0s. Thought was already being given to future developments: Prof. Hertel declared that a Jumo 222 with exhaust-driven turbosupercharger would raise the Ju 388's maximum speed by 50 kph, to 700 kph, and *Oberst* Peltz underscored the necessity of having this higher performance Ju 388 by the end of 1945.[83]

It appears that no clear decision in favor of the Ju 388 had yet been reached, however, for on 29 October 1943 discussions were held on development of the Bf 110 and its successor, the Me 210/

The Bf 110 continued in service as a night fighter until the end of the war. Seen here is WNr. 160 791, a late model G-4, photographed in Brunswick. Crow Collection

Envisaged as the German response to the Mosquito, the Ta 154 was a failure. The A-1 seen here was used as a training aircraft late in the war. It is seen here after discovery by Allied forces. Crow Collection

410, with Jumo 213 and DB 628 engines. The Ta 154, He 219, and Ju 388 were brought into the analysis to determine if the costs for the new types were warranted.[84] Halfhearted decisions and the constant resumption of discussions were characteristics of the department head and development meetings.

Between autumn 1943 and spring 1944, the heads of air armaments made a series of official visits to important aircraft manufacturers under the designation "Operation Hubertus." The visit to the Junkers factories by Göring and Milch on 4 and 5 December 1943 was part of this program. The two decision makers wanted to see for themselves how their plans were being turned into reality, examining Junkers' progress in the field of jet engines. Visibly satisfied, the two Nazi bigwigs praised the accomplishments to date of this "model plant," which was giving Germany "the chance to do great things in the field of jet bombers and fighters."[85]

On 12 November 1943 a high-altitude version of the Bf 110 with a lengthened fuselage, extended wingtips, and DB 628 engines was proposed as a stopgap solution until the Ju 188 or Ju 388 with Jumo 213 engines became available in significant numbers. The proposed variant of the Bf 110 would be capable of operating at heights above 10000 m. Speed seemed to be called for, because the B-29 was expected to begin operating over Germany from bases

The Ju 287 V1 "flying mockup," which was built to test the concept of engine arrangement and forward swept wing, seen at the Junkers branch facility in Brandis. Haubner/Misna Collection

in England. The development of high speed bombers was discussed in late November, resulting in a plan for increased production. The current output of 332 aircraft per month was to rise to 1,140 aircraft per month by the end of 1945, with the addition of the Do 335 and Ar 234. The possibility of canceling the He 219 in favor of the Ju 388 was also discussed, because the two types were viewed as being equal. The idea was rejected, as the drop in production associated with the change could have led to a reduction in total output.

Milch obviously placed great hopes in the new Do 335, for on 3 December 1943 he proposed that the new aircraft be employed as a high speed bomber, day Zerstörer, and long-range reconnaissance aircraft(!). The Do 335 was to be one of the main production types, along with the Ta 152, Me 262, Ju 388, and He 177, with the latter two types seen as medium and long-range bombers.

The subject of a replacement for the Ju 388 was discussed on 7 December 1943, even before the first flight of the Ju 388 V7, as it was feared that the type's performance, especially its service ceiling, would no longer be satisfactory in 1945. While a general improvement in performance was expected with the installation of the Jumo 222, with two-stage supercharger or exhaust-driven turbosupercharger, no improvement in high-altitude capability was anticipated. A service ceiling in excess of 13000 meters did not appear achievable, even with the deletion of a wing or fuselage tank at the expense of range. The ultimate solution for achieving the required performance of 700 kph at a height of 13500 meters was therefore seen as a four-engined machine with a larger wing—the Ju 488. It was scheduled to enter production in May 1945, barely half a year after most reconnaissance units converted to the Ju 388. No definitive decision was made on the fate of the Hs 130, whose performance approached that of the Ju 488, but with much less range. Converting the machine to four engines and building small numbers for the reconnaissance role was still viewed as a possibility.[86]

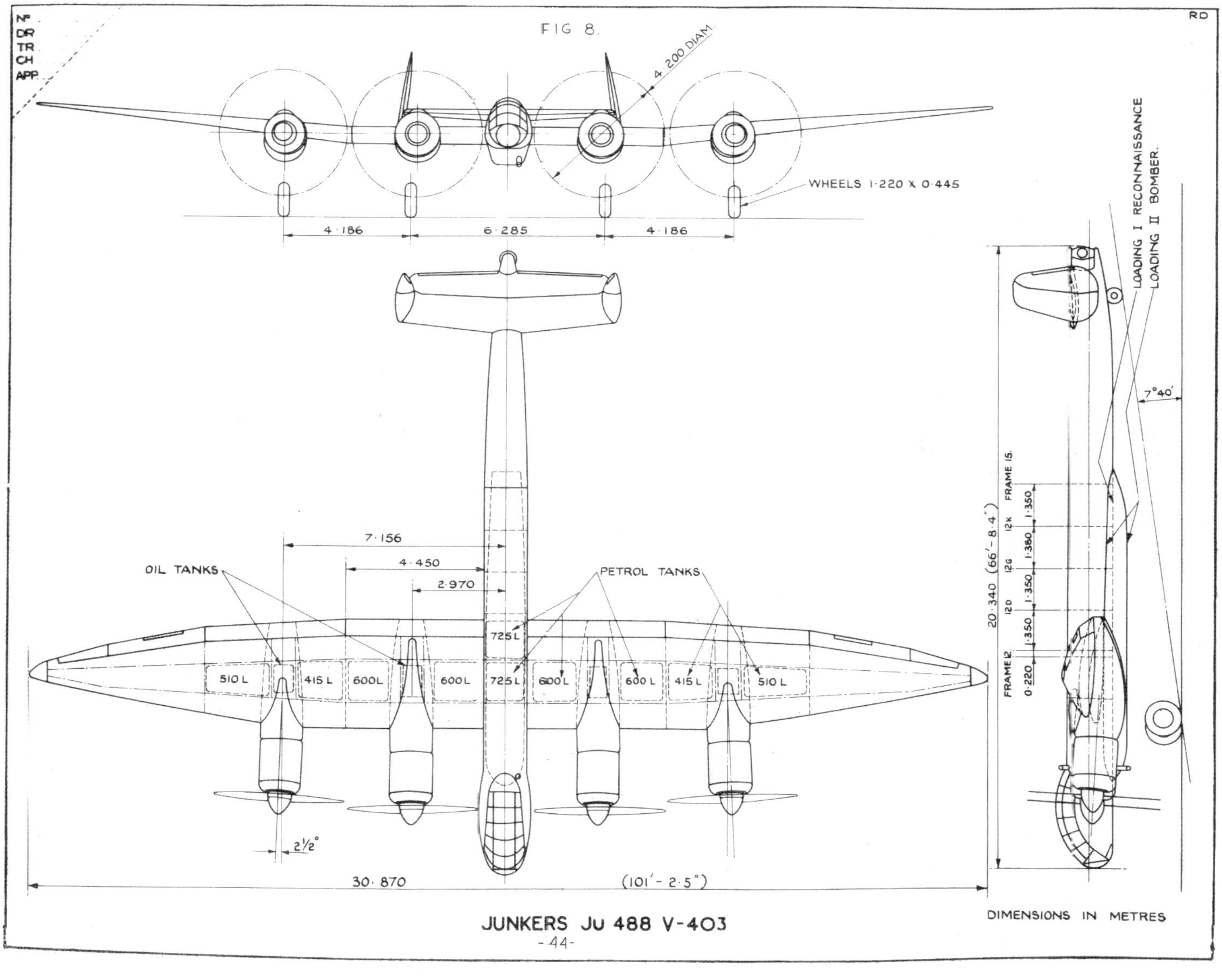

Drawing of the Ju 488 V403 from Allied sources. IWM

The Ju 388 was discussed again on 13 December 1943. Kammhuber and von Lossberg pleaded for various combinations of the planned armament, and discussed the automatic cannon to be installed and their firing ranges. There was disagreement on the merits of a MK 103 or MK 108 installation, and the only aspect that was not in dispute was the installation of two MG 151 cannon in an oblique arrangement.

In mid-December 1943 those in the RLM once again discussed the advantages of the Ju 288 compared to the He 177. It was apparent that the RLM had still not completely discarded the former type.[87] It was claimed that, according to Prof. Hertel, the Ju 288 was 90 kph faster than the He 177 B-5. Knemeyer, head of the subgroup responsible for aircraft development, nevertheless proposed that the Ju 288 be dropped, and be replaced at the end of 1945 by the Ju 488 in the high-altitude role, and a version of the Ju 388 powered by the Jumo 222 for the high speed role.

In the October to December 1943 reporting quarter, the conversion of 10 Ju 388 L-0s was cancelled and converted into "new construction outside the series delivery program," with an identical contract value. In December delivery was taken of the Ju 388 L-0 with the Werknummer *300 001, PG+YA, as the Ju 388 V7. The order for Ju 388 K-0s was increased to 50. Two new contracts for further development of the Ju 188 also appear noteworthy: one was for development of a four-engined Ju 188 (probably the later Ju 488), and the other for development and construction of the Ju 188 G-3.*

The writings of Dipl.-Ing. Reinhart Preuschen, a flight testing engineer in the Dessau flight department (FIA),[88] also make first mention of the Ju 388 L-0/V7 in December 1943. Based on the number of available test reports, this aircraft, which flew long before completion of the V1 to V6 prototypes, was the backbone of the Ju 388 flight test program at Dessau. The V7 was probably also the aircraft that was supposedly inspected by high-ranking members of the RLM while in an advanced stage of construction. As mentioned earlier, Göring and Milch visited the Junkers factories on 4 November 1943.[80] According to Irving, before they left the Dessau factory they were shown a Ju 388 that was under construction. The aircraft was fitted with BMW engines. Göring, who in October directed his state secretary to arrange a demonstration of the latest aircraft and weapons for the Führer at Insterburg airfield, insisted that the machine be shown to Hitler there. It wasn't necessary that it fly, Hitler only had to see it.[90]

Most of the other aircraft from this first Ju 388 pre-production series can be verified by documents and logbook entries, for example, Werknummer *300 009, which was with the* Erprobungskommando *Ju 388 in September 1944,* Werknummer *300 008 (PG+YH) with the Junkers test station at Munich-Riem on 14 October 1944,[91] and finally,* Werknummer *300 004 at Dessau in November 1944.[92]*

Although the Ju 388 V7 had completed only a few test flights by mid-January 1944, at a meeting in the RLM on 14 January it was reckoned that, from the first half of 1945, the "high-altitude

An example of the Ju 88 G-6 with Jumo 213 A engines. This captured example in British hands has a wooden nose cone, and is equipped with FuG 240 radar. The G-6 was treated as an alternative to the Ju 388 for medium altitudes. Haubner/Misna Collection

sector" would be covered by the "Ju 388 and 188 with Jumo 213 E." The high-altitude version of the Bf 110 and the Ta 154 with GM 1 were now given no chance. Construction of the heavy fighter version of the Bf 110 was to continue until January 1945, and the night fighter version until July, after which they would be replaced by the Do 335. At a development conference on 2 March 1944, however, Milch once again considered a higher-performance Bf 110 powered by the DB 605 D, to be kept in production until a successor from the Ju 88/188/388 series with a three man crew was actually available. Reflected in these thoughts was a certain degree of doubt about the availability of the Ju 388 J, or the alternative Ju 88 G-6/G-7.

The situation of the air armaments industry in Germany took a turn for the worse in February 1944, when the USAAF launched an operation dubbed "Big Week." This series of concentrated raids on known aircraft factories inflicted serious damage. One of the consequences was the formation of the "Fighter Staff." From March 1944 it was responsible for the rebuilding of aircraft factories, as well as their transfer to underground installations. The Jägerstab was under the joint leadership of Milch and armaments minister Speer. Karl-Otto Saur was appointed chief-of-staff. With the creation of the Fighter Staff, Speer's armaments ministry acquired greater influence in the decision-making process. Within a few months this body concentrated the available, steadily-weakening forces, while at the same time raising the production quotas of an increasingly limited number of types, mainly fighters for home defense.

Even though the Jägerstab remained fundamentally committed to the Ju 388 as a high-altitude aircraft, the effects of the altered overall situation soon made themselves felt. Despite initial production preparations, at a development conference on 2 April 1944 the fate of the Ju 388 K hung in the balance.[93] Generalmajor Vorwald brought up for discussion the cancellation of the bomber version of the Ju 388 in favor of the heavy fighter variant. Koller[94] replied that, as the only high-altitude aircraft, the Ju 388 also formed the basis for the future reconnaissance aircraft, and therefore they could scarcely do without it. Milch therefore decided to leave the aircraft in the program as before. It was then discussed whether, for the sake of type simplification, the Ju 88 could not be built instead of the Ju 388 J, provided they could "conjure up improved altitude performance." This manner of speech, the use of the word "conjure," casts a revealing light on the distorted view of reality that prevailed in the minds of some of the decision makers. Knemeyer pointed out that this was feasible from the performance aspect, but that there was no pressurized cockpit. Finally, Koller made reference to the Ju 388's priority for the Defense of the Reich, for which an all-weather heavy fighter with an endurance of four hours must remain in the program. Milch agreed, and declared that the Ju 88/188/388 with three man cockpit was the "right" night fighter, thus eliminating the He 219. Koller also took a position in favor of the Hs 130, whose cancellation had already been ordered. He argued that it was needed for development, and because of its pressurized cockpit, as they had to anticipate the Americans coming at an altitude of 14000 meters in half a year. In response, Milch once again stressed that the Ju 388 was "the" high-altitude aircraft. The Hs 130 with Höhenzentrale was of value solely as an experimental aircraft. Frydag admitted that:

> "...the Höhenzentrale is too unproven to send a crew to England with it. The problem isn't the cockpit, which has been solved by Henschel and Junkers, rather, it is the high-altitude engines."

The RLM placed great hopes in the Me 264 for Atlantic and long range reconnaissance. Haubner/Misna Collection

The Ar 234 was the most successful German reconnaissance aircraft. Seen here is an Ar 234 B-2 (8H+DH), which was captured by the British in Stavanger, Norway, and later tested in England. Handig Collection

Speer, who shared leadership of the Jägerstab with Milch, did not attend a meeting until 26 May 1944, when Milch showed him the work that had been done in the previous three months.[95] They had worked out Delivery Plan 226 (1st version), which immediately reduced the number of production types to 32. A further reduction to just 16 types was to follow at the beginning of 1946. The Ju 88/388 continued to fill the "bomber sector," pending the availability of the Ju 287. This so called "Reduced Fighter Staff Aircraft Program" was finalized on 3 June 1944, a few days before the Allied invasion of Normandy. For the heavy fighter role it envisaged the Ju 88, Ju 388, He 219, and the Do 217, plus various versions of the Me 110 G, Me 210 A/C, and the Ta 154 A and C.[96]

Prior to the delivery of the first production Ju 388 L-1, delays in the availability of the BMW 801 forced a rethinking of operational planning. On 10 June 1944 there was a meeting in the office of the Quartermaster-General (Gen.Qu. 6. Abt.) between representatives of the Gen.Qu. 6.Abt., the GL/C-E, and the General of Reconnaissance. The highest-ranking representatives were General Staff *Oberst* Eschenauer, *Oberstleutnant* Knemeyer, and Major Rehm.[97] It had turned out that instead of the 30 Ju 388 Staffeln called for in Delivery Plan 226, only about 11 could be equipped with the type for the long-range reconnaissance role. It was therefore decided to cover the remaining 19 *Staffeln* by equipping six with the Do 335, six with the Ju 388 (Jumo 213 E with MW 50 boost), and the rest with the He 177. The number of He 177 units was expected to be smaller, however, on account of materials shortages, resulting in a corresponding reduction in the total number of Staffeln. Furthermore, it was agreed that the Ju 388 Staffeln earmarked for the medium-level role would be equipped with aircraft powered by the Jumo 213 E, and that the He 177 *Staffeln* would later be equipped with the Ju 488. It was assumed that the Ju 388 "would probably not be kept at the front for very long," and that a "replacement aircraft type" would have to be reckoned on in 1946.

It is interesting that just three Staffeln equipped with the Ar 234 had been put into service by that date, and it is not entirely clear whether this was because not enough aircraft could be pro-

Taken in July 1943, this photograph depicts a Ju 88 prototype (SL+PC), probably the V60. The aircraft was used to test the bulged bomb bay beneath the fuselage. Griehl Collection

Another view of the same aircraft. Note the sensor probe on the starboard wing. Griehl Collection

duced, or because the authorities had too little confidence in this revolutionary jet aircraft.

Reconnaissance over the Atlantic was a real problem child. The old tactic of flying low and shadowing enemy shipping had to be abandoned, because shortages of production capacity and raw materials made it impossible to equip large formations with "large aircraft possessing adequate structural strength." The following proposal was therefore made:

"A tactic will have to be adopted in which single high-performance aircraft capable of about 700 kph at maximum boost altitude will fly above the weather at heights of approximately 12000 to 13000 meters, carrying out the reconnaissance and shadowing roles with radar."

In Knemeyer's view an aircraft suited to this new tactic could be developed from the Me 264. Until then (roughly the winter of 1945-46) they would have to get by with a single Staffel of Ju 290s, which would later be replaced by a Staffel of He 177s. The air armaments situation ruled out production of the Ju 390. These limited resources would be bolstered by prototypes and pre-production versions of the Me 264 (old version), the Ju 488, and an unnamed Hütter aircraft (probably the Hü 211).

It is also remarkable that consideration was being given to a standard reconnaissance aircraft that was supposed to cover all areas (day, night, and high-altitude reconnaissance). It was also decided that interchangeability of normal and high-altitude engines would have to be a design feature of such an aircraft.

Although it was already obvious that a successor to the Ju 388 was needed, the type soon became one of the main pillars of air armaments planning. First, on 29 June 1944, it was decided that the increase in production of the Ju 88/388 to 280 aircraft per month in September 1944 would be followed by a further increase to 420 per month in autumn 1945, while production of the He 219 would continue at the rate of 50 per month.

On 1 July 1944, at a meeting held by Göring, the organizational talents of "*Hauptdienstleiter*" Saur, who was obviously taking over more and more of Milch's duties, received lavish praise. Before the meeting ended, a long list of aircraft were struck completely from the production program.[98] In the new aircraft program agreed upon on 8 July 1944 (Delivery Plan 226, 2nd Version), the Ju 388 was one of five main types—together with the Me 262, Ar 234, Do 335, and Ta 152—which would fill all roles. All other types were to be cancelled.[99] Approximately 550 Ju 388s and 525 Do 335s were to be built monthly. The bulk of these, together with 500 Ar 234s to be completed each month, were to go to the bomber arm, resulting in a significant rise in the quality and quantity of its equipment.[100]

Chapter VI of this book is dedicated to the extremely ambitious production plans for the Ju 388 and their actual outcomes.

The second part of this chapter outlines the most significant design innovations, the pressurized cockpit, and remotely controlled tail turret. Testing of the new bulged bomb bay, also planned for the Ju 188 G, began in the summer of 1943 using the Ju 88 V60 and the Ju 188 V11 (obviously a prototype of the Ju 188 G-2). At the beginning of August 1944 both machines were used in preliminary experiments for the Ju 188 G-2 and Ju 388 K-0/K-1, their purpose to determine at what angle 250-kg and 50-kg bombs left the new bulged bomb bay on release. A Siemens movie camera was mounted beneath the starboard wing to film the dropping of the bombs.[101]

Ju 388 Design Development

The formal changeover from the Ju 188 J/K/L to the Ju 388 J/K/L took place between May and September 1943. An order for 10 zero series, or pre-production aircraft, was issued on 24 May 1943. A series of drawings from the Amsterdam design bureau depicting the forces on the rear fuselage have survived in the special collection of the Deutsches Museum. This document, dated 18 May 1943, already makes reference to the Ju 388 K, as does an overview of the dimensions of the horizontal stabilizer dated 28 May 1943. Here are two almost identical drawings; one overwritten with "Ju 188 E," and the other titled Ju 188 J/K/L, with the "1" crossed out and replaced by a handwritten "3." It is unclear, however, if the new RLM number 8-388 had in fact already been issued, or whether the

The transfer of technology from the Ju 288 to the Ju 388 is obvious in this view of the pressurized cockpit. Deutsches Museum

document was corrected at a later date, for the aircraft division's monthly report makes no mention of the Ju 388 until September.

The Ju 388's speed and altitude capabilities were supposed to enable it to operate in well-defended airspace. From the concept of a high-altitude, high-speed aircraft came the requirement to subject some components to a process of drag and weight reduction. The crew compartment in particular, and of course the engines, had to be adapted to high-altitude flight.

It was not without justification, therefore, that the armament of the Ju 388 K and L was limited to a remotely-controlled tail turret. While this equipment was technically demanding, its location and dimensions resulted in a very low-drag installation.

Development of the Pressurized Cockpit

A significant factor in the development of the new three-man cockpit, in addition to designing it as a pressure capsule, was drag reduction. The crew compartment was reduced in size compared to that of the Ju 188, and drag was reduced by eliminating the dorsal turret. The aerodynamics were further improved by narrowing the crew compartment and deleting the ventral bath.

In designing the pressurized cockpit, Junkers could draw on a wealth of experience gained with the Ju 49, EF 61, Ju 86 P and R, and the Ju 288.

The pressurized cockpit developed for the Ju 49 high-altitude research aircraft was the first ever built in Germany. It had a cockpit dome with portholes, which from a static point of view were extremely favorable. To provide the best possible view from the small portholes, these were situated directly in front of the crew at eye level. The portholes were double-glazed. The Ju 49's pressure capsule could be lifted up and out of the fuselage. This was largely for construction and technical-production reasons, for sealing and repair proved much simpler if an expensive and sensitive component like a pressure capsule could be built and serviced separately. For small crews of two to four men, Junkers clung to the idea of the

The pressurized cockpit of the EF 61 with Elektron cupola. Griehl Collection

The Ju 86 P/R pressurized cockpit with glazed "beetle eye" nose and "bay" window in the canopy glazing. Griehl Collection

The cockpits of the Ju 288 and Ju 388 were also similar in detail: in this photo the moisture traps in the double glazing are visible. Several panels also have a round glass insert for astral navigation. Deutsches Museum

pressure capsule as a separate component until the end of the war. The pressurized cockpits of the EF 61, Ju 86, Ju 288, and Ju 388 were all attached to the fuselage using threaded ball-socket connectors, and were removable. They were of course aware of the fact the entire fuselage of a commercial aircraft had to be pressurized.

Admittedly there was another reason for the separate pressurized capsule of the Ju 49: to prevent leaks by eliminating the forces and loads produced in the fuselage during flight.

The pressurized capsule of the EF 61 was the first to have full glazing. The pressurized cockpit was initially produced as a single-wall plexiglass dome. This plan had to be abandoned, however, because the rather brittle plexiglass, with its high thermal expansion and sensitive outer service, was not equal to the demands of a pressurized cockpit. As well, there were insurmountable difficulties with resistance to penetration and misting from inside. The interchangeability of the dome, and the installation of weapons and other equipment caused other problems. In the end, prior to the commencement of flight trials the EF 61 was temporarily equipped with a cupola of welded Elektron, and an offset cupola with portholes.

The Ju 86 P was the first to have a fully-glazed pressurized cockpit. By then Junkers had amassed sufficient experience for the fuselage forces to be transmitted through the walls of the pressurized capsule without affecting its integrity. The glazed portion of the Ju 86 P's pressurized cabin was divided into a fully-glazed spherical cap in front, and the cockpit canopy directly behind it. The panels next to the heads of the crewmembers allowed them to see vertically downwards. Loads were absorbed both by the canopy framing, and the outer glazed panels. For the first time, the rear bulk-

The significant innovation in the further development of the Ju 188 was the adoption of a pressurized cockpit, for which several mockups were built. This mockup was photographed in early February 1943, and the photo, which is part of the company album held by the Deutsches Museum, bears the caption "Ju 188 high-speed bomber." Note the periscope for the FA 15 (PVE 11) behind the cockpit. Deutsches Museum

Frontal view of the same mockup. In all significant aspects it is identical to the ultimate Ju 388 high-altitude cockpit. The only differences are the absence of a frame on the triangular side panel next to the nose cap, and the presence of the side window, which was later deleted. Deutsches Museum

This photo of a somewhat later mockup was taken at the end of February 1943. It may have been a modification of the mockup in the previous photo. In addition to the PVE 11, there is also a rearview periscope. Also interesting is the bulged bottom, which was not present on the production version. In this photo it is hidden by the wooden frame on which the mockup is sitting. Deutsches Museum

This forward view shows the division of the triangular side panels and the rearview periscope for rearwards firing armament. Note the Ju 88 control yoke. Deutsches Museum

The universal usability of the pressurized cockpit is illustrated by this mockup of the cockpit of the "Ju 188 V403" which, as the four throttle levers reveal, was a prototype of the Ju 488. True double glazing with moisture traps has already been installed. Deutsches Museum

At the beginning of 1943 the mockup was modified for evaluation of a heavy fixed armament (probably 37-mm). Deutsches Museum

This photograph documents the mockup's gradual metamorphosis from a cockpit for a high speed bomber and reconnaissance aircraft to a night fighter cockpit. Antennas for the FuG 212 have already been installed on the still uncovered nose. Through the cockpit frames one can see the "armor," which is represented by a piece of chipboard. Griehl Collection

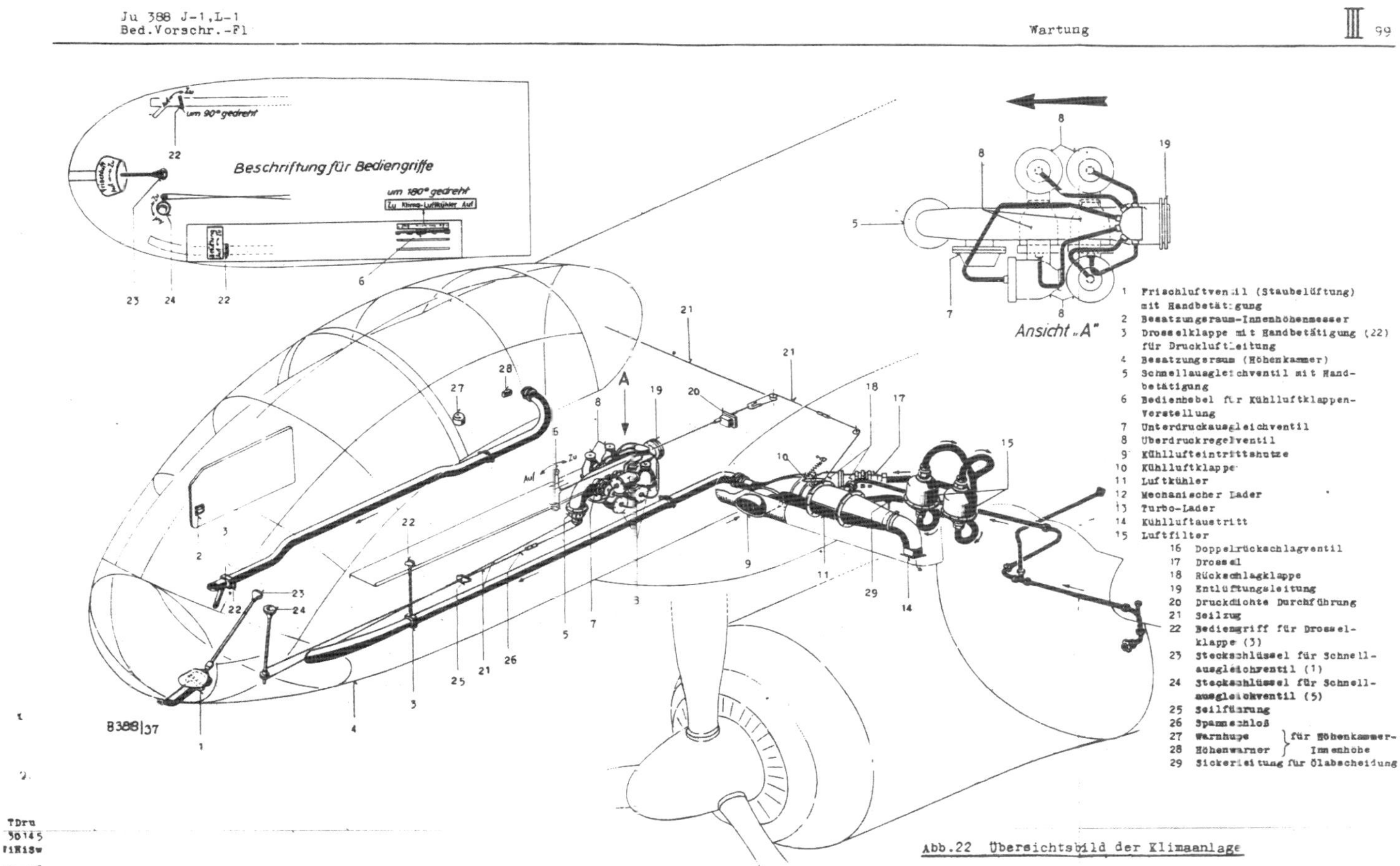

Drawing of the ultimate production version of the climate control system from the pilot's operating notes (BDV).

head was a completely flat, rigid wall. The side walls of the pressurized capsules in the Ju 288 were also almost flat, the cockpit cross section resembling a rectangle with rounded corners. The slightly curved walls were stiffened by means of stout formers and tension rods.[102]

Taken a short time later, this photo shows the mockup still with antennas for the FuG 212. One can even see where the opening for the 37-mm cannon was puttied over. Griehl Collection

On 8 September 1942 the RLM issued an order for the development and mockup construction of the new three-man cockpit for the night fighter and high-speed bomber variants of the Ju 388.[103] At the end of the year several full scale mockups were built for the usual purpose of evaluating the layout, and the arrangement of equipment in the new cockpit. Six fixed weapons (MG 151 or MK 108 cannon) were supposed to be installed in the long-range night fighter version. These mockups were extensively photographed for the factory photo album, thus documenting the development of the later Ju 388 pressurized cockpit. The transfer of technology from the Ju 288 to the Ju 388 was obvious, as the shape and layout were essentially identical.

Another photograph of the mock up taken in February 1943 shows a defensive position at the rear of the canopy, with a handheld machine-gun (as on the later pre-production machines) and radar antennas. The first cockpit mockups for the night fighter and Zerstörer versions appeared in March 1943, and were derivatives of the design for the bomber and reconnaissance variants. While the exterior shape of the canopy was the same, there were significant differences in the forward canopy framing. The fully-glazed rounded nose was replaced by near vertical glazed panels for the pilot, and a stepped nose similar to that of the Zerstörer and night fighter versions of the Ju 88. The modified nose of the mockup accommodated large-caliber (probably 37-mm) fixed weapons and radar antennas.

Incidentally, in the factory album the photos of the cockpit mockups were usually captioned "Ju 188 high speed bomber," and in one case "Ju 188 S."

The double-glazing of the Ju 388's canopy was formed from 8- and 6-mm plexiglass sheet, and the distance between the panels was 10 mm. The intermediate frames between the plexiglass sheets were obviously made of scrap material; the edges of the thicker panels were double milled, while the thinner ones were single-milled. Rectangular rubber strip (4 x 8 mm) was used to seal the double panels.[105] In order to prevent condensation from forming in the intervening spaces, moisture-absorbing capsules were installed in the double panels.

In developing the climate control system, necessary for maintaining pressure and heating, the designers adopted a policy of simplicity from the beginning. Instead of a constant internal pressure, equivalent to a height of 3000 meters for example, they aimed for an over-pressure of 0.23 atmospheres. Production of such a pressurized cockpit was much simpler, especially with regard to sealing. At 14000 meters the internal pressure, meaning the air pressure inside the cockpit, was equivalent to a height of 8000 meters. While this meant that the crew had to wear oxygen masks, oxygen consumption was much lower than it would have been without pressurization. For production reasons, it was specified that as much of the equipment needed for the climate-control system as possible should come from existing stocks. The lowest cockpit temperature during high-altitude flights at night in winter was to be -8° C, for which heated flight suits were specified.

Pressurized air from the engines, cleaned by an active charcoal filter, was used to maintain pressure in and heat the cockpit; this method is essentially the same as that used in present day commercial airliners. In the beginning, the pressurized air was taken from directly behind the turbosupercharger, but later this was changed to behind the intercooler (and behind the mechanical supercharger of the BMW 801 TJ). Non-return valves were installed in the lines that carried pressurized air from the power plants to the cockpit to prevent a loss of pressure in the system following the failure of an engine. Provision also had to be made for manual closing of the lines to prevent the possible entry of carbon monoxide in the event of engine trouble. As well, in the beginning a flow regulator was interposed to minimize the effect of engine speed. This resulted in an almost constant flow into the cockpit of 150 kg/hr per engine, but once the source of pressurized air was switched to behind the intercooler the regulator was dispensed with.

During high-altitude flights cockpit pressure was regulated by pressure relief valves. Because the existing pressure relief valves were extremely volume-sensitive, six were arranged in parallel to regulate cockpit pressure. A rapid pressure equalization valve was used for manual decompression and flights with no pressurization. When it was open, the pressurized air served solely to heat the cockpit. Alternatively, especially in summer and at low altitudes, fresh air could be fed into the cockpit by a forced air ventilation system.

The following receptacles and valves were present in the high-altitude cockpit of the Ju 388:

1. Receptacle for forced-air ventilation
2. Two receptacles for pressurized air from the power plants
3. Valves for maintenance of constant internal pressure
4. Safety valve against external over-pressure
5. Rapid pressure equalization valve for manual decompression of the cockpit (combination with No. 3 possible)

In the first prototypes and pre-production aircraft the climate-control system was controlled by means of a so-called "Climate Switch" located near the radio operator's position. This rotary slide could be used to select the system's various operating modes, for example "Forced Air" and "Depressurize Cockpit."

Testing revealed that the climate switch was unsuitable for practical use. In new condition it was difficult to operate, with too much play it became leaky, and after a moderate amount of use it usually jammed, because sand and dust entered through the forced-air system. Another source of criticism was the complicated housing, which was found to "complicate construction."[106,107] The climate switch was therefore replaced by individual valves (see below), with which the crew had to select the desired operating status. It was also found that the existing filters ceased to be able to absorb oil vapor at temperatures above 80°C. As a result, oil fumes often entered the cockpit. Intake air coolers therefore had to be installed in the leading edge of the wings between the engines and fuselage. The intakes that had been incorporated into the wing design for the installation of "Kärcher Stoves" were used for cooling air inlets. The air exited the bottom of the wings through openings that were also already present.[108] The cooling air inlet had to be modified for aerodynamic reasons, however. The intake air cooler was ultimately made adjustable, to provide sufficient heating capacity for cold weather operations.

Development work on the cockpit continued, even after completion of the first pressurized cockpit at the end of November; the first flight of the Ju 388 in December 1943.

A design error was discovered when the first periscope was installed in a cockpit canopy. A lug cast on the periscope housing for attachment of the bracing strut had been overlooked. Tests were carried out to determine

"whether a lower installation of the periscope is possible with respect to eye position and relocation of the FA-15 shafts, or if modification of the periscope is unavoidable."[109]

Between 26 and 28 February 1944 stress tests were carried out with a Ju 388 canopy in the presence of Director Zindel, in which the canopy was subjected to loads simulating a high-G pullout. The canopy was certified with and without constant internal pressure. A specially made crew compartment, equivalent to the lightened production version, was used for the tests. There were no internal fittings, because the forces they exerted on the canopy during a pullout (4.7 g) were simulated with the aid of a system of levers. Without internal over-pressure a small wrinkle became visible between Frames 7 and 8; however, this disappeared again when a constant internal pressure of 0.25 atm was applied.[110]

At the beginning of May 1944 development of the pressurized cockpit, especially with respect to the climate and pressure control systems, was still not complete. The Dessau design bureau summed up the outstanding points in a memo. The most striking aspect of this document dated 6 May 1944 is that many of the complaints later expressed by the *E-Stelle* may be found there: the recesses in the right instrument panel were not deep enough, the pilot's oxygen hose was too short, the observer's seat could not be folded out of the way if the pilot's seat was too far back (in the last two stops), the sliding panel in the canopy (for use in bad weather) warped easily and leaked, etc. Also seemingly worthy of mention is that the RF 2B rearview periscope hampered the pilot, and when not in use was turned out of the way. The ratchet mechanism in the device was criticized as being inadequate.

In addition to the problems with the climate switch already mentioned, its procurement was also anything but simple. For this reason the switch from the V7 had to be fitted in the Ju 388 L-0 WNr. 300 003 (PG+YC) located in Merseburg. Soon afterwards, the switches were replaced by four separate valves. The 19-3416 A pressure equalization valve made by Schaeffer and Budenberg was chosen for the forced-air intake and cockpit air outlet. The company also designed the air regulator flaps for closing of the supercharger air lines, based on a Junkers pattern. Production was to be undertaken by the airframe manufacturer, however. Prototype installations were to be carried out in the Ju 388 V6 and V7.

The noise created by the forced-air entry came under much criticism. During flight trials at Merseburg a metal or wooden scoop over the intake was tested successfully, and was installed on subsequent aircraft.[111]

Despite the construction of mockups, many cockpit-related problems were not discovered until testing of the first prototypes and pre-production aircraft. The crews complained about the absence of a handgrip on the left side of the back rest on the pilot's or radio operator's seat; moreover, the head armor of the pilot's seat struck the EZ 6 receiver when the seat was tilted prior to an emergency exit. The armor of the observer's seat also had to be modified to make it possible to get by the folded seat with a parachute on. The pilot's oxygen hose proved to be too short if an oxygen parachute (high-altitude parachute) was not used.[107] In May 1944 production was already under way of a pre-production series of 1,000 of these high-altitude parachutes. The new parachute could be worn as a seat, back, or chest pack, and was to be used as stan-

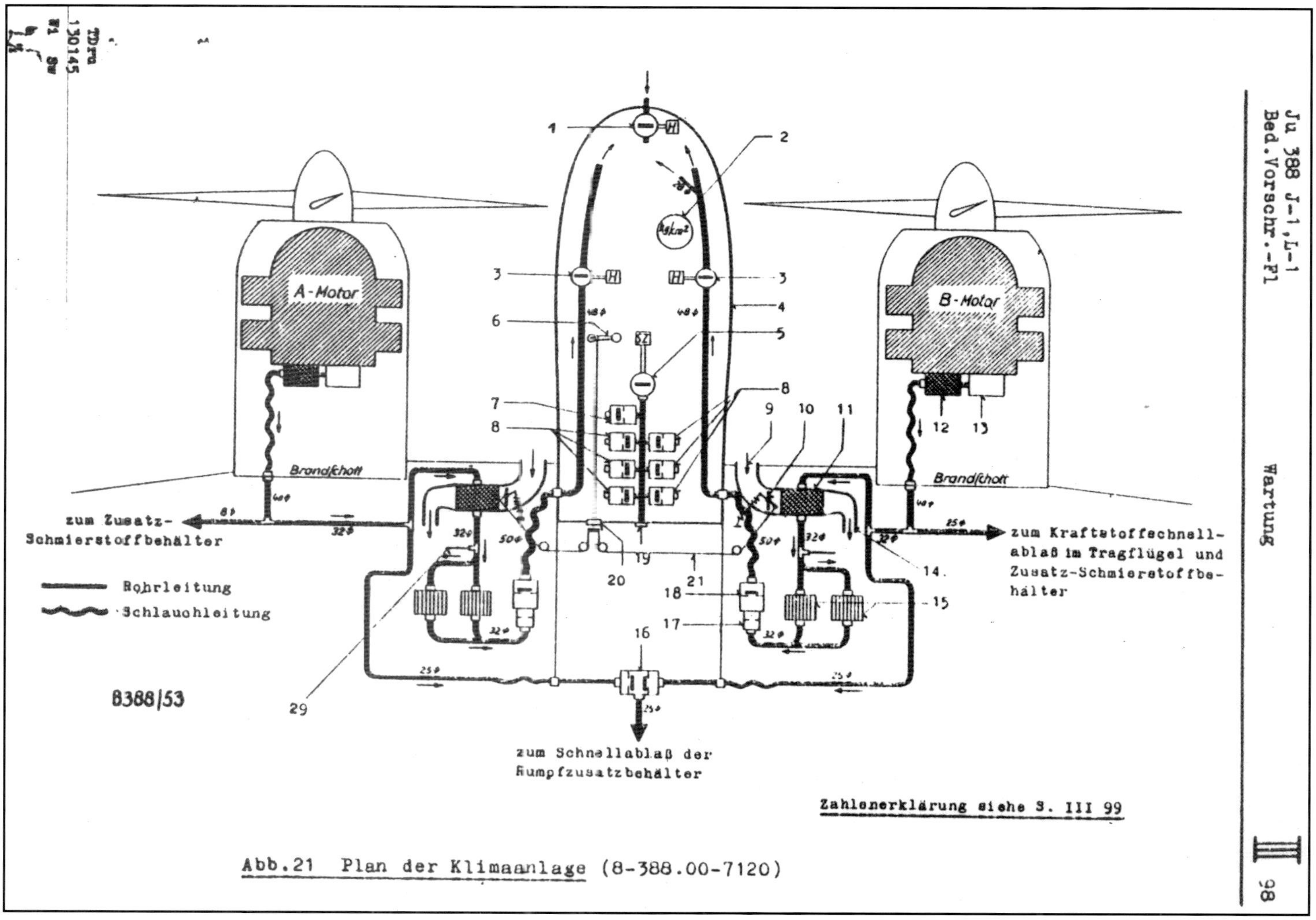

Schematic drawing of the climate control system, also from the pilot's notes.

The pressurized cockpit section used for breaking point tests.

dard equipment on all missions flown at heights greater than 8,000 meters.[112] It is amazing that these problems were not identified during seating trials in the mockups. In the pre-production aircraft, when the radio operator's seat was in a certain position it blocked the handle for the emergency jettisoning by compressed air of the entry hatch, which then had to be jettisoned manually. The associated compressed air system often leaked. The so-called "bad weather window" for the pilot also proved to be poorly designed. The plexiglass frame warped, causing the window to leak. As a result, the frame was later enlarged and made of metal, especially since the gluing of the plexiglass frame significantly restricted visibility anyway.[107]

The rear section of the cockpit glazing, which could be jettisoned in an emergency, also caused numerous problems. It tended to leak, and the poorly-designed jettisoning mechanism worked badly. Finally, in August 1944 the *E-Stelle* Rechlin cabled the *Luftwaffe* construction supervisor (BAL) at Junkers in Dessau, with instructions that "...effective immediately, jettisoning of the canopy is to be demonstrated to the construction supervisor during the handover of all aircraft."

As well, in most cases the tilting of the pilot's seat ("standard pilot seat"), which was necessary for emergency exit, did not proceed as planned. The handle designated for this purpose was too small, and in most emergencies it was not found, and a man wearing fur-lined gloves could not operate it, even if he could locate it. Once again the design of the mechanism was poor, locking bolts jammed and Bowden cables broke. Since tilting of the seat was a safety requirement of the *E-Stelle* Rechlin, the design bureau tried to resolve the problem quickly. Dipl.Ing. Dzulko of the Series Production Design Bureau (Kobü-Rh) also feared that "minor points, about which the soldiers continually complain, might perhaps contribute to an aircraft being discredited."

On 26 May 1944 the high-altitude capsule of a Ju 388 was compressed to destruction in the presence of Prof. Hertel. Because of increasing leakage in the canopy, especially the jettisonable section, only 0.48 atmospheres of internal pressure could be reached, although no significant breakage was found.[115] At that time tests were being carried out by Arado in Brandenburg, involving 20-mm ammunition fired at plexiglass panels in extreme cold. During his visit on 9 May 1944, Dr. Stahl asked if a pressurized cockpit from the Ju 388 might be made available for these experiments.[112]

Development of Tail Turrets for the Ju 188/Ju 388

As the inadequate defensive capabilities of the Ju 88 and Ju 188 came under increasing criticism from the front line units, Junkers responded by building prototypes of the Ju 188 with manned tail turrets.

In late December 1942 the company began building mockups of future manned tail turrets. A series of original Junkers factory photographs in the special collection of the Deutsches Museum reveal that two lines of development existed at that time.[116] One mockup shows a manned tail turret with two manually-operated MG 151s, while another armed with a single machine-gun bears a marked resemblance to a later tail turret. It also had a periscopic sight for the rear gunner, obviously to compensate for the restricted view caused by the relatively large rudder.

According to former Junkers employees, work on installing a tail turret in a Ju (1)88 fuselage began in 1942 at the Junkers branch plant in Bernburg. A rumor circulated among the factory's personnel that it was a conversion for a famous bomber pilot, the highly decorated Hauptmann Werner Baumbach. At the turn of the year 1942-43,[117] an attempt was made to copy the ammunition feed from a crashed enemy bomber. There were initial problems with chafing in the ammunition belt feed, however, these were eventually solved.

Both manned and unmanned turrets were tested on the Ju 288. Here a manned turret with four MG 131 machine-guns. Such a four gun turret was also to have been used on the Ju 290 B. Haubner/Misna Collection

For cg reasons the ammunition had to travel a relatively long distance from the magazines to the turret, and a belt brake was needed to prevent the belted ammunition from sliding backwards in unusual flight attitudes (especially in a dive).

Originally five fuselages without skinning from the Aschersleben production line were supposed to be delivered to Bernburg for conversion. Former Junkers employees[118] remember the conversion as follows: when the fuselage arrived in Bernburg, not all of the fuselage skin had been removed. As the facility had been designed for the large-scale production of components, special requests such as this probably strained its capacity to the limit. For this reason, completion of the work was handed over to apprentices.

Former apprentices Karl-Heinz Pitschler and Jürgen Bressem related that, "as a special project, in Bernburg the machine was on the paving stones in front of the hangar, but at that time [the tail turret] still had no weapons (MG 131) [Ö] and was fitted with segmented glazing."

Photos of a wooden mockup of the entire Ju 188 fuselage taken in late February 1943 show an altered rear fuselage that matches

A wooden mockup of the future tail gun position was produced in February 1943. This early mockup still displays an angular shape Griehl Collection

The Bernburg Ju 188 V2, WNr. 260 151, was fitted with a tail turret with bulged plexiglass side windows. Note the thickened rear fuselage and shortened rudder without notch in this September 1943 photo. Griehl Collection

Rear view of the Ju 188 V2's tail turret with two MG 131 machine-guns. Griehl Collection

this description, the armor-glass side panels giving it a rather angular appearance.

In autumn 1943 Kurt Rutsch, also an apprentice with Junkers at that time, saw a Ju 188 with tail turret during taxiing trials. He believed that the tail turret looked "rounder" at that time. After these taxiing trials, Junkers company pilot Steckhan allegedly declared, "the aircraft is tail heavy, you can fly it by yourselves!" The aircraft in question was probably WNr. 260 151, a machine built in Bernburg and given the designation Ju 188 V2. According to a Preuschen memo, on 7 July 1943 the Ju 188 was "ready for collection" at Bernburg. The aircraft was ferried to Dessau, and was tested there in the summer of 1943. Another aircraft, the Ju 188 V1, had previously been fitted with the HL 131 Z/1 manned tail turret, and was test flown by Wendel on 12 June 1943[119]; after one flight at Bernburg, in the evening it was flown to Dessau. The aircraft was subsequently tested at Rechlin from August to October 1943. The tests were obviously halted because of space limitations. On longer missions the turret would have been barely acceptable, even for the smallest gunners.[120]

As company photos show, the versions of the tail turret differed clearly in the arrangement of the side windows and shape of the rudder. One turret, photographed in June 1943,[121] has smaller side windows and a shortened rudder. By September there was a turret with much larger side windows, and a rudder similar in shape to that of the Ju 388. In the meantime, in August a wooden mockup of the HL 131 turret in the tail section had been built. In addition to a reduction in the size of the rudder, the weight of the HL 131 Z turret shifted the center of gravity. This was partly offset by the bulged bomb bay between Frames 6 and 20 planned for the Ju 188

Unlike the Ju 188V2, the Ju 188V1, which was fitted with a manned tail turret in Dessau, had a notched rudder like the later Ju 388. Griehl Collection

A wooden mockup of the future tail turret with a representation of the FHL 131 was built in August 1943. Once again, the bottom of the rear fuselage was bulged. Griehl Collection

G/H series (similar to that of the Ju 88 V60, prototype for the Ju 88 A-15[122]), and the gun pack of the night fighter/heavy fighter version located between Frames 6 and 9.

The manned tail turret had to be pressurized for a high-altitude aircraft, which brought other problems in addition to the unfavorable weight distribution. Eventually, everyone—not just Junkers—reached the conclusion that remotely controlled turrets were the solution to all these problems. Junkers therefore developed the FA-15 hydraulic remote control system. The PVE 11 periscopic sight was developed in cooperation with Zeiss especially for this system. At the same time Rheinmetall-Borsig developed a remotely controlled variant of the HL 131 Z, designated the FHL 131 Z.

To give the reader some idea of the design and operation of the FA-15, a brief extract from the "Brief Technical Description of the FA 15 Hydraulic Remote-Control System" of May 1944 is reproduced on the following pages, along with a function diagram:

The installation of the PVE 11 and the associated fairing is clearly visible in this close-up photo of the Ju 388 L-1 WNr. 340 084 (RT+KD). The two openings in the front of the bulged bomb bay served as air intakes for the hydraulic cooler. In the wing root leading edge is the intake for the cockpit air cooler. EADS

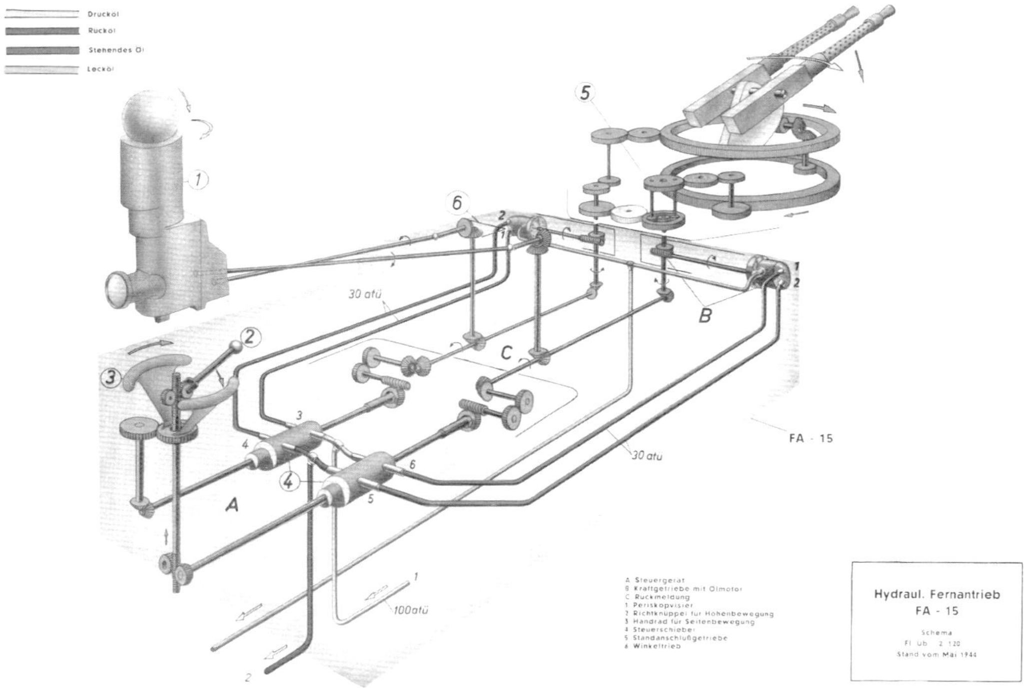

Schematic drawing of the FA 15 remote control drive from the short technical description of May 1944. EADS.

The FA 15 remote-control system is a hydraulic drive for revolving turrets with pure directional control. The pressure oil required for operation is taken from the aircraft's hydraulic system.

The main components of the system are:
A) the control unit
B) the power drive
C) the feedback

The control device A consists of the control lever (2) for elevation control, and the hand wheel (3) for azimuth control, plus the sliding control valve (4), the worm gear, and the connecting drive for the response shafts.

Pressure oil lines connect the sliding control valve and servo motor, which powers the power drive, in this case a worm drive, which has one drive with a splined shaft to the turret connecting drive, and one splined drive for the feedback C. The pressure oil flows through the sliding control valve to the servo motor, and through the same valve back into the return line.

The feedback serves on the one hand to set the periscopic sight (1), and on the other to return the outer plug (see illustration) of the sliding control valve (4). Commands are issued by moving the elevation control lever (2) or the hand wheel (3). These movements cause the associated control valve (4), consisting of the command and feedback valves, to move, allowing the appropriate amount of oil to pass through the valve opening and activate the servo motor. The latter moves the turret by means of the power drive, and by way of the feedback shafts with angle drive (6), as well as the sight (1) and the feedback valve (4). It runs until the feedback valve has turned far enough to restore equal pressure to both sides of the motor.

The weapon is always moved by the same value as the aiming lever or the hand wheel; thus, the weapon and the sighting controls always point in the same direction. The aiming lever and hand wheel can be used together as a coarse sight to place the target within the periscope's field of view.

If the pressure oil system is switched on and the sliding control valve (4) is in the rest, or hydraulic center, position, then, with flow pressure at 100 atmospheres, each side of the servo motor is prestressed to approximately 30 atmospheres. Thus, in the rest position, the connecting lines from the sliding valves to the servo motor are kept under constant pressure at all times. The idling amount flowing through the sliding valves creates an oil circulation to the valve, making the system ready for action at all times.

Movement of the control valve (4) connects one lateral control section with the plus line, and the other with the minus line. The resulting pressure differential starts the servo motor, which by way of the feedback system turns the feedback valve (4) until it reaches the hydraulic center position. The speed at which the turret moves is dependent on the rotation of the control valve compared to the feedback valve.

With the sliding control valve in the center position, there is a circulation of oil from the plus side via the control edges to the minus side. Thus, when the pressure oil system is turned on, there is a continuous flow through the control system. The return oil and the overflow oil return through separate lines."

A later tail turret with one remotely controlled MG 151 cannon installed in a Ju 288. This was very similar to the turret later used by the Ju 388. Griehl Collection

As with the pressurized cockpit, there was a transfer of tail turret technology from the Ju 288 to the Ju 388. Junkers gained its first experience with remotely-operated tail turrets with the FHL 151/20 in the Ju 288 C-1. Ultimately, the installation of the FA 15 was planned for the Ju 288's nose, dorsal, and tail barbettes, and the tail turret of the Ju 388. This followed a rearrangement of the control runs after the periscopic sight, control system, and barbettes had been modified. The dwindling supply of ball bearings following the attack on Schweinfurt proved to be a problem, for just the installation of the FA 15 in the Ju 288 required 440 ball bearings per aircraft, while the figure for the Ju 388 was 274. Consequently, in mid-November 1943 Junkers pressed development department GL/C-E 2 of the RLM's technical office to raise the priority level for the procurement of ball bearings for the pre-production and production versions of the Ju 288 and Ju 388. This was to be handled by special committee SW-3. At the same time the designers tried to reduce the number of ball bearings required for the FA-15. Experiments were conducted with homemade ball bearings, bicycle ball bearings, and slide bearings made from a material called "Presskö."[124] Then, in the spring of 1944, they came up with the idea of using capsule ball bearings, which were to be ordered from a company in Würzburg, in place of normal ball bearings. The outcome of the resulting experiments is not known.[125]

View of the rear cockpit bulkhead with radio equipment (mainly the FuG 10) and the controls for the FA 15. To the right is the PVE 11 periscopic sight. Taken in September 1944, the photo is captioned "Ju 388 L-1 620 305" in the Junkers company album (see *Werknummer* summary). Griehl Collection

According to a message from the design bureau dated 8 November 1943, it was originally planned to deliver the Ju 388 J-1 without a tail turret. Logically only the four prototypes for the Ju 388 K-1 and L-1, at that time still designated the V6/V7 and V8/V9, were to receive the FHL 131 Z. For supply reasons, it was also requested that the FHL 131 Z/1 B production version be used instead of prototype turrets (FHL 131 Z/1 A), as the two versions were not interchangeable. As well, it was felt that the production turret should be tested in at least one or two prototype aircraft.[136]

The highest levels of the *Luftwaffe* command reacted to remotely controlled turrets with disinterest or ignorance. As proof of this, the following extract from the stenographer's record of a meeting held by the Reichsmarschall in the new adjutant's office on the Obersalzberg in Berchtesgaden on 9 July 1944:

Göring: If bombs are the responsibility of the General of Bombers, which I agree with, then the General of Fighters should be mainly responsible for guns.
Galland: Yes, fixed guns. My weakness is movable guns and gun turrets, in which, I must admit, I have very little interest. The remote-control system is also rather dreadful.
Göring: What future aircraft will require remote control?
v. Brauchitsch: The 287!
Schubert: The Ju 388 also has remote control, the 234 not.
Marienfeld: It is being equipped with two fixed rearward-firing guns. It is exactly the same installation as on the 287 and 388.[127]

Göring's question shows that he lacked even a rudimentary knowledge of the potential of remotely-controlled gun turrets. It also demonstrates a shocking lack of detail knowledge about two of the five aircraft types he had decided should remain in the delivery plan just the day before. What then had he based his decision upon? Galland obviously had no qualms about using personal disinterest as a criteria for the allocation of responsibilities.

Numerous changes to the FHL 131 Z caused output to drop in October 1944. To counter this, on 30 October a meeting was held at the Rapido-Werken in Radebeul, where the FHL 131 Z was being built under license. Dr. Stahl insisted that "the Ju 388 absolutely must be equipped with the tail turret," which had to operate with complete reliability. It was requested that the first examples of the latest model turret be delivered to the units on 15 November. Aircraft equipped with the old tail turrets were to be retrofitted under the direction of Rheinmetall-Borsig. The first five to 10 machines were to be retrofitted by 5 November. Beginning in February 1945 the Rapido company was to deliver 160 tail turrets per month. Audi was also bought into production of the FHL 131 Z.[128]

Concerning the tail turret, EKdo 388's final report[129] observed:

"The defensive armament (2 MG 131s in remotely-controlled tail turret) can be bolstered by a dorsal gun position, of which there is a prototype installation (field of fire from 90 degrees to starboard to almost the longitudinal axis to the rear and above). As it is likely that attacks at high-altitude will likely be made by single aircraft only, the defensive armament appears adequate. The tail turret is still prone to breakdown and must be well maintained."

The pre-production aircraft, on the other hand, were equipped with production-style lens mounts at the rear of the canopy. These are visible in numerous photos, but none show a weapon actually fitted. The most likely reason for this is failure to resolve the prob-

Remotely controlled tail turret with MG 131 Z under construction sometime in mid-October 1944. One of the two guns has been fitted in one turret, while on the other the turret cover has been installed temporarily. Griehl Collection

lems associated with the opening for the gun in the pressurized capsule.

At the beginning of 1944 the KdE even considered the addition of a ventral gun position in the Ju 388 to defend against fighters with upwards-firing guns and jet fighters. What they had in mind was a ball turret designed by Dir. Blume (Arado). Junkers had made similar studies, but concluded that such a turret was unacceptable for cg reasons.[130]

In August 1944, during development of the Ju 388, an improved FHL 151 Z remotely-controlled turret, which was envisaged both for the Ju 388 and the EF 122, had been tested in the wind tunnel. Like the FHL 131, it had two vertically-staggered guns. Dr. Strauss, who carried out the experiments on 24-25 August 1944, found that drag with the guns in the normal position had improved by 45% compared to an earlier design.[131]

The Ju 388's remotely-controlled tail turret (barbette) brought Junkers a patent dispute with the Arado company just before the end of the war. Arado first approached the Junkers "Licenses and Patents" department on 16 December 1944, referring to the secret patent "Aiming Arrangement for Gun Turrets" (A 89 133 XI/72f), and a supplement entitled "Periscope Arrangement with Several Lenses" (A 97 818 XI/72f). This letter apparently never arrived, or was quietly ignored, and exactly two months later Arado restated its demands in another letter:

"We...regret that as of today we have not received any confirmation from you that our copyrights are being used in your Ju 388 design or the weapons control used with it. We herewith request that you have this confirmation sent to us immediately and indicate the numbers used."[132]

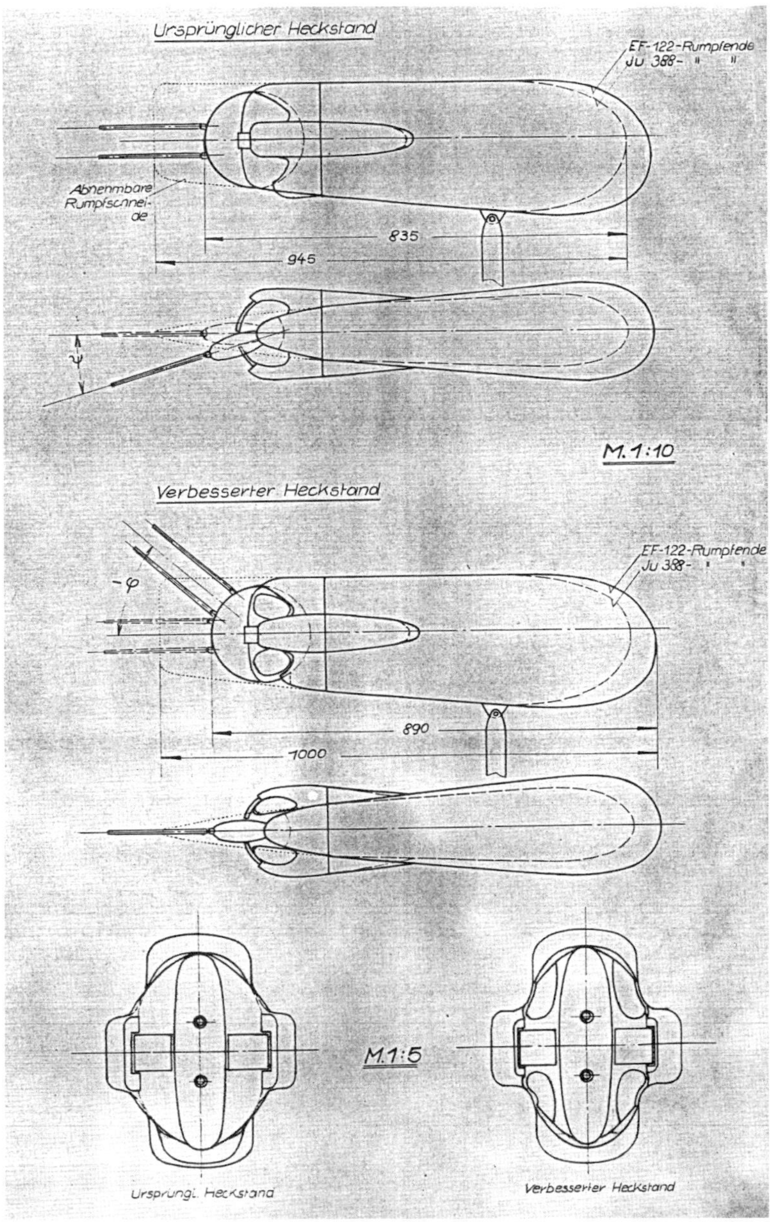

This turret, with obvious aerodynamic improvements, was to have been installed in the Ju 388 and EF 122 together with the FHL 151 Z.

In addition to the new pressurized cockpit, this forward view shows the absent ventral bath and rearwards facing WT 131 with rearview periscope for aiming. The two BMW 801 TJ-0 engines have three-blade propellers, but their smooth cowlings and larger diameter clearly differentiate them from other versions of the BMW 801. Also note the two windows in the cockpit floor, and the arrangement of camera windows behind the wing root. Ott Collection

A glance at a drawing in the Arado patent document reveals a marked similarity between the Arado product and the arrangement selected for the Ju 388 with PVE 11. No additional documentation on this patent dispute is available, and with the war nearing its end the matter was probably dropped.

In closing, it should be noted of the remotely-controlled tail barbette that this concept was entirely futuristic. Both superpowers successfully employed similar defensive systems in their strategic jet bombers (B-52, Tu-22, Tu 28) well into the 1980s.

To conclude this chapter, here is a summary of the most significant differences between the Ju 188 and the Ju 388, based on photographs of aircraft DW+YY:

1. The streamlined canopy of the Ju 388 reconnaissance and bomber variants, part of the pressurized cockpit with double-glazed windows. The heavy fighter/night fighter variant had a slightly-modified, stepped version of this canopy.
2. The BMW 801 TJ-0 (9-8801 J-0) power plant, normally in combination with four-blade propellers (concerning the alternatives, see Chapter IV). The early version J-0 lacked the prominent external ram-air intakes, circular air inlets on the underside of the cowlings. The TJ engine also differed from the BMW 801 D/G versions used in reconnaissance aircraft and bombers in its "bulge-free" cowling, always combined with a central exhaust pipe ("exhaust scoop") on the upper surface of the wing, and cooling gills at the rear end of the engine cowlings.
3. Air intake and exhaust vent halfway down the port side of the fuselage associated with the Kärcher heater for wing deicing.

A number of aircraft also exhibited one or both of the following features:

1. A revised, more angular tail with a notch at the bottom of a shortened rudder. Beneath this sat the rearwards-firing armament in the form of a FHL 131 Z tail turret (with bulged rear fuselage) remotely controlled by means of a PVE 11 periscope and FA 15 remote drive.
2. Long bulged bomb bay for carriage of bombs, cameras, and larger fuel tanks in the K (bomber) and L (reconnaissance) variants.

Chapter 3: The Variants of the Ju 388 Multipurpose Aircraft

Overview

The Ju 388 was to have been built in three main versions: the Ju 388 J heavy fighter/night fighter; the Ju 388 K high-altitude bomber; and the Ju 388 L high-altitude reconnaissance aircraft. There were three sub-variants of each type (-1, -2, and -3) that differed in the type of engines installed. As the only high-altitude engine immediately available, the BMW 801 TJ-0 was to power the initial series. The second series was to be powered by Junkers' own high-performance engine, the Jumo 222. This variant would have an impressive performance, even compared to its Allied counterparts. The third series was to be powered by the Jumo 213 E, also produced by Junkers. Further details may be found in the variant summary reproduced here. It is based on original Junkers Flugzeug und Motorenwerke (JFM) documents (F-388.00-02) found in the Deutsches Museum. The information contained therein should be treated with caution, as it is a summary of planned variants, and all performance figures are estimated.

Numerous technical descriptions and operating manuals (January 1945 edition) for the Ju 388 J-1 and L-1 have survived, making if possible to gain a very accurate picture of the equipment and performance figures for these types. The design of, and differences between, the various versions are described in the text that follows.

Ju 388 J *Zerstörer*/Night Fighter

Type	Engine and Power Plant Designation	Maximum Speed	Range at Max. Cruise	Service Ceiling	Equipped Weight	Gross Weight
		Altitude 11.6 km	Altitude 11 km			
Ju 388 J-1 with BMW 801G	9-8801 J-0 Day: 621 km/h Night: 589 km/h	Day: 2200 km Night: 2160 km	Day: 12850 m Night: 12550 m	Day: 10135 kg Night: 10230 kg	Day: 13275 kg Night: 13310 kg	
		Altitude 7.2 km	Altitude 8 km			
Ju 388 J-2 with Jumo 222 A/B/3	9-8222 A/B Day: 650 km/h Night: 626 km/h	Day: 2000 km Night: 1850 km	Day: 11900 m Night: 11600 m	Day: 11120 kg Night: 11235 kg	Day: 14177 kg Night: 14362 kg	
		Altitude 7.2 km	Altitude 8 km			
Ju 388 J-2 Jumo 222 E/F	with 710 km/h Night: 685 km/h	Day: 1680 km Night: 2570 km	Day: 13600 m Night: 13350 m	Day: 11540 kg Night: 11565 kg	Day: 14510 kg Night: 14690 kg	Day:
		Altitude 10.2 km	Altitude 9.2 km			
Ju 388 J-3 with Jumo 213 E	9-8213 D Day: 617 km/h Night: 585 km/h	Day: 2230 km Night: 2040 km	Day: 12500 m Night: 12050 m	Day: 10050 kg Night: 10140 kg	Day: 13110 kg Night: 13220 kg	

Version differences:

Day: 2 MK 103 or 2 MG 151 in addition to FHL 131 Z

Night: 2 MK 108 and 2 MG 151 and 2 obliquely-mounted MK 108 in addition to FHL 131 Z

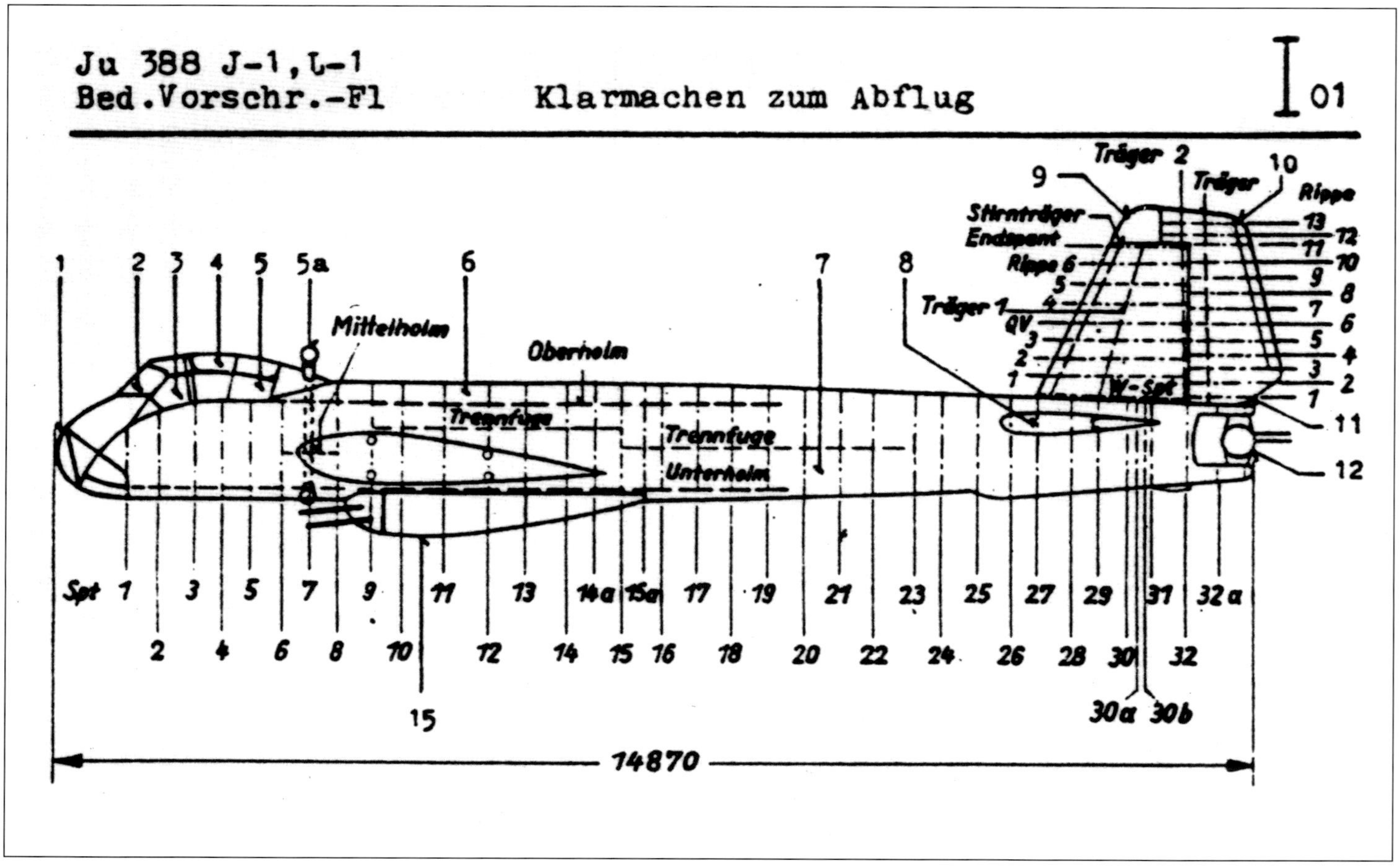

Drawing of the Ju 388 J-1 from the pilot's notes of January 1945.

Ju 388 K Bomber

Type	Engine and Power Plant Designation	Maximum Speed	Range at Max. Cruise	Service Ceiling	Equipped Weight	Gross Weight
Ju 388 K-1	9-8801 J-0 with BMW 801 G	altitude =11.6 km 610 km/h	altitude =11 km 1770 km	12850 m	10250 kg	14275 kg
Ju 388 K-2	9-8222 A/B with Jumo 222 A/B	altitude =12 km 635 km/h	altitude =8 km 2080 km	11700 m	11215 kg	16000 kg
Ju 388 K-2	with Jumo 222 E/F	altitude =11.5 km 695 km/h	altitude =11 km 1800 km	13500 m	11545 kg	15930 kg
Ju 388 K-3	9-8213 D with Jumo 213 E	altitude =10.2 km 593 km/h	altitude =9.2 km 2160 km	12450 km	10085 kg	14400 kg

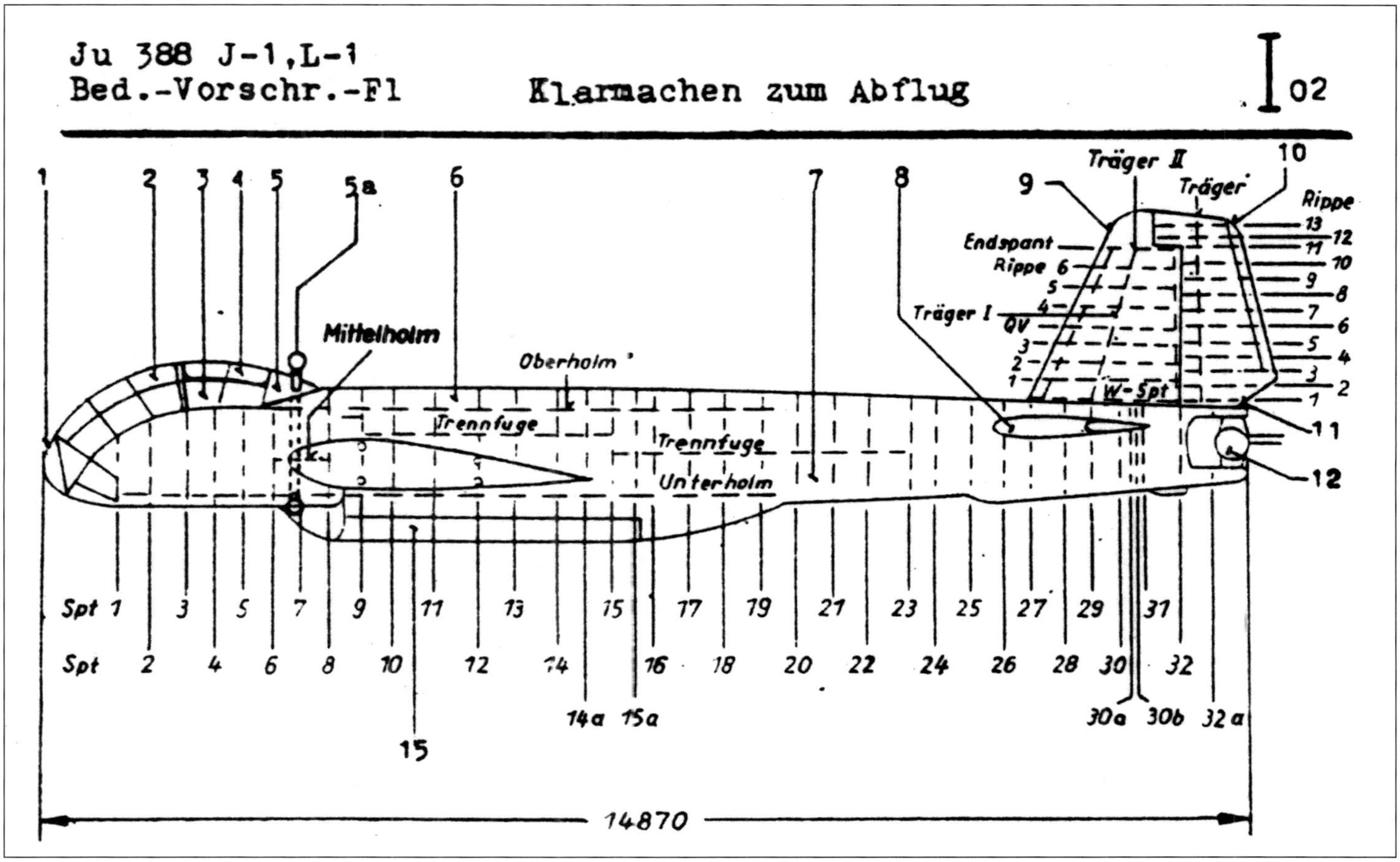

Drawing of the Ju 388 L-1 from the pilot's notes.

Ju 388 L Long-Range Reconnaissance Aircraft

Type	Engine and Power Plant Designation	Maximum Speed	Range at Max. Cruise	Service Ceiling	Equipped Weight	Gross Weight
Ju 388 L-1	9-8801 J-0 with BMW 801 G	altitude = 11.6 km 620 km/h	altitude = 11 km 3100 km	12800 m	10150 kg	13890 kg
Ju 388 L-2	9-8222 A/B with Jumo 222 A/B/3	altitude = 7.2 km 650 km/h	altitude = 8 km 2860 km	11800 m	11225 kg	14840 kg
Ju 388 L-2	with Jumo 222 E/F	altitude = 11.5 km 712 km/h	altitude = 11 km 2450 km	13500 m	11565 kg	15180 kg
Ju 388 L-3	9-8213 D with Jumo 213 E	altitude = 10.2 km 608 km/h	3150 km	12400 m	10060 kg	13675 kg

Ju 388 L-1 RT+KD, WNr. 340 084, an early production machine, photographed at Merseburg in September 1944. One of the horizontal antennas for the FuG 217 may be seen above the starboard wing; vertical antennas were used on most other production aircraft. The engines are still fitted with internal ram air intakes. EADS

Ju 388 L High-Altitude Reconnaissance Aircraft

The Ju 388 L could be built in two basic variants for the day and night photo reconnaissance roles.

The fuselage was of stressed skin construction. Trapezoidal in cross-section, it had four main stringers and flush-riveted skinning. The frames were arranged perpendicular to the longitudinal axis. The fuselage was split lengthwise into upper and lower shells. Provision was made two fuel tank bays between Frames 9 and 15, while a bulged fairing beneath the fuselage could accommodate additional fuel tanks and cameras. The remotely-controlled FHL 131 Z barbette was located in the rear fuselage behind Frame 15. The rear fuselage was accessible by means of a hatch behind the second

Junkers Ju 388 L	**Data from Brief Technical description No. 110145 and Concept Design Bureau 3682**
Purpose:	twin-engined day or night long-range reconnaissance aircraft of Load Group H, Purpose group 3
Wingspan:	22.00 m
Wing area:	56.00 m^2
Engines:	L-1: BMW 801 TJ in power plant 9-8801 J-0 L-3: Jumo 213 E in power plant 9-8213 D-1
Engine output:	L-1: takeoff power 1,615 h.p., climb and combat power 1,472 h.p. at ground level and 1,430 h.p. at 12.3 km (maximum boost altitude). C3 fuel
Propellers:	L-1: 4-blade VDM Dural variable-pitch propellers with automatic control (diameter 3.726 m) L-3: 4-blade VS 19 propellers with wooden blades (diameter 3.6 m)
Armament:	FHL 131 Z with FA 15 and PVE 11, 600 rounds per gun
Max Gross Weight:	14 700 kg
Max Landing Weight:	12 000 kg
CG Limits:	0.78 to 1.20 m behind wing leading edge.

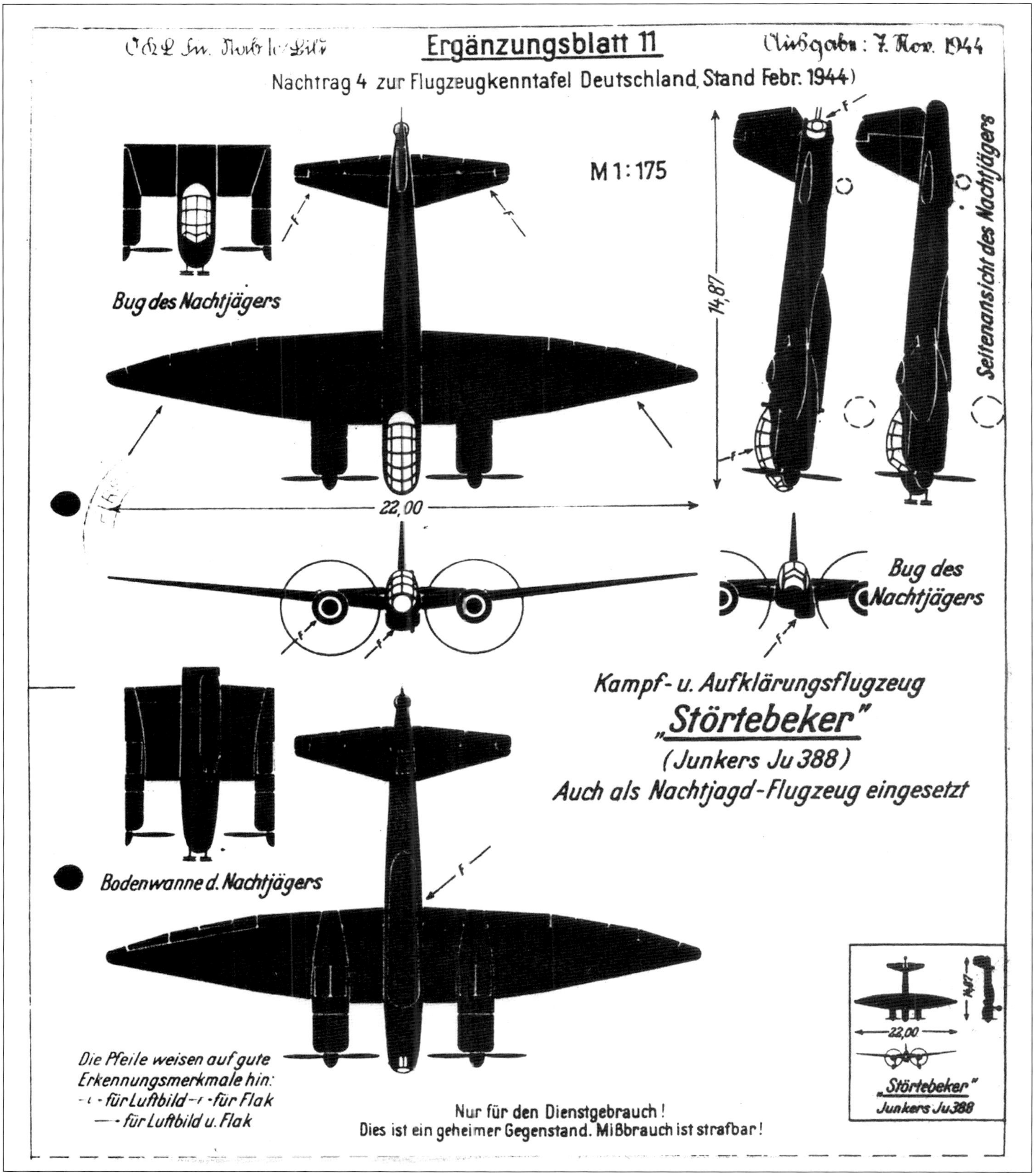

Official Ju 388 recognition sheet.

fuselage bay. The turret provided an uninterrupted field of fire, extending 45 degrees upwards and 60 degrees to either side. 600 rounds of ammunition were provided for each gun, half contained in ammunition boxes behind the second fuselage bay, and the other half in the belts from the boxes to the turret.

The cockpit was designed as a pressure-tight high-altitude chamber for an internal overpressure of 0.23 atmospheres. At an altitude of 13 km, therefore, pressure inside the cockpit was equivalent to an altitude of 8 km. As with modern airliners, a certain amount of supercharged air was bled from the engines to maintain internal pressure. This passed through a purification filter, and also heated the cockpit. The cockpit also had fittings for electrically-heated flight suits and an oxygen system. Four oxygen bottles for each crew member were housed in the starboard wing. The upper part of the pressure capsule was fully glazed. The glazing consisted of double-glazed plexiglass panes with moisture-absorbing cartridges to prevent fogging. Because of their slight curvature and the quality of the glass, these plexiglass panes were not suitable for astral navigation. They were therefore partly replaced by double-glazed silica-glass[133] windows (astro-windows). There were two possible methods of evacuating the cockpit in an emergency. In the event of a belly-landing or ditching, the rear section of the canopy could be jettisoned. For a bale-out, the entry hatch could be opened by compressed air (150 atm) and then jettisoned. An inflatable life raft was stored in the upper fuselage behind the cockpit.

Crew seating had the observer positioned to the right of and slightly behind the pilot. The pilot sat in a standard seat with head and back armor, while the observer's seat had back armor. The radio operator sat with his back to the pilot, facing the radio equipment, and was protected by an armored panel in the rear of the canopy, and armor plate in the left upper part of the cockpit bulkhead. The radio operator also operated the remotely-controlled tail turret. He did this by means of a Junkers-developed FA 15 hydraulic remote control system. For aiming, a PVE 11 double periscope was installed in the rear of the cockpit on the left side. It could be used to simultaneously scan behind and below the aircraft.

The retractable undercarriage, which was housed in the rear of the engine nacelles, had oil-air oleos, and was virtually identical to that of the Ju 188, which in turn was a strengthened version of the Ju 88 undercarriage. The 1140 x 410 mm mainwheels were each provided with a braking cylinder, and braking was controlled by the rudder pedals. The tailwheel (560 x 200) was fully-castoring, and was fitted with a KPZ spring-oleo.

Fuel System (Daylight Reconnaissance Aircraft)

Relief Tank in Inner Wing	port	415 liters
	starboard	415 liters
Auxiliary Tank in Outer Wing	port	425 liters
	starboard	500 liters
Auxiliary Tank in Fuselage	forward	1680 liters
	rear	500 liters
Fuel Capacity		3935 liters

All versions of the Ju 388 were fitted with the Ju 188 wing, a cantilever structure attached to the fuselage by four threaded ball-socket connectors on each side. The "interior life" of the wing did, however, differ in several respects from that of the Ju 188.

The horizontal tail was also taken from the Ju 188. The elevators were internally-balanced with coupled Flettner tabs controlled from the cockpit. The all new vertical tail was centrally arranged, and the rudder and Flettner tab were used together to trim the aircraft in flight. The Ju 388 was also equipped with slotted ailerons and flaps.

Elevator and aileron control was by means of a movable control column on the fuselage centerline with a swing arm for the control yoke. The rudder was controlled with adjustable foot pedals. Small hand wheels in the portside console controlled the aileron, elevator, and rudder trim tabs. Elevator trim could also be adjusted electrically. A Patin PDS 11 two-axis autopilot was also planned.

As in the Ju 188, wing deicing was achieved by warm air from the engines. A Kärcher gas-operated heater housed in the fuselage provided warm air for deicing of the tailplane and heating of the camera bay. Propeller deicing was by means of deicing fluid, which was housed in an 18-liter tank in the wing. The relief tanks in the wings and the fuselage tanks were each equipped with two pumps, one of which was a backup. The backups were selected by means of a toggle switch on the starboard side of the cockpit. Because of this backup system, no provision was made for an emergency hand pump.[132]

The fuel tanks in the fuselage, and the unprotected tank (500 l) in the starboard wing were equipped for fuel jettisoning. The unprotected tank in the starboard wing could also be flushed with supercharger air to eliminate potentially dangerous gasoline fumes.[133] Design and testing of the fuel dumping and tank flushing systems

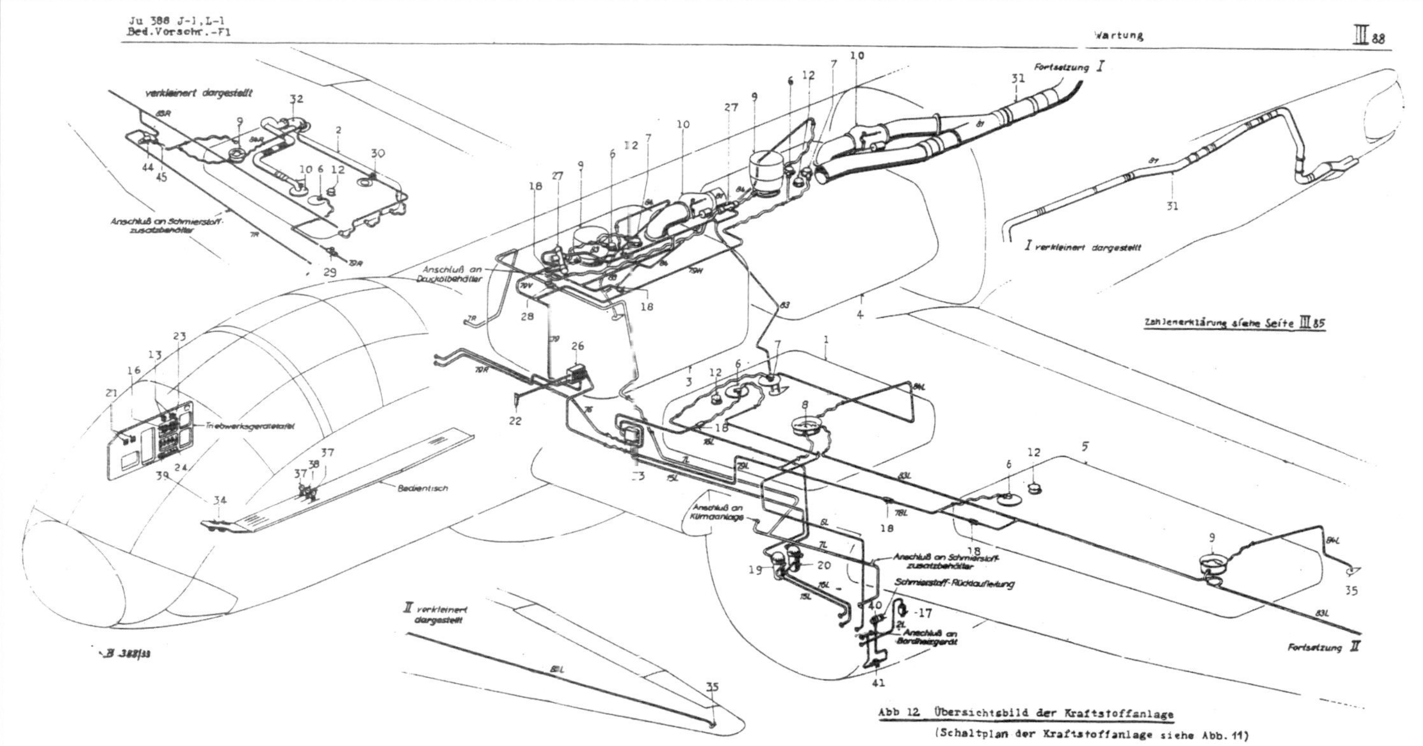

Drawing showing the location of the fuel tanks from the pilot's notes.

1. Entnahmebehälter (ER und EL mit je 405 l)
2. Zusatzbehälter rechts (AR)
3. Zusatzbehälter im Rumpf vorn (RV)
4. Zusatzbehälter im Rumpf hinten (RH)
5. Zusatzbehälter links (AL)
6. Behälterkopf mit Pumpe
7. Nebenbehälterkopf m. Pumpe
8. Entnahmebehälter-Füllkopf
9. Zusatzbehälter-Füllkopf
10. Kraftstoff-Schnellablaflventil
11. Kraftstoffbehälterpumpe
12. Vorratsgeber
13. Tankanzeige
14. Kraftstoff-Entlüfter
15. Kraftstoff-Förderdoppelpumpe
16. Schnellablaflschalter
17. Kraftstoffdruckgeber
18. Rückschlagventil
19. FB-Armatur (zu P1)
20. FB-Armatur (zu P2)
21. Kraftstoffdruckmesser
22. Hebel f. Ventilbatterie
23. Schalter f. Tankanzeige
24. Schalter Reservepumpe
25. Reservepumpen
26. Ventilbatterie Spant 9
27. Regelventil für Ablafl
28. Rückschlagventil
29. Ventil f. Umpumpleitung
30. Absperrventil
31. Schnellablaflleitung
32. Schnellablaflleitung
33. Ventilbatterie im Flügel
34. Ladedruck-Drehzahlmesser
35. Sickertopf
36. Entfällt
37. Bedienhebel FB-Armatur
38. Hebel f. Behälterschaltung
39. Schalter f. Schnellablafl AR
40. Ringdüse (Kaltstart)
41. Mischhahn (Kaltstart)
42. Zusatzbehälterpumpe
43. Reservepumpe

Ju 388 L-1 (F) without flame dampers **State III, 4 wing tanks** **16 Nov. 1944**		**Concept Design Bureau** **Performance Data** **No.: 388/844/2a**	
Engines:		801 TJ-0 with Fimag intercooler	
Takeoff power	h.p.	1,615 + 11 kg	
Combat power at 0 km	h.p.	1,472 + 8 kg	
Combat power at 12.3 km	h.p.	1,430 + 62 kg	
Condition:		with external tank	without external tank*
Equipped weight	kg	10100	10100
Crew	kg	300	300
Fuel	kg	3015	3715
Oil	kg	260	260
Life raft	kg	65	65
Ammunition	kg	102	102
External tank + rack	kg	—	70
Gross weight	kg	13842	14612
Wing loading	kg/m≤	247	261
Average flying weight	kg	12335	12735
Max. speed at 0 km	kph	416	415
Max. speed at 12 km	kph	617	610
Estimated range	km	2275	2780
at cruise speed at 11 km	kph	560	557
Estimated range at	km	2960	3457
economical cruise at 11 km	kph	480	490
Initial climb rate	m/sec	6.75	5.80
Time to climb to 11 km	min	30	38
Ceiling on 2 engines at takeoff weight	m	12800	12250
Ceiling on 2 engines after climb	m	13050	12750
Ceiling on 2 engines at average flying weight	m	13450	13250
Ceiling on 1 engine at average flying weight	m	6900	6750
Initial climb rate, single engine	m/sec	1.55	1.50
Takeoff distance	m	780/1180	900/1350

*after conversion **with 1255-liter non-jettisonable fuel tank

State III: production state without leading edge cable cutters, with adjustable cooling gills, two cameras. Fimag intercoolers. Standard fuel weight: 0.78. Armament: FHL 131 Z with 2 x 600 rounds of ammunition.

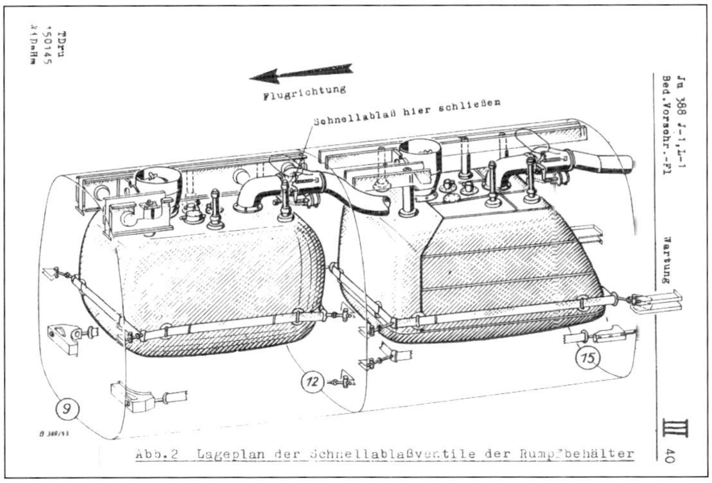

Overview of the fuselage tanks and fuel jettisoning systems

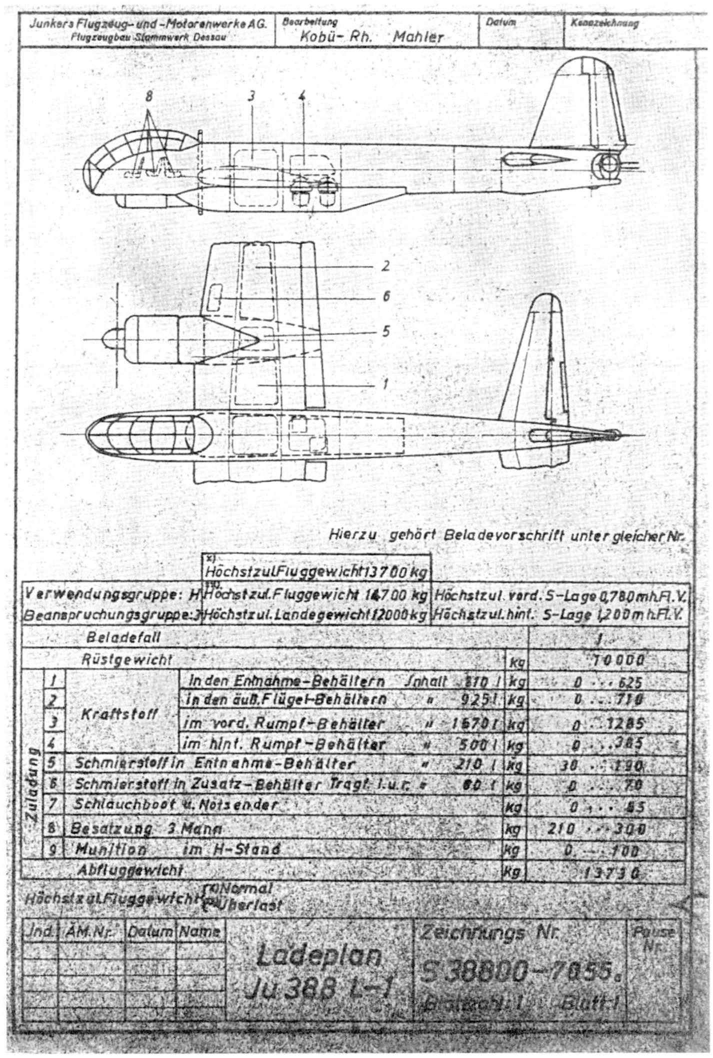

Ju 388 L-1 loading diagram.

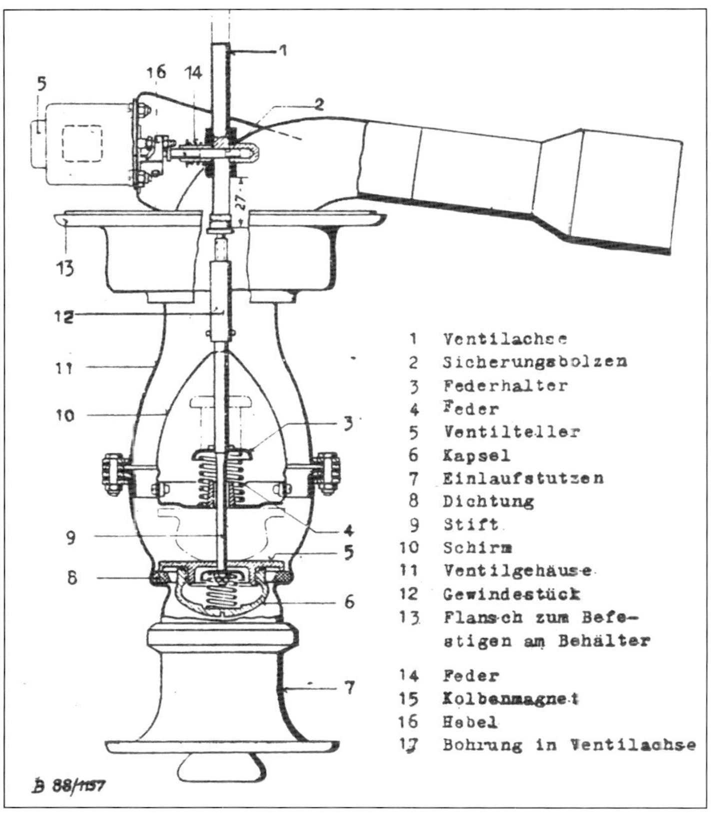

Cross section of the fuel jettison valve of the starboard auxiliary fuel tank.

was a cooperative effort between the *E-Stelle* Travemünde and Junkers. The test station had previously developed a fuel jettisoning system for the Do 217 E in 1941. The corresponding order was completed in December 1944 with the Junkers report Kobü-TN[134]/V353, which unfortunately is not available to us.[135] The fact that the closing report was completed by Junkers suggests that the *E-Stelle* probably performed an advisory role in this project because of its previous experience.[136]

In order to increase fuel capacity, and thus range, provision was to be made for carriage of an externally-mounted 900-liter tank. The tank would be mounted on a jettisonable framework attached to External Load Bearer I. Consideration was also given to carriage of a bomb instead of the fuel tank.

The cameras were installed beneath the rear fuel tank in Fuselage Bay 2. They could be mounted vertically, or angled to the side at 10, 15, 20, or 30 degrees. The day photo-reconnaissance version was to have two Rb 20/30, Rb 50/30, or Rb 75/30 cameras, while the variant intended for nocturnal operations was to carry two NRB

35/25, NRB 40/25, or NRB 50/25 cameras. As well, the large forward fuselage tank of the day reconnaissance version was replaced with a smaller 725-liter tank to create space for an L-frame with eight illumination bombs. Four more illumination bombs could be carried on two racks. In order to make room for the illumination bombs, the two cameras had to be mounted side by side, with a reduction in fuel capacity to 2,980 liters.

Each engine had its own lubricating system. An armored oil tank was mounted in each wing between transverse members I and II. The tanks had a capacity of 136 liters, however, the maximum allowable content was 105 liters. There was also a 40-liter auxiliary tank in each wing. Current for the aircraft's electrical system was supplied by a 3,000-Watt generator on each engine. While generator output was 6,000 Watts, average sustained consumption was approximately 2,200 Watts. Batteries rated at 48 ampere-hours were provided to cover peak loads.

The radio system consisted of a FuG 10 short- and long-wave set with Peil G 6 direction-finder, FuG 25a, FuG 101a radio altimeter, and FuBl 2 F instrument landing aid. A FuG 16 Z(Y) radio was used for station to station communications. Also provided was a FuG 217 Neptun II tail warning radar.

The performance figures for the Ju 388 L-1 were impressive: during testing in the USA, WNr. 560 049 achieved a maximum speed of 383 mph (616 kph) at an altitude of 40,300 feet (12 283 m). This was close to the 620 kph cited in the technical specification of 2 August 1944.[137]

Ju 388 L-0 *Nacht* (Night)

At the Fighter Staff meeting on 6 June 1944, General Staff Major Eschenauer acknowledged the need for night reconnaissance aircraft:

"For lone reconnaissance aircraft to be effective in 1945, they will have to sneak in at high-altitude and use illumination bombs or possibly operate at medium altitudes under cover of darkness. The lone daylight reconnaissance aircraft is no less effective in 1945."[138]

On 23 October 1944 Junkers' technical department issued Construction Directive 20, which directed that all Ju 388 K-0 aircraft under construction were to be completed as L-0 Nacht night-reconnaissance aircraft, with equipment for the carriage of illumination bombs.[139] On 16 November 1944, the Steering Committee for Aircraft Production directed that the 20 Ju 388 L-0 Nacht aircraft under construction (former K-0s) were to be equipped with a night camera system. Drawings for the airframe installation and associated wiring, plus an explanatory note were to be completed by 7 December 1944. Prior to delivery, the tail turrets of these machines were to be brought up to the latest standard.[140] The spent casing bag was to be modified based on the prototype installation in WNr. 230 156 and information provided by *Herr* Kemkemer (Kovermi). It was envisaged that a Ju 388 L-0 Nacht would be selected for testing of the turret. With the tail turret modified to the latest standard, but with no cameras, it was to be flown to Tarnewitz. Once trials had been completed, it was to be flown back to Merseburg for installation of cameras. All aircraft were to be provided with infrared lighting for the instruments as per the master plan.[141] In July 1944 trials had been conducted involving a He 111 (SU+MY) equipped with infrared lighting and searchlight batteries in the Berlin area. It was found that infrared lighting and red filter glasses developed by Prof. Knothe provided effective shielding against glare. As a result of these trials, Department E5 of the *E-Stelle* Rechlin was prepared to recommend the general adoption of the procedure.[142]

Ju 388 L-3 with Kalkert Ka 430

The Ju 388 L-3 powered by Jumo 213 E engines was envisaged as a tug for the Kalkert Ka 430 battle zone transport. In September 1944 Mitteldeutschen Metallwerken began preparing a report on the tug-glider combination.[143] It concluded with an extremely positive assessment of the Ju 388 L-3 as a tug (taking into consideration takeoff distance, service ceiling, range, and tow speeds). The only area that came in for criticism was the high towing speed of 350 kph, which exceeded the Ka 430's maximum speed. Looking back, the idea of using a dedicated high-altitude aircraft in the role of glider tug at low altitudes (it is unlikely that the Ka 430 could have been equipped for high-altitude operations) and speeds appears highly suspect. One must take into consideration, however, the fact that the He 111—the most widely-used glider tug—was to go out of production for good in the summer of 1944. It was for this reason that a proposal was made at the Fighter Staff meeting on 3 July 1944 to investigate use of the Ju 88 in the role of glider tug.[144] As the Ju 88 was to have been replaced by the Ju 388, it is entirely plausible that the successor type was also considered. Furthermore, the Ju 388 was the only remaining conventional bomber type in Delivery Plan 226. Jet-powered types were deemed unsuitable, as was the Do 335 on account of its push-pull configuration.

Ju 388 K High-Altitude Bomber

Ju 388 K-0

Since about mid-1943, plans had existed for a pre-production series of 20 Ju 388 K-0s to be built at Merseburg before the Ju 388 K-1 entered mass production. At the beginning of May 1944 the pre-production series was increased to 50 aircraft. The aircraft were to be "converted" at Merseburg using major components of the Ju 188 E-1 and A-2. This does not mean that largely complete aircraft were flown to Merseburg and converted into Ju 388s there. Instead, Ju 188 E-1 and A-2 fuselages (without cockpits, electrical systems, and radio equipment), tail sections (without automatic pull-out systems), and wings with undercarriage were taken from the production line and transported to Merseburg, where they were modified in accordance with "Component Plan Ju 388 K-0." Major components of two Ju 388 K-0s, WNr. 230 152, and WNr. 230 154[145] were to be shipped to Toulouse for the construction of "two flying mockups of the special version of the Ju 188 (V401 and V402)."

The Ju 388 K-0 was very similar to the Ju 388 L-1 (and thus also the Ju 388 K-1). The most obvious external difference was the absence of a tail turret and the standard Ju 188 tail surfaces. The cockpit was similar to that of the Ju 388 L-0, with a dorsal gun position (B-Stand) mounting one MG 131. 500 rounds of ammunition for the belt-fed weapon were stored in an ammunition box on the starboard side of the cockpit. Unlike the Ju 388 L-0, the Ju 388 K-0's cockpit had several glass panels for astral navigation.

The aircraft's fuel system was similar to that of the Ju 388 L-1 built to night photo-reconnaissance standard. Two auxiliary oil tanks were to be installed as standard equipment (Equipment Set B13), as was an auxiliary fuel tank in the port wing.[146] The Ju 388 K-0 had the bulged ventral bomb bay of the Ju 388 K-1/L-1; however, two MG 131s with 400 rounds per gun were installed in the rear of the bay (ammunition boxes on the right and left sides of the bomb bay beneath Frame 16). Like the Ju 388 L-0, a cockpit-mounted Rf 1 or Rf 2 B rearview periscope was used to aim the guns.[147] Also like the L-0, the radio operator's seat in the Ju 388 K-0 was not rigidly mounted, and had to be swiveled to the left in order to abandon the aircraft.[148] For the carriage of stores, provision was made for the following rack mounts:

4 Schlösser 500 XII (Fl 50 310)
2 Schlösser 2000/XIII B-1 (Fl 50 199)
1 L-Gerät 8 Schloss 50 B-1
4 Träg 2 Schloss 50/X B-2

Deliveries of fuselage stores carriers totaled just two Type 2000/XIII B-1 mounts. All of the other mounts were "holding devices" or components of *Rüstsätze* (equipment sets).[149] The bombsight was a Lotfe 7H. The bulged bomb bay necessitated new bomb door actuators, a new electrical arming mechanism, and new emergency pull rods. The aircraft were supposed to be delivered with four Type 500 XII mounts.[147] In addition, a non-jettisonable rack was to be installed on Load Bearer II (*Rüstsatz* M14).[150]

The only known photographs of a Ju 388 K-0 were taken in late June-early July 1944 at Merseburg. They depict KS+TA, WNr. 230 151. The most noticeable difference from the Ju 388 L-1 is the absence of a tail turret. The Ju 388 L-0 Nacht would have been similar to KS+TA, as it was converted from the Ju 388 K-0. EADS

Ju 388 K-1

The first description of the Ju 188 K-1 bomber, later to become the Ju 388 K-1, appeared in Amendment 1 of the construction directive for the "Construction of two Ju 188 K-1 Bomber Prototypes" dated 31 May 1943.[151] This amendment more accurately described or corrected the basic design of the pattern aircraft, which was first mentioned in March 1943. Surprisingly, apart from minor deviations the wording is identical to that of the construction directive for the 10 Ju 388 L-0s (see above). In addition to the changes necessitated by the tail turret, there were some frame changes in the area of the bulged bomb bay. As well, two new fuel tanks were envisaged for the fuselage. For the bomber role, the Ju 388 K-1 was to have had the necessary fittings and wiring for a Lotfe 7 D bombsight, arming mechanism, and emergency pull rods for the bombs. The use of *Rüstsatz* M8 for tank jettisoning was also planned. In addition to the radio equipment envisaged for the Ju 388 L-0, the K-1 was to have the FuG 129a Bernhardine[151a] and FuG 217 Neptun II.

The subject of equipping the Ju 388 with guided weapons was the subject of discussions at the highest level. As early as 1943, when plans were being made for attacks on England using V1 rockets fired from fixed ramps, the search had already begun for even cheaper alternatives. Guided weapons could be launched from beyond the range of the English defenses. Dr. Vogt designed the BV 246 Hagelkorn (Hailstone) as a "cheap variant," a glide bomb of the simplest design. An aircraft would bring it close to the target, from where it would be released to glide the rest of the way. It was

Close-up of KS+TA. The only external difference between the Ju 388 K and L was the glazed triangular bulge for the bomb sight under the cockpit. The opening beneath the nose cap is the forced air intake. Note the opened bad weather window. EADS

essentially a winged bomb, with a guidance system simpler than that of the V1. On 17 December 1943 a GFM/GLM conference was held in Berlin[152] to initiate series production of the BV 249, Fi 103, and A4. General Peltz wanted to use KG 101 at Greifswald for testing of the BV 246, and asked which aircraft apart from the Ju 188 were suitable as carriers. Knemeyer subsequently named the Ju 388, and Milch threw in the He 177 as well. *Oberst* Petersen brought up the still unanswered question of how to determine the correct release point, from which the bomb was to cover the remaining 200 km to the target from an altitude of 7000 meters at a glide angle of 1:30. No solution was found. Milch subsequently gave the green light for 500 examples of the BV 246.

The BV 246 was not the only guided weapon considered for use with the Ju 388, however. The Reichsmarschall's conference, which began at 1130 hours on 9 July 1944 in the new adjutant's office on the Obersalzberg, concerned itself with the implications of the new aircraft program adopted the day before. Among the topics discussed was guided weapons (Hs 293, Hs 298, Fritz X, and A4). Göring expressed the opinion that these weapons were not worth the cost. General Koller, on the other hand, wanted to use up the existing stocks at least, but Göring resisted the idea:

Göring: But you will need too much junk from me.
Koller: I don't need much.
Göring: I have no carriers for these missiles.
Koller: I'm hoping for the 410.
Göring: The 410 is out. Instead you'll be getting a lot of fighters until the switch to the 388 and 335 heavy fighters.
Koller: Of course we can carry on with that. The 388 will work fine.[153]

This brief excerpt from the minutes reveals the alarming lack of expert knowledge on the part of the participants. Göring spoke vaguely of "junk"; he obviously didn't know what additional equipment had to be installed in the aircraft for the use of guided weapons. Koller, too, should have known from previous meetings that the Me 410 was one of the aircraft that was to be dropped following the adoption of the new aircraft program. Unfortunately, it is impossible to determine from surviving documents to what extent his vague ideas about using the Ju 388 as a carrier of guided weapons were translated into concrete plans.

A high-altitude aircraft armed with a stand off weapon, such as a remotely controlled missile that could be released from long range (or high-altitude), would surely have been an operational success. A lengthy period of testing would probably have been required, however.

As a combat aircraft the Ju 388 surely represented a stopgap solution. By the end of May 1944 the *Luftwaffe* command was agreed that in the Ju 287 the *Luftwaffe* would again have an aircraft that would be a "really modern medium bomber."[154] At the "Designers' Meeting with the Reichsmarschall," which was held on the Obersalzberg on 25 May 1944, Göring declared that Junkers was to use the capacity resulting from the cancellation of the Ju 290 and Ju 390 to "launch into the Ju 287 with a vengeance."[155]

It is therefore not surprising that the armaments staff canceled the Ju 388 K-1 in September 1944.[156] According to Delivery Plan 226/2A of 28 August 1944, ATG alone was to have produced 741 Ju 388 K-1s and 120 K-3s. When Delivery Plan 226/2B appeared on 16 September, only the J and L series remained. The only production Ju 388 K-1 ever built was delivered to JFM Prague on 31 July 1944, where it was to be tested to destruction. The *Werknummer* of this aircraft, which was built by ATG, is unfortunately not known.[157] All other Ju 388 K-1s under construction in August/September 1944 were converted into reconnaissance aircraft.

In spite of the early cancellation of the Ju 388 K, Junkers still held out some hope for production of the aircraft at a later date. At the beginning of January they were still considering a human labor expenditure of 400 hours, the cost of updating the equipment lists "in the event that construction of the K-1 series should be planned again."[158]

The Blohm & Voss BV 146 glide bomb. The example seen here is preserved in the NASM. Handig Collection

Ju 388 M Torpedo Bomber

The most exotic variant of the Ju 388 was the Ju 388 M torpedo bomber, about which almost no documentation exists. It was envisaged as a successor to the Ju 188 LT (aerial torpedo) variants.[159] The Junkers company albums contain a number of photographs of WNr. 340 084, RT+KD, bearing the caption "Ju 388 M-1." It was, however, clearly an L-1 production aircraft built at Merseburg. Unlike most other Ju 388s, which had vertical FuG 217 antennas beneath the wings, this machine had the other antenna arrangement, with horizontal antennas above the wings. It is almost certain that the aircraft was not an M-1 conversion or a forerunner of that variant. In its monthly reports, the FTA[160] section attached to *Erprobungskommando* 388 always referred to the aircraft as a Ju 388 L-1. The much-photographed RT+KD was one of the first Ju 388 L-1s from the Merseburg production line, and it is quite possible that in this case the "M" in M-1 might also stand for "pattern aircraft." Incorrect captions are sometimes found even in the factory albums, and a mistake by the company photographer is the most likely explanation.

One of the most reliable sources on the Ju 388 M-1 is an almost complete summary of the components and equipment of the various members of the Ju 88 family compiled by von Bonin and Ort of the Design Bureau's Project Department on 10 August 1944.[161] According to this, the Ju 388 M-1 was to have had the cockpit of the Ju 188 E-1 and the fuselage of the Ju 188 G-2. Because of the defensive position in its tail (manually operated MG 131 Z), its tailwheel was enlarged from 560 to 630 mm. Defensive armament in the cockpit was similar to that of the Ju 188 A-3, with single MG 131s firing fore and aft; however, the rotating turret with MG 131 of the Ju 188 A-3 and G-2 was dropped. The B2 position was to have been equipped with an MG 151. Radio equipment consisted of FuG 10 P with APZ 6, FuG 16 Z(Y), Fu Bl 2H, FuG 101, FuG 217, and FuG 200. The records contain no clue as to the type of engines used, however, the fact that the Ju 388 M-1, like the Ju 88 A-17 and Ju 188 A-3, was to have no oxygen system suggests that high-altitude engines were not used. Fuselage fuel tank arrangement was identical to that of the Ju 388 L, with a forward tank holding 1680 liters, and a rear tank with a capacity of 510 liters. As the large forward tank (1680 l) of the Ju 388 "Daylight Photo-reconnaissance" variant occupied part of the bulged bomb bay, the Ju 388 M-1 obviously had the bulged bomb bay of the Ju 388 K/L and the Ju 188 G. It was more similar to the Ju 188 G than any other variant of the Ju 388. According to a Preuschen memo dated 1 August 1944, the Ju 188 V11 was supposed to serve as prototype for the Ju 388 M-1 with L10 and LT1B. Even within the Junkers company itself, this series was sometimes referred to as the Ju 188 M1, and sometimes as the Ju 388 M1. The most obvious case is in the title of Change Notification 5033/44 of 26 June 1944: "Ju 388 M-1 (Ju 188 M-1)." In another change notification from the Dessau Production Department concerning changes to the fire extinguishing systems for Ju 88/188/388 aircraft dated 11 September 1944, a "Ju 188 M1" is mentioned together with the Ju 88 S-3, Ju 88 G-6, and the remaining Ju 388 variants. The torpedo system—in particular the setting of the launch angle and running depth[162]—was similar to that of the Ju 88 A-17 and Ju 188 A-3. A change notification (20924/44) dated September 1944 confirms that the ETC 2000 rack was also used.

Another photo of RT+KD from the same series of photographs. The caption accompanying this photo in the company album reads "Ju 388 M-1," but this is most likely an error on the part of the photographer. In every other document RT+KD, which later went to EKdo 388 on Rechlin, is identified as an L-1. EADS

The last production torpedo bomber to reach front-line service, in particular with KG 26, was the Ju 188 A-3. Externally, the machine differed from the standard bomber in having ETC 2000 bomb racks, which were clearly larger than standard racks, beneath the wings, and the fairing on the starboard side of the forward fuselage over the control linkages. Deutsches Museum

One of the two ETC 2000 racks, and the fairing over the installations needed to set the torpedoes are visible in this photo taken in September 1943. The equipment for the torpedo version was taken from the Ju 88 A-4 torp or the A-17. Deutsches Museum

On 8 September 1944 a "Construction Directive for the Development and Prototype Installation of the L 10 in the Ju 388 M-1" was issued. It was based on an advance order from the OKL-TLR, Fl/2/A, dated 26 August 1944. The directive called for the development, design, and testing of an equipment set for the Blohm & Voss L 10 torpedo-glider to be mounted beneath the fuselage of a prototype aircraft. The aircraft's *Werknummer* was not recorded in a subsequent amendment. As well, PVC subcarriers and fairings were to be modified for carriage of the ETC 2000. A subsequent amendment dated 14 September 1944 brought an end to the work, however: "Because of the new program, the work referred to above will not be completed."[163] This program change probably ended the entire Ju 388 M-1 program as well.

In November 1944 the *E-Stelle* Travemünde carried out a preliminary study on equipping the Ju 388 with mine detecting equipment.[164] Like a torpedo bomber, a minesweeping aircraft had to fly low over the water; therefore, the Ju 388 M would probably have been predestined for the role.

Why an aircraft lacking the Ju 388's characteristic pressurized cockpit and high-altitude power plants was ever designated a Ju 388 and not a Ju 188 remains a mystery. Was the use of the successor design's serial numbers intended to suggest modernity to the issuer of federal contracts? There are certainly clues that this was the case. Written in mid-March 1944, Junkers production plan E 189 envisaged the production of a total of 687 Ju 188 G-3 torpedo bombers by VW. The first aircraft, powered by two Jumo 213 engines, were to be delivered in November 1944, while the target of 50 aircraft per month was not to be achieved until April 1945. The location for final assembly of the planned 682 examples of the Ju 188 G-2 bomber (Jumo 213) had yet to be determined, however. Following production startup in October 1944, 50 machines per month were to be built from May 1945. Only the J, K, and L versions of the Ju 388 appeared in this plan.[165]

In mid-May 1944, however, Delivery Plan 226, Version 1, reduced the number of Ju 188 G-2s to just 71 aircraft, to be built by VW between August 1944 and January 1945. Now, however, the Volkswagen factory was to build 655 "Ju 388 M-1 cuto" with BMW 801 TG power plants. Like the earlier Ju 188 G-3, 50 were to leave the production line each month. Production startup was scheduled for January 1945.[166] Instead of the Ju 188 G-3, therefore, VW was to build the Ju 388 M-1, which differed only in its BMW 801 TG engines, in almost exactly the same numbers. The only difference was the later production startup for the Ju 388 M-1, which was not to reach the production target of 50 aircraft per month until June 1945.

Both plans ended after the first quarter of 1946, and there were no plans beyond that. Did the Ju 188 G-3 therefore extend its career with under engines under the name "Ju 388 M-1"? There is some evidence that it did, however, definitive proof is lacking.

The Ju 388 J High-altitude Day Zerstörer, Bad Weather Fighter, and Night fighter

The oldest available information on the Ju 388 J comes from Amendment I of the construction directive for the "Construction of two Ju 188 J-1 Experimental Night fighters" dated 26 May 1943.[167] The wording of the amendment is largely similar to that of the construction directive for 10 examples of the Ju 188 L-0 issued two days earlier (see copy in following pages). The Ju 188 J-1 was to have the tail section of the Ju 188 F, although modifications were required on account of the tail turret. There were more significant differences in the cockpit canopy. The forward section was to be armored, with electrically-heated armor glass panels for the pilot and observer. The armor glass panels were to be provided with fresh air flow to prevent misting. The control yoke was to be designed so as not to obscure the pilot's view of the navigation equipment. It was also planned to install a large backup turn-and-bank indicator above the combined turn-and-bank indicator and artificial horizon, and a coarse altimeter for the observer. Plans called for the following to be added to the standard radio equipment of the Ju 188 L-0: FuG 212 Lichtenstein SN2 (FuG 220); Uhu (probably FuG 135 "Uhu II") or Fuchs (Fox) (probably FuG 134); plus FuG 217. The FuG 216 was replaced by the FuG 16 ZY.

The most significant differences from the Ju 388 L-0 concerned the aircraft's armament. For rearward defense, a remotely-controlled tail turret consisting of a FHL 131 with two MG 131 machine-guns was envisaged. The guns would be aimed by means of a periscope mounted on the port side of the fuselage in front of the observer's seat, providing a view above and beneath the fuselage. This was identical to the tail turret later installed in the Ju 388. The Ju 188 J-1's offensive armament was to consist of six MG 151/20 cannon with 250 rounds per gun. The six cannon were arranged in two rows beneath the fuselage (each with three guns side by side). Provision was to be made for replacement by two MG 151/20s and four MK 108s (100 rounds per gun), or four MG 151/20s and two MK 103s (100 rounds per gun). It was envisaged that the guns in the first row would be fitted with flash suppressors. Separate triggers were to be provided for the MG 151s and the automatic cannon (Maschinenkanonen or MK). The reflector gunsight was to be seated in a quick-release mount, making it possible to fold the sight out of the way, or remove it prior to an emergency landing. In the end none of the prototypes carried an armament of six cannon beneath the fuselage (see Chapter V); instead, this was limited to four machine-guns or cannon.

The carriage of bombs by the Ju 188 J-1 was not anticipated, but a linkage for the jettisoning of fuel tanks was to be fitted for Equipment Set M8. At the end of 1944, however, a change notification made it clear that the Ju 188 J-1 was to have neither air-dropped weapons or jettisonable fuel tanks.[167a] Initially it was planned to equip the Ju 188 J with a life raft, however, this was also later abandoned, probably because the military situation made Zerstörer missions over water very unlikely, making the raft unnecessary ballast.[168]

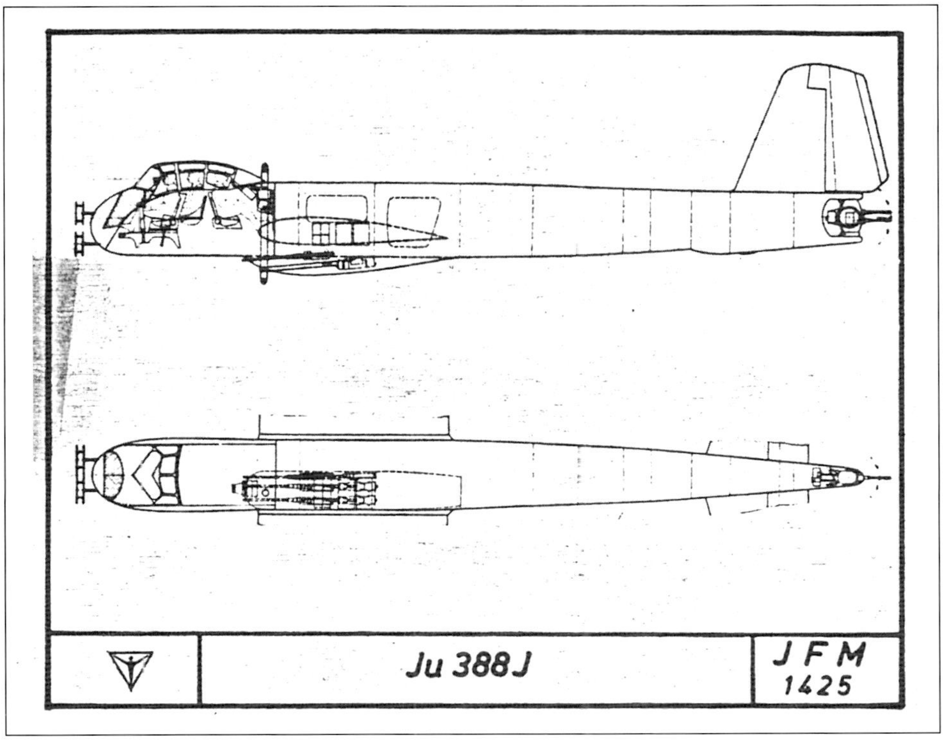

Early drawing of the Ju 388 J with FuG 212. Note the absence of oblique armament.

In November 1943 it had been assumed that the Ju 388 J-1 would be delivered without a tail barbette,[124] but in July 1944 the Technical Office decided that the Ju 388 J-1 be equipped with this after all.[168a]

The equipment for the J-1 and J-3 versions was finalized in a Junkers technical description from late 1944,[169] by which time only the strategic reconnaissance and night fighter versions remained in the production program. Offensive armament consisted of two MG 151 cannon with 180 rounds per gun, and two MK 108s with 110 rounds per gun, the latter housed in a bulged ventral bay between Frames 9 and 12. As well, two MG 151s with 200 rounds per gun were mounted behind Frame 15 at an angle of 70 degrees.

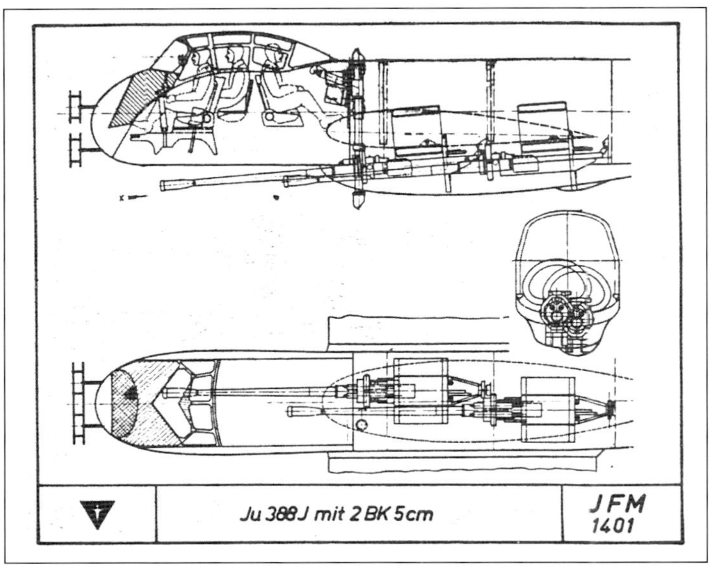

Drawing of the Ju 388 J with BK5 cannon. Design work for the installation of large caliber weapons was completed very early on.

Finally, like all versions of the Ju 388 powered by the Jumo 213 and the J-2, the Ju 388 J-3 was to have been equipped with a Dachlaurin fire extinguishing system.[170] The plan called for three extinguishing zones in each wing, and just one in the fuselage, encompassing the two stores bays. The engines were incorporated into the associated extinguishing zones in the wings.[171]

On 26 November 1944, during a telephone conversation with Dipl.-Ing. Kaspar of Aircraft Procurement and Supply, Junkers designer Oberingenieur Schilling learned that Galland was opposed to the use of water-methanol injection by Ju 88 and Ju 388 night fighters envisaged for the anti-Mosquito role. Galland's argument was that neither type would be fast enough, even with MW power boosting. As well, methanol was in short supply and was needed by the day fighters. Junkers intended to equip both the Ju 88 G-7 and the Ju 388 J-3 with two 150-liter tanks that could hold either fuel or the methanol-water mixture. The only additional equipment needed was one electric valve per side.

Kaspar promised to obtain approval for this solution so that Junkers could order the necessary equipment.

During the conversation Schilling also learned that efforts were again under way in the Technical Office to further reduce monthly production targets for the Ju 88 family. Consideration was obviously being given to dropping the Ju 388 J-1.

Furthermore, the night fighter units were demanding that the mounting angle of the oblique weapons be increased from 70 to 80 degrees, as the existing angle exposed them to enemy fire. Because these changes were to be incorporated into the Ju 88 and Ju 388 production line, Kaspar asked Schilling to advise him of the design and modification costs.[172]

In order to be able to engage enemy aircraft with greater effectiveness and at greater ranges, consideration was given to a "heavy fighter armament." The ability to open fire from greater range meant a reduced risk of being hit by defensive fire. On 23 September 1944 the Armaments Staff issued a "collective report" on aircraft construction. In it the staff envisaged the following development of the Ju 388 J:

"Continued development of the Ju 388 with 50-mm long-barreled gun for daylight missions for the time being and detailed report on: accuracy, stabilization questions, possible installation of Jumo 222."

The desire for heavy weapons had supported from the highest levels. On 4 November 1944 Hitler issued orders for the following immediate measure:

Ju 388 L-1 (F) without flame dampers **State III, 4 wing tanks**		**Concept Design Bureau** **Performance Data**	
Ju 388 J-1 (N) **State III, with tail turret** **16 Nov. 1944**		**Concept Design Bureau** **Performance Data** **No.: 388/846/2a**	
Engines:		801 TJ-0 with Fimag intercooler	
Takeoff power	h.p.	1,615 + 11 kg	
Combat power at 0 km	h.p.	1,472 + 8 kg	
Combat power at 12.3 km	h.p.	1,430 + 62 kg	
Purpose:		Night fighter with heavy armament , flame dampers and airborne radar	
Equipped weight	kg	10485	
Crew	kg	400	
Fuel	kg	2490	
Oil	kg	260	
Ammunition	kg	340	
Gross weight	kg	13965	
Wing loading	kg/m≤	250	
Average flying weight	kg	12730	
Max. speed at 0 km	kph	401	
Max. speed at 12 km	kph	583	
Endurance at altitude	hr.	3.02 (11 km)	3.03 (8 km)
at cruise speed at 11 km	kph	538	520
Endurance at altitude at	hr	3.69 (11)	3.45 (8)
economical cruise at 11 km	kph	480	405
Initial climb rate	m/sec	6.3	
Time to climb to	min	34.5 (11)	21 (8)
Ceiling on 2 engines at takeoff weight	m	12500	
Ceiling on 2 engines after climb	m	12750	
Ceiling on 2 engines at average flying weight	m	13000	
Ceiling on 1 engine at average flying weight	m	6000	
Initial climb rate, single engine	m/sec	1.2	
Takeoff distance	m	800/1200	

*including 1 hr. at combat power
**with 1000 liters of fuel and all ammunition for tail turret

State III: production state without leading edge cable cutters, with adjustable cooling gills, two cameras. Fimag intercoolers. Standard fuel weight: 0.78, capacity 3280 l. Search equipment: SN-2 radar with Morgenstern antenna and Naxos. *Zerstörer* armament with 2 MG 151 and 2 MK 101 plus 2 MG 151 oblique. Defensive armament: FHL 131 Z tail turret.

"The immediate installation of heavy long-range cannon and special weapons, like the Panzerschreck, rockets of all kinds, etc., in aircraft to be committed (against enemy bombers)."

When it came to the allocation of the necessary resources, these measures were given "absolute priority." Speer derived the following order from this general directive[173]:

"In addition to installation of the 50-mm [MK] 214 in the Ju 88 and Ju 388, a feasibility study is to be begun immediately for the installation and testing of one or two such weapons in the Ar 234. Similar studies must be undertaken for the 335 and the 262..."

It is impossible to say whether these plans were ever realized. The following directive, issued in January 1945, suggests that work on them did not get very far:

Ju 388 J-3 (N) with flame dampers 24 Nov. 1944		**Concept Design Bureau Performance Data No.: 388/851**	
Engines:		9-213 E	
Takeoff power	h.p.	1,750 + 103 kg	
Combat power at 0 km	h.p.	1,580 + 85 kg	
Combat power at 10.2 km	h.p.	1,270 + 140 kg	
Condition:		without external tank with MW 50	with 1 external tank with MW 50
Equipped weight	kg	10410	10410
Crew	kg	400	400
Fuel	kg	2410	2950
Rack + tank	kg	—	60
Oil	kg	260	260
Ammunition	kg	340	340
Gross weight	kg	14100	14700
Wing loading	kg/m≤	252	262
Average flying weight	kg	12750	13055
Max. speed at 0 km	kph	439	438
Max. speed at 10.2 km	kph	576	573
Emerg. speed with MW 50 and flame dampers at 8.2 km	kph	626	624
Emerg. speed with MW 50, no flame dampers at 9.1 km	kph	647	645
Endurance at 9.2 km	hr	3.74	3.65
at max cruise speed	kph	516	509
Endurance at 9.2 km	hr	4.65	4.30
at economical cruise speed	kph	438	446
Initial climb rate	m/sec	7.85	7.00
Time to climb to 9.2 km	min	27.7	34.0
Ceiling on 2 engines at takeoff weight	m	11100	10550
Ceiling on 2 engines at average flying weight	m	11800	11650
Ceiling on 1 engine at average flying weight	m	2300	2000
Initial climb rate, single engine	m/sec	*, **	*, **
Takeoff distance	m	670/920	750/1010

"Chances of and requirements for success, and installation possibilities for the 50-mm rapid-fire cannon, primarily the 214 at first, are to be thoroughly investigated and described for the Types 262, 234 and 388. Particular attention is to be paid to the time-limit situation, comparison of the possible and required number of weapons per aircraft, anticipated performance losses, etc. Project under the control of Knemeyer."[174]

These questions were to be resolved by 18 January 1945 by *Oberst* Knemeyer, Blume, and *Hauptdienstleiter* Saur.

There are photos of the mockup 37-mm cannon installation, but so far none have surfaced of the 50-mm BK 5. It is astonishing that consideration was still being given to large caliber cannon, for the participants must have been aware of the lack of success by aircraft so equipped (especially the Me 410, but also the Ju 88 P with a 75-mm gun). The ventral housing for the 50-mm KwK 39 had already been tested on the Ju 88 V57/1-(P-4), thus saving development time. The arrangement of the two 37-mm cannon was also very similar to the early Ju 88 G, which had two MG 151/20 cannon in the nose. In operational use this arrangement produced an uncomfortable blinding effect, and it was soon dropped from production aircraft.

Planned Developments

In August 1944 thought was already being given to increasing the range of the Ju 388. Such a machine was to be produced in mockup form using the fuselage of a Ju 388 L-1, WNr. 340 083, a new production aircraft built by ATG. The contract also included the experimental installation of two obliquely mounted weapons behind Frame 15 in the fuselage torso of the Ju 188 V2 (WNr. 260 151, landing accident on 2 June 1944, 60% damage), and the preparation of drawings for series production.[175]

As well as lengthening the fuselage, the designers took another path in increasing the range of the Ju 388. In a message dated 31 October 1944, the Prague design bureau envisaged the creation of a third bay between Frame 15 and Frame 20. The equipment installed there, such as the transformer frame, Kärcher heater, equipment board for the FuG 217, Leonhardt transformer, and GDU had to be relocated. The ammunition boxes for the tail turret extended back to just in front of Frame 20, and the cable balancing box for the direction finding equipment also had to be moved (because of the weapons). In the bomber versions a specially designed 550-liter fuel tank was to be installed in the new bay between Frames 15 and 20. This made it necessary to relocate the drive shaft for the bomb doors, and created a new a completely new hatch on the bot-

A large caliber weapons installation was examined in March 1943 using the Ju 388 J cockpit mockup. Griehl Collection

tom of the fuselage for installation of the tank. As well, new openings were required in the upper shell for a tank filler point and fuel pumps. In the reconnaissance version, a new bay between Frames 15 and 18 was to house either two daylight cameras (50 x 30 or 75 x 30) or two night cameras (NRB 35/25). These were to be installed in such a way that they could be positioned vertically, or tilted to the side at various angles. Instead of the new hatch installed in the bomber version, the reconnaissance was to have a hatch with cutouts for the cameras. As to the fighter version, the *Kobü-Waffen* (Design Bureau Weapons) would decide how the additional space was to be used.[176]

In addition, while production of the Ju 388 was getting under way, thought was again given to rationalization. On the one hand there were production "refinements," which meant simplifying the design of individual assemblies or components. For example, refinement of the Ju 388's electrical system meant that all lines were installed as single cables with the aircraft's aluminum skin acting as ground wire. On the production side, there were efforts to reduce as much as possible the large number of standard parts, semi-finished components, and other small parts. Such a mandatory parts selection list was issued for the Ju 88 family on 3 August 1944. From then on, only the semi-finished and standard parts contained in the list could be used in design changes and innovations. It is interesting to note that the materials lists for the Ju 88 G-1 and Ju 388 J were based on this mandatory parts list for the Ju 88 family.[177]

Despite the striking of the Ju 388 K-1 from the aircraft procurement program, there was still a requirement for bomber aircraft. In November 1944, therefore, thought was given to converting the Ju 388 J and L for use as bombers. For this purpose, the Ju 388 J-3 was to be fitted with an auxiliary bombing system. Two bombs would be carried at the wing load points, and dropped by means of a toggle switch. No bomb sight was envisaged, however. The Ju 388 J-1, which appeared at about the same time, also had these wing load points, but no bombing system was envisaged for this subtype.

The Ju 388 L-1 *Nacht*, deliveries of which were supposed to commence in the spring of 1945, could carry 12 50-kg bombs. As these were actually intended to be illumination bombs, no bomb sight was provided. Increasing the small number of Ju 388 L-1

Nacht photo reconnaissance machines, or converting the Ju 388 J required very many "external parts," meaning parts from outside subcontractors.

The Junkers engineers were of course aware that a bomber with no bomb sight could only be viewed as an interim solution, and therefore the Ju 388 K, which differed from the Ju 388 L only in its equipment, was again taken into consideration. The existing documentation was, however, not up to the latest standard, and "reintroduction" would also require numerous external parts (subcarriers, ZSK, ZBK, bomb sight, bomb doors drive, etc.), only some of which were present in the Ju 388 L-1 Nacht. As well, a completely new

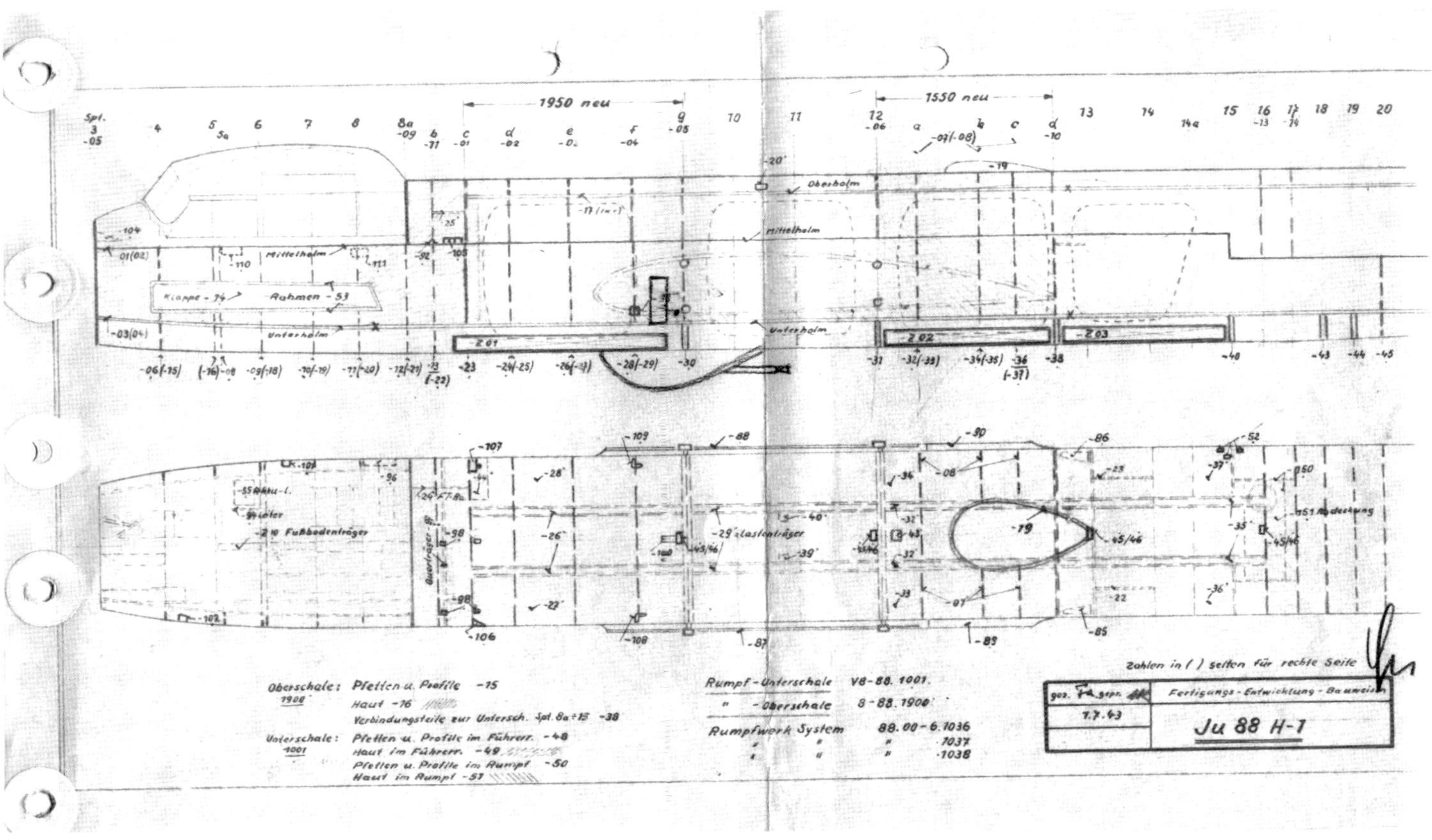

Junkers had gained experience in lengthened fuselages with the Ju 88 G-10 and H. The dimensions of the new sections are indicated in this factory drawing of the Ju 88 H. Consideration was also given to extending the fuselage of the Ju 388 for a *Zerstörer* version. Vernaleken Collection

electrical system would have been required. The L-1 Nacht was only equipped to carry 50-kg bombs, while the K-1 had to have a system capable of dropping heavy bombs.[178]

Finally, because of the shortage of aluminum a number of assemblies were to be converted to wood. As a first step, the landing flaps, vertical stabilizer, rudder, horizontal stabilizer, elevators, and extended wingtips were redesigned for wooden construction. Series documentation was to be delivered in early 1945 after the completion of load testing.[179] Experimental components were already in use in October 1944. As the second stage, it was planned to transition to wooden wings, ailerons, and landing flaps beginning in May 1945. The monocoque construction planned for the wings and control surfaces had already been tested on the vertical stabilizer of the Ju 352.[180]

Armament and Equipment Sets (*Rüstsätze*)

Initially, the following table of equipment sets were envisaged for the Ju 388.[181]

These equipment sets were later changed, however, and their designations were completely changed. In the following chapters we attempt to describe in greater detail the equipment sets envisaged for the type at the end of 1944. No equipment states

The following equipment sets (*Rüstsätze*) were initially envisaged for the Ju 388:[183]

B13	Two auxiliary oil tanks, installed in the port and starboard wings. Planned and installed as standard in Ju 388 K-0/K-1 and L-0/L-1.
M 8	Two wing racks, Load 1, non-jettisonable. Envisaged for Ju 388 L-0/L-1. Not installed as standard. Procurement by RLM special order only.
M 14	Two wing racks, Load 2, non-jettisonable. Envisaged for Ju 388 K-0/K-1 and L-0/L-1. Installed as standard in Ju 388 K-1. For other variants procurement by RLM special order only.
M 20	1 "Rack 8 Clasp 50" and 4 "Bearer 2 Clasp 50" for carriage of 16 x 50 kg bombs in fuselage. Envisaged for Ju 388 K-0/K-1.
B 12	Fuel tank in rear fuselage bay, 1050 l, envisaged and installed as standard in Ju 388 J.

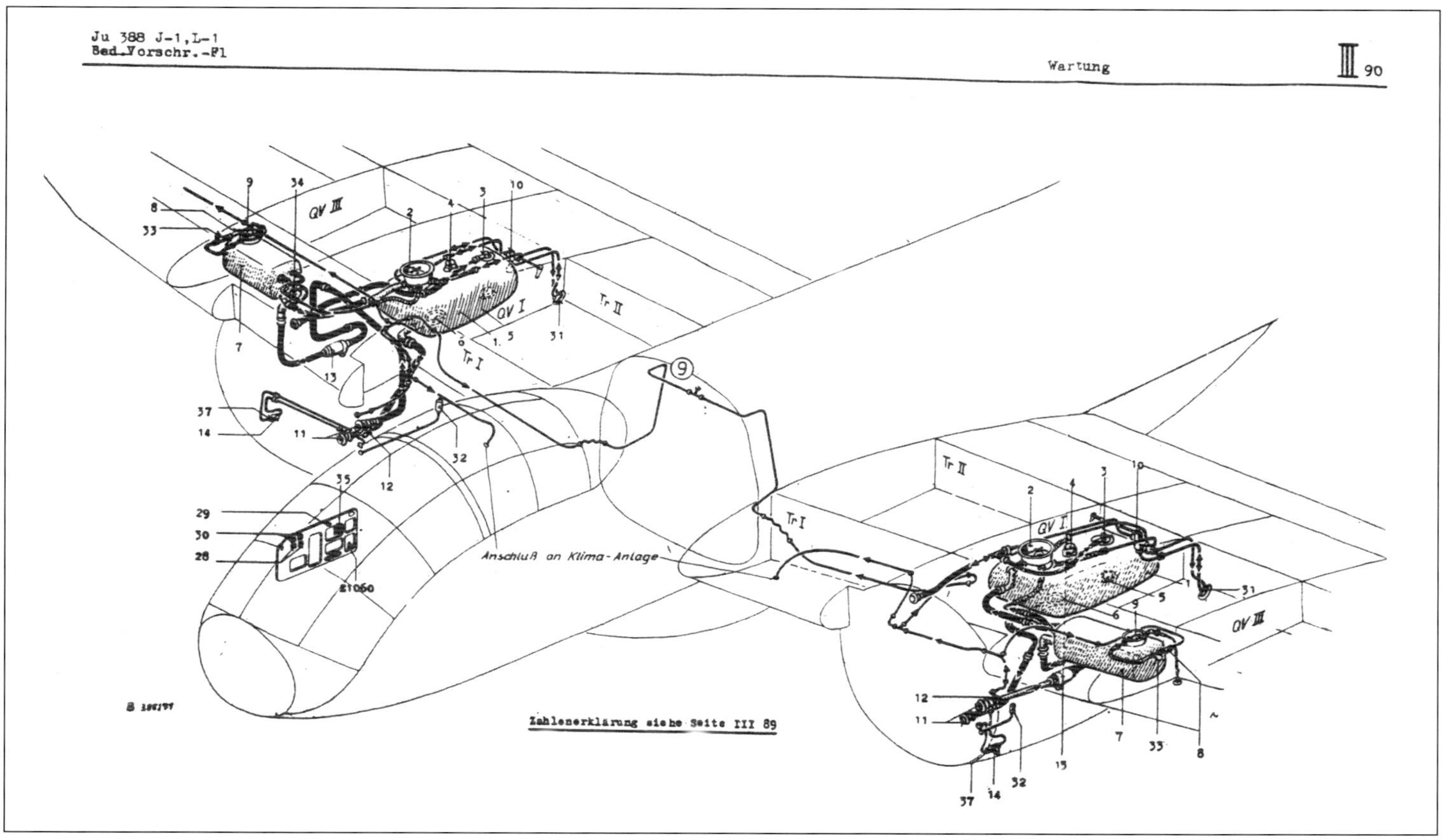

Arrangement of lubricant and fuel tanks in the wings. Transfer of fuel from the auxiliary tanks had to be accomplished using bleed air, as manual or electrically driven pumps failed at low temperatures (high viscosities).

(Rüstzustände), identified by a latter after the type designator, had been specified for the Ju 388 by the end of 1944. Minutes in which an equipment state "D" was mentioned for the Ju 388 J-1, and an equipment state "A" for the Ju 388 L-1, were immediately contradicted by the design bureau.[182]

Defensive Armament

Dorsal armament of the Ju 388 K-0 and L-0 consisted of an MG 131 mounted in an armored lens incorporated into the structure at the rear of the canopy. This armament was installed and tested in Ju 388 L-0 WNr. 300 005 (PG+YE) at Merseburg on 19 and 20 May 1944, resulting in a number of difficulties. The machine was probably one of the first in which the MG 131 was installed. There were clearance problems with the sight bracket—too little space between the weapon mount and the rear of the radio operator's seat, which was also too narrow. If the weapon was swung more than 20 degrees to either side the aiming line was blocked by the canopy framework and weapon mount. The padded headrest of the radio operator's seat, which was felt to be too tall, proved bothersome during firing, and had to be removed. Finally, there were also complaints that the spent casing chute was too long. The most serious problem, however, was that the weapon installation compromised the pressurization system:

"After a successful firing test, during which no further problems arose, the crew compartment was placed under pressure. The play between the weapon and buffer is so great that no pressure at all can be maintained. Installation of the dorsal gun position thus makes it impossible for the cockpit to be pressure sealed. Use of the climate control system at higher altitudes is therefore not possible. Modification of the pressure-tight weapon opening is necessary."[183]

This problem could not be solved in the short term, however, and so the aircraft had to be delivered without a usable dorsal gun position.[184] At the same time that the MG 131 was being installed in the dorsal position, weapons were obviously also installed in PG+YE (C-Stand, or ventral position) and tested. In order to make possible a rapid replacement of the rearview periscope after firing the ventral weapons, it was suggested that alignment markings be applied to the vertical stabilizer. These would make it possible to do without the adjustment fire that would otherwise be necessary.[183]

Report No. 7/44 by *Herr* Schweitzer of the FTA-Stelle dated November 1944 describes the testing detachment's involvement with other armament variations:

"The E-Kommando (388) is working on a weapons installation of its own (MG 131) in the right side of the rear upper window. The purpose is to provide a defense on both sides. The radio operator can comfortably operate the machine-gun while resting on the armature control rod."

Taken in the autumn of 1943, this photograph shows the cockpit of a Ju 388 K-0 with jettisonable cockpit roof removed. In the bottom left corner may be seen the covered socket for the rearview periscope, in the upper right corner the bolted mount for the dorsal machine-gun. Griehl Collection

Name	Designation Old No.	New No.	Remarks
Weapons Set	388 W10 (388F90)	8-388 F 821 2 x MK 103	*Zerstörer* installation 2 x MG 151, Ju 388 J
Weapons Set	388 W11 (388F91)	8-388 F 822	*Zerstörer* installation 4 x MG 151, Ju 388 J
Weapons Set	388 W12 (388F92)	8-388 F 823 2 x MK 108	*Zerstörer* installation 2 x MG 151, Ju 388 J
Weapons Set	388 W 13	8-388 F 824	Oblique installation 2 x MG 151 behind Frame 9

The Ju 388 J's Offensive Armament

The design bureau's equipment set overview of 18 October 1944 listed the following fixed weapon installations (see table avove):

The oblique weapons were originally to have been installed behind Frame 9, however, a decision was later made to position them behind Frame 15. This location was initially rejected by the Technical Office for cg reasons, which restricted the amount of ammunition that could be carried for the tail turret. In association with the installation of the oblique weapons behind Frame 15, a tank with a capacity of 475 liters was installed in the forward fuselage bay. As a result, the weapons arrangement beneath the fuselage had to be modified. The tank made it necessary to lower the ammunition boxes and design new ammunition feeds for the weapons.[184]

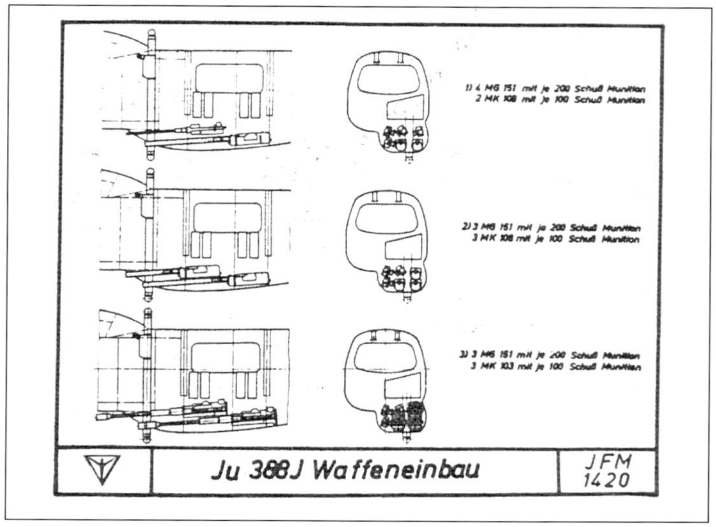

Overview of armament variations.

Air Dropped Weapons

Rüstsatz M 18 requires some explanation. On the Ju 88/188, attachment points for stores carriers ("Last 1") were located between the wing root frame and cross-member I. Additional stores carriers ("Last II") could be attached further outboard, between cross-members IV and V. All stores carriers could be jettisoned in the event of emergency (for example, the loss of an engine). On the two inner stores carriers, which were designated *Rüstsatz* M 8, the Ju 188 could carry between 500 and 3600 kg of bombs, or drop tanks with a capacity of 900 liters (wooden version) or 1200 liters (metal version).[185]

On the Ju 388—with the exception of the M-1 series—the attachment points for the inner stores carriers ("Last 1") were moved outboard by 260 mm to create additional space for the bulged bomb bay. The equipment set for the stores carriers on "Relocated Last I" was now designated M 18, and was no longer jettisonable. Access panels and fittings were moved accordingly. The mechanical bomb arming system was replaced by an electric one made by the Esbe company, and all lines for the jettisonable tanks were moved to the starboard wing. It was anticipated that the external stores carriers ("Last II" or *Rüstsatz* M 14) would be completely eliminated from the Ju 388.

From the first machine, variants powered by the Jumo 222 or Jumo 213 (Ju 388 J-2, J-3, K-3, and L-3) were to be equipped with *Rüstsatz* M18, to allow carriage of a 900-liter external tank beneath the starboard wing.[186] For cost reasons this equipment set was not envisaged for the Ju 388 J-1, K-1, L-1, and K-0.[132]

Name	Designation Old No.	Remarks New No.	
Bomb equipment in fuselage	M 20	8-388 F 810	Ju 388 K
Bearer for wing loads Load I	M 18	8-388 F811	Electric armed-safe block as holding device, not jettisonable, for displaced load with new fairing

Though of limited quality, this photo is noteworthy, as it shows a Ju 88T with long range tanks on the wing racks. The outboard racks were very rarely used. Haubner/Misna Collection

Cameras for the Ju 388 L

Name	Designation Old No.	New No.	Remarks
Daylight Photography	—	8-388 F 830	Two daylight cameras behind Equipment Set Frame 12 Ju 388 L
Night Photography	—	8-388 F 801	Two night cameras behind Equipment Set Frame 12 Ju 388 L

Engine Fuel and Oil Supply System

Name	Designation Old No.	New No.	Remarks
1050-l Auxiliary Tank	—	8-388 F 800	Tank with two fuel pumps, in rear Fuselage otherwise as 8-388 F 802
Auxiliary Oil System	B 13	8-388 F 801	Ju 388 L-1, J-3, L-3

Rüstsatz B13 was an option for the Ju 388 J-3, as the ability to carry a 900-l external tank for extended range was necessary.[182]

Factory photo of an NRb 50/25 night camera, taken in November 1944. At that time Junkers was converting Ju 388 K-0s into the "L-0 *Nacht*" version.

Chapter 4: Power Plants and Propellers

The purpose of this chapter is to describe the engines and propellers projected for the Ju 388, and actually installed in prototypes and production aircraft. Information on the type, design, and performance of the various power plants is followed by a brief summary of their development, and an examination of the engine related problems encountered during testing of the Ju 388.

The Ju 388 was mainly flown with the BMW 801 TJ-0, which was also the standard engine of the L-1 series. All Ju 88/188/388 airframes were designed for 40-ton standard engine mounts, which could accept BMW and Jumo power plants.[187]

Several prototypes were constructed with Jumo 213 E engines, and at least one with Jumo 222s. They were used to test the engines chosen for the -2 and -3 sub-variants. As well, there were plans to equip the Ju 388 with the DB 603, but when the war ended this was probably still in the experimental stage. Mixed power was obviously also in the planning, and theoretical calculations were made in this direction.[188]

First, however, we will examine the general considerations associated with the selection of power plants for the Ju 388.

The Power Plant Question

Originally the Ju 388 was supposed to be powered by the BMW 801 TM engines (probably Project P8035). In the period from 4 to 6 October 1943, Herr Wilhelm of the Junkers Concept Design Bu-

BMW P8035 engine project, with the engine and turbosupercharger as individual, easily separated components. MTU

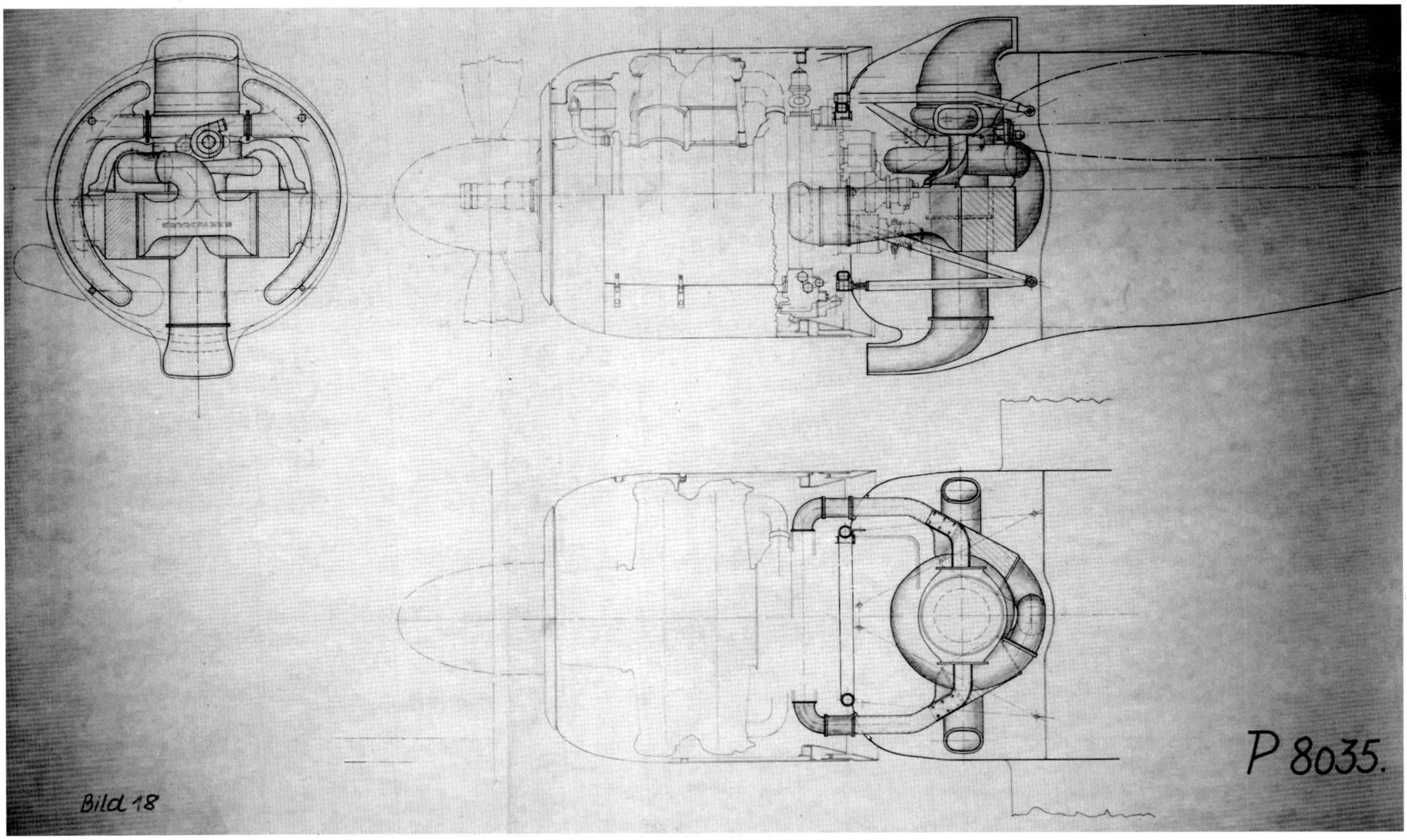

This drawing illustrates the modular design of the BMW P8035. MTU

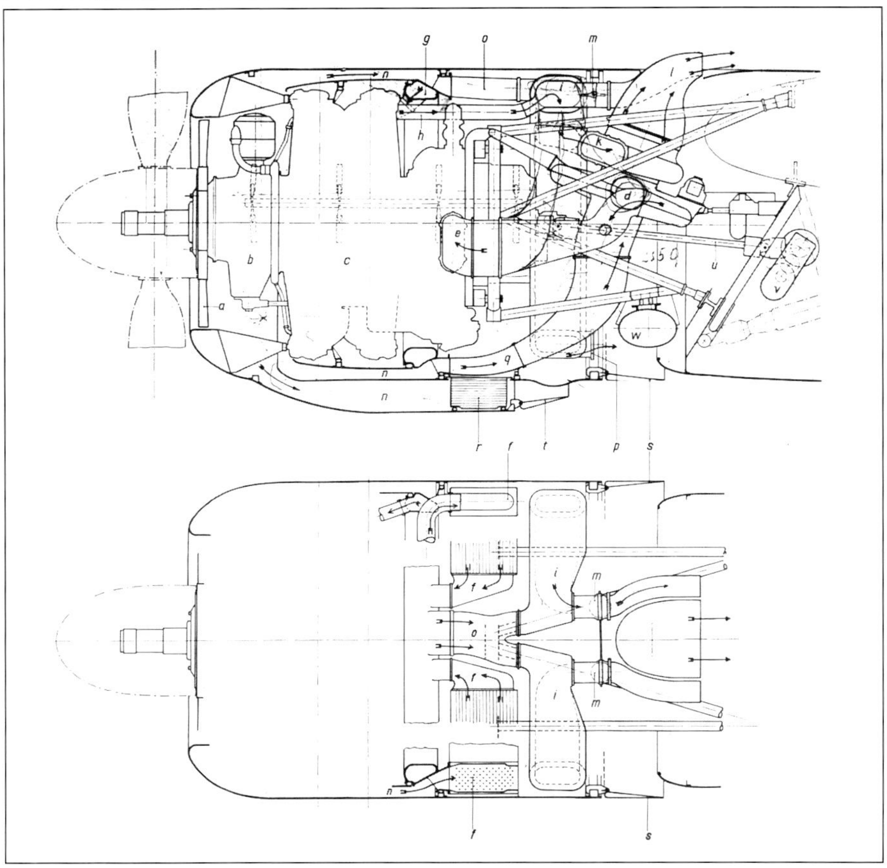

In contrast, the compact design of the BMW 801 TJ—here in an early drawing—meant that the turbosupercharger was closely integrated with the engine. MTU

reau and the Amsterdam Design Bureau discussed various concepts for the BMW 801 TM and its mockup.[189] On 15 September 1943, at a conference held by BMW, it had been decided to produce the flame damping system (FlaV) as a common equipment set for the BMW 801 TM and TJ. As the TJ power plant was to be built in greater numbers than previously thought, it was also decided to design just one intermediate nacelle for both engines. It was also decided that the sliding rings meant to serve as compensation for construction tolerances would be deleted, significantly improving the aerodynamics. The concept design bureau was demanding a 10 percent increase in the size of the intercooler. As this would require an additional 2,000 to 2,500 hours of design work, it was decided to initially postpone this measure. A communication from Prof. Hertel dated 25 October 1943 reveals that the BMW 801 J power plant was increasingly coming to the fore for the Ju 388,[190] because this power plant's numbers were steadily rising, while the BMW 801 TM project faded completely into the background. At a meeting of department heads on 12 October 1943, which was also attended by Hertel, Generalingenieur Mahnke suggested increasing production of the BMW 801 TJ by 500 engines to gain another three months for testing of the BMW P8035 turbosupercharged engine. Milch agreed with the proposal, and ordered that priority be given to the *General der Aufklärer* (General of Reconnaissance) for BMW 801 J engines for 30 Ju 388 T-1s.[191]

It is interesting that Hertel already foresaw very clearly the maintenance (and thus the reliability) problems that would result from poor accessibility. There was no alternative to the BMW 801 TJ in sight, however, and in spite of everything the power plant had to be accepted, as the series solution for the Ju 388. Hertel asked his associates Cruse, Wilhelm (concept), Klein, Du Bois (design), and Herzog (production inspection) to prepare a report on the BMW 801 TJ in order to exert as much pressure on BMW as possible. Through this he hoped to get a successor power plant as soon as possible. The report was to be broken down into the following categories:

a. Requirements which absolutely must be raised and absolutely have to be met.
b. Requirements which are important and could be met
c. Requirements which are important, but whose infeasibility are obvious[190]

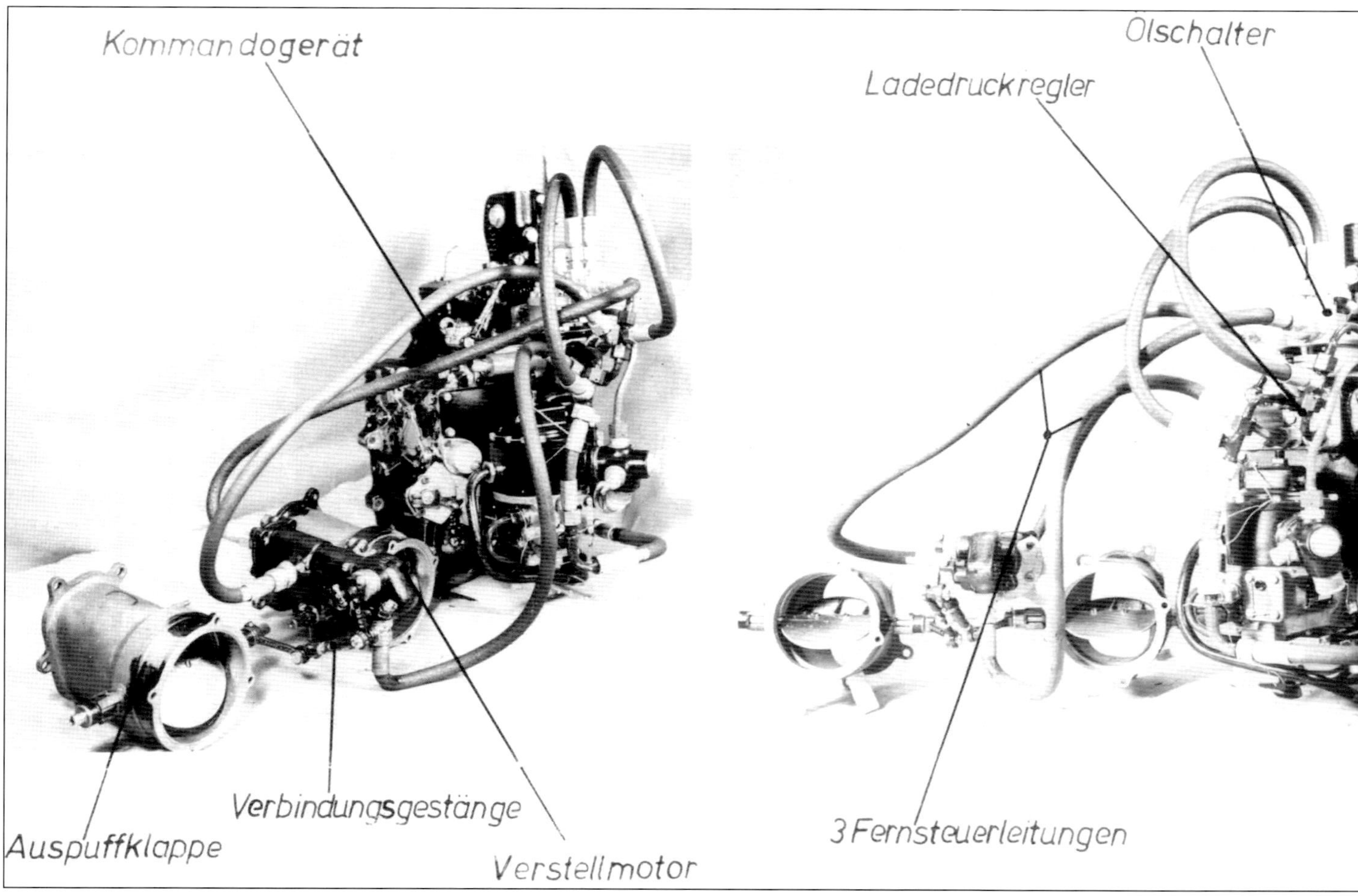

Automatic control of the turbine by the engine control system was tested, as this photo confirms; however, the designers were unable to make the system work well enough to go into production. MTU

Hertel was undoubtedly a clever tactician and farsighted engineer, for one of the main criticisms by *Erprobungskommando 388* in its final report was the aircraft's high maintenance requirements.

"The high incidental maintenance time per flying hour shows that during frontline operations in winter the aircraft serviceability rate would depend largely on whether work areas and hangars were available. Despite careful maintenance by well-trained personnel, serviceability rates will scarcely be the same as those of other aircraft."[192]

A few days later, on 1 November 1943 it was confirmed that the BMW 801 TM would be abandoned. The Ju 388 would have to be built with the BMW 801 TJ. The engine still had numerous shortcomings, however, which BMW had so far refused to address because of the small number produced. This justification disappeared, now that the Ju 388 was to be built in quantity with the engine. The Junkers design bureau subsequently seized the initiative, and on 1 November 1943 summarized the most important shortcomings. Flight tests, probably with a Ju 88, had revealed an unexplained pressure loss in the ram air intake that was equivalent to a reduction in maximum boost altitude of 400 meters. As well, the supercharger air was heated by 20°C, which cost another 100 meters in maximum boost altitude, and more than 4% of performance. Flight tests also revealed that, at altitude, the effectiveness of the cooling fan fell by 10% compared to its performance at ground level. Junkers therefore requested that the improved cooling fan under development by BMW begin flight trials as soon as possible. As well, Junkers was basically in favor of the development of an adjustable cooling fan.

According to statements by BMW, the planned turbine controller was not yet ready for service; therefore, the BMW 801 TJ's turbines had to be manually controlled. Those at Junkers asked themselves if this manual control could be retained without endangering the engines, as the high-altitude fighter in particular would be making rapid changes in altitude with no time for resetting of the turbines. Pending the availability of the turbine controller, the production engines would be fitted with a sensor for the supercharger air temperature. As well, the propeller pitch control drive of the BMW 801 D basic engine was too weak, the reliable pitch change speed being too low even for propellers with counterweights. It was not just for this reason that Junkers desired the installation of the BMW 801 E as a further development of the engine.

At that point in time all BMW 801 engines were suffering piston problems; however, Junkers categorically refused a solution involving a reduction in performance of 100 hp, as this would have meant a speed loss of 14 kph for the Ju 388. The cowling, latches, and cowling attachments were all seen as insufficiently strong, and it was requested that towed tests be conducted under water to test the cowling's strength.

As the BMW 801 TJ power plant was still fitted with the old Junkers attachment fittings, a request was also made for a change to the 40-ton standard mounts for the Ju 188 wing. A reminder was issued of the need for improved accessibility "to the vital parts" of the engine.[193]

On 15 November 1943 designers Cruse and Wilhelm, together with *Herr* Danner of Jumo, attended a meeting with Dr. Schilo, Plenipotentiary for Power Plant Development (BTE) in the RLM. Also present were representatives from Borsig, Daimler-Benz, and Hirth-Motorenbau. Dr. Schilo's remarks revealed that the BMW 801 TJ was in fact the only available power plant with an exhaust driven turbosupercharger. The use of an exhaust driven turbosupercharger with an air cooled engine had proved more problematic than with a liquid cooled power plant, however. The RLM therefore planned to halt production of the BMW 801 TJ in February 1945, and replace it with an entirely new exhaust driven turbosupercharger engine based on the DB 603. The turbines and cooling system, however, were to be designed in such a way that an alternative solution based on the Jumo 213 could be produced quickly if required. Jumo would therefore have to assist Daimler-Benz with the design. On account of the increasing urgency attached to the Jumo 222, Jumo was also ordered to build an exhaust driven turbosupercharger engine based on the Jumo 222.

Cruse responded with the argument that an exhaust-driven turbosupercharger engine based on an air cooled engine was "right" for high-altitude aircraft, which could not be armored for weight reasons. According to Dr. Stieglitz (Hirth-Motorenbau), performance calculations for the 9-2426 (TK 19) turbosupercharger resulted in a maximum boost altitude of 13 km for the DB 603 A and 14-15 km for the DB 603 G, and a maximum continuous power of 1,400 hp. Use of the Jumo 213 A resulted in a maximum boost altitude of between 14 and 15 kilometers. No precise estimates had been made for the Jumo 213 E, for the turbosupercharger would have to be enlarged for use with that engine. It was anticipated, however, that maximum boost altitudes in excess of 15 km could be achieved. When questioned by Schilo, Prof. Van der Nüll confirmed that the figures provided by Hirth were obtainable, as some of the calculations had been done at his institute. He rejected a larger turbosupercharger for the Jumo 213 E, as it would then be too large for the DB 603.

From Junkers' point of view, maximum continuous power was of primary importance. The Ju 388 required maximum continuous power of 1,400 hp at a height of 13 km, and 1,500 hp at 14 km. For the Ju 488 these values, which included the thrust produced by the turbines, were about 100 hp lower. The first flights with the BMW 801 TJ resulted in an estimated thrust figure of 75 to 80 kg.

To create a uniform basis of computation, it was decided to make the calculations for the 603/213 turbosupercharger and the Jumo 222 turbosupercharger (which was to be a new design, not a compromise solution) in cooperation between the turbine and engine companies under the direction of Prof. Van der Nüll. For this purpose representatives from Borsig, BMW, Daimler-Benz, and Jumo were to meet with Prof. Van der Nüll at the Hirth offices in Stuttgart on 22 November 1943.

Borsig's suggestion to arrange the turbine behind the engine with a vertical axis was rejected by Junkers, because this would have required that the engine be moved forward 450 mm. The company also feared negative effects on directional stability.

The alternative was an installation with an horizontal axis, which still required that the engine be moved forward 350 mm.

Use of the Jumo 222 as a power plant for the Ju 388 was discussed as early as the end of 1944. This Jumo 222 E, which has a mechanically driven supercharger, is on display in the Deutsches Museum. Development of a high-altitude version with turbosupercharger was also planned. Handig Collection

Even a minor relocation would have significant effects on the structure of the engine nacelles and leading edge, which could easily lead to wing vibration. Wind tunnel experiments and flight trails were needed to make a final determination. According to Junkers and Borsig records, *Lufthansa* was supposed to convert a Ju 88 for flight trials, with an intermediate trestle and cowling mockup simulating the new engine position. A final decision on which configuration would be used would be made based on the results. An extension of the February 1945 deadline was foreseeable. Herr Mann therefore proposed that the Ju 388 be immediately redesigned to take the Jumo 222, and that the turbosupercharged engine based on the DB 603 be used as forerunner to the Jumo 222 engine with turbosupercharger. Prof. Hertel also favored this engine as the ultimate solution for the Ju 388 and Ju 488. *Herr* Friedrich of Daimler-Benz again raised the topic of the DB 622 and the selected arrangement of the turbines, because no forward relocation of that engine was necessary, just a modification of the intercooler.

After the conference Cruse met once again with Major Knemeyer and pleaded—in his view successfully—for the retention of the BMW 801 TJ until the Jumo 222 with exhaust-driven turbine was ready for service. At that point it was obviously planned to produce only some of the 500 Ju 388s per month as pure high-altitude machines.[194]

BMW 801 TJ-0

The BMW 801 TJ Engine

The BMW 801 was the most powerful German radial engine to see service during the Second World War. Fourteen cylinders were arranged in two rows. Bore and stroke were both 156 mm, and all

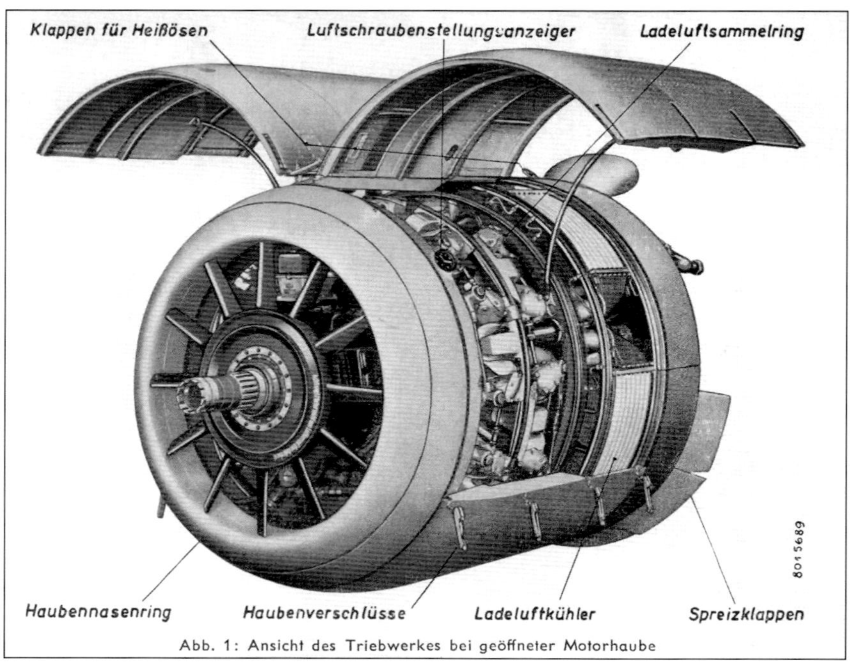

Overall view of the BMW 801 TJ-0 engine from the engine chart. Clearly visible are the single exhaust pipe and the intercooler. MTU

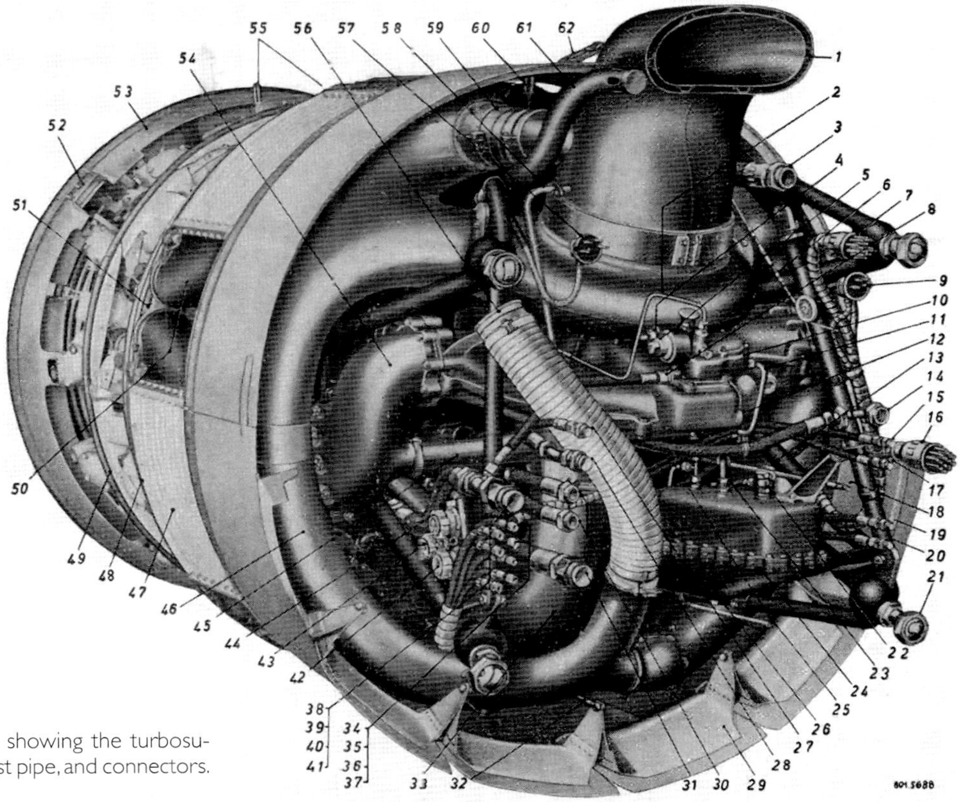

Abb. 2: Ansicht auf Anbauseite mit Anschlüssen

Rear view of the engine showing the turbosupercharger, central exhaust pipe, and connectors.

1. Exhaust nozzle
2. Pressurized oil line to turbine
3. Oil tank ventilation in engine
4. Turbine oil relief valve
5. Turbine oil filtering screen
6. Actuator for free exhaust gill
7. Electrical socket for rpm indicator, intake oil temperature, control mechanism
8. Turbine oil return line
9. Electrical socket for generator
10. Turbine oil pump
11. Turbine oil header from radiator
12. Sediment drain
13. Engine control device (*Kommandoger%ot*) oil filler point with dipstick
14. Pressurized air for fuel jettison
15. Hydraulics, cooling gills "OPEN"
16. Electrical socket for starter, mixture indicator, lighting, propeller pitch indicator
17. Hydraulics, cooling gills "CLOSED"
18. Throttle lever actuator
19. Hydraulics, cowling nose ring "OPEN"
20. Hydraulics, cowling nose ring "CLOSED"
21. Engine attachment point
22. Oil return line from intercooler
23. Header to intercooler
24. Engine control device oil tank
25. Engine control device and tank ventilation
26. Engine control device tank sediment drain
27. Hydraulic fluid from hydraulic pump
28. Header to hydraulic pump
29. Cooling gills
30. Fuel header lines
31. Oil header
32. Cooling gill actuator
33. Supercharged air suction line to turbosupercharger
34. Fuel ventilation return line to tank
35. To boost pressure gauge
36. Propeller deicing
37. To oil pressure gauge
38. Oil return line
39. Cold start fuel
40. To fuel pressure gauge
41. Starter fuel injection
42. Hot oil to engine
43. Hot oil to cooler
44. Hot oil to tank
45. Turbine bearing ventilation
46. Exhaust collector with heat exchanger
47. Intercooler
48. Cowling seal, pressurized air collector ring
49. Cowling seal, cylinder baffle
50. Compressed air from mechanical supercharger to cooler
51. Compressed air collector ring
52. Propeller pitch indicator
53. Forward cowling section
54. Compressed air to mechanical supercharger
55. Forward lifting eyes
56. Hot air connection
57. Electrical socket for starter
58. Turbine housing vent
59. Free exhaust pipe
60. Rear lifting eye
61. Engine housing vent
62. Lifting eye access hatch

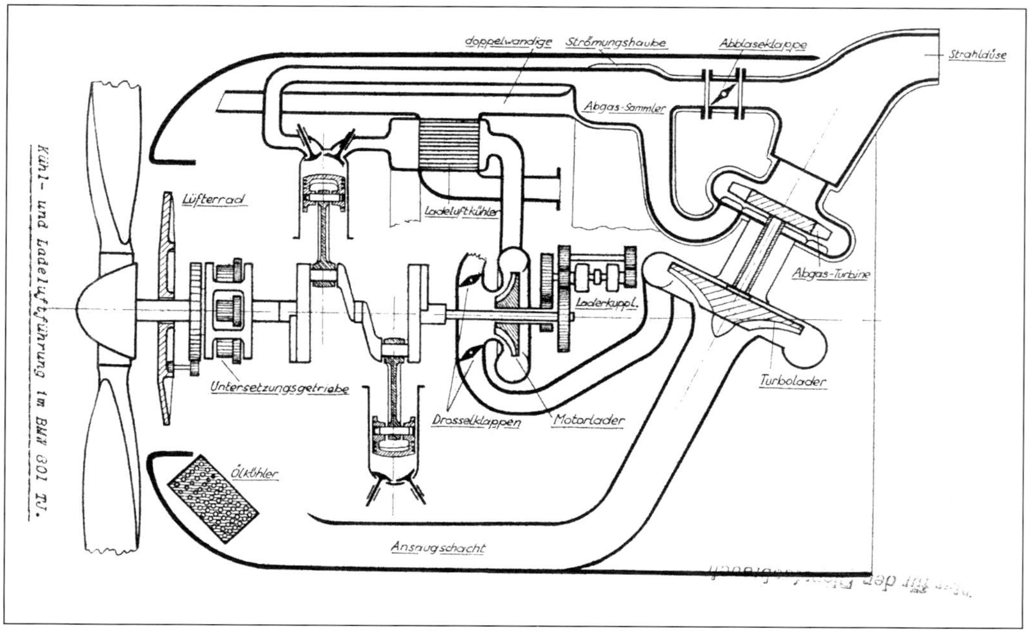

Schematic drawing showing the path followed by the supercharger air in the BMW 801 TJ. MTU

versions of the engine had a displacement of 41.8 liters. The air cooled engine was equipped with a cooling fan, which ensured sufficient cooling while stationary or in low-speed flight. Part of the cooling air flowed through the oil cooler, and negative air pressure at the front of the engine forced it out through the rear of the nacelle. Like all German high performance engines, the BMW 801 was equipped with direct fuel injection. The engine was equipped with an automatic engine control system (Kommandogerät), which made it possible to control the engine with a single control (throttle lever). The "T" attached to the engine designation stood for the complete engine installation, including cooler, cowling, and exhaust system. The next letter identified the actual variant; for example, the J-0 version was the BMW 801 TJ-0. On the other hand, the letter "M" before the engine version signified the "naked" version of the engine.

Early on, BMW developed plans for a BMW 801 with an exhaust-driven turbosupercharger under the project designation P8001. The engine's mechanically-driven two-speed supercharger was to form the first stage. From this project ultimately came the BMW 801 J and K. The only difference between the BMW 801 K, which was later abandoned, and the J version lay in the propeller drive, which was designed to rotate in the opposite direction of the crankshaft.[195]

The first experimental versions of the BMW 801 J were based on the BMW 801 A and had a maximum boost altitude of just 8 km. Eventually Junkers switched to the BMW 801 D, whose mechani-

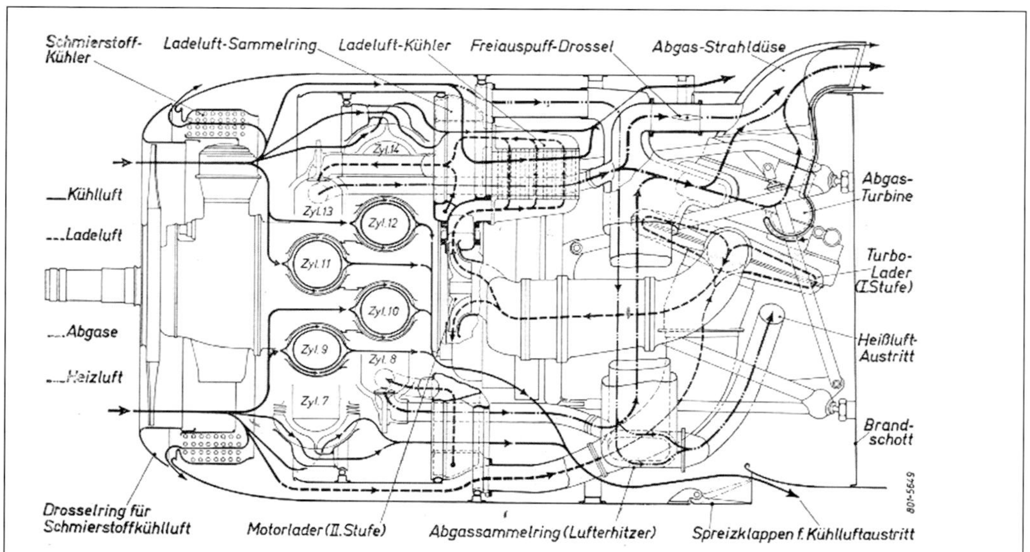

Abb. 3: Ladeluft — Kühlluft- und Abgasführung

The flow of cooling air in the engine. MTU

Front view of the BMW 801 TJ-0 engine from the installation guide. MTU

Of note in this side view from the installation guide are two of the huge intercoolers, and the split valve cover of the BMW 801 D. MTU

cally-driven supercharger possessed a higher pressure ratio. In response to a requirement for a maximum boost altitude of 11.8 km, in early 1942 the turbosupercharger of the BMW 801 had to be completely redesigned. The impeller circumferential speed increased from 365 to 400 m/sec.[196] The exhaust turbine, with solid chromium-manganese steel blades, completed a 100-hour endurance run in January. In the BMW 801 J the exhaust gases were fed into the turbine uncooled. The high exhaust temperatures (800 to 850°C) and the high circumferential speed (250 m/sec) produced high thermal and mechanical stresses that could not be handled by solid blades. Hollow blades with internal air cooling were therefore introduced.[197]

Exhaust collector, turbosupercharger, and central exhaust pipe of the BMW 801 TJ-0. The pale hose below the central exhaust pipe is the hot air attachment. This photo shows that the accessories, such as injection pump, starter, etc. were completely covered by the turbine installation, making servicing considerably more difficult. MTU

Development of other high-altitude engines based on the BMW 801 proceeded slowly; therefore, the BMW 801 TJ-0, which used 95-octane C3 fuel and was based on the BMW 801 D (Series 2), was the only available high-altitude engine for a long time. Deliveries of the BMW 801 Q with the Hirth 9-2426 exhaust-driven turbosupercharger (TK 19) were not expected to begin until May 1945.[198] In November 1944, therefore, high priority was given to the BMW TQ/2 with the 9/2426 turbosupercharger based on the BMW 801 E.[199]

The main difference between the BMW 801 J-0 and the BMW 801 D was the turbosupercharger installed behind the accessories tray. The BMW 801 J-0 thus had two supercharger stages: the first was a continuously variable exhaust-driven turbosupercharger, the second a mechanically-driven two-speed supercharger for ground and high-altitude power. Before the compressed supercharger air was fed into the collector ring, it passed through four air cooled supercharger air final coolers arranged in parallel. If the maximum boost altitude was to be raised beyond that of the BMW 801 D, the diameter of the cooling fan would obviously have to be increased to raise cooling capacity. Cooling air was regulated by hinged shutters.[200]

The reduction drive for the propeller, with a ratio of 0.54 of engine rpm, was designed as a synchronized drive for clockwise-rotating VDM variable-pitch propellers. On several experimental examples of the BMW 801 J-0, the previously-mentioned control system for single lever operation also controlled the exhaust gas turbine via actuators on the waste gates, which operated in parallel with the exhaust turbine, and consequently expelled most of the exhaust gas when in the open position. Gradual closing of the waste gates made the turbine infinitely variable. In spite of the obviously good experiences with the automatic control system, on production engines the admission of exhaust gas into the turbine had to be controlled manually. For manual control, it was recommended that

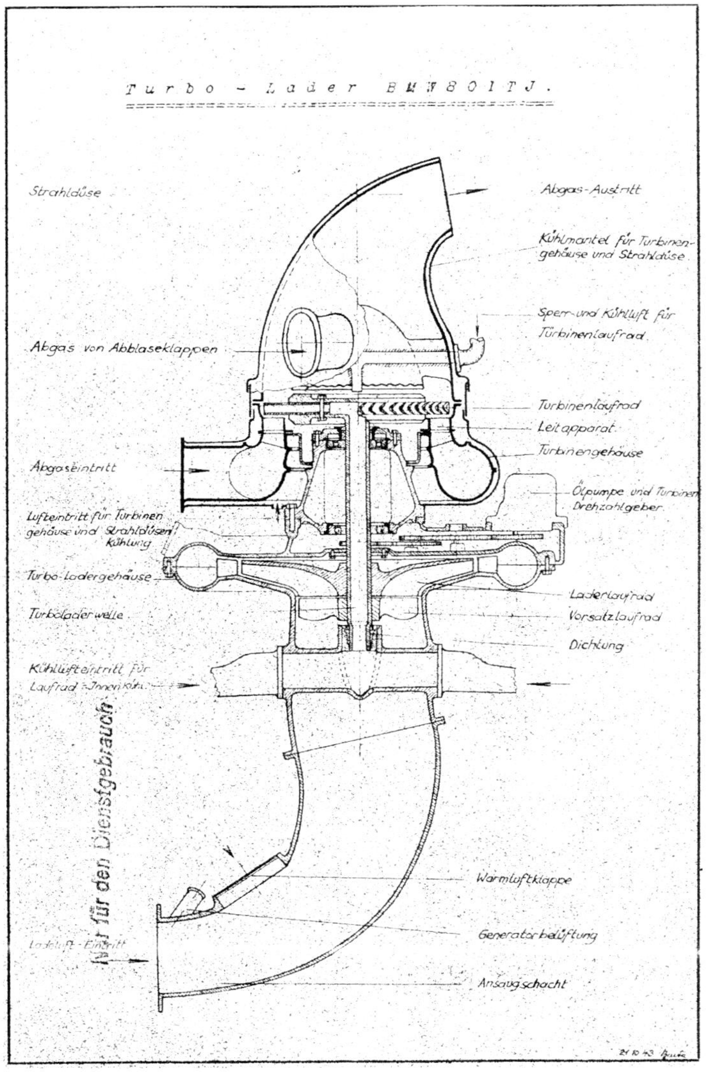

Detail drawing of the turbosupercharger installation in the BMW 801 TJ.

Early version of the exhaust driven turbosupercharger.

Impeller (front) and turbine rotor (rear). At the bottom left are two of the internally cooled turbine blades.

This perspective provides an exceptionally good view of the huge exhaust collector and turbosupercharger installation. On this early engine the free exhaust next to the central exhaust pipe is noteworthy; one of the two waste gates is also clearly visible.

The two waste gates used to regulate the turbine with control linkages.

All photos on this page: MTU.

boost pressure be allowed to fall by increments of 5/100 atmospheres[201] until the mechanical supercharger's maximum boost altitude of 9 to 9.5 km had been reached—then the waste gates closed until the original value had been reached again. These constant adjustments meant that the waste gates were repeatedly closed. After the high-altitude supercharger was engaged, the waste gates were to be opened just enough to prevent boost pressure from dropping. This prevented the unnecessary venting of excess air. The waste gates were then supposed to be gradually closed—like the low-level supercharger—until maximum boost altitude was reached.[202] Takeoff power by the BMW 801 TJ-0 (engine 9-8801 J-0) was 1,615 hp, and available climb and combat power was 1,472 hp at ground level and 1,430 hp at the maximum boost altitude of 12.3 km.[203] The developed version (BMW 801 TJ-1) produced 1,810 hp at ground level and 1,490 hp at 11000 m. It was based on the BMW 801 E/S and had one piece valve covers.

Note: In an effort to produce high-altitude engines without the modifications to the actual power plant system required for the BMW 801 J, it was also proposed that the turbosupercharger be integrated into the airframe as an "integral turbosupercharger" or "displaced turbosupercharger." To test these concepts, in early 1942 the 323 P turbosupercharger was developed from the turbosupercharger of the BMW 801 J.[204] In mid-1943 this turbosupercharger was installed in the fuselage of the Hs 126 D-ISDR, which was powered by a Bramo 323 R. The engine exhaust was fed through a 3.4-meter long pipe on the right side of the fuselage to the exhaust-driven turbosupercharger. The compressed air from the turbosupercharger passed through an intercooler on the right side of the fuselage, and then along a 3.4-meter pipe back to the engine. About 39 hours of flight time were logged without major problems, and the maximum boost pressure with the low level supercharger in operation increased from 2 km to 8.7 km as calculated. The drag producing components of the "displaced turbosupercharger" had a negative effect on lift, and the Hs 126's reachable service ceiling was limited by the airframe (and not the engine), and was equivalent to "just" 10.9 km.[205]

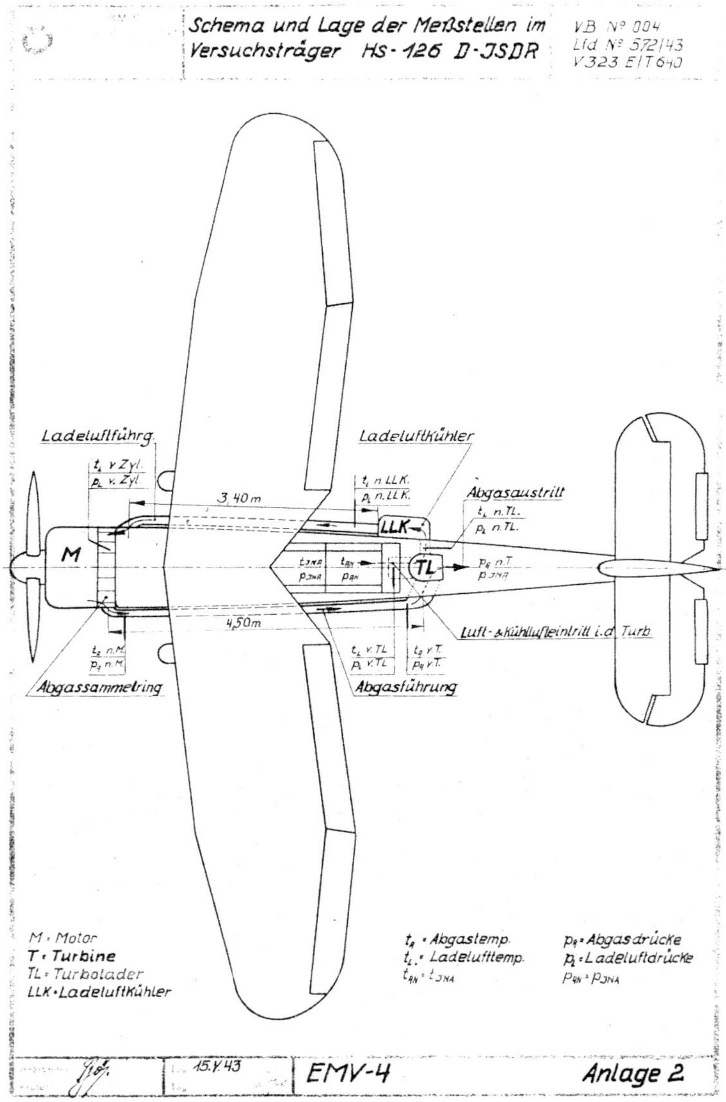

Installation of a "displaced" turbosupercharger in a Hs 126. MTU

In this unusual view of the engine from above, the waste gates with actuators and free exhaust may be seen clearly. MTU

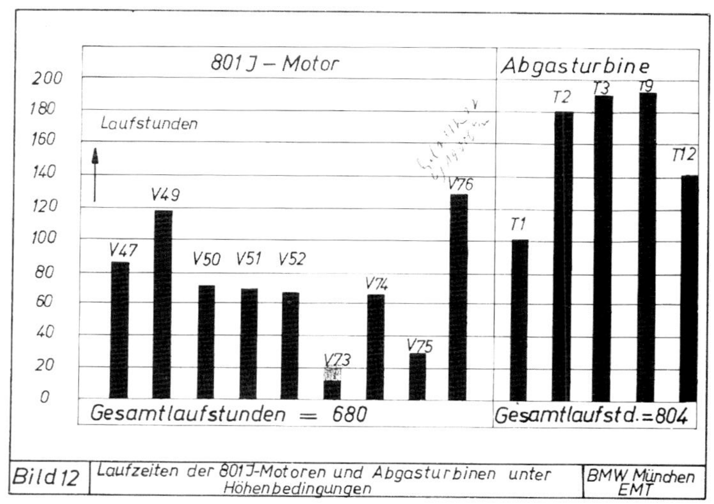

As this graphic shows, several of the prototype engines had a significantly higher operating life than the first pre-production engines. MTU

Borsig-built turbosupercharger for the BMW 801 TJ in the Deutsches Museum. Handig Collection

In spite of these favorable test results, BMW soon abandoned the integral turbosupercharger after it was found that the airframe design changes for the Ju 88/188, for example, were much greater than expected, and much too expensive.[206]

In 1942, in an attempt to achieve a better separation of engine and turbosupercharger while avoiding the disadvantages of the integral turbosupercharger, BMW created Project P8035 based on the BMW 801 E. The turbosupercharger was located in an insert between the engine and firewall.

The BMW 801 TJ-0 was installed in several prototype aircraft. Already in 1942 the Hs 130 C V4 was fitted with experimental engines based on the BMW 801 A, as was a Ju 88.[207] A Ju 188 (WNr. 260 308[208]) is also known to have been fitted with BMW 801 TJ-0 engines, and was tested at Rechlin.[209] It seems noteworthy that in May 1943 *Oberst* Rowehl of VfH Mitte reported to the RLM that "the testing of a Ju 188 with BMW 801 Js is proceeding surprisingly well."[210] In spite of this, the RLM initially backed the BMW 801 TM, another high-altitude power plant with turbosupercharger. In June 1943 *Generalmajor* Vorwald (representing the RLM GL/C-E-3) decided that plans to increase the BMW 801 J pre-production series by 100 engines should be canceled, in order to ensure the start-up of the BMW 801 TM.[211] In the summer of 1944 he wanted to add 38 engines to the BMW 801 TM pre-production. Earlier plans to convert 20 Ju 88s to the BMW 801 J in the *Luftwaffe* maintenance facility in Staaken were abandoned. The engines were now to be assigned to the KdE, while the VfH was to install two in a Ju 188.

Production of the BMW 801 was hardly sufficient to meet the needs of the aircraft manufacturers. As a result, from December 1943 to May 1944 Engine Program 325 was short 865 engines, and the number of spare engines was only 800 to 900.[212]

The state of deliveries was discussed at the RLM development conference on 14 April 1944, the main topic being the BMW 801 TJ.[213] Director Buchs of BMW-Spandau gave an account of the difficulties that were preventing the company from adhering to "Engine Program 324." The principal one was the unavailability of exhaust driven turbines resulting from bombing of the Borsig factories, but the Kolb firm was also having problems welding the exhaust collectors. Meanwhile, a remedy had been found in cooperation with Eberspächer, Junkers, and BMW. Of the 20 to 30 BMW 801 TJ-0 engines delivered to Junkers so far, 20 had the poor quality collectors. These were approved for test purposes, provided special safety measures were adhered to. For the reasons given, there were no deliveries in March. Deliveries were supposed to resume in May in a scaled down program, provided the exhaust collectors were delivered as planned. Because there were too few engines and airframes available for testing at that time, *Oberst* Petersen urgently demanded the allocation of additional power plants. As it stood, testing could not begin until June. Vorwald subsequently ordered that the delivery of power plants for testing be accelerated; the GL-C specialist department was to arrange the allocations together with the type supervisor. At that time, little flying was being done by Junkers, on account of bad weather and difficulties with the oil cooler. Production startup of the pre-production series was being hampered by errors in drive production and reduced supercharger performance caused by manufacturing errors.

At the beginning of May, the engine program had earmarked just 66 BMW 801 TJ-0 power plants for prototypes and pre-production machines. A third of these were supposed to be spares, leaving just 44 active power plants for 22 aircraft. Six of these had already been earmarked for the creation of three Ju 88/188 engine test beds by DLH, which meant that Junkers could only equip 19 Ju 388s, six prototypes, and 13 pre-production aircraft with the BMW 801 TJ-0. At first, therefore, there were no power plants available for 55 of the 68 pre-production machines planned at that time.[214] Junkers did, however, have slightly more BMW 801 TJ-0 engines than it needed for the production Ju 388, so that by the end of 1944 another 38 pre-production aircraft could be equipped with the power plant. With advance orders already on hand for a version of the Ju 388 powered by Jumo 222 engines, and inquiries from the RLM

The fuselage of Ju 388 WNr. 560 049 in the annex of the NASM in Silver Hill in April 1993. At the joint between the cockpit and fuselage is the periscope (PVE 11), which was used to train and aim the guns of the tail turret. To its right is a half-open servicing hatch. The outline of the wing is clearly visible, probably because the aircraft was overpainted at least once in the USA. The finish is weathered, because the aircraft was initially stored outside at Silver Hill. Within the outline of the wing may be seen two of the threaded ball socket connectors, and connections for the controls, fuel, oil, etc. RLM 02 gray-green was applied as an interior color on part of the wing-fuselage joint, while part is natural metal. Vernaleken

The cockpit of the aircraft from the other side, photographed in 1999. Note the flare port in the starboard side of the cockpit, which—because of the pressurized cockpit—was sealed from the inside by a rubber plug. The canopy framing is painted in a typical dark gray finish, RLM 66. Although the upper fuselage appears to be a brown color similar to RLM 81, this Weser-built machine was in fact painted with RLM 82 dark green. Vernaleken

Looking forward at the central instrument panel. To its left is the control yoke, to its right the radar scope for the FuG 217 (Neptun II). The camera control panel should appear beneath it, but instead there is an American radio. Arranged on the starboard side of the cockpit are electrical switches and engine instruments.

Close up of the blind flying panel. Beneath the combined artificial horizon and turn-and-bank indicator is the repeater compass, and directly above it the airspeed indicator and vertical speed indicator. At the very top of the panel are the pitot heat indicator lamp and the AFN 2 homing indicator for the Fu Bl 2. In the row below are the indicator for the radio altimeter (AFN 101, left) and the barometric fine-coarse altimeter. Vernaleken

FuG 217 radar scope with hood and centrally-arranged control switch. The control knob was used to adjust brightness. The recessed adjusting slots could be used to adjust focus, sweep origin, sensitivity, and range. Vernaleken

Directly above the Patin-built DF repeater compass is a UV lamp. Left above it is the observer's O_2 gauge.

View from the pilot's seat of the portside console with large throttle levers and, on their right, smaller levers for controlling the turbines. To the left of the throttle levers are the ignition switches. Farther forward beneath the two visual indicators for the rich-poor switch are the combined boost pressure-rpm indicators, arranged vertically. Vernaleken

Close-up of the side console with aileron and elevator trim wheels. In the recess above the rudder trim control knob are the switch for the FuG 101, the transmit selector switch, socket, and control for the heated flight suit, and the intercom controls. The small blue handle to the right of the rudder trim was used to operate the control shutter for the pressure ventilation with supercharger air from the port engine. Vernaleken

Starboard cockpit sidewall with instruments for monitoring oil pressure, temperature, and fuel pressure. In the center of the photo beneath the so called "twelve-lamp device" are the landing flap and undercarriage switches. To the right, beneath the yellow outlined fuel and oil contents gauges are the fuel transfer and fuel jettison switches, as well as the fuel consumption selector switch. In the recess below are various switches, including those for the identification lights and auxiliary pumps. Vernaleken

Rear cockpit bulkhead with FuG 10, PVE 11 periscopic sight, and the control device for the FA 15. Beneath the FuG 10 is the FuG 16. The small red handle next to the PVE 11 was used to release the life raft. Beneath it is the emergency hydraulic switch. Vernaleken

View from the second fuselage bay into the rear of the fuselage. Behind the ammunition boxes for the tail turret is a rack with the radio equipment's transformers, transmitters, and receivers. Behind the wooden boards may be seen the master compass. The receiver, transmitter, main distributor, and transformer for the FuG 217 may be seen in the right foreground, mounted on the fuselage wall. Vernaleken

The gasoline powered K%orcher Blo. 15 heater was used to heat the horizontal stabilizer and camera equipment. Vernaleken

Left: Antenna reel (AH 10) for the FuG 10's trailing antenna with antenna shaft (ASch 10 A) and severing device. Next to it on the right is the AAG 3 antenna tuning device. Vernaleken

The fuselage on a newly constructed metal support in 1999. Clearly visible are the four threaded ball socket connectors for attaching the wings and the open entry hatch. The likewise disassembled Do 335 is visible in the background. Vernaleken

Close up of the entry hatch with the compressed air cylinder for emergency opening in flight. Given the entry hatch's complicated design, it is not surprising that the Allied intelligence teams initially took it to be a hydraulic speed brake. Vernaleken

The black lever was intended for "regular" opening of the entry hatch, the red recessed handle for emergency opening and jettisoning. Vernaleken

As the fake cross applied by the Americans peels off, the original transfer markings become visible. In this view one can see the trailing antenna, and the original splinter camouflage scheme in the late war RLM colors 82 and 83. Only the color of the RLM 76 has been changed by overspraying with a light gray. Handig/Skol

Right: Operating diagram of the FA 15 from the brief technical description. EADS

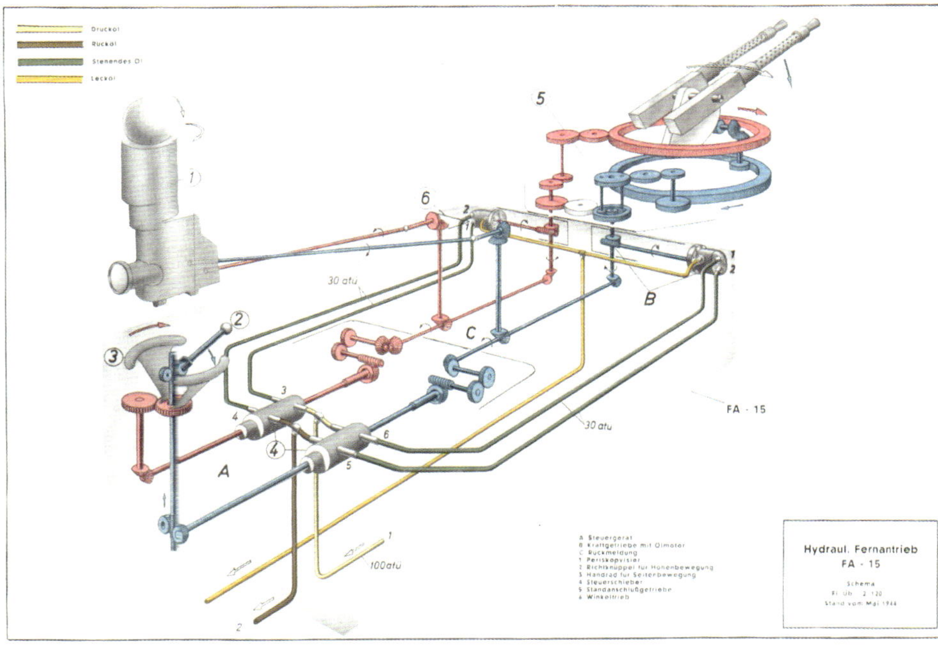

Clockwise from below left: Close up of the PVE 11 on the upper fuselage. Note the armor plate on the right of the photo. Vernaleken

Rear fuselage with FHL 131 tail turret without guns. This view reveals that the aircraft has a base coat of RLM 02 beneath the two gray shades. The tailwheel coors were made of wood. Handig/Skol

Bottom right: Control device for the FA 15 and PVE 11 periscopic sight in the cockpit of the Ju 388 L-1. The control device conceals the FBG 3 remote control device for operating and tuning the trailing antenna, as well as the switch for selecting the fixed or trailing antenna. Vernaleken

Junkers Ju 388 L-0/V8
WNr. 300 002
Upper surfaces RLM 70/71
Undersides RLM 76

Junkers Ju 388 K-0
WNr. 230 151
Upper surfaces RLM 02*
Undersides RLM 01

*may also have been RLM 65

about test beds with Jumo 213 E engines, Zindel made a virtue out of necessity, and proposed that the remaining 17 aircraft be offered as test beds for the Jumo 213 and Jumo 222 (see below).[215]

In mid-July 1944 output by the BMW factory in Munich-Allach, where the BMW 801 TJ-0 was produced, was suffering because of missing rail deliveries caused by Allied bombing. Instead of the customary 15 to 19 tonnes of express goods, just 2 to 3 tonnes had reached the factory. As well, many trains carrying materials for BMW were destroyed in the Munich area. Laim station, which was responsible for deliveries to the factory, was especially hard hit. The second level of the underground production facility at Allach was inaccessible, as there was a shortfall of 200 security personnel. As a result, in the period from 1 to 18 July 1944 just 45 engines were delivered from Allach, which was supposed to produce 800 BMW 801 engines per month. By mid-July all factories had delivered only 243 of these engines—far too few for the 1,200 Fw 190s planned monthly.[216]

Propellers for the BMW 801 TJ

It was envisaged that production aircraft would be equipped with the VDM 9-12 188 four-blade propeller with Dural blades, with a diameter of 3.726 meters and a chord/diameter ratio of 8%. The blades (9-12 152.10), which were equipped with counterweights, were from the production unit used by variants of the Ju 88/188 powered by the BMW 801 D, mounted in a four-blade hub. On 20 July 1944 six such propellers were in use on three prototype aircraft. At the same time three other prototypes, including the V3 and V7, were equipped with the backup type, the four-blade VDM 9-12 178 propeller. It also had Dural blades, but lacked counterweights, and had a diameter of just 3.6 m (chord/diameter ratio 7.9%). The blade tips of the two propellers on the V7 had been shortened by 2 cm on account of damage.

VDM Type 9-12 188 A[217] Settings	
Basic Setting:	25° = 12^{05}
Takeoff position:	26° = 12^{00}
Limits:	
Level flight:	21.3° = 12^{35}
Feathered:	92° = 6^{00} ± 5'

Due to a shortage of four-blade propellers, Ju 388 L-0/K-0 machines delivered by Merseburg were temporarily fitted with the production propeller from the BMW 801-powered Ju 88/188, the VDM 9-12 152 three-blade Dural propeller (diameter 3.7 m).[218]

At the beginning of March 1944 a BMW modification team carried out modifications to the reduction drives for the propellers of those BMW 801 TJ-0 engines delivered earlier. Two complete drives per engine were delivered on 9 and 10 March. At Dessau and Merseburg, work began on the removal of the 16 drives still requiring modification.[219]

Initially, spinners for the VDM 9-12188 A four-blade propellers were hard to come by. When, in mid-August, 10 spinners (9-12110.32) were delivered to the main factory in Dessau, two were immediately reserved for the first ATG production aircraft in Merseburg. A fitter was to be sent there immediately. As well, the Ju 388 V1, V5, and V7 were now to be equipped with propeller spinners. Another spinner was to be installed on the Ju 388 V3, which already had one. The last remaining spinner from this delivery was held in reserve.

While the Ju 388 V4 already had spinners, the Ju 388 V2 and V6 and the other production aircraft in Merseburg and Rechlin were supposed to receive spinners from the next delivery. According to VDM, 80 spinners were on the way.

Early Ju 388 L-0s like DW+YY were equipped with three-blade propellers because of a shortage of four-blade units. Ott Collection

It was anticipated that the prototypes (with the exception of the V6) would be refitted with production propellers or blades as counterweights were received.[220]

According to a communication from VDM in November 1944, the counterweights could be eliminated from the four-blade propeller. This made it possible to use the same deicing method as conventional three-blade propellers, in which deicing fluid was sprayed through a small tube ("trombone tube") onto the blades. Junkers now pushed for the earliest possible delivery of propellers with these tubes, so that the aircraft might be delivered with functioning propeller deicing systems.[221] In the summer of 1944 there was still no tested system for propellers with counterweights. Tests were therefore scheduled using a Ju 88 G-1 at the testing station in Munich-Riem.[222]

Finally, in a letter to Junkers dated 1 December 1944, the OKL ordered that the counterweights for the VDM 9-12.188 A propeller were to be omitted effective immediately. The reason for this measure was probably the condition of the propellers delivered to Junkers and ATG.[223] These had a basic setting of 26 instead of 25 degrees, and instead of the usual 12 mg unbalance, up to 170 mg was measured. For those propellers already delivered or in storage it was not simply a matter of removing the counterweights, however. After the counterweights had been removed the blades had to be completely rebalanced.[224]

Overview: Experiences with the BMW 801 TJ in the Ju 388

The first BMW 801 TJ-0 engines delivered to Junkers exhibited numerous shortcomings. Many of the complaints voiced in earlier letters (from 1st to 11th November 1943) had not been resolved by December, when engines 385 134 and 385 166 were delivered. The list of design flaws included 16 points, the list of construction errors 17.[225] For example, the exhaust fairings on the two power plants differed in height by 17 mm, and many hoses and lines were loose, rubbed against other parts, or were bent. In order to obtain reliable data on hose and cable lengths for quantity production of the BMW 801 TJ-0 engine, the mockup engine and engine number 385 122 were mounted on production left and right wings at Dessau. Prior

On the engine cowling of the BMW 801 TJ, facing the cockpit, was the propeller pitch indicator. Handig Collection

to installation, the wings were checked to ensure that their firewalls were up to production standard. On 23 December 1943, this prototype engine installation was inspected by a group led by Director Zindel, *Herr* Reichel of BMW, and Oberingenieur Klein. It was not possible, however, to clear up all the disputed points. For example, the status of the supercharger air bleed for the cockpit ventilation system could not be ascertained until flight tests of the Ju 388 L-0 began. There was determination, however, to deal with most of the remaining points by noon of 24 December. The engine would then be documented photographically and returned as quickly as possible to Munich, where a second prototype engine was to be built. Its parts would serve as patterns for the production engine.

The Ju 388 production aircraft had already been designed for the BMW 801 TJ-1.[227]

The cooling fans were another source of frequent problems. The 801.557-734 and 801.557-033 fans earmarked for the 801 TJ-0 were not available at first; consequently, the company was forced to fall back on the type 801.557-032. The cooling fan was a relatively fragile component. As a rule, it was forbidden to turn over the engine by hand using the fan because of the possibility of damage to the blades.

The cooling fan was powered by a reduction drive, and reached speeds of about 5,000 rpm during takeoff. There were several known cases of cooling fans breaking or shedding blades with the engine running.[228] In one instance at Ifa-Merseburg, a section of the cooling fan (801.557-033.11) with four blades broke away from one of the engines on Ju 388 L-0 WNr. 300 293 (DW+ZA). The pieces were sent to BMW for examination, and it was determined that there were flaws in the casting where the break had occurred. At that time the cooling fans were checked by "centrifuging," or running them at 125% of actual rpm. Up to 25% of cooling fans failed to pass this test. In actual service, the loads placed on the fans were only about 40% of those during centrifuging (according to another source, however, the forces during centrifuging were 45% more than normal operation). According to BMW, failure should actually have occurred during the test, and not after three hours of operation, for the fan had been homogenized, and bore a symbol indicating that it had been run at 5,600 rpm for one hour. The company

This perspective illustrates the size of the cooling fan, and makes apparent the immense danger posed by breakage in flight. Handig Collection

believed that it was possible to avoid damaging the fan when it was overstressed during centrifuging.[229] Junkers, however, raised the suspicion that overstressing the fans during centrifuging had caused minor flaws in the material to become fatally weakened, making failure inevitable.[228] BMW categorically rejected the notion of a design error, because forces at the break point were much less than at the location under maximum stress. Nevertheless, as a precaution the hub, which was subjected to great forces, was reinforced. The loss of a blade at Rechlin was also traced to a casting flaw. Microcavities (small cavities or pores created during casting by contraction of the material during cooling) just beneath the outer surface were the most probable cause of both failures. To avoid such incidents in the future, BMW decided to step up quality control of the cooling fans by carrying out the following checks:

1. Sandblasting, to remove the outer layer over cavities just beneath the surface and reveal them.
2. Spot probes to determine whether the cooling fan was tempered.
3. Centrifuging
4. Visual inspection
5. Lime water test or fluorescence method (check for cracks).

Further, only high grade material was to be used for cooling fans. As well, it was anticipated that, beginning in October 1944, there would be a change from Elektron sand casting to iron mould casting.[229] After completion of the necessary mould, all cooling fans were to be replaced by the improved 801.557-036 version, which also had a deicing grille for propeller deicing. Initially eight of the new cooling fans were delivered to Junkers; however, they were used up by the end of September 1944. The remaining stocks of the "older and oldest style" cooling fans also fell to zero, and flight testing was interrupted by a shortage of cooling fans.[231] Deicing fluid was carried to the propeller blades by telescopic tube. In addition, Junkers was to refine the design produced by the *E-Stelle* Rechlin's deicing detachment in Riem and make it ready for production.[232]

On 20 October 1944 BMW hosted a meeting with the Special Committee for Power Plants and the *E-Stelle* Rechlin.[233] The theme was the BMW 801 TJ for the Ju 388. Attending from the special committee were Director Schilo, Dipl.-Ing. Jellinghaus, and Dipl.-Ing. Lussnat. The *Luftwaffe* was represented by *Oberstabsinginieure* Hoffmann (TLR/Fl-E 3) and Hasche (*E-Stelle* Rechlin E 3), as well as an unidentified gentleman from the BAL (*Luftwaffe* organization that oversaw the construction of aircraft). Speaking for Junkers were Director Zindel and Dipl.-Ing. Wilhelm, while in addition to Dipl.-Ing. Schabert,[234] who was absent for part of the conference, BMW was represented by Directors Bruckmann and Beifang, plus Dr. Ammann. Director Schilo began by remarking that the purpose of the conference was to bring about clarity concerning the state of the engine with respect to performance, complaints and engine failures, reliability, and operational readiness (or the measures required to bring this about). He also stated that in flight testing it had not been possible to completely match the performance estimates that Junkers had made using BMW 801 TJ performance figures provided by BMW.

During development BMW had promised a maximum boost altitude of 11.8 to 12.4 km without dynamic pressure. Based on this, Junkers calculated a maximum boost altitude of 13 to 13.5 km, taking into consideration dynamic pressure. This was not possible with the internal ram-air intake used initially, however: at the end of 1943 the BMW 801 TJ-0 was only achieving a maximum boost altitude of 11.6 km on the test bench. With increased engine temperature this value dropped to 11.3 km. During flying operations with production engines, the aircraft only achieved about 11 km, and sometimes they failed to reach even this figure. Because a reduction of 2 km in maximum boost altitude meant a speed loss of 40 kph, a solution had to be found quickly. Junkers therefore called for the introduction of an external ram-air intake. As well as a 1500-meter increase in maximum boost altitude, the external intake also resulted in a reduction in supercharger air and cylinder temperatures of 30°C and 10°C, respectively. The increase in engine performance resulted in a speed increase of 16 kph at the "old" maximum boost altitude. It was decided to introduce the external ram-air intake into production as quickly as possible, and to modify all previously delivered engines. The external intake did increase drag, however, resulting in a loss of 8 kph at low level, but only 4 kph at maximum boost altitude due to reduced air density. This was accepted because of the gains in altitude and performance.

Beginning in September 1944, engines were delivered with the external ram-air intake. The majority of the engines delivered to date were retrofitted with the external intake in a joint modification program by Junkers and BMW.

The BMW 801 TJ suffered a further loss of performance because supercharger air temperatures measured ahead of the cylinders were still 15°C higher than calculated, even with the external intake. With each degree rise in supercharger air temperature, engine performance fell by 3.5 to 4 hp, resulting in a total loss of 50 to 60 hp. The supercharger air temperatures predicted by BMW were only achieved in high-speed flight at 600 kph.

Fortunately, it was discovered that a larger Fimag intercooler could be fitted without modification, bringing a significant improvement in supercharger air cooling, and increased engine performance at maximum boost altitude. BMW put the performance increase at approximately 150 hp, and also predicted a reduction in cylinder temperature of 8 to 10°. At the conference on 20 October 1944 there was agreement that testing and introduction of the new intercooler should proceed as quickly as possible.

Only on the BMW 801 TJ-0 with internal ram-air intake were the cooling gills hydraulically operated. In order to increase the cooling effect, on engines with the external intake, blocking was used to set the gills at 5° outwards. Not until the introduction of the BMW 801 TJ-1 were adjustable cooling gills to be reintroduced as a standard feature. All Ju 388 airframes were designed to allow the adjustable cooling shutters of the BMW 801 TJ-1 to be connected immediately.[235] Production of this engine was not scheduled to begin until January-February 1945, however. It was anticipated that it would take three to four months to complete the switch from the BMW 801 TJ-0 to the TJ-1. The BMW 801 TJ-0's fixed cooling gills and increased outflow of supercharger air resulted in a speed loss of 16 kph at low level, and 26 kph at maximum boost altitude.

The adjustable cooling gills of the TJ-1 could be closed at high speed, eliminating part of this loss (13 kph at an altitude of 5 km, and 12 kph at 11 km). To improve cylinder cooling, the TJ engines were also fitted with a larger cooling fan, which resulted in an increase of 50 hp at ground level and 20 hp at altitude.

The original central exhaust pipe of the BMW 801 TJ-0 had a cross section area of 3.2 dm^2, however, this was changed several times, and the production version of October 1944 had an exhaust cross section area of 4.2 dm^2. This significant increase resulted in a theoretical loss of 20 kg of exhaust thrust at maximum boost altitude, equivalent to an engine performance loss of 60 hp at 600 kph. Based on new performance measurements, Junkers feared that the exhaust thrust loss might be as great as 30 kg.

Recent tests by the DVL with the Ju 388's VDM four-blade metal propeller at high mach numbers caused Junkers further concern. It was feared that there would be a significant loss of propeller effectiveness, as the propellers operated at very high mach numbers in high-altitude, high-speed flight. Propeller blade width was restricted by the VDM pitch control mechanism installed in the BMW 801 TJ-0, the largest usable VDM metal propeller having a blade width of 8% and a diameter of 3.7 m. The high c_a values in high-altitude, high-speed flight, and the associated loss in propeller effectiveness therefore had to be accepted. The profile of the propeller blades, which was not suitable for high mach numbers, was also criticized. VDM had just completed a hub capable of accommodating four wooden propeller blades. This would finally have allowed Junkers to use broader wooden propeller blades with the BMW 801 TJ-1, combined with a more robust pitch control mechanism.

The airframe, especially the leading edge balloon cable cutters, also contributed to the overall speed loss of 30 kph at low level, and 90 kph at maximum boost altitude; however, at 6 and 9 kph, respectively, this was rather minor.

Though the intercooler was huge in comparison to the actual engine, ideal supercharger air temperatures could only be achieved in high speed flight at 600 kph. At lower speeds the air stream and cooling fan provided insufficient cooling. MTU

Performance Summary Ju 388 with BMW 801 TJ-0 – State I

	at ground level				at maximum boost altitude			
	A	**B**	**C**	**D**	**A**	**B**	**C**	**D**
Calculated maximum speed	1.44		425			651		at altitude of 13 km
Spec. Sheet 388/377 for Ju 388 L-1 of 16 Dec. 1943								
Maximum boost altitude reduced from 13.0 to 11.0 km						- 40		
Loss caused by inadequate supercharger air cooling					- 80	- 17		+ 30°C
Loss of 30 kg of exhaust thrust						- 18		
Cooling gills	+ 0.22		- 16		+ 0.20	- 26		
Cold air exit loss								
Larger cooling fan		- 50	- 5			- 20	- 4	
Reduced engine performance at 500 m		(- 75)	(- 7)			(- 40)	(- 8)	
Leading edge cable cutters	+ 0.04		- 4		+ 0.04		- 6	
Chamber cooler	+ 0.02		- 2		+ 0.04		- 3	
State I: L-1 series with internal ram-air intakes	1.72	- 50	398		0.26	- 100	537	at altitude of 11 km

State I was equivalent to an unmodified 388 L-1 production aircraft with internal ram-air intakes. The speeds of 398 kph at low level and 535 – 550 kph at 10.5 – 11 km are very close to the estimated figures above.

Performance Summary Ju 388 with BMW 801 TJ-0 – State II

	at ground level				at maximum boost altitude			
	A	**B**	**C**	**D**	**A**	**B**	**C**	**D**
State I L-1 series with internal ram-air intakes	1.72	- 50	398		+ 0.26	- 100	537	at altitude of 11 km
Improvement: deletion of leading edge cable cutters	- 0.04		+ 4		- 0.04		+ 6	
Cockpit air cooler intake	- 0.01		+ 1		- 0.01		+ 2	
Aileron gap cover	- 0.03		+ 3		- 0.03		+ 4	
Drag caused by external ram-air intakes								
Reduction in supercharger air temperature caused by external ram-air intakes						+ 80	+ 17	
Increase in maximum boost altitude caused by external ram-air intakes							+ 30	at altitude of 12.5 km
Loss of propeller effectiveness based on new DVL tests							- 12	
Production State II L-1 with 801 TJ-0 with external ram-air intakes	1.72	- 50	398		+ 0.21	- 20	580	

State II was equivalent to the modified series state Ju 388 J-1 with external ram air intakes. Leading edge cable cutters not to be introduced until the end of the year.

Legend

A: Drag-producing area
B: H.P.
C: Speed in kph
D: Remarks
TLL: Supercharger air temperature

A Merseburg production aircraft (WNr. 340 083) was tested in this series state on 30 August 1944; the aircraft still had leading edge cable cutters and no covers over the aileron gaps, plus an additional horizontal antenna for Neptun R, cockpit air cooler without improvements, and no periscope fairing. Speed losses at maximum boost altitude of 12.5 km were:

Speed loss	Cable cutters	-6
	Aileron gap covers	-4
	Cockpit air cooler	-2
	No periscope fairing	-5
	Additional horizontal antenna	-3
Total speed loss		-20

The estimated value of 580 kph must therefore be reduced by 20 kph to 560 kph, which is equal to the figure achieved at an altitude of 12.5 km.

Ju 388 V4 Comparison

Flown	560 kph at altitude of 12 km
Speed difference for tail turret	-6
Speed difference for periscope	-3
Speed difference for cable cutters	+6
Speed difference for aileron gap covers	+4
Speed difference for cockpit air cooler	+2
Total speed difference	+3 kph
Max. speed at altitude of 12 km	583 kph
For 500 m maximum boost altitude	+7 kph
Max. speed at 12.5 km in Series State II	590 kph

As the figure of 580 kph was achieved at a reduced weight (500-1000 kg less), this reduces maximum speed by 10-20 kph; the figures for the V4 are thus also quite close to the estimated value for State II.

Performance Summary Ju 388 with BMW 801 TJ-0 – State II

	at ground level				at maximum boost altitude			
	A	B	C	D	A	B	C	D
Series State II L-1 with 801 TJ-0 with external ram-air intakes	1.72	- 50	398		+ 0.21	- 20	580	
Improved intercooler		+ 55	+ 6	TLL = -15 °C		+ 150	+ 30	TLL = -50 °C
Adjustable cooling gills on 801 TJ-1			+ 12				+ 12	
Loss in maximum boost altitude caused by improved intercooler							- 6	- 300
Series State III 388 L-1 with external ram-air intakes and improved intercoolers and 801 TJ-1		+ 5	416			+ 130	616	altitude = 12.2 km

A further speed increase could be achieved through increased propeller effectiveness. The introduction of an improved propeller is possible with the more robust gearing of the TJ-1, and at present is being looked into.

After Junkers initiated flight trials, there were complaints and mechanical problems involving the BMW 801 TJ-0, some of them serious. Engine failures seriously delayed factory testing. The search for the causes of the engine's reduced performance and the related measures to improve performance resulted in many man hours of additional work.

The 59 BMW 801 TJ-0 power plants used in the Junkers flight test program logged a total of 248 hours by 1 October 1944 (just 4.2 hours per engine on average). During this time 30 engines had to be removed from the aircraft because of serious faults, or damage that could not be repaired in place. One engine did, however, reach 37.5 hours before it had to be changed.

It is noteworthy that 21 of the engines had to be removed because of problems that had nothing to do with the power plants themselves. As well, a significant percentage of the failures in the high-altitude components had nothing to do with the high-altitude drive itself. On three occasions a turbine seized because an oil line had broken.

All of this led Dir. Schilo to ask a provocative question: did Junkers fear that production and/or operational use of the Ju 388 with the BMW 801 TJ engine would lead to a disaster? Director Zindel replied that such a disaster need never happen, provided BMW did everything necessary to avert it by eliminating the remaining problems associated with the engine, which were well known.

BMW's representatives (Dir. Bruckmann and Dr. Biefang), were forced to admit that most of the problems affecting the 801 TJ engine had also appeared in the standard D-series engine without high-altitude accessories, and therefore had to be addressed as quickly as possible. Obviously only some of the problems could be fixed by technical changes in the engine, for BMW blamed some of the problems to start-up difficulties at the Allach factory near Munich, and inadequate acceptance checks. According to BMW, the serious problems encountered by Junkers, in particular, could be attributed to the fact that it had used mainly TJ-0 pre-production engines and early production engines from Allach in its test program.

This may explain why the *E-Stelle* Rechlin achieved better results with later production BMW 801 TJ-0 engines. *Oberstabsinginieur* Hasche reported that the *E-Stelle* and the test detachment logged a total of 840 flying hours with 40 engines, an average of 21 hours per engine. Only 13 engines had to be removed because of problems—essentially the same as Junkers. Four engines even achieved 90 flying hours without incident. Performance was found to be 75 hp below the specification at a height of 500 m, however, and compared to the value achieved on the test bench it was 100 hp lower. As well, effective engine performance was reduced by 50 hp on account of the larger cooling fan. High-altitude performance had not been checked to date. The *E-Stelle* Rechlin declared "that the 801 TJ power plant in the Ju 388 is fit for front-line service," with the proviso that BMW eliminate the remaining problems, and that the necessary high standard of maintenance by the units be assured. Dir. Schilo again saw the need to stress to BMW that it must take all steps needed to improve the reliability of the 801 TJ engine. For winter operations, it was imperative that the pipe that served as attachment point for the wing deicing be retrofitted. This was to be undertaken together with modification of the outboard intake. As of recently, all engines delivered by BMW were equipped with complete propeller deicing systems, further enhancing the aircraft's all weather capabilities.

One particularly serious complaint was the tendency of the engines to leak oil, combined with an unacceptably high oil consumption rate of 20 to 30 liters per hour. Together with engine and component failures, this was the most common cause for engine removal. It also made servicing difficult, and caused numerous breakdowns. BMW believed that it had discovered the cause, tracing it to inadequate sealing of the supercharger bearing bushing. By correcting this fault, it believed that it could eliminate the oil leakage and high oil consumption problems.

The BMW 801 TJ-1 differed from the BMW 801 TJ-0 in the following areas:

- New improved exhaust collector with significantly better accessibility to the engine mount, resulting in simplified maintenance of the TJ-1.
- Controllable cooling air gills, and a new hinged mounting strip.
- Improved control of the 801 E engine.
- Strengthened VDM propeller pitch gearing.

A BMW 801 TJ-1 is displayed in the aviation department of the Deutsches Museum in Munich. The external ram air intake, which was retrofitted to the BMW 801 TJ-0, is notable, as is the one piece valve cover of the BMW 801 E/S. Handig Collection

- Cylinders with 18-mm Bosch high-altitude sparkplugs with significantly improved reliability over the Siemens plugs of the TJ-0, which had to be replaced after 20 hours of operation.
- Improved seating of the exhaust turbine's waste gate to prevent jamming.
- Modified engine control system enabling the use of MW 50. MW 50 injection not possible with the TJ-0.

At the conference it was also decided to give the 801 TJ priority in the allocation of improved connecting rods, in an effort to solve the problem of connecting rod failures. In response to the problem of cooling fan failures, it was decided to implement a more rigorous testing procedure, pending the change to iron mould casting. *Oberstabsinginieur* Hasche subsequently pointed out that with the approach of winter, improvement in cold weather starting was of particular importance. BMW therefore scheduled starting trials in its cold chamber. As well, the users were to be urged to adhere to the starting procedure contained in the manual issued by BMW.

In an effort to master the countless complaints about the BMW 801 TJ-0 and enable future problems to be discovered and dealt with by the manufacturer, Junkers, at the express wish of BMW, sent an engine inspector by the name of Keller to BMW from 6 to 20 November 1944. At Dessau the required modification work had grown to such an extent that it could not be handled by the BMW field team, or Junkers. Work errors accounted for the majority of complaints. Finally, at the beginning of December 1944, the more than 30 shortcomings found by Keller were discussed with representatives of BMW. Junkers immediately submitted a proposed solution to each problem.[236]

The objective of development was to increase engine output from 2,200 to 2,400 hp. Initially this was to be achieved through the use of 100-octane C3 fuel, while B4 (87-octane) fuel with power boosting was to be used later. It was anticipated that methanol-water injection would provide an increase in performance, and BMW planned to deliver more precise information in early 1945.

Data concerning measures for possibly increasing the performance of the BMW 801 TJ became available in mid-December 1944. In addition to the adoption of the Fimag intercooler and the external ram-air intake, these included increased boost pressure and power boosting. Based on the supercharger's maximum boost altitude, it was calculated (using all available test data) that combat power could be increased by 200 hp without power boosting. The increase in takeoff and emergency power was 320 hp. Engine output could be increased by 420 hp using 150 kg of MW 0 (pure water injection for supercharger air and internal cooling), or 250 kg of MW 50. This "special emergency power" could only be used for 15 minutes at low level, and 30 minutes at high-altitude. It was calculated that this extra power would result in speed increases of 40 kph (combat power), 60 kph (takeoff power), and 68 kph (special emergency power).

The allowable cylinder head temperature would be exceeded while climbing at the increased combat power, but only marginally.

Side view of the Deutsches Museum's BMW 801 TJ-1, probably the only one in existence. In this view of the partially cut away engine one can see that there were no fundamental differences in the arrangement of the exhaust driven turbosupercharger and intercooler compared to the TJ-0. Unlike the BMW 801 TJ-0, the TJ-1 was only built in very small numbers. Handig Collection

At increased emergency power, however, a rise of 13° was calculated, making necessary the use of MW 0 or MW 50 to lower temperature. The required quantities were relatively small, as the design of the BMW 801 TJ had done away with the ignition advance of the BMW 801 S and T in favor of lower cylinder head temperature.

These performance estimates were to be verified by bench and endurance tests, after which the achievable speed gains would be measured in flight tests.[237]

It is not known if the BMW 801 TJ-1 was ever installed in a Ju 388.

Junkers Jumo 213 E

The Jumo 213 Engine

The Jumo 213 was developed from the Jumo 211, which was used mainly to power the Ju 87, Ju 88, and He 111. Like its predecessor and the Daimler-Benz engines, it was of inverted-vee configuration, with 12 cylinders in two banks. With a bore of 150 mm and a stroke of 165 mm, it had a displacement of 34.97 liters. There were two intake valves and one exhaust valve per cylinder, plus two sparkplugs that sat in the water-cooled cylinder head. The pistons traveled in cylinder linings that were inserted singly into the engine block, where they were pulled firmly into the cylinder head by four con-rods per lining. Coolant flowed around these cylinder linings. Two rubber gaskets shut the crankcase off from the coolant circulation system, while in the cylinder head there was a soft metal gasket between the cylinder lining and combustion chamber. The six-throw crankshaft rested in seven slide bearings and one roller bearing, while the camshaft was seated in seven slide bearings. The engine was equipped with direct fuel injection, and a so-called "pressurized water cooling" system. The coolant was half water and half Glycol antifreeze, which protected the engine to temperatures as low as minus 35°C. For anti-corrosion protection, Fl 44400 preservative oil was added to the mixture (1.5%). A so called "motor operating device" (MBG), similar to the Kommandogerät of the BMW 801, controlled engine functions, allowing the pilot to control the engine with a single lever, the "performance selector lever." The MBG adjusted injection rate, boost pressure, and engine rpm as required.[238]

Development of the Jumo 213 E was nearing its end in the spring of 1944. Based on the Jumo 213 A, it was equipped with a two stage supercharger with mechanical three speed gearing, an intercooler, and a high-altitude ignition system. Thus modified, the engine achieved a maximum boost altitude of 9 km.[239] After a 50-hour function test, the first Type 9-8213 D power plant with Jumo 213 E completed its 100-hour endurance test on the OMW diesel engine test stand between 23 April and 17 May 1944. The test program followed the Rechlin model: the hour was divided into 30 minutes at combat and climb power, 5 minutes at takeoff power,

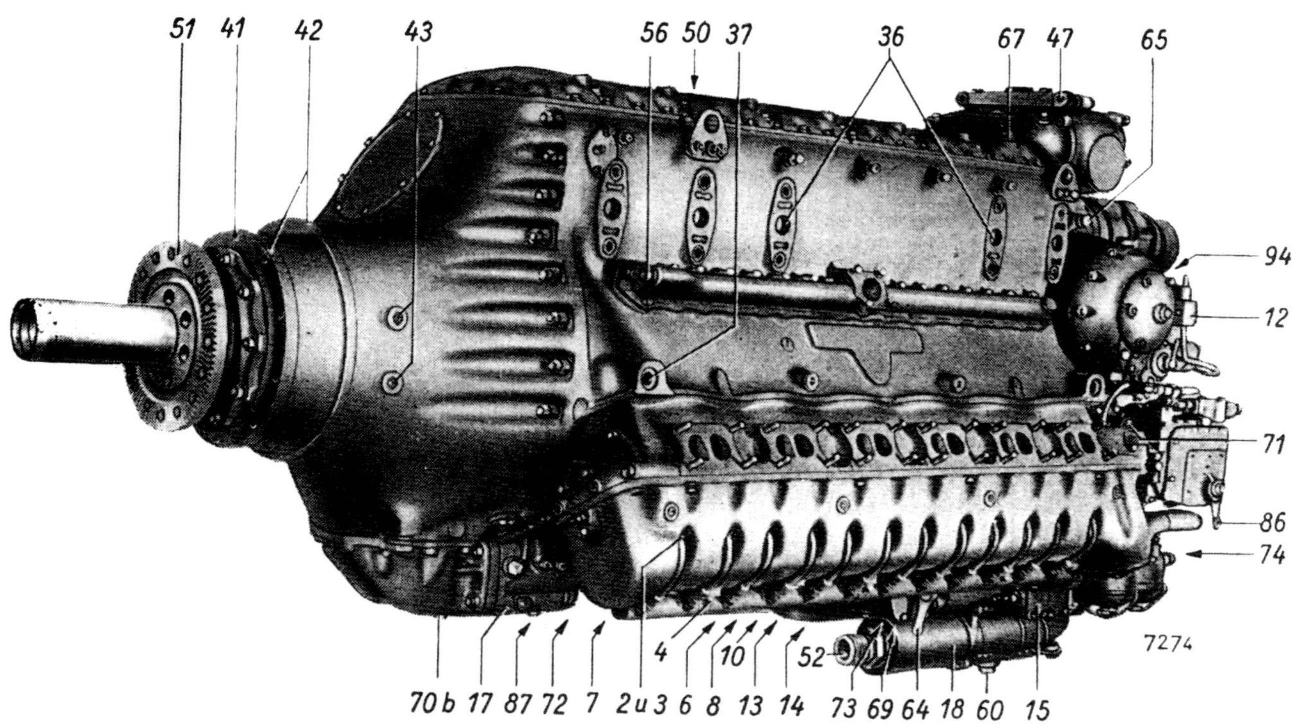

Abb. 1. JUMO 213 A-1. Ansicht auf linke Motorseite

This illustration of the Jumo 213 A is from the handbook. The Jumo 213 E differed from the engine shown here in having a two stage supercharger with three speed gearing.

and 25 minutes at maximum allowable continuous power. The endurance run had to be interrupted a total of 17 times due to problems, the first after 9 hours, the last after 94 hours. The installed engine was replaced after 84 hours. Overall there were no serious problems, only the exhaust pipes caused persistent problems and proved to be unsuitable.[240]

The Jumo 213 E was powered by 87-octane B4 fuel, and had a takeoff and emergency power rating of 1,750 hp at 3,250 rpm. Power output at 9000 m was 1,420 hp. "Special emergency output," or power with water-methanol injection, was 2,050 hp for takeoff and 1,700 hp at altitude. The Jumo 213 EB with increased boost pressure weighed 20 kg more, but it promised 1,900 hp for takeoff at the same rpm (1,520 hp at 9000 m), and a special emergency rating of 2,250 hp (1,840 hp at 9000 m). The increased performance compared to the Jumo 213 E did, however, result in greater thermal loads. While the heat abduction of the Jumo 213 E at climb and takeoff power (1,600 hp, or 3,000 rpm) was 480,000 kcal/h, while the same value for the Jumo 213 EB was 530,000 kcal/h. According to Jumo records, the Jumo 213 AB was at the point of entering production when the war ended.[239] Jumo 213 E/F engines fitted with a modified guide vane ring to overcome supercharger surging, and an exhaust valve controlled by the MBG were designated 213E1/F1 by Jumo.[241]

There was also experimentation with GM1 power boosting. While flight testing of the Jumo 213 E/F at Rechlin was halted by fuel shortages in January 1945, on the high-altitude test bench one engine produced an amazing 3,120 hp at an altitude of 12 km. The subject engine was equipped with a GM1 boosting system that injected 150 g/sec directly ahead of the supercharger.[242]

The Jumo 213 was delivered as a "power egg," and was sometimes used as a "standard power plant" for twin-engined aircraft. These engines differed from the fighter version in having different attachment fittings and cowlings. With standard fittings, the engine was interchangeable with other engines, such as the DB 603, for example. The Jumo 213 had a dry weight of 940 kg; with accessories such as oil and supercharger air heat exchangers, while installed weight was 1020 kg.

The following brief excursus will give a better picture of the scope of power plant testing done by Junkers Motorenbau. In October 1944 roughly 40 Junkers aero-engines in approximately 30 test beds logged a total of 130 flying hours. The bulk of this effort was allocated to testing the various versions of the Jumo 213. Some were installed in new prototypes of the Ju 388, Fw 190 D-9 and D-12, and the Ta 152. As well, the Jumo 222 was tested in the He 219 V16 and V19. Fifteen engines (or 37.5%) broke down, usually after a brief operating period, due to technical defects or manufacturing errors. By comparison, in November 1944 just 13 engines (31.5%) broke down. 160 flying hours were logged in spite of bad weather and overwhelming Allied air superiority. Contrary to what has been written elsewhere, the Jumo 213 F, essentially a Jumo 213 E in which the intercooler was replaced by an MW 50 power boosting system with an injection rate of 90 l/h, was built and tested. From September 1944 it was installed in the Fw 190 V55 (D-11), V56, V57, and V59. Six engines broke down during testing in October and two in November.[243,244] Development of the Ju 388 continued until the end of the war. The Jumo 213 J, which featured a newly developed four-valve cylinder head and increased bore of 155 mm (displacement 37.5 liters), with a compression ratio of 1:7.5, was bench tested, though without the ultimate supercharger. Sixty kilograms heavier than the Jumo 213 E, the Jumo 213 J was capable of producing 2,250 hp for takeoff at 3,700 rpm. At its maximum boost altitude of 11 km the engine was still capable of 1,620 hp. "Special emergency power" was a proud 2,600 hp at ground level, and 1,900 hp at an altitude of 9.8 km.

But back to the Jumo 213 E. In October 1944 the engine is known to have been used in the Ta 152 V29, the Ju 188 (WNr. 160 082), and the Ju 388 (see below).[245] In November 1944 these were joined by the He 219 V4 at Heinkel's Vienna facility, and the Ju 52 V203 and the Ju 188 WNr. 160 098 at the *E-Stelle* Rechlin.[246,247] While in October the performance of the Ju 388 and Ta 152 at maximum boost altitude was below expectations, in November they did succeed in attaining their calculated maximum speeds. The biggest problems were caused by the annular radiator segments. In just 12 flying hours by the Ju 388 V21, eight of these segments had to be replaced after cracks developed in the chamber bases. Replacement was difficult and time consuming. Spare parts did not fit because of production deficiencies, and in several cases the entire radiator section had to be removed. The Jumo 213 E's inadequate cooling performance was probably attributable to the surprisingly low coolant circulating capacity of just 9.5 l/sec. Beginning in November 1944 new D-Type radiator assemblies with so called "small radiators" (production version) were tested in the Ju 388. With these it was found that cooling capacity was just adequate. While under interrogation by the British in June 1945, *Oberst* Petersen, Commander of Test Stations, stated that work on the Jumo 213 E had continued until the approaching front forced an end to further testing. The objective had been to increase output from 1,900 to 2,500 hp using methanol-water or water injection.[248]

The Ju 388 with the Jumo 213

Equipping the Ju 388 with the Jumo 213 was not a spontaneous emergency solution; rather, it was the result of a lengthy deliberate process. In February 1944 the Fluid Mechanics Department conducted experiments with a 1:7.5 scale model of the Ju 388 with Jumo 213 engines in the wind tunnel at Dessau. The model had powered propellers and variable position landing flaps. The objective of the experiments, which ran from 1 to 5 February 1944, was to determine the aircraft's flight characteristics about the vertical and lateral axes. The experiments were based on an engine output of 1,600 hp at 1,250 rpm. As well, a favorable location was to be found for intake scoops on the side of the engine cowlings. Dr. Strauss found that the Ju 388 had much improved directional stability at full power, compared to the Ju 88 A-4, although he could not explain exactly why.[249]

Inquiries by the Technical Office (GL/C-E) about test beds or prototypes with the Jumo 213 E, which was being discussed as an alternative to the BMW 801 TJ, were very convenient for Junkers, as the planned number of BMW 801 TJ-0 engines was insufficient for all of the pre-production aircraft. On 3 May 1944, therefore, Zindel proposed the construction of five Ju 388 K-0 test beds with Jumo 213 E engines, simplified versions with no bulged bomb bay (and equipment). One aircraft would be used by Jumo as an engine test bed, and two by Junkers for general performance trials (and as reserve aircraft). Two other machines would be used for engine trials at Rechlin. Delivery of the aircraft was to take place between July and October 1944.[215]

S model of the Ju 388 with Jumo 213 E engines (scale 1 : 7.5) in the wind tunnel in February 1944. Griehl Collection

A construction directive for five Ju 388s with Jumo 213 E engines was issued on 23 May 1944. Initially, the five aircraft were designated V11 to V15.[250] Only about one month later these designations were changed to V21 to V25.[251] As well, in mid-July the construction plans were changed. Two aircraft would be built as engine test beds, and three as L-0 operational aircraft.

The first aircraft to be equipped with the Jumo 213 E was the Ju 388 V21, WNr. 230 157. This matches an entry in Reinhart Preuschen's notebook on 24 July 1944, where a Ju 388 V21 suddenly appears with the note "1st aircraft with Jumo 213 E." Further information on the testing of this machine, which flew for the first time at Merseburg on 17 August 1944, may be found in Chapter V.

The data in the May construction directive for the five test beds and the July amendment were modified, and stated more precisely in another amendment dated 12 August 1944. As a result of decisions by the RLM and the type supervisor, not only were the Ju 388 V21 to V25 to serve as test beds for the Type 9-8213 D (Jumo 213 E) power plant, but some were also to be prototypes for the Ju 388 J-3.[253]

The type supervisor, Hauptmann Dr. Stahl, also decided to produce two further engine test beds for the Jumo 213 A and Jumo 213 E from the Ju 388 L-0 series. On 3 September 1944 the Chief of Aircraft Procurement and Supply directed *Lufthansa* to undertake conversion of the two aircraft. Junkers protested, for it already had a verbal agreement with the type supervisor that the conversions would be carried out in Dessau (FSD).[254] According to a construction directive dated 25 September 1944, Ju 388 L-0 WNr. 300 294 was to be fitted with Jumo 213 A power plants, while the Jumo 213 E was intended for Ju 388 L-0 WNr. 300 293.[255] Preuschen later designated these machines the V30 (Jumo 213 E) and V31 (Jumo 213 A-1).

The above-mentioned monthly report confirms that Ju 388 L-0 WNr. 300 293 (V30) was equipped with the Jumo 213 E. The Ju 388 V31 was most likely the aircraft equipped and flown with Jumo 213 A engines in Merseburg. On 26 October 1944, however, an amendment was issued to the above cited construction directive. The V31 was to be equipped with Type 9-8213 D power plants (Jumo 213 E) at Merseburg (FSD-M). After delivery of the engines

The Ju 388 L-0V31 (DW+ZB), one of at least four machines equipped with Jumo 213s. Especially noteworthy is the sleek engine cowling with relatively small annular radiator. The damage evident here appears to have been inflicted intentionally. Ott Collection

from Rechlin, the aircraft was also to be ferried from there to Merseburg. A decision on the use of the engines removed from the V31 was to be made later.[256]

The OKL's experimental unit also tested a Ju 388 with Jumo 213 E engines (Type 9-8213 power plant) and VS 19 propellers at Oranienburg. The aircraft was the Ju 388 L-0 V32, with the *Werknummer* 300 295 and code T9+DL. The Jumo 213 E engines were fitted in the maintenance hangar at Oranienburg,[257] and the first static test was carried out on 12 September 1944. The aircraft made 23 flights before crashing on 5 January 1945. It logged a total of 9.40 hours, reaching a maximum altitude of 10200 m, and a maximum speed of 565 kph.[258] Weighing 12685 kg, the aircraft was only flown at maximum continuous power; therefore, the projected maximum speed of 608 kph did not appear too unrealistic. Testing of T9+DL will be described in detail in Chapter VII.

As a result of a decision by the Development High Commission, and according to a telex sent by the type supervisor on 10 November 1944, six prototypes of the Ju 388 J-3 with MW 50 power boosting were to be built for the anti-Mosquito role, probably in addition to the previously named prototypes.[259] The aircraft were to be fitted with wooden nose caps for the Morgenstern radar antennas.[260]

Propellers for the Jumo 213

In mid-June 1944 the Junkers design bureau in Dessau ordered 14 VS 9 variable-pitch propellers with spinners for five Ju 388 test beds powered by the Jumo 213 E. These propellers, four of which were intended as spares, were to be fitted with wooden blades. Overall diameter was to be 3.6 meters with a blade width of 12.2%. The model of blade with respect to the shape of the engine cowling (power plant 9-8213 D-1 or 9-8213 B-2 V) was to be determined at a later date. Two propellers were to be delivered every two weeks from mid-July to mid-October, while six propeller blades were to be provided by the end of September 1944 for use as spares.[262]

The propeller blade requirements were spelled out in a memo by the design bureau's propeller department on 29 June 1944. The blades envisaged for the 9-8213 D-1 power plant would have to have a cutout at the root of the blade to ensure clearance from the engine cowling with the blades in the feathered position.[263]

Contemporary cutaway model of a Junkers variable pitch propeller.

A more effective four-blade VS 19 propeller was envisaged as standard for the Jumo 213 E, in order to take advantage of the engine's increased power. Consequently, all prototypes with Jumo 213 E engines were also to receive the VS 19. On 30 August, therefore, an earlier order was canceled, and a new order was placed for 10 three-blade VS 9, and 10 four-blade VS 19 (9-21 019 A-0) variable-pitch propellers.

Half of the VS 9 propellers were to be fitted with Heine tempered-core blades (Type 9-30 337.16, diameter 3.5 m, chord/diameter ratio 11.52%), the other half with standard Heine blades (Type 9-30 372.11, diameter 3.6 m, chord/diameter ratio 12.2%). As well, one set of each type was to be kept in reserve. Eight of the VS 19 propellers were also to receive the Heine tempered-core blades. Envisaged for the remaining two were Type 9-41 304.11 blades made by the Schwarz company, with a diameter of 3.6 m and chord/diameter ratio of 10%. Two sets were to be held in reserve. Four VS 9 propellers had already been delivered when the contract was issued (15 July and 15 August 1944). The rest were to follow in pairs at 14-day intervals until the end of September. The VS 19 propellers were to be delivered in the same fashion from mid-September to mid-November.[264] Further information as to which aircraft were to receive these propellers is not to be found in the corresponding document. The delivery dates given suggest, however, that the V21 must have flown with VS 9 propellers, at least in the beginning. According to the technical description of December 1944, VS 19 propellers were also envisaged as standard equipment for the J-3 and L-3 series.[265]

Junkers Jumo 222

The Jumo 222 Power Plant

In April 1937 a team under Dipl.-Ing. Ferdinand Brandner of Jumo began working on an engine for the Ju 288. The engine was first installed in the fifth prototype of the Ju 288 in 1941. The Ju 88 V5 made its first flight with the engine on 8 October 1941.[266] In autumn 1941, however, the RLM ordered the engine, which was in the vital performance class, dropped from the production program, which so hindered its development that the Jumo 222 was never built in significant numbers. Development and testing of the Jumo 222 continued, but in the shadow of the Jumo 213, and various jet engines. Not until late in the war, when it was selected to power almost every piston-engined aircraft remaining in the delivery program, did the Jumo 222 approach production maturity.

The Jumo 222 was a liquid-cooled four-stroke engine with direct fuel injection, and a compression ratio of 1:6.5. Its 24 cylinders were arranged radially in six rows of four. While the crankshafts of all versions turned clockwise, the propeller shafts turned in different directions. The A, C, and E series had propeller shafts that turned counterclockwise with a reduction ratio of 2.727:1, while the B, D, and F series had clockwise-rotating propeller shafts with a reduction ratio of 2.73:1. All versions of the engine were designed to run on B4 (87 octane) fuel.

The individual variants and sub-variants differed, not only in the type of supercharger used, but they also had different displacements. The significant differences between variants are listed in the following table:

Jumo 222 A/B	Jumo 222 A/B-1	Jumo 222 A/B-2	Jumo 222 A/B-3
Bore	135 mm	135 mm	135 mm
Stroke	135 mm	140 mm	140 mm
Displacement	43.3 l	49.9 l	49.9 l
Supercharger	single-stage with two-speed gearing	single-stage with two-speed gearing	single-stage with two-speed gearing
Takeoff and Emergency Power at Ground Level	2,500 h.p. at 3,200 rpm	2,500 h.p. at 2,900 rpm	2,500 h.p. at 3,000 rpm
Climb and Combat Power	2,200 h.p. at 2,900 rpm at ground level	2,250 h.p. at 2,700 rpm at ground level, 2,020 h.p. at alt. of 5 km	2,230 h.p. at 2,700 rpm at ground level, 1,900 h.p. at alt. of 6 km
Remarks		Intake valves enlarged from 48 to 51, exhaust valves from 48 to 58 mm.	Maximum boost altitude raised from 5 to 6 km by new supercharger.
Development Status	pre-production series built.	pre-production series built.	Production planned, but no production engines completed

At the beginning of 1944 Ifa saw an immediate requirement for the Jumo 222 A/B-2 for the Ju 290. The Jumo 222 A/B-2 was the only version then considered by Jumo to be ready for production. At that time it was still assumed that the Ju 290 with Jumo 222 engines would enter production at the beginning of 1945, and that output would reach 50 aircraft per month by the end of the year. The designers would have liked a higher maximum boost altitude for the Jumo 222 A/B-2, and would have preferred to use the Jumo 222 C/D. Because the Ju 290 also had to have a long-range bombing capability, emphasis was placed on developing an engine with reduced fuel consumption.[267]

The Jumo 222 C/D was a development of the Jumo 222 A/B-3 with increased performance, which was to be achieved by increasing engine speed to 3,200 rpm and displacement to 55.44 liters. Design work was complete by the end of the war, and a prototype with a bore of 145 mm, and a stroke of 140 mm was under construction. The engine was designed to produce 3,000 hp for takeoff, while at low altitude it was capable of producing 2,600 hp at combat power (2,900 rpm).

The Jumo 222 E/F, which Junkers envisaged for the Ju 388—in particular the Zerstörer version—beginning in mid-1945,[267] was based on the Jumo 222 A/B-3, and had a two-stage supercharger for improved maximum boost altitude. Except for the accessories housing and additional parts in the cooling system, it was largely identical to the A/B. The end cooler was designed as a heat exchanger (air-water). The first supercharger stage was powered by a gear transmission, while an infinitely-variable fluid drive was installed for the second stage. This hydraulic clutch caused difficulties in the beginning. At the end of 1943 there were three of these engines in the Otto-Mader plant, and six pre-production engines were completed by the end of the war, with another four under construction. A maximum boost altitude of 9 km was measured on the high-altitude test stand, and development of the supercharger had to continue to achieve the desired maximum boost altitude of about 11 km. At 3,000 rpm the aircraft had a takeoff and emergency power rating of 2,500 hp, and 1,920 hp was produced at an altitude of 9 km. Climb and combat power at low altitude was 1,780 hp at 2,900 rpm.

On 19 November 1943 Brandner and Schorlemmer met in Berlin at a meeting with *Fl.Oberstabsinginieure* Cuno and Fischer of the *E-Stelle* (Dept. E3). The Jumo representatives described the technical state of the Jumo 222 to the *E-Stelle* engineers, who had no personal experience with the engine, and asked E3 to include it in testing as soon as possible in order to allow them to form a qualified opinion on the engine. Up to that time the Jumo 222 had obviously not been tested on the bench, nor in the air at Rechlin, and

A Jumo 222 E/F on display in the Deutsches Museum as a "bare" engine, mounted on the original engine bearers. The BMW 801 TJ-1 engine is visible in the background. Handig Collection

Cuno was therefore forced to rely on second hand information in making his assessment of the engine. Cuno subsequently received an inquiry from Dir. Francke of Heinkel, who asked if they could count on mass production of the Jumo 222 beginning at the end of 1944. As the BAL was still reporting significant problems with the connecting rods (failures related to production), Cuno was forced to respond in the negative.

At that time Junkers only had available A/B-1 engines capable of producing 2,500 hp. Work was under way to increase output to 3,000 hp by boring out the cylinders. Five engines with a larger bore were already complete, with work in process on three more. The increased displacement meant that the A/B-2 could produce 2,500 hp at 2,900 rpm instead of 3,200 rpm, reducing the load on the bearings. As E3 had available testing capacity and shared the opinion of the Jumo 222, one or two Jumo 222 A/B-1s were to be bench tested, and later replaced by A-B/2s. As well at that time DLH in Staaken was in the process of installing a Jumo 222 in a Ju 52 for E3. DLH was also converting five Ju 188s as test beds for the Jumo 222 on behalf of the KdE. The first two aircraft were to receive A/B-1 engines, which would later be exchanged for A/B-2 or E/F engines. Consideration was even given to fitting the last three aircraft with Jumo 222 A/B engines with exhaust-driven turbosuperchargers (see following text). The *E-Stelle* assumed that the Jumo 222 E/F with two-stage supercharger would be the main production variant. It was Cuno's belief that variants with exhaust-driven turbosuperchargers would be used mainly for special roles (high-altitude and long-range flight). According to Jumo, Kommando Rowehl also planned to use several Ju 288 prototype airframes with Jumo 222 A/B engines for special missions. Cuno urged that these machines be tested by Jumo and Rechlin before they were used operationally.[268]

As a result of these negotiations and inquiries by airframe makers, soon afterward, on 30 November 1944, Brandner felt the need to take a position on the Jumo 222 in order to allow the office's planning authorities "to reach a quick, purposeful decision." He did so in a memorandum to his superiors, Prof. Mader and Director Madeis. Because of the limited number of pre-production engines, Brandner proposed that testing of the Jumo 222 be limited to aircraft considered important by the *Luftwaffe*. According to Brandner, it was envisaged that Dessau would produce 265 engines in 1944, while 95 engines were needed for the various requested test beds. The problem was, however, that the engines had to be assembled into complete power plants, and it was still not certain whether this could be done. Brandner felt that the Jumo 222 A/B-2 was already fit for general use. Two examples of the Jumo 222 E/F were already running on the test bench in OMW, and there was confidence that the September target date for the first two engines could be met. Orders were issued in Oberursel for the additional components for 12 more engines. After evaluating all installation proposals and the development status of the various engines and power plants, Jumo came up with the following sequence—beginning with the least risky installation:

Ju 290	4-engine transport	standard A/B-2 engine
He 219	2-engine night fighter	E/F with mechanically-driven 2-stage supercharger
Ju 488	4-engine high-altitude bomber	A/B-2 with exhaust-driven turbosupercharger

Engine Requirements were as follows:[269]

1944 Engine Requirements for Development of the Jumo 222*

	Jan	Feb	March	Apr.	May	June	July	Aug	Sept	Oct	Nov	Dec
Factory Testing	3	2						2	2			
Heinkel He 219			4	2					**2**			
Staaken Ju 252					*1*		*1*					
Staaken 4 Ju 188		2	2	2	2	*4*	*2*	*2*				
Jumo 4 Ju 288			4	4	4							
Ifa Ju 488				4	4	4	4	*2*	*2*	*4*	*4*	*4*
Ifa Ju 290					4	4	4	4				
A/B-1 67	3	4	10	12	14	8	8	6	2			
Total *Turbo 26*					*1*	*4*	*3*	*4*	*2*	*4*	*4*	*4*
E/F 2									**2**			

* The original color coding based on engine type has unfortunately not survived and was reconstructed.

As for implementation of these plans, according to JFM's first quarterly report of 1944 the Jumo 222 was only belatedly included in engine program 325a, but without the inclusion of production numbers. The factories of Special Committee T1 were fully occupied with the Jumo 213 (engines and power plants), and with accessories, propellers, and jet engines; therefore, just 36 Jumo 222 engines were built new or repaired.

A total of six Jumo 222 E/F prototype engines were completed, and these were used for test bench experiments. Performance measurements on the high-altitude test bench were in preparation. Dornier was considering installation of the Jumo 222 in the Do 335, however, there were difficulties with the ventral exhaust. Design work on the Jumo 222 production engine was passed on to Henschel, however, Jumo retained technical responsibility for the engine.[270]

The ultimate engine program 326/III became available in the second quarter of 1944. In spite of a lack of production capacity, the RLM demanded the monthly production of 190 Jumo 222 engines for the Ju 388 and Me 264 beginning in February 1945.

In fact, just 34 engines were completed in the second quarter, plus nine repaired engines. Because of enemy bombing, production of the Jumo 222 in Dessau had been completely halted, and was being dispersed. The dispersed operation was not expected to complete the first engines until autumn 1944. The available stocks of engines were maintained, however, and testing went on without interruption. Development work on the Jumo 222 E/F was given the highest priority. From the available engines of various versions, 25 were made available to the main aircraft plant in Dessau (FSD) for installation in the Ju 388. Ten engines went to the *E-Stelle* Rechlin for installation in the Ju 388 and Ju 52, and three were issued to Heinkel for installation in the He 219. Junkers' engine division supervised the installation of the Jumo 222 in the Ju 388.[271]

The new RLM program 326a was issued on 1 September 1944, toward the end of the third quarter. The start of Jumo 222 quantity production had meanwhile been put off until August 1945, and following cancellation of the Me 264 the only remaining aircraft was the Ju 388. Although there was still no available production capacity, a suggestion by Jumo to engage the Pommerschen Motorenwerke (PO-MO) was rejected. Because of the dispersal of production, only seven new engines were built and eight repaired.

Testing of higher-performance engines and the high-altitude engine were given priority. Assembly of the E/F version was delayed because of losses to enemy bombing. As a stopgap measure, about 20 A/B-1 and A/B-2 engines were converted to E/F standard in Oberursel. By the end of the quarter five conversions had been completed, and the first three power plants assembled. One of these engines was envisaged for installation in a Ju 352 by DLH, and two more were to be installed in a Ju 388 by FSD. The engine division also continued to oversee installations of the Jumo 222 with standard power plant by Heinkel (He 219), Hütter (Hü 211), and Junkers (Ju 388). Development of a fighter engine was also begun.[272]

The standard power plant 9-222 A/B V1 completed its acceptance test on 2 October 1944, and was subsequently delivered to Staaken.[273]

Jumo 222 Engine Production (1944)[272]

Jan	Feb	March	Apr.	May	June	July	Aug	Sept	Oct	Nov	Dec
17	13	6	11	12	11	—	5	2			

View of the right side of the Jumo 222 E in the Deutsches Museum in front of a French built example of the Ju 52. With its three engine layout, the Ju 52 was an ideal engine test bed. One machine was used for flight tests with the Jumo 222 in January 1945. Handig Collection

Because of the continuing fuel shortage, in the last quarter of 1944 testing of the Jumo 222 was limited to short test runs and function tests. In spite of this, function testing of the Jumo 222 engine control system was concluded. Measures were also initiated to prevent blade vibration, and improve the mounting of the second stage impeller. Experiments with the two-stage E/F supercharger under high-altitude conditions continued on the exhaust test bench. Enlarging the diameter of the first stage intake delayed the onset of the feared supercharger surging without negatively affecting the other values. Tests with just the first stage with the housing reversed confirmed suspicions that it was responsible for the complete supercharger's poor performance.[274]

Flight testing of the Jumo 222 also took place in late 1944 and early 1945, but to a much lesser extent than with the Jumo 213. By the end of November a Jumo 222 A/B III had flown just 15 hours at Rechlin, while in Merseburg the first Ju 388 with A/B III engines, probably the Ju 388 V9, completed static tests. The aircraft was then supposed to be ferried to Rechlin. In Vienna, that same month Heinkel completed a He 219 with Jumo 222s in preparation for transfer to Dessau.[275]

In January 1945 a Jumo 222 (variant not known) was tested in a Ju 52 for at least 26 hours with no serious problems.[276]

As early as the beginning of 1943 Jumo planned to make the Jumo 222 into a high-altitude engine by adding an exhaust-driven turbosupercharger. A cooperative effort with Borsig was begun at the behest of the RLM under the leadership of Prof. Dr. Van der Nüll. Borsig's task was development of the turbosupercharger. To speed development, existing components from the BMW 801 TJ's turbine were used, but maximum boost altitude without ram-air was just 10.9 km. The RLM ordered 50 of these turbines, however, enemy action probably delayed their delivery until February 1944. Jumo was supposed to design the associated engine, but little was done because of the high priority attached to the Jumo 213. Completion of the engine was not expected until early summer at the earli-

The left side of the Jumo 222 E-U2 in the Deutsches Museum. Development of the engine began in 1937 under the direction of Dipl.-Ing. Ferdinand Brandner. Handig Collection

est, after which it was to be installed in single examples of the Ju 252 and Ju 288 prepared by Staaken. Whether these plans were carried out is uncertain.

Borsig also began development of an exhaust-driven turbosupercharger designed specially for the Jumo 222. Ifa envisaged that the resulting high-altitude engine would be used in the Ju 388 beginning in October 1944.[267] The first engines were to be based on the Jumo 222 A/B-1 or -2, and the Jumo 222 A-B III was also to be used later. Takeoff and emergency power was to have been 2,400 hp at ground level, and 2,070 hp at 12.3 km (both at 3,200 rpm). Climb and combat power was 2,170 hp at ground level, and 1,900 hp at 12.3 km at 2,900 rpm. Jumo began preliminary testing of the basic engine at the end of 1943-beginning of 1944, but by the end of the war testing of the Type 222 B-2-22 engine had proceeded no further than the low-altitude test bench.

Testing of the turbine controller design was completed in the last quarter of 1944, however, tests with the Rheinmetall-Borsig turbine were delayed by a missing turbine rotor.[274]

At the beginning of February 1945 work on an experimental Jumo 222 power plant with one of the named turbosuperchargers reached the stage where tests on the high-altitude test stand could begin. The following passage is found in the monthly report by OMW Test Stand II for January 1945:

"Provided there is fuel available and sufficient interest, in the coming days Borsig could begin high-altitude test stand experiments with a turbosupercharged engine."[277]

In the opinion of Dr.-Ing. A. Müller, the official responsible for turbine engines, because of the Ju 388's unfavorable gliding power requirements, a performance increase "in the area of 12-13 km" was very significant. As Dr. Müller thought he had recognized a "special inclination toward the Jumo 222" by JFM, in a letter to the Hirth-Motoren company he asked that it get a version of its 9-2426 (TK 19) turbosupercharger for the Jumo 222 to the test bench as quickly as possible. Dr. Müller took this action, even though he was convinced of the greater development potential of the Daimler-Benz 9-8603 F engine, which was equipped with the same turbosupercharger.[278]

On the subject of turbine-equipped engines, EHAG's Hirth-Motoren company subsequently turned to *Fl.-Oberstabsinginieur* Mann of the OKL (TLR/Fl - E3) on 4 November 1944:

The Ju 288 V9 (VE+QE) carried out a test flight with Jumo 222 A-2/B-2 engines on 6 January 1945, long after the type was cancelled. Testing was subsequently cancelled in favor of the He 219. Seen here is the first Ju 288 powered by Jumo 222 engines, the V5. Haubner/Misna Collection

"In the turbosupercharger field we are currently working on three different versions of the Type 9-2426, for the DB 8603 F, BMW TQ/2, and Jumo 222 A/B 3 engines, and two versions of the Type 9-2279 for the DB 603 U engine. As these are very complex projects, which in the case of the 801 TQ/2 and 222 A/B 3 are just beginning, it is necessary to establish their priority relative to one another, and to the Type 109-011. We would very much like to discuss this and the overall state of turbosupercharger development, and we therefore request that you visit us during your next visit to southern Germany."[279]

After the Allies crossed the Rhine, Junkers employees hid two E-type engines, along with other Jumo 222s in a cave in Oberursel. They were later handed over to the Americans. Following a detailed examination, they ended up at the National Air and Space Museum in Washington. In 1978 the Jumo 222 E-U2 was given back to the Deutsches Museum.

The Ju 388 with Jumo 222 Engines

On 24 April 1943 the *Technische Amt* ordered three prototypes with Jumo 222 engines. Just under a year later, on 27 March 1944, following a meeting between Junkers and the *Technische Amt* (on 17 March), an advance order (SS 4940-5101-0576/43) was issued for 20 twenty aircraft. It was obviously envisaged at first that all of these aircraft would be built with a lengthened fuselage, but a few weeks later, on 17 April, the *Technische Amt* advised Junkers that the aircraft were to be completed without the fuselage extensions "to avoid any delays in delivery."[280] In addition, the three planned Ju 188 experimental aircraft with Jumo 222 engines (probably the three aircraft mentioned previously, ordered in April 1943) were canceled. Six prototypes were envisaged later, however, for a "final solution with extended fuselage," three as bombers, and three as heavy fighters. The fuselage was to be designed in such a way as to allow either the manned tail turret with four MG 131s, or the unmanned turret with two MG 151s (heavy fighter) to be fitted without major modifications. Work could not have progressed very

In flight photo of the Ju 288 V9 taken in June 1942. Like the Ju 288 V9 the Ju 388 V9 also flew with Jumo 222 engines in early 1945. Griehl Collection

far, however. At the end of November 1944, work on increasing the range of the Ju 388 J-2 was not planned until the beginning of February 1945, because prior to this there had been little reason to expect that a version of the Ju 388 with the Jumo 222 would materialize.[281]

In view of this advance order (SS 4940-5101-0576/43) and the shortage of BMW 801 TJ-0 power plants, on 3 May 1944 Zindel proposed building a total of 12 Ju 388 K-0s as test beds for the Jumo 222 A/B. Initially, two of these were planned engine test beds for Jumo, and two for the *E-Stelle* Rechlin, to be delivered in September and November 1944. The remaining eight machines, to be delivered between September and October, were supposed to serve as "series pattern machines," and for operational testing of the Ju 388 J-2 Zerstörer with Jumo 222.[215] The very next day the JFM engine division agreed to deliver the engines, provided the RLM approved the installation. The stated deadlines could only be met, however, if Henschel delivered according to program. None of the engines had yet reached the test stand; therefore, changes (and thus delays) had to be expected. Because Jumo could not deliver the first four engines until September, instead of August, consideration was even given to taking back four of the Dessau-built prototype engines promised to *Lufthansa* for the Ju 188s.

Finally, the ratio of J-2 prototypes to test beds was reversed, probably because testing of the Jumo 222 engine had not progressed very far. With reference to advance order SS 4940-5101-0576/43, on 11 May 1944 a construction directive was issued for the "construction of eight test beds from Ju 388s with 9-8222 A/B (Jumo 222) engines."[283]

The design bureau in Dessau released the first information on the eight engine test beds on 23 May 1944. Provided the fuselages to be used had not already been prepared for the K-0 version, they were essentially to be built to Ju 388 L-0 standard (no bomb bay), and were only to have the K-0's radio equipment. Apart from the MG 131 in the cockpit, no armament was envisaged. Precise information as to deicing, auxiliary conduits, and control and hydraulic systems was to follow later.[284] One moth later the information was revised. The aircraft were to be built to L-0 standard in any event, and be fitted with the wings of the Ju 388 J-2.[285]

More detailed information about the Zerstörer prototypes powered by the Jumo 222 was also issued at the end of May. Apart from their power plants, these aircraft were to be similar to the Ju 388 V2 and, in addition to an offensive armament consisting of two MG 151s and two MK 108s (weapons set W12), they were also to have the remotely controlled tail turret. The cockpit shell, undercarriage, and climate control system could be taken from the Ju 388 K-0/L-0. Bombing equipment was not envisaged, and information concerning the wings, control system, and electrics was not yet provided.[286]

The planned number of prototypes with Jumo 222 engines subsequently fluctuated greatly. On 11 July 1944, the Concept Design Bureau (Kobü-Entwurf) ordered VS 19 propellers for seven Ju 388 test beds with Jumo 222 A/B engines.[287] The first amendment to the original construction directive issued on 12 August reduced the number of test beds from eight to three.[288] On the same day there was an amendment to the construction directive for the 50 Ju 388 K-0 aircraft, seven of which were planned as prototypes (V9 to V15) with Jumo 222 A/B engines. The Ju 388 V9, V11, and V13 were envisaged as engine test beds, the remaining four were initially to serve as prototypes for the Ju 388 J-2.[289] In September 1944, however, the type supervisor decreed that the first three of these machines should be delivered with BMW 801 TJ engines. The aircraft, whose completion was to be moved forward "in keeping with the change in the series program," were to be equivalent to the Ju 388 V2, and were also to have an oblique armament mounted behind Frame 15 (weapons set W13).[290] Thus, the Ju 388 V15 remained the only example of the Ju 388 J-2 Zerstörer.[291]

Finally, on 13 November 1944, Junkers issued a construction directive for six testbeds consisting of Ju 388 airframes with Type 9-8222 A/B (Jumo 222) engines. It was anticipated that the aircraft would use VS 19 four-blade propellers with a diameter of four meters. The testbeds were to aircraft of the K-0 series. It is likely, however, that the intention to build the aircraft to L-0 standard was maintained, for the document lists the significant differences between the K-0 and L-0 series. Apart from the changes necessary to accommodate the Jumo 222, these included a strengthened undercarriage with 1400 x 410 wheels and lower retraction struts made of steel, one Kärcher heater in each wing for wing deicing, additional lubricant tanks (*Rüstsatz* B 13), and a jettisonable fuel tank (900 l) beneath the starboard wing. The only armament was to be one MG 131 in the dorsal position. The installation of cameras was not envisaged at first, therefore, the magazine doors could be permanently sealed.[292]

A few days later, however, after discussions with the type supervisor, it was decided that cameras would be installed in four of the testbeds. In order to obtain the first two engine testbeds as quickly as possible, only the last four machines were considered. Infrared lighting was not envisaged for any of the aircraft.[293] It was found that for cg reasons ballast would have to be added to the vertical stabilizers of the testbeds. The ballast was to be of cast concrete.[294]

On 28 November 1944 JFM submitted an offer for "Development and Construction of Six 8-388 Test-Beds with Jumo 222 A/B Engines" to the Chief of Aircraft Procurement and Supply. Including development, which was set at 5,000 design hours, the six aircraft would cost 1.15 million Reichsmarks. Envisaged were the Ju 388 V9, V11, V13, V18, and V20. The Ju 388 V9 was to be the first aircraft, with delivery to the KdE in January 1945. Also included in the letter of offer was an offer for "Development and Construction of an 8-388 J-2 Operational Trials Aircraft with Jumo 222 A/B Engines" for 195,000 RM. The project would require 2,000 design hours, and the aircraft envisaged was the Ju 388 V15, WNr. 230 903, also to be delivered in January. For construction of the aircraft major components would be taken from the Ju 188 production line.[295]

Work on realizing these plans was centered in Merseburg, where the first Ju 388 with A/B III engines, probably the Ju 388 V9, completed static tests in December. The aircraft was supposed to be ferried to Dessau.[296] On 19 December 1944 representatives of Jumo negotiated with the Raspe company in Berlin for operational tanks for the lubrication system of the Jumo 222 for the Ju 388 and Hü 211.[297] Function tests with the 9-222 B 0 1 standard power plant on the test bench were temporarily halted after a fire in the exhaust tub. The engine was released for flight testing of the Ju 388.[298] It is not clear whether one of the available testbed's engines was replaced, or another machine modified. The last proof of use of the Jumo 222 in the Ju 388 appears at the beginning of February 1945. Jumo 222 B-2 No. 48, which had been converted to B-3 standard, exhibited insufficient oil pressure while installed in a Ju 388 at Junkers in Merseburg (probably the V9). During static tests to determine the cause of the low oil pressure, the gearing was completely destroyed as the result of an assembly error. The engine was stripped and reassembled, and was ready for collection on 5 February 1945.[277]

Propellers for the Jumo 222

On 11 July 1944 the propeller section of the "Concept" Design Bureau in Dessau ordered propellers and spinners for seven Ju 388 testbeds equipped with Jumo 222 A/B engines. The propellers were four blade VS 19 variable pitch units, with a mounting hole diameter of 220 mm and 16 attachment bolts. A total of 18 VS 19s were ordered, including four spares, half of them lefthanded and half righthanded. As the blades envisaged for the Jumo 222 A/B and E/F (4 m diameter, 10.5% blade width) would not become available until October 1944, a stopgap solution was adopted for the first 10 propellers: 20 Schwarz Type 9-41 285.13 V (righthanded) and 20 Type 9-41 286.13 V (lefthanded) blades were ordered. These had a blade width of just 9%. The Heine and Schwarz companies each received an order for 16 blades of the ultimate version. Delivery of the first set was planned on 25 August, with an additional set at the middle and end of each month until 30 December 1944.[287]

The number of testbeds with the Jumo 222 was reduced from eight to three on 12 August 1944, and on 20 September the propeller section amended the order. Just six active and two spare blades were now to be delivered. While the number of final version blades ordered remained the same, the number of interim blades was reduced accordingly.[299]

The Ju 388 with Daimler-Benz DB 603 Engines

With a displacement of 44.5 liters, the DB 603 V-12 engine was essentially an enlarged development of the DB 601. Unlike the DB 601, the DB 603 had two sparkplugs per cylinder, a combined engine propeller control unit and pressure water cooling with temperatures up to 115°C. The supercharger was located on the side of the engine; therefore, the compression of the two cylinder banks was different (left 1 : 7.3, right 1 : 7.5) to compensate for the pressure difference caused by the different lengths of the supercharger air lines. There was a total of 36 different versions of the DB 603, often differing in details, such as type of fuel, compression, and modifications for special accessories.[300] Efforts to equip the Ju 388 with the DB 603 can be traced back to November 1943. On 22 November of that year a meeting was held involving the Hirth-Motorenwerke, BMW, Junkers, and Daimler-Benz companies, its purpose being to establish basic assumptions for calculating the performance of the Hirth 9-2426 (TK 19) turbosupercharger. On 30 April 1944 Hirth wrote to Daimler-Benz requesting installation details and more precise information about the engine, so that it might redo its calculations specifically for the Ju 388. The engines under consideration were the DB 603 U and DB 603 E. The DB 603 G was not considered because of its higher compression.[301]

On 5 April 1944, there was a meeting in Dessau between *Herr* Löhlein and *Herr* Kirbach of Daimler-Benz AG, and *Oberingenieur* Klein, *Herr* Schnelle, and *Herr* Maier of JFM. The purpose of the meeting was to clarify technical detail questions concerning the firewall of the Ju 388 for the 9-8603 F (DB 603 U) engine. According to a specification sheet dated 4 August 1944, the DB 603 U with Hirth 9-2426 (TK 19) turbosupercharger had a takeoff and emergency power rating of 1,600 hp at 12.8 km, and a climb and combat power rating of 1,410 hp at an altitude of 13.3 km.[302] As the engine was to have a Riedel starter, a suitable place had to be found for the fuel line; the heat output for wing deicing was also discussed. As the Daimler-Benz engine had a greater diameter, the Ju 388's frame diameter would have to be increased. Details concerning the electrical system could not be clarified at the meeting, as JFM had not had time to obtain a competent consultation partner.[303]

Another reference to the Ju 388 and the DB 603, this time in another version, is found in connection with the Do 335. On 5 May 1944 the Plenipotentiary for Power Plant Development in the RLM, Dir. Dipl.-Ing. Schilo, advised Daimler-Benz by registered mail of *Oberstleutnant* Knemeyer's desire that the first two DB 603 L engines, delivery of which was planned for June, should be delivered as quickly as possible without further testing by Daimler-Benz for immediate installation in the Do 335. In order to accelerate the work, they intended to omit the installation of intercoolers and use water-methanol injection for cooling. A Ju 388 was to be made available on demand for flight testing, and further testbeds were envisaged "to broaden the testing base."[304] Daimler-Benz categorically rejected the idea of releasing the first two ultimate DB 603 L engines without a minimum of its own testing, however, and proposed instead that it deliver the interim engines V131 and V132, which had originally been earmarked for Rechlin.[305]

In October 1944, obviously at the urging of Dipl.-Ing. Lusznat (OKL/TLR/FL-E 3 VI),[306] there were serious plans to build a bomber power plant for the Ju 388 based on the DB 603, and use the Jumo 213 Es envisaged for the Ju 388 to increase output of the Ta 152 H. The DB 603 E/F engine was to be used initially, to be supplemented by the DB 603 L later. Deliveries of the Ju 388 with DB 603 power plant were envisaged as follows:

The fuselage of the BV 155 in the old NASM annex in Silver Hill in the summer of 2000. In this aircraft the TK 15 exhaust driven turbosupercharger was installed "displaced" behind the cockpit. Exhaust and supercharger air pipes ran along the side of the fuselage. Handig Collection

Supercharger air pipe connection on the DB 603 U-0 in the BV 155 V3. Handig Collection

March 1945	5 aircraft
April 1945	17 aircraft
May 1945	47 aircraft
June 1945	100 aircraft
July 1945	140 aircraft
as of August 1945	200 aircraft

Consideration was also given to equipping the Ju 88 G-7, of which somewhat more than 1,000 were to be built with Jumo 213 engines, according to existing plans, with the DB 603 E/F. As production of the Ju 88 G-7 was supposed to end in July 1945, the changeover did not appear worthwhile for time considerations alone.[307] A tabular survey of the various night fighter types, which unfortunately is not dated, seems to reflect these considerations, for as well as the Ju 88 G-7 and Ju 388 J-3 with Jumo 213 E engines, it also includes the same two types with DB 603 E/F engines. According to this survey, use of the DB 603 E/F would have resulted in significant speed losses (47 kph for the G-7 and 51 kph for the Ju 388 J-3), which were surely due in part to the reduced maximum boost altitude, from 9.1 to 5.7 km. Only in initial climb rate did the DB 603 E/F bring about a marginal improvement of 0.15 and 0.5 m/sec.[308]

Based on this performance data alone, it would have been a mistake to continue pursuing a version of the Ju 388 with the DB 603 E/F engine. Furthermore, it appears to make little sense to equip a high-altitude aircraft with a technically expensive pressurized cockpit with engines whose maximum boost altitude was just 5.1 km.

Daimler-Benz carefully examined the possibility of developing a bomber power plant for the Ju 388 based on the DB 603 E/F or DB 603 L at the beginning of November 1944; however, it reached the conclusion that such a development could only come at the expense of the Type 9-8603 B2 power plant which, based on the DB 603, was under development for the Ta 152 C and Do 335. Given

"At present and in the months to come, the Ju 388 aircraft type will be used primarily as a night fighter, day *Zerstörer*, and bad weather fighter. The required altitude for these roles is normally no more than 10 km. After deployment of the B-29, which is expected sometime between the end of this year and the beginning of the next, the Ju 388 will be used mainly to intercept these aircraft, as they will probably operate at an altitude of 12 km, and the Ju 388 is the only aircraft with sufficient armament and radius of action capable of operating as a fighter at those altitudes. The B-29 is not expected to operate at altitudes above 12 km, as damage to the pressurized cockpit would pose a serious danger to the crew. Operations at extreme altitude are thus not anticipated, even by the Ju 388.

Until the B-29 is committed, in the event of material shortages the Ju 388 will have to give way to the rest of the fighter program, and is currently assigned the relatively low Priority Level I. If deployment of the B-29 is expected, the Ju 388 program will be given priority, for the reasons listed above.

The Ju 388's performance is lower than previously estimated, although recently the BMW 801 TJ engines have performed according to specs. An increase in maximum boost altitude compared to the present engines would bring only a minor performance increase, which would not justify switching to another engine. Consequently, for performance reasons there is presently no requirement to install another engine, for example, the 8603 F, in the Ju 388, unless such a requirement arises as a result of procurement difficulties. Installation of an engine that offers significantly improved performance over the entire performance range, such as the Jumo 222 E, would be desirable. Based on the latest performance figures, even the use of the Jumo 222 engine with exhaust driven turbosupercharger would not promise any particular advantages, unless it is envisaged for small numbers of special purpose aircraft.

From the above it is apparent that, from the performance point of view, there is no present or future requirement to install an exhaust turbine high-altitude engine with our 9-2426 device in place of the BMW 801 TJ engine presently in use, as the anticipated performance increase is too small for the Ju 388's envisaged operational role. The change to another power plant will take place if the BMW 801 TJ can no longer be obtained, or if a power plant based on the Jumo 222 becomes available."[310]

the importance of this power plant, Directors Nallinger and Mann were unwilling to take responsibility for delays caused by development of a new power plant. Lusznat was therefore advised that Daimler-Benz declined to develop the new power plant, at the same time pointing out that, in addition to the BMW 801 TJ-0, the Type 9-8603 F based on the DB 603 U engine with exhaust-driven turbine would be available for the Ju 388. This power plant had already been designed, and prototypes were being built.[306] Schilling raised the same idea during a telephone conversation with Dipl.-Ing. Kaspar of the OKL TLR C-E on 26 November.[309]

A report by the Hirth-Motoren G.m.b.H. (Zuffenhausen) dated 13 November 1944, which concerns several meetings with representatives of the TLR, offers a probably authentic insight into the considerations by the *Luftwaffe* concerning the Ju 388. We have chosen to reproduce it here, as it makes reference to almost all of the engines discussed so far (begins p 121).

As Hirth had long belonged to Ernst Heinkel AG (EHAG) by this time, the statement about the alleged poorer performance of the Ju 388 airframe is to be treated with the appropriate caution. The He 219 was, after all, a competing night fighter design from the Heinkel stable. Comparing the extensive performance data in the previously cited document of 20 October 1944 concerning a meeting between BMW and JFM, it quickly becomes apparent that this statement was rather probably a side blow, for it could be proved that the BMW 801 TJ was responsible for the greatest speed loss by the Ju 388.

Development of the Type 9-8603 F engine and associated supercharger was pursued with great vigor, probably on account of the BV 155. As the increased boost pressure of the DB 603 had been approved for the turbine engine, in November the Hirth-Motorenwerke recalculated the performance of the 9-8603 F engine with the Type 9-2426 (TK 19) turbosupercharger, with and without MW 50. By increasing boost pressure from 1.48 to 1.8 atm (1.95 atm with MW 50), takeoff and emergency power was now calculated to be 1,980 hp at 11.5 km (2,260 hp at 11 km with MW 50). Increasing boost pressure from 1.35 to 1.45 atm for climb and combat power (1.65 atm with MW 50) resulted in a power output of 1,530 hp at 12.8 km (1,790 hp at 12.1 km with MW 50).[311]

With the Type 9-2279 (TK 15) turbosupercharger and the Blohm & Voss intercooler then envisaged, it was estimated that the DB 603 U would have a takeoff and emergency power rating of 1,610 hp at the maximum boost altitude of 15.4 km.[312] Admittedly, in December 1944 the 9-2279 turbosupercharger, which was envisaged for the BV 155 C, was still in the experimental stage. All of the outstanding installation questions concerning the 9-2279 V21 were supposed to be clarified between Hirth and Blohm & Voss on 12 December 1944.[313] The turbosupercharger was far from a lightweight, the V1 to V10 prototypes each weighing 173 kg.[314] before January 1945 was over the DB 603 U with the Type 9-2279 turbosupercharger was supposed to be tested on the DVL's high-altitude test stand in the configuration anticipated for the BV 155 C, and its performance measured.[315] Testing apparently progressed rather slowly, however, as the heat exchanger for the intercooler was still not ready at the end of January.[316]

To what extent were these plans implemented? The Ju 388 V27 (WNr. 230 902) and V28 (WNr. 230 906) were designated test beds for the DB 603 U in a construction directive amendment dated 12 August 1944. The type supervisor and the RLM decided that the aircraft should come from the batch of 50 Ju 388 K-0s. The major components of both aircraft were to be delivered to DLH in Staaken at the beginning of September. Conversion to the DB 603 U and final assembly were to take place there.[289]

Perhaps indicative of the state of the work is a document from January 1945. On 14 January 1945 Daimler-Benz employee Stotz reported to his director that the "388 intermediate framework" being built by *Lufthansa* in Staaken was "not yet finished." As yet it is impossible to say if a Ju 388 powered by DB 603 engines flew prior to the surrender.

Propellers for the DB 603 U

In April 1944 the "Head Department for Propeller Theory," which had been relocated to Erbach, in the Odenwald, ordered calculations from VDM for propellers for the Ju 388 with DB 603 U engines (engine type 9-8603 F).[317] Detailed plans based on this information are found in a Junkers design bureau file memo dated 25 August 1944 that makes reference to a meeting with VDM on 8 August 1944. There was a choice of two alternatives:

1. Two six-blade Dural propellers, each consisting of two three-blade hubs one behind the other (weight 365 kg without counter weights) with a diameter of 3.8 meters, and a chord/diameter ratio of 7 to 7.5%.
2. Two four blade propellers of reinforced wood (weight 235 kg without counterweights) with a diameter of 3.6 meters, and a chord/diameter ratio of 10.5%.

Use of the six-blade propellers would result in a change in the center of gravity, making it necessary to add 200 kg of ballast, making the aircraft 460 kg heavier than the version with wooden propellers. As well, the production version of the DB 603 U was supposed to be equipped with the strengthened VDM pitch control gearing (9-13030), with 800 Watt electric motor and 800 Nm pitch moment, which was also envisaged for the BMW 801 TG. It proved possible, however, to use the old gearing (9-13009), with just 500 W engine power and 420 Nm pitch moment, in the prototype engines. However, as the propellers' pitch moment was 720 Nm and 520 Nm (four-blade), the propellers would have to be fitted with counterweights, raising the weight of the wooden propeller to 267 kg. Apart from this, the higher production and assembly costs weighed against the six-blade propeller. It was therefore decided to use the four-blade wooden propeller. The Ju 388 with DB 603 L or DB 603 E/F was to have the same propeller as the Ju 388 with BMW 801 TJ, the four-blade Type 9-12188 Dural propeller with a diameter of 3.726 m. Counterweights would no longer have been required.[318]

Close-up of the turbosupercharger installation behind the cockpit of the BV 155. Handig Collection

Chapter 5: Factory Testing of the Ju 388

Prototypes of new designs were tested extensively by the manufacturer before being sent to the *Luftwaffe*'s testing center in Rechlin for additional testing. There they underwent further testing to determine if they met all performance requirements, and their general usability was examined, with an emphasis on technical questions.[319]

Later, or in some cases at the same time, operational trials were carried out by a specially formed trials unit (Erprobungskommando, or EKdo). Such units consisted mainly of personnel from front line units assigned to the trials unit to evaluate a new aircraft type under operational conditions.

As far as the Ju 388 was concerned, the fundamental question was whether the aircraft should be thoroughly tested before the start of production, or whether testing and production should take place simultaneously. In order to begin production as soon as possible, the decision was made to initiate production while testing was under way, beginning in mid-1944. The fact that many components of the Ju 388 had already been tested in the Ju 88 and Ju 188 probably influenced this decision. Modifications and improvements that resulted from testing were successively "infiltrated" into ongoing production of engines and airframes. Early production aircraft were subsequently modified to the latest production standard at the modification facility in Welzow.

This chapter is dedicated to the Ju 388 test organization in the main facility at Dessau, whose flight department was responsible for flight testing new types, and to the branch plant at Merseburg.

This photograph was probably taken in Dessau at the beginning of October 1943, and probably shows the fuselage of the Ju 388V7 under construction. One of the wings is visible in the left foreground. The lighter metal panels in the fuselage center section suggest that the Ju 188 fuselage was converted for installation of the bulged bomb bay. On the right the first panels covering the fuselage cockpit attachment points have just been fitted. Deutsches Museum

Prototypes of the Ju 388 K and L

Ju 388 L-0/V7, Backbone of the Factory Test Program at Junkers-Dessau

As stated previously, the first Ju 388 to be completed was the L-0 V7, which was built at Dessau. The aircraft's *Werknummer* was 300 001, and it bore the manufacturer's code PG+YA.

On 11 December 1943, even before its first flight, Junkers sent a telex directly to Göring, asking that it be allowed to retain the aircraft for thorough testing. A few days later, on 14 December, Junkers again turned to the Reich Minister of Aviation and Commander-in-Chief of the *Luftwaffe*, requesting that the V8 also be assigned to the company. The justification for the request reveals that Junkers also wanted to use the V7 for engine trials:

"The components installed in the Ju 388 series, such as the pressurized cockpit with climate control system and BMW 801 TJ-0 high-altitude engines, must undergo thorough examination, as they have not been sufficiently tested to date. We must point out that the 801 TJ engine has so far only been flown at normal altitudes, and will reach its maximum boost altitude for the first time in the Ju 388. Behavior and performance of the engine at that height are of great interest."[320]

Furthermore, according to Junkers, finalization of the Ju 388 L-0 and K-0 variants depended to a great extent on the results of these tests. As well, the company felt that using the two prototypes as operational trials aircraft was not advisable due to insufficient experience with high-altitude engines and pressurized cockpits.

While the V7 was nearing completion, the flight department, in close cooperation with BMW, determined which measuring instruments should be installed in the V7 and V8. The instruments were to be installed by the time flight testing began but, if this could not be done without delaying the first flights, installation was to take place later.[321]

By mid-November 1943 final assembly had reached an advanced stage. Both BMW 801 TJ-0 engines had been fitted when this photo was taken. The pressurized cockpit has yet to be fitted, and one can see hydraulic installations and elements of the control and electrical systems on Frame 9. By this stage the first system tests had already been carried out. Deutsches Museum

On 15 December 1943 the Ju 388 V7 was inspected by the BAL (Ing. Friedrich), in Assembly Hall 203. Elements of the autopilot, and the compressed air cylinder for the entry hatch emergency opening system were still missing, but otherwise the aircraft was obviously complete. Even the wiring for the camera installation had already been installed.[322] The next day the machine underwent a much more detailed inspection by Examiner Wähler of FSD Production Inspection. Included in his report to the design bureau were 65 complaints, and recommendations for improvement in the areas of the cockpit, fuselage, and engine nacelle extension, with another 20 for the electrical system.[323]

Flugingenieur Preuschen, head of flight testing at the Dessau works, also encountered the Ju 388 V7 (L-0) for the first time on 16 December 1943, without comment, however.

At 1154 hours on 22 December 1943 Flugkapitän Matthies took the Ju 388 V7 into the air for the first time, accompanied by flight engineers Körner and Heintz. The aircraft still lacked an autopilot, and the radio equipment was incomplete. On takeoff, the port engine failed to develop full power because of fouled plugs, reaching just 2,150 rpm instead of 2,450. While flying at 380 kph in level flight at a height of 3200 meters, oil temperature rose to 90°C. This exceeded the short term allowable temperature of 85°C (5 min.), and the continuous allowable of 70°. Fuel pressure was too low, and oil pressure was also at the lower limit. The cooling gills lacked sufficient rigidity and could not be controlled. There was vibration in the aircraft's tailplane. When the throttles were closed, the starboard propeller feathered itself. Other complaints included noise interference in the intercom system,[324] and too little pressure buildup in the pressurized cockpit.

After the maiden flight, which lasted 40 minutes, the oil coolers, cooling fans, and propellers were replaced. The measuring system was not yet complete, and the sensors installed for measuring cylinder head temperatures were not functioning correctly. As well, various instruments were missing, such as an oil pressure gauge with a 15 atmosphere measuring range and rpm indicators for the turbines.[325] On 3 January 1944 Preuschen wrote in his notebook that he expected the aircraft to be ready to fly again by Epiphany (6 January). In fact, it did not make its second flight until 9 January 1944, and once again it was with Flugkapitän Matthies at the controls. The maintenance test flight, which began precisely at noon, lasted 45 minutes. During this flight the port engine lost oil, the mixture indicator failed to work, and the starboard engine's coupler-curve was incorrect. Almost immediately after landing, at 1249 hours, Matthies took the machine up again on a short four minute flight.

The Ju 388 V7 took off on its third flight just three days later. Again there were problems with the engines. At a height of 2000 meters the starboard engine's boost pressure dropped off, and the turbine could not be engaged. There were also problems with the propeller pitch control mechanism. There was severe fouling of the port engine, and fuel pressure was too low. The cockpit interior altimeter (Fl 22325) was still missing, even though it was a vital piece of equipment that enabled the crew to determine exactly when to go on oxygen, and what flow rate to use. The forced air ventilation system was also found to be too loud. The line for tapping supercharger air for the pressure ventilation system had yet to be connected on the starboard engine. Experiments were supposed to be carried out involving connections in front of and behind the in-

This just completed pressurized cockpit, marked "5/17," was used for one of the prototypes, as the mounting on the end of the cockpit typical of the pre-production machines is absent. Griehl Collection

tercooler. The measuring equipment also failed to perform properly during the flight. Several thermal elements failed, and no current reached the sextuple temperature recorder.[326]

The aircraft's fourth flight on 14 January had to be broken off at a height of 3.2 km, because the engines vibrated badly while climbing at combat power, and flames were observed coming from the exhausts. During takeoff engine rpm had been down by 100. As a result, the next day the BMW field representatives in Dessau made a thorough check of the port engine, examining the spark plugs and fuel injectors. It was found that several plugs were loose, a total of four plugs in two cylinders were wet, and the insulation of one ignition cable was torn. All spark plugs and injectors were replaced. Immediately fouling was reduced significantly. The engine was closely monitored, and at high power the mixture was set 1% leaner. Bad weather prevented any high-altitude flights from being made for the time being.[327]

What was obviously the first high-altitude flight took place on 20 January 1944. At an altitude of 5000 meters there was a great deal of noise in the radio receiver, probably because of static charge buildup. Oil mist appeared in the cockpit at high-altitudes. As well, the hardened glass windows displayed a tendency to fog up. Sealing between the plexiglass and hardened glass windows (different dimensions) was ineffective, especially on the exterior, resulting in misting during flight. The superchargers shifted gears at 9800 m (port) and 10100 m (starboard).[328] Soon after the high-altitude superchargers were engaged there was a sudden drop in boost pressure at combat power at 10.2 km. After landing it was found that an oil line had burst, resulting in the destruction of the turbine bearings. The port engine, which to date had logged 5 hours and 49 minutes (including a 2 hour acceptance run), had to be removed. One of the nine engines delivered for use in other prototypes was installed in its place.[329] While the engine was being changed, work was also carried out on the airframe. The autopilot mounting was strengthened before the unit was calibrated, and the supercharger air feed for the pressurized cockpit was connected to the starboard engine behind the intercooler. Also criticized were the leaky compressed air system (entry hatch) and the poor delivery of fuel from the fuselage tank.[328]

On the afternoon of 31 January 1944 Matthies took the aircraft into the air for a 17 minute test flight with the new engine. The turbine became fully engaged, because the shaft for controlling the exhaust flap had seized. Probably for this reason engine speed on takeoff was 100 rpm below maximum.[330]

Until the end of January the Ju 388 V7 was definitely the only aircraft flying. Flight characteristics were found to be largely similar to those of the Ju 188.[231] Elevator travel was less than on the Ju 188; consequently, control forces were greater and sensitivity reduced. Restrictors had to be installed in the landing flaps hydraulic system to produce operating times equal to those of the Ju 188. Operation of the oil cooler ring and engine cooling gills was still unrestricted, and they deployed so abruptly that it was feared the system might be damaged. As a result, BMW was called upon to install restrictors there as well.[332]

Vibration of the tail surfaces was observed during the Ju 388 V7's first flights. It was suspected that the jet of exhaust gas produced by the BMW 801 TJ-0 was causing airflow separation. In

The exhaust system of the Ju 388V7, which was completed and test flown at the turn of the year 1943-44. This variation of the exhaust system caused the stream of exhaust gases to strike the wing, causing severe vibration. The exhausts were modified repeatedly until an aerodynamically suitable form was found for the production aircraft. Deutsches Museum

late January 1944, therefore, wind tunnel experiments were carried out on the central exhaust port at Dessau. No significant flow separations were discovered. The flow of exhaust gas was angled about 5 degrees above the wing's datum plane, and it was thought that this might be capable of causing flow separation; however, the tests were conducted with total flow through the turbine (exhaust flap closed). With the exhaust flap open, the exhaust gases went straight into the exhaust at significantly higher speeds, causing it to lose its directional effect. In the opinion of the aerodynamicists this was responsible for the vibration in the tail surfaces. This scenario could not be simulated, however, on account of the "pulsating character" of the exhaust.[333]

A stationary engine run was carried out on the afternoon of 1 February, during which a pressure integrity check (depressurization) of the pressurized cockpit was carried out. It was found that the bad weather window leaked badly, and it had to be adjusted during the night. Another depressurization test was carried out the next day, and the window sealing was found to be good. The climate control switch, on the other hand, was not to be used in flight, as it was almost impossible to adjust. No replacement was immediately available, because all remaining climate control switches were to be installed in the pre-production machines, so as not to delay their completion.

An additional panel with engine instruments was installed in the cockpit of the Ju 388V7 for the flight engineer, who sat in the observer's seat. Two data recorders were installed below the panel. The arrangement of the instrument panel on the right side of the cockpit differed greatly from that of later machines, as is evident in this photo taken in January 1944. Griehl Collection

During the pressure integrity checks, it was found that the entry hatch was very difficult to close, and could only be accomplished with the assistance of two men pushing against the outside of the door with their shoulders. It was impossible to close the hatch from inside. Tests to determine the leakage rate had to be abandoned after a piece of testing equipment broke down; however, it was then discovered that the entry hatch could not be opened from inside or outside. Finally, the compressed air emergency opening system had to be employed. Several rivets on the hatch mounting were sheared off in the process. It was discovered that an aluminum part in the closing mechanism had broken. If no one had been in the cockpit, they would likely have had to punch a hole from the outside to open the hatch.[334]

The V7 made another maintenance test flight on 5 February. After two minutes in the air, at combat power the port engine began leaking oil badly through the ventilation port, whereupon the breather tube was shortened by 50 mm. The objective of the next flight on the following day was to determine the rate of climb between 1300 and 2200 meters. At a speed of 260 kph and a takeoff weight of 10.3 tons, this was found to be 10.95 m/sec. While at an indicated airspeed of 270 kph, the starboard engine suddenly began vibrating heavily and smoking, and engine speed dropped rapidly. The entire airframe exhibited strong transverse vibrations, probably caused by excessive play in the port landing flap. After a flight that lasted just 12 minutes, Matthies landed safely back at Dessau.[335] It was discovered that fouled sparkplugs had caused the engine problems. To avoid problems during takeoff, the high-altitude sparkplugs (Bosch 240 ET 7 I B) were replaced with those from the BMW 801 D (240 ET 7 I A). As a result of the change, subsequent takeoffs were trouble free. At 1225 hours the next day Matthies took off again in the V7. The flight lasted 45 minutes.

The next flight began at about 1300 hours on 9 February. The flight plan called for a climb to high-altitude at combat power, but after reaching 3700 meters the flight had to be aborted when the starboard engine's forward control mechanism failed. The engine was subsequently removed after an operating time of 7 hours, 27 minutes (including 2-hour acceptance run). As the machine was equipped with numerous measuring devices, all of the leads had to be reattached. The enforced pause caused by the engine change was also used to work on the aircraft. The pressurized cockpit connectors (threaded ball-socket mounts) were retightened, and the valve cluster and climate control switch were removed and checked. As well, air lines and filters were checked for oil, and the supercharger air bleed on the port engine for the pressurized cockpit was connected. In the process, BMW had to remove the cooler.[336] The exhaust gas counter-pressure indicator on the port engine was not functioning, consequently, the measuring line had to be washed out with solvent (carbon tetrachloride). The propellers were set at 92°, because when the engine had failed it was discovered that the engine was not completely stopped with the propeller in the gliding (feathered) position. Rubber seals were installed between the engine cowlings and nacelle extensions to improve the aerodynamics.[337]

During the subsequent static test the new starboard engine had no turbine oil pressure due to a missing component. As a result, the first maintenance test flight (without climate control switch!) with the new engine probably did not take place until 15 February. During the 104-minute flight the starboard engine's oil pressure was found to be strongly dependent on engine speed, and was just 1.8 atm at 1,700 rpm. Fuel pressure in the starboard was also too low, just 1.35 atm. After two minutes of flight at combat power, the port engine again began leaking oil heavily through the breather. A 15-mm longer thin plate orifice was subsequently installed on the breather line. During the flight a hydraulic line burst in the airframe in front of Frame 9, forcing the crew to use the emergency undercarriage and flap lowering system for landing. Matthies nevertheless managed to land the aircraft safely at Dessau at 1630.[338]

Two days later he was airborne again at 1553 hours. The speed course flight was made at combat power to test the thin plate orifice. It worked perfectly, and no oil was seen. During the 42-minute flight, however, the aircraft lost one of the rubber seals installed in the seam between the engine cowlings and nacelle extensions. A cover made of metal strip had to be fitted to prevent this from happening during future flights. The climate control switch was still absent. If no new switch was available, prior to the next flight the supercharger air bleeds were to be capped at the engines, and the associated lines connected to the forced-air ventilation system.[339]

A second high-altitude flight by the Ju 388 V7 was carried out on 19 February 1944. While climbing at maximum allowable continuous power, the starboard engine's high-altitude supercharger refused to shift into high gear until 10000 meters was reached, while use of the throttle lever was needed to get the port engine's to engage at 10700 meters. During the subsequent climb to 11.3 km, the boost pressure dropped a total of six times because the engine supercharger switched back into low gear. The flight was subsequently aborted. Once again an elbow in the hydraulic system had broken, and the crew were almost forced to make a belly landing. It took seven minutes to lower the undercarriage using the emergency system. As a consequence, the hydraulic hand pump, initially installed behind the radio operator's seat, was moved to a position next to the pilot's seat, where it could be operated more effectively. As a result of the supercharger having shifted gears much too high, after the flight BMW checked the barometric cells in the supercharger's controlling mechanism.[340]

Problems with the BMW 801 TJ-0 engine's exhaust collector now appeared for the first time. After 4 hours, 26 minutes flying time, on 22 February 1944 exhaust gas began coming from the starboard engine's warm air scoop due to a leak in the starboard collector.[341] The collector was subsequently replaced.

Another high-altitude flight was begun on 25 February. This time, while climbing at maximum allowable continuous power, the superchargers shifted into high gear at an altitude of 7.9 km, which was too low. As the machine continued to climb, the boost pressure on the port side again dropped repeatedly as the high-altitude supercharger disengaged. The flight was therefore aborted at an altitude of 9300 m. The aircraft vibrated, especially while in descent with power off. Loose connections were found on several measuring devices.[342]

On the afternoon of 26 February 1944 BMW replaced the barometric cells in the port engine. The aircraft was able to complete its next high-altitude flight, once again at maximum allowable continuous power, on 27 February. This time the superchargers shifted into high gear at 8100 m (port) and 8850 m (starboard), but in spite of the new barometric cells, the port engine again exhibited variations in boost pressure. The flight was aborted at an altitude of 9.1 km. Prior to returning to Dessau, however, the aircraft was flown at combat power in level flight at altitudes of 8.0 km and 6.5 km. Boost pressure fluctuations reappeared at maximum allowable continuous power at 8000 meters. The technicians were unable, however, to reproduce this behavior on the ground with the high-altitude superchargers engaged. To avoid further delays in testing the port engine was immediately replaced. By that time engines 385 167 (port) and 385 129 (starboard) had logged 9 hours, nine minutes and 7 hours, 31 minutes, respectively, including a two hour test run.[340] The smell of oil continued to penetrate the cockpit of the Ju 388 V7. The first rust spots had appeared, and had to be painted. There were strong fluctuations in the starboard rpm indicator, and it was decided that it should be mounted less rigidly.[343] The rpm and boost pressure indicators would be mounted on rubber cushions like those of the Ju 188, especially since the BMW 801 TJ was found to run much rougher than the BMW 801 D.[344] Furthermore, fuel was leaking from the fuel tank fittings, with the tank in the port wing leaking worse than the starboard one.[343] In addition to numerous minor modifications, the fuel transfer pump system and autopilot were to be checked, and all the screws in the canopy framework tightened. As well, the cooling fans were to be painted with colorless Zapon paint (corrosion protection). The transmitter transformer had to be adjusted, as it was interfering with the intercom system. And finally, the cockpit was to be made ready for inspection to verify that it was ready for night operations, and a check was to be made to determine if covers had been placed over the nose ring seams.[345]

On 29 February there was an inspection of the V7's illumination system. Present were two members of the "Electrics" design team (Kobü-Elt.), Flugkapitän Matthies and pilot Lehmann. After the canopy had been covered, positions were selected for the UV lamps to provide optimum illumination of the instruments. After the inspection was over, further complaints were raised about the electrical system. The external thermometer and position lights would have to be moved away from the entry hatch to prevent it from becoming tangled in electrical wiring when jettisoned. As well, to gain access to the autopilot resistor box in flight the pilot had to stand up.[346]

As the Ju 388 was designed to operate at high-altitudes, the propeller pitch control mechanisms of the Ju 388 V7 were provided with cold protection boxes. Because of supply problems, however, equipping subsequent aircraft was problematic. Following the installation of a new port engine (385 169) in early March, the first flight to reach maximum boost altitude was carried out on the 6th. At a height of 12 km with an external air temperature of -63°C, maximum boost altitude was determined to be 11320 meters for the port engine and 12020 meters for the starboard engine. The flight had to be aborted at 12200 meters when the starboard engine's oil temperature began to rise. The aircraft vibrated very badly while in a power off descent. Ground runs were subsequently carried out to determine the cause of the port engine's low maximum boost altitude, and the starboard engine's high oil temperature. As a precaution, it was decided to ready two spare engines.[347]

Maximum boost altitude is dependent on boost pressure, which builds up behind the turbosupercharger. When the engines were run with the exhaust flaps closed, it was found that pressure behind 385 169's turbosupercharger was also lower on the ground. During the search for the cause, it was discovered that one of the seals between the supercharger air elbow and the inlet housing had not been installed at the factory—it was found at the bottom of the cowling. The seal was installed, and another test run was made on 11 March 1944, a Saturday. The difference in boost pressure between the two engines was reduced, but they still only produced 1.49 and 1.62 atmospheres at combat power. Turbosupercharger revolutions were measured at 9700 and 10800 rpm. The search for the cause continued. No foreign bodies were found in the air shafts leading to the engine, and all lines were tight, leading to the conclusion that the turbosupercharger itself must be the problem.

In the search for the cause of 385 129's elevated oil temperature, the first thing checked was the oil cooler safety valve, but it was found to open at 12 atm as prescribed. On 8 March there was an air raid alert, and at 1328 Matthies took off to fly the V7 from Dessay to Alt-Lönnewitz. The flight, which lasted 12 minutes, was made at low level, and once again oil temperatures were high. When the all clear was given, Matthies flew the aircraft back to Dessau, landing there at 1705 hours. March 10th saw the installation of new propeller pitch drives mounted on ball bearings. In the process one forward engine cowling (with cooler) was replaced after the starboard engine's higher oil temperatures were traced to a defective seal in the oil cooler. After removal the forward section was checked, but all the seals were found to be in order.

On 11 March 1944 the Ju 388 L-0 V7 was equipped with engines 385 169 (2 hrs, 23 min flight time) and 385 129 (7 hr, 51 min).[348]

The aircraft was flown again on the morning of 11-12 March. On the day after, there was a seating and evacuation test involving Hptm. Dr. Stahl, the Ju 388 project supervisor, together with Stabsing. Kuhn and Ing. Pehlke of the *E-Stelle* Rechlin. Friedrich, a representative of the BAL attached to Junkers, was present. It was found that the three crew members had complete freedom of movement, even in winter flight suits with parachutes on. The entry hatch was large enough to allow the crew to enter with parachutes on, although the pilot and radio operator were not supposed to put theirs on until they were in their seats. The radio operator's seat came in for criticism. When slid to the left it could block the entry hatch, and the headrest had to be modified to prevent the parachute harness from becoming caught during evacuation. An evacuation test showed that it was possible for the observer to get out in four seconds, even though he first had to release his harness and turn left toward the pilot. On the other hand, the pilot's seat was criticized, as it made it difficult for the pilot to get out after the canopy had been jettisoned. The armored seat back would have to be modified to allow it to fold back, otherwise the pilot's parachute pack might become caught on the upper edge. The arrangement of some radio equipment also had to be changed. It was recommended that the jettisonable canopy section be lengthened by 15 to 20 cm at the front; however, this would require the relocation of a fuselage frame.[349]

Another high-altitude flight was made on or about 13 March, and the port engine's turbosupercharger was still not working properly. Above 11 km the starboard engine's oil temperature rose again as oil pressure fell sharply. There were also problems with the autopilot, which kept veering the aircraft to port. After the flight it was decided to install restrictors in the hydraulic lines serving the annular oil coolers, so that they would open and close more slowly. Orders were issued for the installation of a FuG 16, but for the time being the aircraft was to be kept flyable and weighed with full tanks.[350]

After 4 hours and 5 minutes flying time, there were more problems with the exhaust collector, this time on the port engine.[341] The cause was a faulty weld seam. The total hours suggests that this was engine 385 169. On 20 March the aircraft was at BMW, and one day later the port engine's right collector had to be replaced because of leakage. While attempting to install a new collector a serious problem was discovered: not only were there problems with

In this side view of the BMW 801 TJ-0, the supercharger air elbow and its attachment to the intake housing are visible next to the quick release fastener for the upper cowling section (upper left). MTU

the weld seams, the diameter of the entire replacement part was too great, and it fit poorly. Both collectors—the new one and the last one removed—were subsequently sent to Kolb, the manufacturer. It in turn shifted some of the blame onto BMW, which had failed to provide adequate documentation or a suitable mockup. This further complicated the already difficult manufacturing process. Deformations posed another problem. The manufacturer used water under pressure to check for leaks in the collectors. On its own, Kolb had increased the pressure of the water from 2.5 to 3.5 atmospheres to improve reliability, but in the end this only resulted in deformations. BMW reacted about a month later. X-ray examinations had revealed welding defects in early production collectors, including those containing nickel; consequently, these were only approved for use in flight testing, and only for a maximum of 50 hours.[351] They were not to be used for service engines, and thus were not to be delivered to the air force. All new engines were to be equipped either with the ultimate operational version of the collector built under improved quality control procedures, or older collectors that had been X-rayed and were still subject to the 50 hour restriction. As well, as of 15 April 1944 all engines were to be delivered with the strengthened engine mounts with 40-tonne attachment points (instead of 31.5 tonne units).[352]

On 23 March the problems with the V7's autopilot had still not been solved, and during a 16-minute test flight by Flugkapitän Matthies the port engine leaked oil, and the starboard engine's idle speed was too high. Another test flight was obviously imminent, for Heintz ordered that five of the total of seven minor complaints were to be resolved by 0900 hours (the following day?). As well, the aircraft's fuel tanks and two 110-l oil tanks were to be filled.[353] It appears that the aircraft was flown on 24 March, for there is a brief complaint report dated 24 March 1944, and a more detailed one dated 26 March.[354] During the flight the aircraft was unstable while climbing, and at 400 kph it lost a hatch cover from the port engine cowling. The antenna mast vibrated with an amplitude of about 20 cm. It appeared that the pumps in the fuselage tanks were not delivering enough fuel at altitude.

The starboard engine's idling speed was still too high. When switched to M2 engine rpm dropped by 100, and the same thing happened at altitude. Turbine oil pressure was also too low. The port engine vibrated at altitude, especially with weak fuel mixtures. The port propeller pitch mechanism failed to function, as did the electric elevator trim. With the undercarriage retracted the starboard undercarriage position indicator flickered, and the tailwheel did not appear to be locked. Both oil coolers were now working too slowly.

The huge exhaust collector and its attachment to the turbine on an early BMW 801 TJ with a tubular engine mount, probably no more than a mockup. MTU

Cutaway exhaust collector of the BMW 801 TJ-1 in the Deutsches Museum. The round openings mark the spots where the exhaust pipes from the individual cylinders join the exhaust collector and are welded to it. The weld seams had to be of very high quality because of the extreme thermal loads to which they were exposed. Handig Collection

There were also problems with the FuG 10 and the DF. The recently installed FuG 16 still had to be tuned to the flight department's ground station. After the most serious problems had been addressed, the aircraft's fuel and oil tanks were filled again. The oxygen supply also had to be replenished.

On 6 April 1944 the V7's propellers were damaged in a collision with a vehicle. At Dessau the aircraft taxied off the runway from the northeast in the direction of Acceptance Test Hangar 225. A Jumo truck leaving Jumo Flight Operations Hangar 234 suddenly turned onto the taxiway leading to the acceptance test hangar, and was struck by the propeller blades, the tips of which suffered damage. The propellers (VDM Type IV 9-12178) bore the serial numbers 6753 and 7197. The driver who had caused the accident was given a warning.[355,356]

At a meeting held on 14 April 1944, consideration was given to converting the aircraft into a fighter; however, this plan was obviously not carried out. On 25 April one of the aircraft's turbine oil lines broke.[357]

In April 1944 the V7 also served the design bureau's Physics Research Department. Fogging of the double-glazed cockpit panels was a serious problem that had to be rectified. A recess was therefore made in each field for a moisture trap filled with a hygroscopic substance.[358] This was to absorb the moisture and dry the air in the spaces between the glazed panels. To date, relatively expensive and very fragile silicagel moisture traps had been used, either in the form of filter stone or filtering cardboard containers, which were made by Junkers in Dessau.[359] Laboratory tests carried out in October and November 1943 revealed, however, that another substance, kolphosal—a mixture of phosphorous pentoxide and aluminum powder—was much more effective, had a longer life, and was also cheaper. In April 1944 tests were carried out to evaluate the new kolphosal moisture traps during flight operations and determine their lifespan. The V7's canopy was therefore fitted with both conventional silicagel and kolphosal moisture traps. Although a misunderstanding between the pilot and flight engineer resulted in the series of tests being terminated before the kolphosal mois-

Right instrument panel of a prototype or pre-production machine in May 1944. At the bottom left a cockpit interior altimeter (Fl 22325) had already been installed. Flight trials with the V7 had revealed the need for the instrument. Griehl Collection

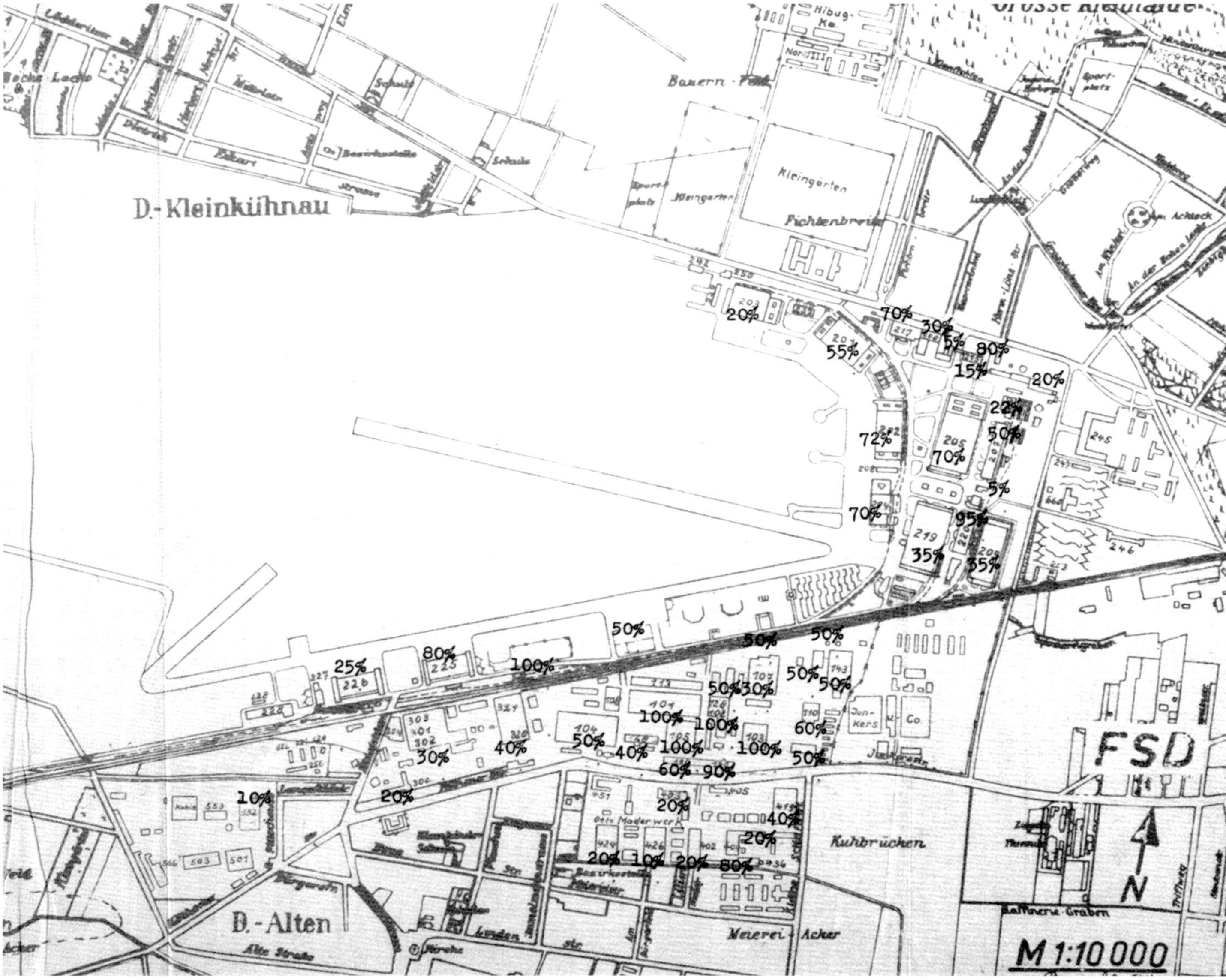

Map of the Dessau plant. Each building was marked with a figure indicating damage percentage at the end of the war.

Left: One of the insets for the moisture traps in the cockpit of Ju 388 L-1 WNr. 560 049 in the NASM. Vernaleken Collection

ture traps were saturated, the new material clearly demonstrated its superiority. The kolphosal moisture traps were in the double-glazed panels for 26 days. During that time the Ju 388 V7 made 10 flights, including six high-altitude flights above 11 km.[360] In the same period the silicagel moisture traps with filter stone containers had to be replaced six times, and those with filtering cardboard containers four times. Silicagel turned a reddish color when saturated with water. Unlike kolphosal, however, silicagel could be regenerated by heating it in an oven.[361] When the kolphosal was "spent," as evidenced by the appearance of brown flecks in the trap's cardboard container, the trap had to be replaced with a new one.

At the beginning of June modifications were made to the exhaust fairings of the Ju 388 V7. BMW installed bearer arms for a flame damping system, after which the installation underwent a final check. According to a file note entitled "Ju 388 Prototype Aircraft: Work Scheme" dated 11 June 1944, after initial tests of a new thicker exhaust fairing, the flight test program was to be transferred to Prague. The flight test program was to explore the following areas:

- thicker-gauge exhaust fairing
- flame damping system
- external ram-air intake
- climate-control system with air cooler
- fuel pumps (together with DBU)
- fuel leakage

Preuschen had expected the aircraft to be ready to commence trials with the flame damping system on 10 June 1944, and 14 June was given as the ready-to-fly date in the file note. But on the following day Preuschen found it still in Hangar 201, probably because conversion work was still going on. He logged a flight on 18 June, and mentioned the machine several more times in June with the note "Flame damper System." From 19 to 24 June the aircraft may have been in Berlin, where aircraft 300 004 and 300 005 were being tested.[363] A high-altitude flight at 11600 meters with the flame damper system is known to have been made on 29 June. Tests carried out between 25 and 30 June revealed that airflow over the upper surface of the wing in the area of the flame dampers was still

Pilot Wendel, *Flugkapitän* Kindermann, and *Flugversuchsingenieur* Hotopf (from left) pose for the press in front of the Ju 90V1, with which they set two international altitude records with 5000 and 10000 kg payloads on 4 and 8 June 1938. Vernaleken Collection

poor. Night stationary ground runs were also carried out. Tests were carried out with and without flame dampers, and the effects of the exhaust screens and mixers were checked with the dampers in place. It was found that with the turbine engaged, flame damping was most effective with just the mixer; it was thus possible to greatly simplify the actual flame damping system.[363]

In addition to minor complaints, on 1 July there were problems with the measuring devices. While the aircraft was ready to fly on the morning of 9 July, after one day of testing (12 July) it had to return to the hangar on the 13th. It was probably there to have its starboard engine replaced, for when the port engine exhibited output fluctuations due to falling boost pressure after 7 hours and 46 minutes of flight time on 15 July, the starboard engine (385 256) still had no hours in the air. One week later the port engine had to be changed after just 8 hours, 42 minutes flying time, as the supercharger was leaking more oil than the sparkplugs could tolerate.

Before the new port engine (serial number 385 220) could be used, on 25 July 1944 both propeller pitch control systems (VDM) had to be replaced because they could not be switched from automatic to manual.[364] Two days later Flugkapitän Rupprecht Wendel carried out a test flight that lasted from 1636 to 1720 hours. The following day, 28 July 1944, Wendel took off from Dessau again, at 1512 hours. He climbed the machine to 13150 meters and did not return until 1729 hours. The aircraft again appears in Preuschen's records on this day with the notation "oil cooler." On 1 August 1944 he noted that flight testing of the Ju 388 V7 was continuing.

On 4 August 1944 modifications were made to the cylinders, perhaps the installation of measuring devices for cylinder head temperature experiments. The machine was actually supposed to resume flying on 7 August, but after 8 hours, 42 minutes the starboard engine's VDM propeller pitch control mechanism had to be replaced, and for the same reason as the port engine. As a result, the V7 did not fly again until 0745 hours on 9 August. This is verified by an entry in Wendel's logbook dated 9 August 1944, which states that an altitude of 10300 meters was reached in a flight that lasted from 0748 to 0857 hours. On that day the Askania boost pressure regulator on the starboard engine had to be replaced after almost 13 flying hours. The V7 did not fly on 10 August, possibly because the supercharger was leaking oil again after just 4 hours and 10 minutes of operating time (though an engine change was not yet necessary). On 11 and 12 August 1944 engine and cockpit pressure measurements were carried out with the Ju 388 V7 for the purpose of revealing the differences between the internal and external ram-air intakes. Wendel was probably at the controls for most of these flights, for he flew the V7 on 11 August (1404 to 1611 hours) and twice on the 12th (0920 to 1027 hours and 1709 to 1909 hours). At that time the machine had four-blade propellers without spinners or counterweights. As a result of the previously mentioned damage to the blade tips, propeller diameter had been reduced to 3.68 m. In preparation for the tests described in the following text, cooling fans with broad metal blades were installed, together with sealing strips in the oil cooler nose ring and original short BMW exhausts. The 60-mm gap between the nacelle extension and the rear of the engine cowling was left uncovered.

The most the cooling gills were opened during measurements with the internal ram-air intake was +5 degrees, compared to 15 degrees during tests with the external intake. Half of each set of flaps, which were covered by the intake scoop, were set at 15 degrees. After the last flight with the external intake, the lower cowling halves of both engines and the intake pipes on the turbosuperchargers were replaced by components with the internal ram-air intake. The comparison flights were made under generally similar conditions on the same day.

Tests to determine the dynamic pressure utilization of the two types of intake had already been conducted in July 1944. During a flight on 7 July the dynamic pressure utilization of the engine with the internal intake fell to 100% at an altitude of 5.5 km. With the external intake, on 11 August utilization was still 130% at a height of 13 km. Both versions achieved about 180% on the ground.

During the flights on 11 and 12 August tests were also carried out to measure the cylinder head temperatures. The hottest cylinder turned out to be the starboard engine's No. 4. All of the others, including the port engines Nos. 4, 9, and 10, proved to be 10 to 30°C cooler.

The comparison between internal and external intakes showed that temperatures differed above 6.5 km. With the internal intake, the maximum allowable temperature of 245°C, which was permitted for a maximum of 15 minutes, was exceeded above 10000 meters, while with the external intake this did not occur until 11750 meters. The maximum allowable temperature for longer periods was 235°C.

Generally, cylinder head temperatures were too high. This was shown during a flight on 11 August 1944, when the aircraft initially climbed to an altitude of 10 km at 2,300 rpm and 1.28 atm of boost pressure, the maximum allowable sustained output. The climb was subsequently continued at combat power and 245 kph to an altitude of about 12800 meters. Cylinder head temperature rose to 310°C. In comparison, at low altitude the aircraft achieved 260 kph while climbing at combat power. The problem of cylinder head temperatures could have been solved by enlarging the gaps in the cooling gills, but possible negative aerodynamic effects had to be examined first.

During these flights similar problems were encountered with oil temperatures. There, too, the engine with the internal intake exceeded the critical mark of 90°C at an altitude of 9000 meters, while the other engine did not reach this temperature until 11.3 km. On this occasion the temperature reached 135°C at 12800 m. The BMW 801 TJ-0's allowable oil temperature was 90° for 15 minutes. The Junkers experts remarked that this oil cooling performance would only be acceptable if it could be maintained for 30 minutes, or as long as combat power was available. Even the supercharger air cooling functioned much better with the external intake. Furthermore, with the external intake supercharger air temperature was the same as outside air temperature. At an altitude of 12 km, therefore, it was as much as 14°C cooler than its opposite number with the internal intake, improving thermodynamic effectiveness. The pre-compression received by the turbosupercharger with use of the external intake allowed the turbines to be switched on about 1000 meters higher than usual. It was hoped that Fimag intercoolers with improved cooling performance would later be installed and tested.

Measurements of exhaust gas temperatures, which were between 550 and 650°C, clearly showed the temperature drop produced by the energy discharged into the turbosupercharger. The switch from low- to high-altitude supercharger took place at an altitude of 8.6 km for the engine with the internal intake, compared to 10 km for the engine with the external intake. This, too, was determined from the exhaust gas temperatures.

The experiments on 11-12 August 1944 also revealed the influence of various configurations on temperatures in the cockpit. After passing through the turbosupercharger and mechanical supercharger, the air for the pressurized cockpit was diverted through a limiting nozzle. Before passing through two active charcoal filters, it was cooled again in a so-called "chamber air cooler," for it had been shown that oil could not be filtered out of the hot supercharger air. In this way, compressed air from the port and starboard engines reached the cockpit via a non-return valve. At lower altitudes the cockpit could also be ventilated (inadequately) by forced air. There was one safety valve for protection against external, and six for protection against internal overpressure. Given the fact that the air for the cockpit was bled from the supercharger air, it is obvious that a reduction in supercharger air temperature with the external intake must also have resulted in proportionately lower cockpit air temperatures. Use of the external intake was found to have one pleasant side effect: the pressurized cockpit remained oil free up to an altitude of 13.1 km. Normally oil fumes appeared at 9 km and grew stronger with altitude. Because of the unfavorable position of the metal tube with which air was bled from the starboard engine (the one with the measuring equipment), however, the air was warmed 20°C by heat generated by the turbine ahead of the chamber air cooler. Cooling performance was of course much improved at higher speeds, as the air intake for the chamber air cooler was located in the leading edge of the wing. A reduction in speed from 260 kph to 245 kph meant that air from the high-altitude supercharger stage was only cooled by 70°C instead of 85-90°C. It was therefore suggested that restrictor flaps be installed in front of the chamber air cooler's air intake in order to achieve better control over temperatures inside the pressurized cockpit.

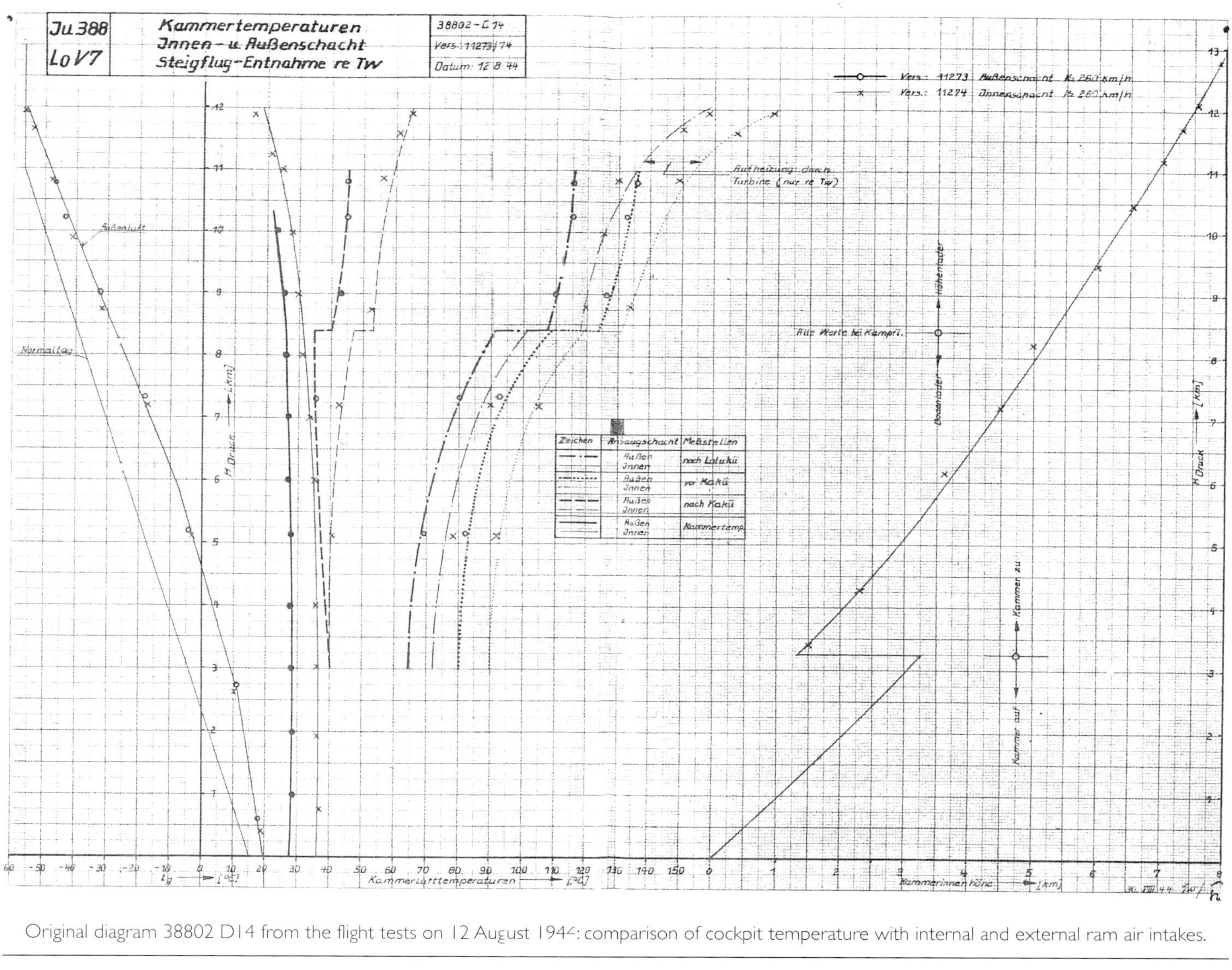

Original diagram 38802 D14 from the flight tests on 12 August 1944: comparison of cockpit temperature with internal and external ram air intakes.

The sun's rays had a considerable effect on temperatures in the cockpit. Tests with a so-called physiological thermometer showed that the effective biological temperature was 8 to 10°C higher than the temperature measured inside the cockpit.[365]

After 9 hours and 26 minutes of operation, on 14 August 1944 the sockets of the port engine's supercharger control switch had to be replaced, as it was no longer functioning correctly.[366] The results of the repair were apparently checked by Flugkapitän Wendel, who took the aircraft up from 1804 to 1835 hours. The next day a static test was conducted with the V7, and on 28 August tests with flame dampers. A test bench run was planned for the following day.

Testing report FlA. 569/44, dated 31 August 1944 summarizes the results of flame damper tests with the Ju 388 L-0/V7.[367] Various configurations had been tested during night static tests. The examples made by the Linde und Eberspächer company were two-part. An exhaust mixer was first attached to the standard BMW exhaust. This divided the flow of exhaust gas, after which it was mixed with fresh air. The second part consisted of a long metal tube that shrouded the mixer and concealed the exhaust flames.

Night static tests with the Linde flame damper revealed that at maximum allowable continuous power (1.28 atm, 2,300 rpm) without the turbine, there was a dark red trail from the upper edge of the outlet approximately 300 mm long. When the turbine was switched on this was reduced to about 200 mm. This was confirmed in test flights to determine yaw sensitivity.[368] As a result the Linde exhaust screen was not installed; instead, an attempt was made to get by with just the mixer until Junkers could design its own. At first a fairing was simply fitted over the mixer, with a metal tube at the rear for the later exhaust screen. With this system the aircraft's handling returned to normal. Even the tail surface vibrations, present with standard BMW exhausts, were somewhat better. In order to get an idea of flame buildup with the standard exhaust system, this was also tested at night. Depending on the power setting, flames from 200 to 750 mm in length were produced. Generally the flames were shorter with the turbine switched on, which made sense, as some exhaust energy was recaptured by the turbine. Junkers continued development of the flame damper system. It even succeeded in developing a system that completely eliminated the flames; however, this design (No. 5) caused tail surface vibrations almost as bad as those from the standard exhaust system. Furthermore, at 18 kg it was also the heaviest, and after a time the corrugated metal shroud began to glow. They finally settled on design No. 3. Weighing 10 kg, it was the lightest, and by diverting the exhaust stream away from the wing it even produced a speed increase of 7 kph, resulting in a top speed of 410 kph at ground level. Tail surface vibrations were significantly better than with the standard exhausts and other flame damping systems up to a speed of 500 kph. The system's only shortcoming was that it did not completely eliminate the exhaust flames. At combat power without the turbine, for example, there was a remnant flame 80 mm long. It was hoped, however, that in flight these flames would not be visible from the designated distance because of increased cooling performance.

As some sort of exhaust fairing had to be applied to the standard BMW exhausts for aerodynamic reasons anyway, for simplicity's sake Junkers considered equipping all aircraft, even the reconnaissance versions, with flame dampers, as they improved handling and performance. Ultimately the V7 was supposed to be converted to the new system so that night flights could be made for systematic performance measurements with and without flame dampers.

The Ju 388 V7 went back to work on 7 September. It was supposed to be finished one week later. On that Saturday (9 September), in a flight lasting a total of 72 minutes, the aircraft made a continuous climb to 11800 meters (Junkers records). According to BMW, it leveled off at that altitude and flew for eight minutes at combat power, during which engines 385 220 (10 hours) and 385 256 (16 hours) used 22 and 18 liters of oil, respectively. BMW's figure for total duration differs somewhat. It claimed that the flight lasted 85 minutes, with 30 minutes each at maximum allowable continuous power and combat power, before descending in a power-off glide and returning to the airfield. The oil tank ventilation system had been changed. It was linked to the engine ventilation tube, and there was no longer any significant ejection of oil.[369] The aircraft flew again on the following Sunday, 10 September 1944. As the day before, it first climbed to 8000 meters with 1.28 atm of boost pressure and 2,300 rpm (maximum allowable continuous power), then continued to 11800 meters with 1.45 atm of boost pressure and 2,425 rpm. The flight lasted 63 minutes. Two more climbs to high-altitude were made on the following day, the flights lasting 61 and 69 minutes. On the second flight the aircraft was climbed to 10000 meters with 1.28 atm of boost and 2,300 rpm before continuing to 11800 meters at combat power. This flight was repeated on 12 September. As well, for 15 minutes it flew a closed course at maximum allowable continuous power, resulting in a total flight time of 84 minutes. Also on that day, the port engine's supercharger control unit had to be replaced after 17 hours, 51 minutes of operation, probably because of slippage in the clutch. One day later the starboard cooling fan lost a blade, the second time this had happened. As well, a valve spring cap failed on an exhaust valve. On that day the machine had completed a climb flight, again to 11800 meters, lasting 77 minutes. Oil consumption based on the flights from 9 to 13 September was between 15 and 23 liters per hour. By 13 September the starboard engine (385 256, a "specially checked" engine) had logged 27 hours and 55 minutes since mid-July. In that time it had made 18 high-altitude flights above 11.5 km. The port engine (385 220, standard engine) had run 19 hours and 13 minutes, and had completed 11 high-altitude flights. The engine's Bosch DW 240 ET (7/1) B (8-4045 K-1) sparkplugs were almost worn out, meaning that the plugs had only lasted for the guaranteed period of 20 hours. The sparkplugs in the bottom cylinders had to be changed more frequently because of fouling. The excessively high fuel consumption was also causing concern. Fuel transfer experiments revealed that when flown at maximum allowable continuous power, the aircraft used about 570 liters of fuel per

Dipl.-Ing. Schinzinger, seen here in conversation with two coworkers, was the longtime head of the flight department in Dessau, and was also responsible for flight tests with the Ju 388. Vernaleken Collection

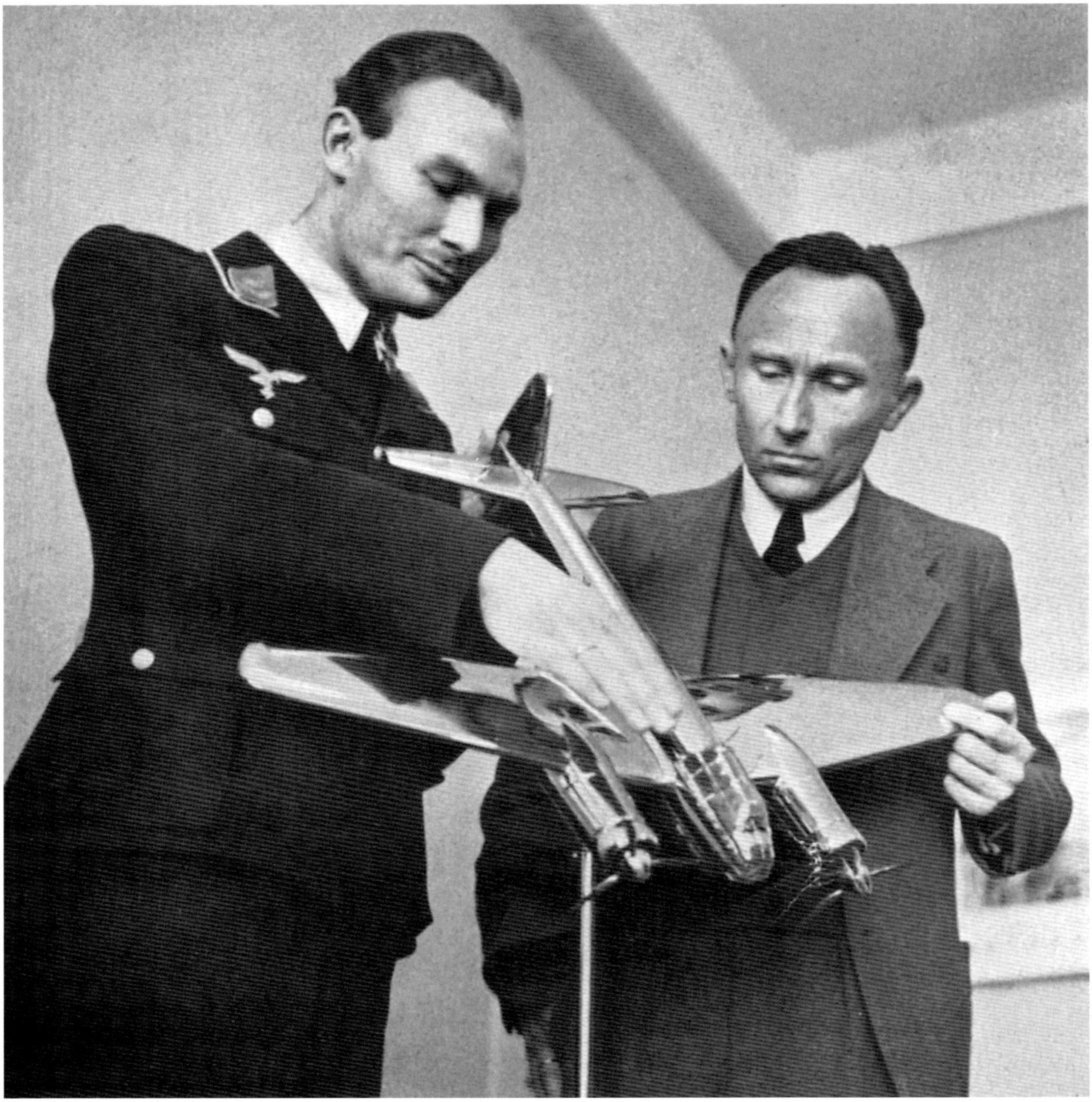

As head of the design bureau, Director Ernst Zindel, seen here in a photo from "Junkers News" posing in front of a model of the Ju 88 with highly decorated bomber pilot Werner Baumbach (left), was also involved in development of the Ju 188 into the Ju 388. Vernaleken Collection

hour, while BMW had promised a consumption of just 430 liters per hour. More precise measurements were ordered.

An oil drain pipe (NW 5) in the supercharger air collector ring also caused problems. Fuel passed through this small tube into the engine cowling, where it usually ignited if the engine failed to start immediately. The drain pipe had to be lengthened so that it passed through the bottom of the engine cowling. On the other hand, advances were made in the oil tank venting system, which had been modified. Air was now vented from the tank into the open, instead of into the engine. Even with full oil tanks (120 l), during a climb to 12 km oil leaked from just one engine, and then in negligible quantities.[370]

On 16 September 1944 the Ju 388 V7 was in the hangar being worked on, and this was continued on the following Sunday. In the course of this work new oil pumps with L and M channels were probably also installed. It is not surprising, therefore, that on 29 September 1944 the engines had still logged just 19 hours (385 220) and 25 hours (385 256), respectively.[371] A planned temperature climb flight to test the new oil pumps was still pending.[372]

In October 1944 the Ju 388 V7 began a series of test flights equipped with the latest intercoolers, which had reduced internal drag, their main purpose being comparative performance measurements. Also to be determined was the amount of speed lost with the ultimate flame damper system installed. Fuel consumption measurements and a range test were also planned, as were engine measurements to investigate the influence of the new oil coolers (improved oil flow rate) on oil temperature. The ultimate exhaust system was to be tested, with measurements of the temperature on the upper surface of the wings and landing flaps. The de-ice air temperature was to be measured with use of the external intake, maximum boost altitude measured with the honeycomb filter in front of the intake shaft iced over, and the cold start system, oil tank venting (with emphasis on cold starts with full tanks), and a flare signaling system tested. These flights were also to be used to determine why the altimeter, vertical speed indicator, and airspeed indicator fluctuated at high-altitude, resulting in significant discrepancies between the calculated and indicated airspeeds. And finally, the flights were to be used to test a Kärcher heater providing heat for the camera and weapons installations, and to determine the range of the FuG 16 at high-altitude.[373]

The next significant comments about the V7 by Preuschen appear in November. The aircraft was flown on 3 November, then again on the 7th. Both flights were plagued by ignition problems. On 7 November the aircraft's port engine (385 220) failed while climbing through 7.5 km. The principal cause is not known. The turbine was removed and sent to BMW. There were problems with the starboard engine on 11 or 12 November. The engine was removed, and its turbosupercharger fitted to the port engine. All of the measuring systems, especially temperature and pressure sensors, subsequently had to be reinstalled,[374] and it is therefore not surprising that the machine was still unserviceable on 16 November. Two days later both engines were removed from the V4, one of

		Economical Power				**Most Economical Power**			
Measured Value	Engine No.	385/275	385/274	387/776	387/837	385/275	385/274	387/776	387/837
Consumption	l/hr	196	224	200	180	260	287	278	256
Revolutions	rpm	1785	1838	1780	1775	2058	2110	2080	1930
Boost Pressure	atm	1.00	1.00	0.98	0.98	1.14	1.14	1.12	1.12
Supercharger Air Temperature	°C	42	46	35	36	44	47	36	37
Exhaust Counter-Pressure	atm	1.07	1.07	1.08	1.09	1.10	1.10	1.12	1.13
Injection Volume	mm≥/stroke	268	290	267	241	302	324	318	307
Rated Value	mm≥/stroke		240 - 255				295 - 305		

		Maximum Continuous Power				**Combat Power**			
Measured Value	Engine No.	385/275	385/274	387/776	387/837	385/275	385/274	387/776	387/837
Consumption	l/hr	464	480	468	466	598	634	649	600
Revolutions	rpm	2310	2325	2320	2320	2460	2470	2430	2430
Boost Pressure	atm	1.28	1.28	1.26	1.265	1.45	1.45	1.43	1.43
Supercharger Air Temperature	°C	46	48	38	38	50	52	43	42
Exhaust Counter-Pressure	atm	1.16	1.15	1.18	1.19	1.22	1.22	1.25	1.27
Injection Volume	mm≥/stroke	478	491	482	478	580	611	636	588
Rated Value	mm≥/stroke		438 - 470				540 - 580		

which (321 219) was subsequently allocated to the V7. On 22 November, however, this engine failed to perform properly, and was replaced by engine 385 266. With the aid of 321 219's turbine, the V7's port engine (385 220) was made whole again.[375] The next day the aircraft took to the air again.

By 24 November 1944 the port engine (385 220) had flown 22 hours and 10 minutes, and the starboard engine (385 266) 26 hours and 19 minutes. The aircraft also flew on 3 and 4 December, reaching an altitude of 12000 meters. The V7 was serviceable on 7 December, but in the evening there were problems with the elevator. A replacement elevator was taken from the Ju 188 V11, which had been grounded, and was now being used as a source of spare parts. The V7 was supposed to be ready to fly again in two days; however, by 9 December the flight times logged by the engines had only risen to 23 hours, 24 minutes (385 220) and 27 hours, 34 minutes (385 226).[376] The aircraft was used by the Dessau Flight Department to measure the cooling performance of the new Fimag intercooler, an early version of which had been installed in the starboard engine.[377] Both engines were now fitted with external supercharger air intakes. Fuel consumption tests were also carried out to determine the aircraft's range.[378] Consumption was measured using a Siemens flow meter. Other values affecting fuel consumption, such as boost pressure and exhaust gas temperature, were recorded using carbon recorders, while supercharger air temperature was recorded by a six-color recorder. Measurements were first taken with the engines in their original state. When consumption rates specified in the installation chart could not be achieved, the engine settings were checked on the ground. The resulting values were outside the tolerance range.[379]

After the engines had been retuned by BMW, further measuring flights were carried out. In the first series of tests, measured fuel consumption by the port engine had been as high as 690/l/hr. After retuning it was still about 450 l/hr. The starboard engine presented a similar picture, with fuel consumption of about 450 l/hr after retuning.[380] The values contained in the installation chart had still not been achieved, however, and even in a power off descent the engines still consumed 120 liters (at 1,360 rpm) to 220 liters (at 1,700 rpm) of fuel per hour. This unexpectedly high rate of fuel

Here the V7's sister machine, the Ju 388 V8, WNr. 300 002, is seen in the third week of March 1944, probably in front of Assembly Hall 203 in Dessau. The rudder still lacks the notch at the bottom, which became standard on later production aircraft. The periscope is also absent. Also noteworthy is the early version fairing over the BMW 801 TJ-0's central exhaust pipe. EADS

consumption would mean a reduction in range of about 10%; BMW was asked to comment on the situation, and an order was issued for all other engines to be rechecked as well.

In December a signal flare system was also tested in the Ju 388 V7. The selected installation site proved entirely satisfactory Only 32 of 80 flares could be fired, however, the rest failing to fire even though the electric percussion pins had been pierced. The system's striking power was found to be insufficient, for all of the remaining cartridges were fired successfully from a flare pistol. A functioning flare signal system would surely have been a useful item of equipment, as it would have eliminated the need for the flare pistol firing port in the canopy, and with it a potential source of leaks.[381]

The V7 was grounded on 13 December 1944 pending the arrival of new engines. The engines with the serial numbers 385 275 (8 hrs, 30 min) and 385 274 (7 hrs, 30 min) had been installed in the Ju 388 K-0 WNr. 230 156, while the brand new engines 387 776 and 387 837 had been mounted on the wings of the Ju 388 V4 a K-1. On 20 December 1944 BMW delivered two BMW 801 TJ engines with the serial numbers 387 976 and 387 970. These were immediately checked against the BMW modification list to ensure that they incorporated all the latest modifications. Instead of the 83 engines called for in the December program, by 28 December 1944 BMW had delivered just 24.[382] On 5 January 1945 the aircraft was still (or again?) grounded, and did not become serviceable until 20 January. After performance measurements on 25-26 January the engines were obviously modified or repaired, for on 29 January 1945 the V7 again appeared in Preuschen's notebook with the comment "supposed to roll out today for a test run." At the end of January the BAL in Dessau handed over to type supervisor Hptm. Stahl a "Test Report on Flame damping Systems on the 8-388 L-0/V7 with BMW 801 Engines." Unfortunately only the cover letter dated 23 January survives in the EADS Archive in Ottobrunn, and it is unclear which was delivered: an updated report, or the one already mentioned.

A defective cylinder grounded the V7 on 10 February 1945. Not until three days later did BMW deliver a new cylinder. The aircraft was supposed to be ready for a test run by 17 February, a Saturday. On the following Tuesday, however, the Ju 388 L-0/V7 was grounded, and does not appear again in Preuschen's writings, which end soon afterwards. A memo concerning a meeting of development heads on 2 February 1945 reveals that the Ju 388 V7, together with the V4 and WNr. 230 151, had been grounded, making test and calibration flights with the new intercoolers impossible.[382a]

Junkers Ju 388 L-0/V8

The Ju 388 V8, which was assigned *Werknummer* 300 002 and the manufacturer's code PG+YB, was built at Merseburg, and initially equipped with engines 385 182 and 385 178. The engines were delivered on 15 February 1944 and installed quite quickly, but there were difficulties when the first attempt was made to start them. For three days (18-20 February 1944) all efforts to start the port engine

This photo of the V8 from the same series shows that the machine was towed out of the hangar by a tracked vehicle with rigid tow bar after a rain. The propeller spinners are still missing. The opening for the absent periscope is visible in the bulged bomb bay. EADS

failed. Changing the spark plugs helped, but when the engine was finally started on 21 February it did not perform properly. The cooling jacket was resting on the turbine, preventing it from moving. The problem was not resolved until 23 February. Harmonizing the fuel, oil, and boost pressures of the two engines, carried out on 25 February, proved very time consuming due to poor accessibility. The next day a propeller pitch control unit was found to be leaking. Finally, on 4 March, both engines were cleared for flight. Based on experience with the V7 at Dessau, the aircraft's hydraulic system was modified, which took four days. On 11 March the propeller reduction drives were changed. Afterwards, when a test run was carried out, it was found that the coupler curves no longer matched.[383,384] This chronicle of the V8 surely sounds rather negative; this is largely due to the fact that it is based on a report entitled "Complaints Resulting from Test Runs," which details problems encountered with the BMW 801 TJ-0 engines.[383]

After these problems had been addressed, on 15 March 1944 the Ju 388 V8 took off from Merseburg on its maiden flight. At the controls was Heinz Lehmenn, making his first ever flight in a Ju 388; he was accompanied by flight test engineer Lange.[384a] The same crew carried out the next documentable flights by the Ju 388 V8.

The first verified "flight complaint" appeared during a test flight on 16 March. On that day the coupler curves and boost pressure fluctuations again came in for criticism. Two days later Lehmann and Lange carried out two more test flights, one of which had to be aborted after the starboard engine's turbine developed problems passing 7500 meters. During a static test on 22 March, probably to check the engines after measures had been taken to alleviate the problem, leaks were found in the turbine lubrication systems.

Another high-altitude test flight on 25 March was broken off after the engines developed problems. At an altitude of 10800 meters the port engine's turbine oil pressure was only 0.3 atm, while the starboard engine's high-altitude supercharger refused to engage. The latter problem was later traced to a stripped screw in the control mechanism. A test flight on 29 March and a maintenance check flight on 7 April 1944 were apparently completed without engine problems, for there were no further complaints until 11 April, when poor accessibility was criticized during a starter change. Another high-altitude flight was recorded on 13 April; however, it also had to be aborted. The port engine's oil temperature climbed to over 90°C, and it was leaking oil badly. In mid-April Preuschen noted that the V8 was supposed to resume flying as soon as possible. In fact, Lehmann carried out a radio test flight on 15 April. On 17 April an autopilot test flight was made, but it had to be aborted after the starboard engine began shaking badly. A single-engine landing was made, after which the starboard engine was examined. It was found that the connecting rod of the 12th cylinder had broken, and the engine was replaced. The aircraft was grounded for a time; however, Lehmann and Lange were able to repeat the autopilot test flight on the 22nd. A maintenance check flight was conducted three days later. The next repair followed an autopilot test flight on 26 April 1944, when the port engine's starter refused to engage and had to be replaced.

Another test flight was carried out on 30 April. During a high-altitude flight on 5 May 1944 problems developed with the port engine's boost pressure regulator. After the high-altitude supercharger had been engaged the regulator failed to maintain the selected value. Furthermore, the engine began to overheat (100°C).

A second series of photos of the Ju 388 V8 was taken at the end of March 1944, just one to two weeks after the first. By that time propeller spinners and weapons (or mockups thereof) had been added. WNr. 300 002's manufacturer's code PG+YB is clearly visible. EADS

During a test flight on 9 May an oil line to the port engine's turbine fractured. Another test flight was carried out two days later, and complaints were filed about both power plants. At an altitude of 10000 m the port engine's oil temperature exceeded 90°C; however, loss of oil from the starboard unit presented a much more serious problem, and the engine had to be shut down. During a test flight the next day (12 May 1944) the starboard engine caused another single-engine landing. There were boost pressure fluctuations, then the engine oiled up during a power-off descent and began emitting white smoke. After landing it was discovered that a line bringing cooling air to the starboard engine's turbine had failed.[385] The turbine had been damaged as a result, and had to be replaced. The starboard engine continued to be a source of trouble, also during a check flight by Lehmann on 25 May. Cylinder No. 7's sparkplugs were found to be fouled, and on 25 May the piston rings were replaced, which required removal of the cylinder. Two days later the engine failed to produce sufficient power, resulting in a check of the spark plugs. The next documented test flight took place on 1 June 1944.

At a meeting held on 6 and 7 June 1944 consideration was given to handing over the machine. Meanwhile, flight testing continued. A maintenance check flight was carried out on 11 June, with BMW engineer Schabert on board instead of Lange. Four days later the "original" crew of the V8 carried out another test flight.

In the period from 19 to 24 June 1944 the Ju 388 V8 was flown at altitudes up to 11000 m to test a cooler for the climate control system. It was discovered that at combat power, temperature inside the cockpit could be reduced by 38°C.[363] On the evening of 7 July 1944 Lehmann and Lange made a high-altitude flight in the V8.

The aircraft was subsequently transferred to Rechlin. It was not part of *Erprobungskommando* 388, however, probably being used mainly for engine testing.

On 1 November 1944 the Ju 388 V8 was flown from Rechlin back to Dessau via Brandenburg. On 10 December Flugkapitän Wendel took the aircraft up on a maintenance test flight that lasted from 1135 to 1208 hours. On 19 December the V8 was flown overland back to Rechlin. It was still there on 19 January 1945, carrying out a brief calibration flight (0946-1003) on that day.[385a]

Junkers Ju 388 K-1/V3

The first prototype of the Ju 388 K-1 high-altitude bomber was created from major components of a Ju 188 E-1 that were modified to Ju 388 K-1/V3, V4 standard. The aircraft had a partial offensive weapons system with no bomb sight, and an experimental version of the tail turret. Junkers Dessau was unable to meet the planned completion date for the Ju 388 K-1/V3, but by the beginning of March 1944 the aircraft was nearing completion with engines 385 127 and 385 179. The supporting framework on the firewall was causing problems. As delivered by BMW it did not fit properly and had to be reworked.[386] On 24 March 1944 the aircraft was accepted by Inspector Wähler (Dessau Production Inspection), as revealed in the Aircraft Final Report dated 30 June 1944. Inspection report memos, with detailed lists of complaints, were compiled by Inspectors Wähler and Klaue on 17, 23, and 31 March.[387] There were fit problems with the pressurized cockpit and fuselage, as well as the bulged bomb bay doors. Also, the gap between the closed tailwheel doors was 20 mm instead of the specified 10. There was also criticism of the design and installation of the climate control system. When delivered to the flight department the aircraft was fitted with four-blade Type 9-12178 propellers, as those envisaged for the aircraft were unavailable. It was anticipated that the correct propellers would be fitted before the aircraft was handed over to a testing station or the air force. Parts of the propeller de-icing system were also missing, and the installation was only complete as far as the cooling fan.[388] On 24 March 1944 the State Aviation Ministry's construction inspection branch (BAL) carried out an inspection of the electrical and navigational systems. No significant faults were found; however, the inspectors did raise several criticisms: the wires running from the distributors between Frames 8 and 9 were unshielded, and DF control switch for use by the observer was poorly situated on the port fuselage wall between the pilot's and radio operator's seats. As well, Inspector Knebel pointed out the lack of UV lamps for the instrument panel on the starboard fuselage wall. For safety reasons, it was also suggested that the undercarriage control switches be made double pole instead of single pole, and that the FK 38 emergency compass be moved, as in its present location it was deflected by approximately 15° when the rudder pedals were used.[389] Preparations for flight were made on or about 27 March. Following earlier engine monitoring experiments, sensors had to be installed for measuring boost pressure behind the turbosupercharger, turbine oil pressure, and turbine flap position.[390]

The Ju 388 V3 was flown for the first time on 2 April 1944. Two days later a planned high-altitude flight had to be aborted at an altitude of 6400 meters when the starboard engine began losing oil. The port engine's waste gate/dump flap would not close properly (only to 30°), and in addition to abnormal behavior when switching from low- to high-altitude superchargers, the port engine intermittently emitted flames at a boost pressure of 1.28 atm (maximum allowable continuous power). Finally, the starboard oil cooler ring opened on its own during flight. Lowering of the landing flaps took 24 seconds too long, and they retracted again at 230 kph, while the starboard landing flap vibrated. After the flight Heintz ordered a measurement of rudder deflection.[391]

On 5 April 1944 the aircraft made a high-altitude flight to 11350 meters. While climbing at combat power, at 11000 meters there was a drop in boost pressure in the starboard engine, while the port engine's turbine again refused to fully engage. The de-icing system air scoop on the port engine was lost during the flight.

As rudder control was too easily jammed and the replacement bomb bay doors limited maximum speed to 350 kph, it was decided to conduct no further flights for the time being and, in addition to replacing the rudder control box, to address those deficiencies found to date. As well as those already mentioned, these included sticking of the port throttle lever. The two engines switched superchargers at different altitudes: 8200 m for the port and 8900 m for the starboard at maximum continuous power, and 8300 m port and 9200 m starboard at combat power. As well, both engines vibrated at weak mixtures, the port worse than the starboard. Two canopy panels were defective and had to be replaced, and because of the loud noise produced by the forced-air ventilation a wooden intake was supposed to be fitted as on the V7. Finally, the remotely controlled tail turret was to be made serviceable.[392]

The presence of an auxiliary fuel tank in the starboard wing only (approx. 430 liters) meant that considerable aileron input (more than half of maximum deflection) had to be applied during every takeoff. Consideration was therefore given to either leaving the tank empty, or installing a counterpart in the port wing.[393] The crew also complained that the weapon control panel of the first K-1 projected too far into the cockpit, making it difficult for the radio operator to exit the aircraft quickly in an emergency.[394] Beginning with the V6, therefore, the panel was to be modified to allow it to fold. Flights by the K-1/V3 and the L-0 aircraft at Merseburg had demonstrated a requirement for a turbine monitor, at least for acceptance and check flights. What was needed was an engine mounted boost pressure gauge. Also raised was the airspeed indicator leakage problem. Once the ultimate bomb bay doors and the production rudder had been installed, it was planned to use the machine for performance flights, and to conduct measurements for the de-ice air and the FA-15 systems. On 16-17 April 1944 Preuschen noted that the V3 was to resume flying as quickly as possible.

In May 1944 the aircraft was used, among other things, for oscillation tests with the PVE 11 A periscopic sight. To help with the design of the PVE 11 A, which was harmonized with the FA 15, Junkers sent designers to Zeiss in Jena. *Herr* Handke of the Zeiss design bureau requested documentation of the oscillation loads at the periscope installation point in the Ju 388, as the BAL representative at Zeiss was demanding the information for acceptance tests. The BAL's demands concerning the periscope oscillation loads were no doubt unrealistically high, for Handke reported that several units had already been damaged in acceptance tests.

To provide Jena with the required data, the Ju 388 V3 was used for vibration measurements, both in the air and on the ground with engines running. A vibration writer-recorder measured actual vibration at the installation point, various places on the fuselage sidewalls, and on the installed periscope itself. During static tests with engine speeds between 1,800 and 2,400 rpm, it was found that

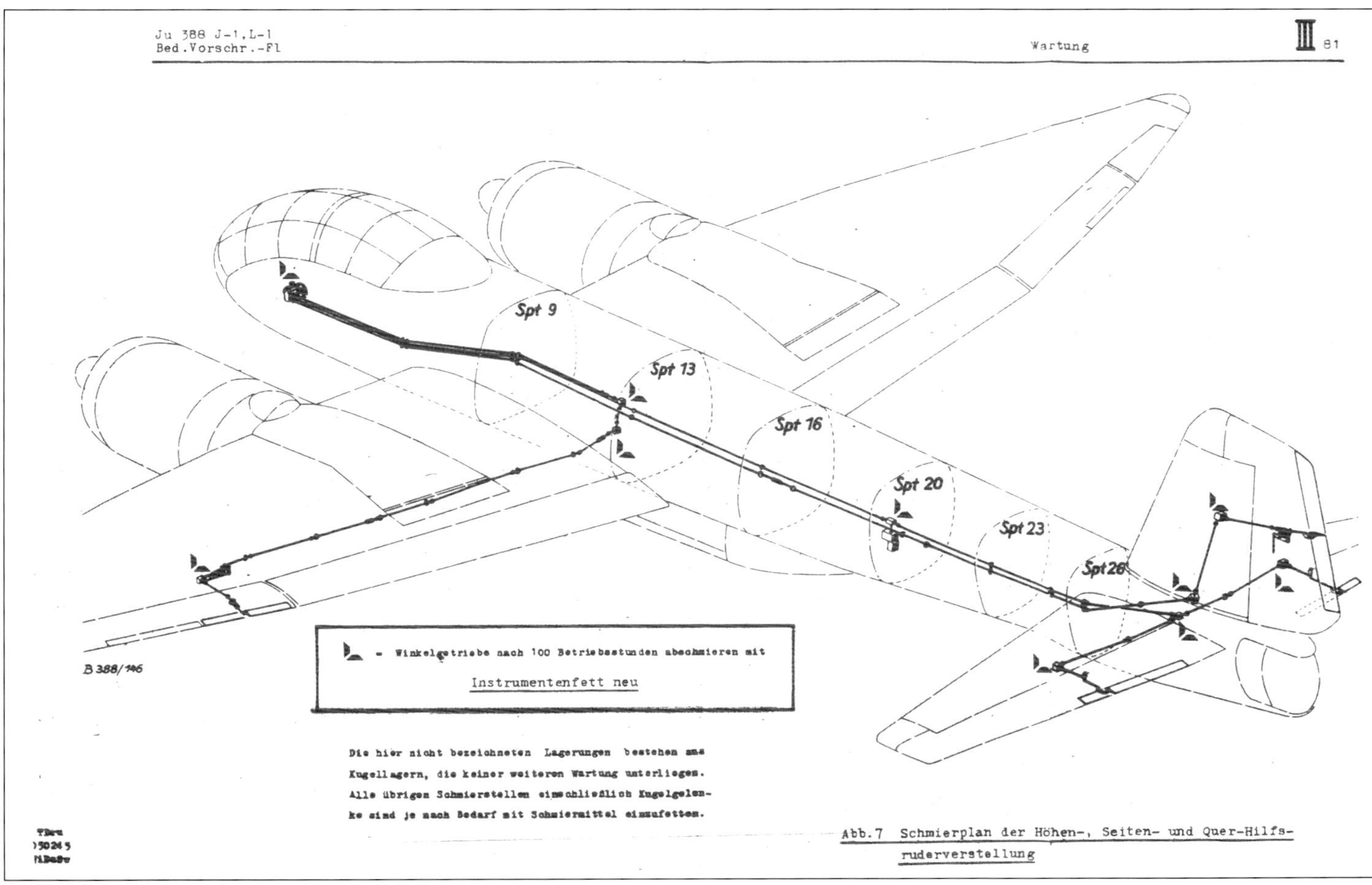

The rudder control lubrication plan from the pilot's notes provide a good overall view of the system.

vibration amplitudes were relatively small, in the worst case +/- 1/10 mm. The recorded vibration frequencies largely corresponded to propeller speed (1,000 to 1,350 rpm). Sometimes there were high frequency overlapping vibrations, whose frequency corresponded to fourfold propeller revolutions (passing frequency of the blades of a four-blade propeller). The amplitude of these vibrations was barely measurable, however, and never exceeded several hundredths of a millimeter. Altogether, the maximum loads calculated from frequency and amplitude were about 0.5 G. Measurements on the Ju 288's pressurized cockpit had produced similar values.

The installation of the periscope in the airframe proved extraordinarily rigid; the natural frequency of its vibrations in installed condition could not be excited even by striking it. It was obviously very high and was strongly damped; therefore, resonance vibrations were not expected.

Earlier systematic experiments with unbalanced propellers using Ju 87 and Ju 88 aircraft with imbalances to the limits of flight safety (unbalance = 120 mg or 2°-3° of individual blade pitch) had resulted in maximum loads of 1 to 1.5 G on rigid airframe components.

The experiment report ended with the conclusion that, with a suitable installation of the periscope in rigid components of the airframe, amplitudes greater than 1/5 mm and acceleration forces greater than 1 g were not to be expected, even with rough running engines or slightly imbalanced propellers.[395]

The failure of a fuel pump on the port engine was reported on 16 May 1944. As well, oil leakage from the port engine (385 127) was unacceptably high. The engine had to be replaced, because every time it was started a significant quantity of oil leaked from the air channel. There were also problems with the turbine waste gate/dump flaps, which refused to close properly. There was thus no standard turbine speed on the ground, and each engine switched from low- to high-altitude supercharger at a different height. Furthermore, all engines experienced problems with sparkplugs (240 ET 7 I-B), which fouled constantly.[396] Measurements of de-ice air temperatures were made on 21 May 1944, and revealed that the leading edges were much too hot. On 24 May a radio check flight was carried out, and the aircraft was flown again on 26 May. Transfer of the aircraft to Rechlin was discussed during a meeting on 6-7 June. The V3 was obviously in Rechlin for performance flights in

Externally the Ju 388 V3 was practically indistinguishable from the V8. The Ju 388 V3 also had just one spinner well into August. EADS

the first half of June, remaining there until about 14 June. The exhaust fairing reinforcements and port wing fuel tank were to be installed when the aircraft returned, and it was also envisaged that the most urgent modifications—the so-called "A-Modifications"—would also be carried out. The plan for the subsequent flight test program included performance flights, wing de-ice measurements, and testing of the FA 15 hydraulic system.[397] Like the V2 and V7, the aircraft had been equipped with the newest, folding version of the FA 15, but unlike the other machines no periscope fairing was envisaged.[398]

The aircraft was still unserviceable on 15 June; one day later there was a second blown tire. Performance flights were carried out between 16 and 18 June. In the period from 19 to 24 June 1944 flights were carried out with the raised fairing behind the engine ("little boat"); however, there was no improvement in the vibration previously reported. Initially none of the measures tried, such as shortening the exhaust pipes, brought any improvement. The exhaust system was therefore redesigned, with metal panels covering the engine compartment.[363]

On 19 June 1944 a flight was made that included a climb to 11000 meters at 260 kph. The flight was also used to evaluate a hydraulic fluid cooler, a heavy metal cooler built by Junkers Kalifor ("Jukalor"), to determine if it met performance requirements. At full counter pressure of 100 atm the fluid temperature was not supposed to exceed the maximum allowable level of 90°C in 45 minutes. With the tail turret hydraulics engaged, this was exceeded within 10 minutes without a cooler. It was found that the heavy metal cooler was almost too large. This was confirmed by a low-level flight at 340 kph on 23 June 1944. Consideration was then given to a light metal cooler with a lower degree of effectiveness (and significantly less weight).[399]

On 21-22 June 1944 the port engine developed a problem. Between 25 and 30 June trials were carried out in an attempt to improve flow over the upper surface of the wing. These included removing the fairing from the exhaust elbow, which resulted in a significant improvement. There was still flow separation on the landing flaps, however. Shortening of the exhaust and removal of the "little boat" fairing resulted in a further deterioration of the airflow. The exhaust system was therefore modified once again. According to *Herr* Wilhelm, this modification included the use of the exhaust pipes previously tested at Oranienburg.[363] According to statements by Preuschen, the work should have been completed on the night of 27-28 June; however, no further performance flights were undertaken with the V3 until 30 June. The aircraft was also cleared for a performance flight on the following day; however, on 2 July the exhaust was destroyed during a performance flight. The aircraft was declared serviceable on 5 July, and was supposed to be ready to fly again at noon on 7 July.

From 10 to 12 July 1944 the V3 was tested daily at Dessau, and was also reported serviceable in the days that followed.

Test flights with the V3 to eliminate the vibration or the airflow separation that caused it were so promising after the conversion that on 7 July Dzulko ordered that all other prototypes were to be modified like the V3 as an interim measure. The aircraft at Merseburg were also to be modified if the opportunity presented itself.[400]

In a memo dated 22 July Dzulko described what was obviously the ultimate solution to the problem, which was to be applied to all previously delivered aircraft as soon as possible. The fairing over the exhaust pipe on the nacelle extension was to be cut off. This modification was relatively simple, and could be performed by fitters of the field service. As well, a completely new exhaust fairing was designed. It consisted of an engine cowling outlet and a cap. For improved servicing, several parts of the nacelle extension that had previously been fixed were made more easily removable.[401]

As previously mentioned, the de-ice air for the wing de-icing system was too hot. To solve this problem, a new Y-branch pipe was developed for the internal air intake with cold air mixing. This was to be introduced into production in August 1944. The new version of the Y-branch pipe was installed in the V3 in July 1944. Temperatures were now found to be right at the upper limit, and a climb at climb and combat power had yet to be made. As speed measurement flights were under way, on or about 25 July a decision was made to postpone these in order to spare the engines. Despite modification of the exhaust fairings and improvement in cooling air flow volume, performance figures for the V3 were worse than those for the V5. To date, however, it had not been possible to find a reason for this.[402] As well as in the V3, the new Y-branch pipes had also been installed in WNr. 300 008. At the end of July 1944 the aircraft was at Rechlin, and the KdE had designated it for de-icing tests at Munich-Riem.[403]

On 15 July the V3 was test flown by von Lossberg, during which it attained an altitude of 8000 meters and a speed of 550 kph.

On 19 July the aircraft's engines were inspected. It was planned to also check the port oil tank, which constantly leaked oil, and the horizontal stabilizer mounts. As the forward fuselage tank was not supposed to be filled(!), it was ordered removed. Finally, it was envisaged that the access hatch in the aft fuselage would be made accessible, and that measuring points for supercharger air temperature be installed in both engines.[404]

The port engine was destroyed during an endurance flight on 22 July, but the aircraft was serviceable again by noon on the 24th. On 29 July, in a flight that lasted from 1830 to 2023 hours, Flugkapitän Wendel reached an altitude of 11000 meters. Wendel flew the aircraft again the next day, from 1058 to 1138 hours. A flight was also logged on 31 July (pilot unidentified).

Further changes were made to the exhaust system from 3 to 7 August 1944. During that time a brief test flight was carried out on the 6th with Wendel at the controls (1623 to 1659 hours), and another the following day (1943 to 2005 hours). The aircraft was supposed to be made ready to fly on 8 August between 1000 and 1030 hours. The only known flight on that day, however, took place from 1721 to 1735 hours, with Wendel flying. The next day Wendel again flew the aircraft at Dessau (1449 to 1600 hours, 9800 m), but one or two days later it was again taken out of service for more modifications. Wendel made another short test flight in the V3 on 12 August (1941 to 1951 hours). The next day the aircraft was fitted with the engines from the V2. Static tests were supposed to be carried out the day after that, and on 15 August it was to take part in a demonstration together with the V1 and V6; however, in the end only the V1 and V6 flew. The next day the V3 developed an engine malfunction.

At the end of August, Dipl.-Ing. Kaspar of the Technical Office advised that the Ju 388 would have to switch from the originally planned Patin autopilot to the Siemens system. The modern Siemens K 12 autopilot had already been tested in a Ju 188 at the *E-Stelle* Rechlin. The system originally caused serious problems until the servo was modified. As the K 12 autopilot was earmarked for the He 219, Ta 154, and Me 262, there was little production experience to fall back on. Installation of the K 12 as a two-axis autopilot (course and lateral control) was completely new, and had never been tested. On 24 August, following consultation with Siemens, the flight department proposed that a K 12 autopilot be installed in a Ju 388 L production aircraft as soon as possible, but with a Patin PKZ 14 directional gyro. The ZK 12 two-axis autopilot, which was intended to be used with the bomb sight during the run-up to the target, was to be installed in the Ju 388 V3 as quickly as possible for testing. Siemens made available an additional fitter, and as well as the two LRM 13 servos (modified) delivered two LDK 11 damping gyros with housings, two LMK 12 E servo amplifiers, an airspeed indicator, one each of LEG 2 and LKW 12 isostatic and booster devices, the LKW 5, and the 19-5821 A servo switch. It was planned to install the aileron servo in the cockpit and the one for the rudder in the fuselage, as usual. Junkers tried to procure a Type 19-5820 A artificial horizon erection switch, an Anschütz 7e artificial horizon, and a Patin repeater artificial horizon from another aircraft or the laboratory. In addition to the directional gyro, the GDU 110 transformer, PSH 17 autopilot control switch, repeater, and master compasses (PfK/f 2 and PfK/m), and the LRg 3 and LRg 15 course selectors were all to be taken from the Patin system already installed in the aircraft. The directional gyro, however, had to be replaced with a B 2 model.[405]

On 24 August 1944 the aircraft entered Assembly Hangar 203, probably for modification or repair work, and was not ready to fly again until the evening of 31 August. Readings were taken from the machine during the next two days, but it is not known if these were in-flight measurements. On 9 September the aircraft was flown to its performance limits. At the beginning of September engines 385 194 (20.5 flying hours) and 385 264 (6.5 hours) were installed in the aircraft.[406] On 12 September static runs were carried out with the Ju 388 V3 and Ju 188 V11 (BMW 801 MG) to check engine performance.[407]

Installation of the flame damping system began on 21 September 1944, and on 25 September both the V3 and V7 were cleared

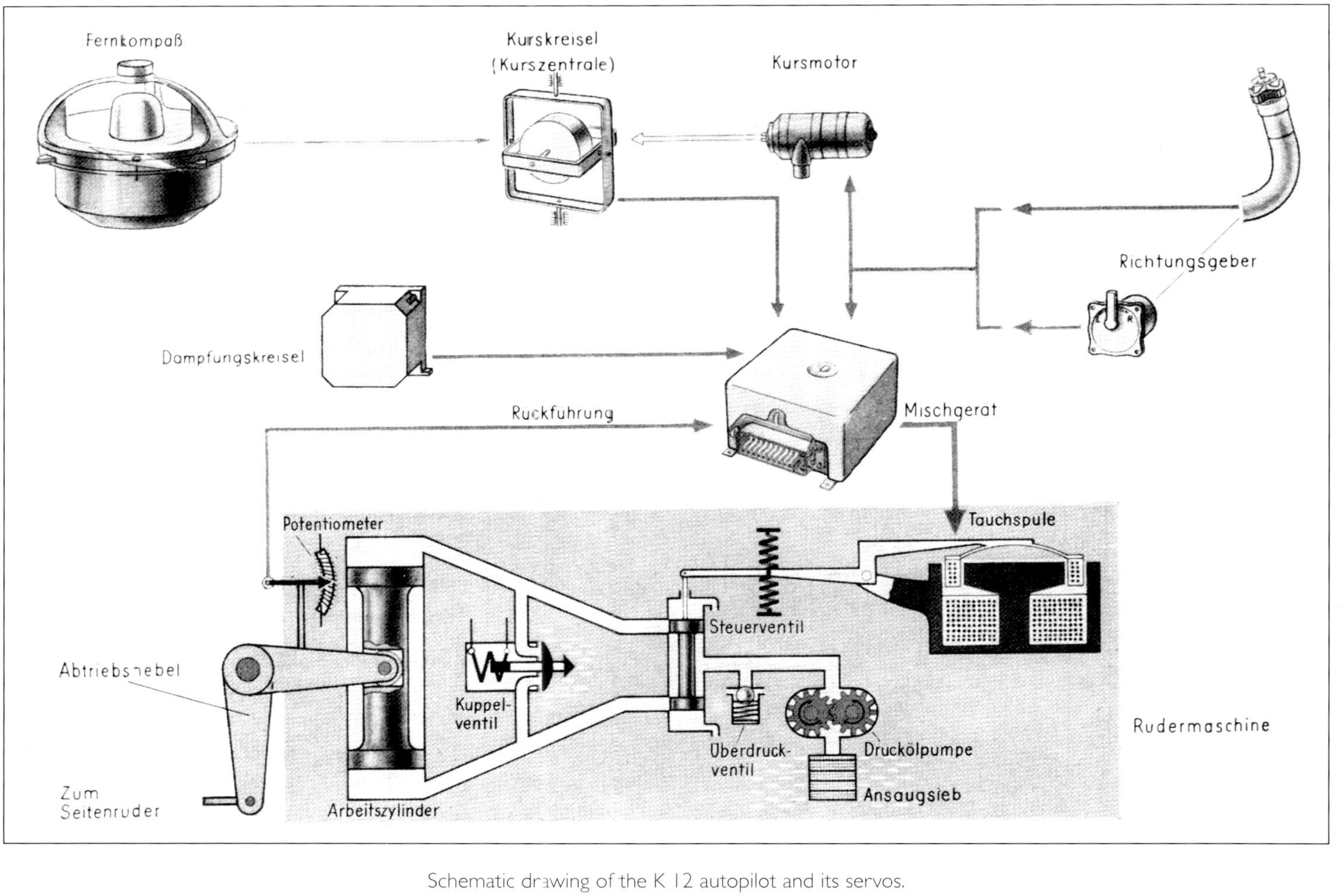

Schematic drawing of the K 12 autopilot and its servos.

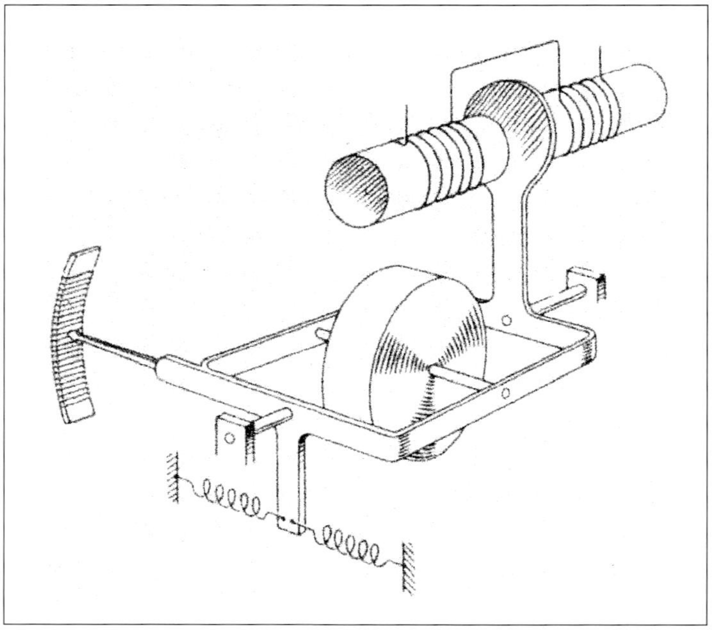

Functional diagram of the K 12's rate of turn, or damping gyro. The gyro itself was damped by an eddy current brake; its signal (rate of rotation about the vertical axis) was measured by a potentiometer.

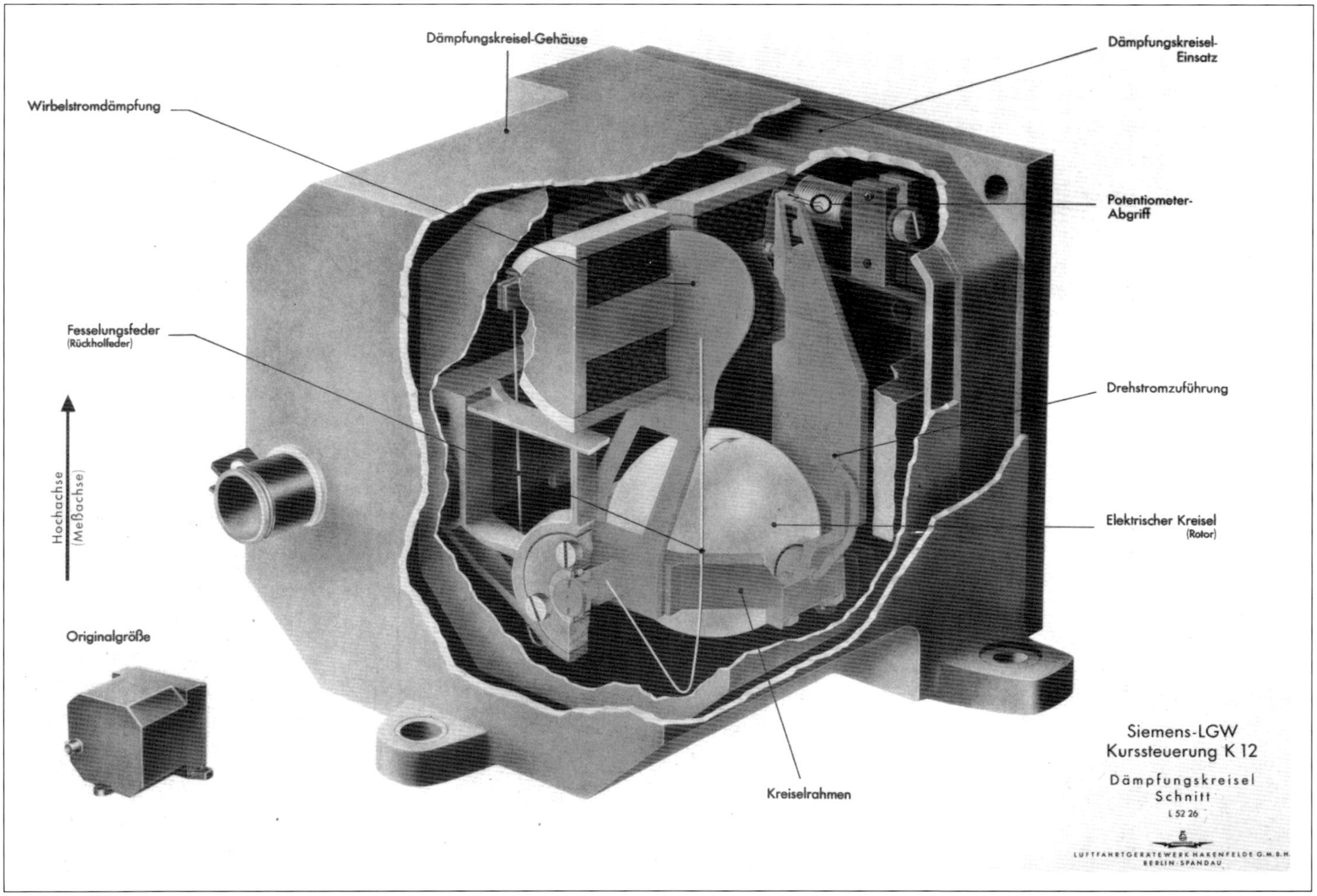

Perspective section through a damping gyro. The spherical rotary current motor is seated in the outer gimbal, at 90° to the gimbal axis. The lever on the right side of the outer gimbal supports the potentiometer arm.

for night flight. The next day a new tank was installed in Fuselage Bay 2. In view of this work, it is not surprising that the flight times for the two engines reported by Ing. Schabert of BMW on 29 September were the same as at the beginning of the month (see above). One must assume, therefore, that the aircraft was not flown during that period.[408]

The flight department's plans envisaged that the Ju 388 V3 and Ju 388 V7 would be used for night trials with the flame damping system beginning in early October 1944. As well, there were still plans to install the Siemens two-axis autopilot, and the aircraft also had to be prepared for the upcoming high-speed trials. In addition to a complete check of the aircraft, a system of sensors had to be fitted, and installation of the tail turret and the FA 15 completed. The planned series of flights was to be used to monitor the performance of the FA 15 system and assess the aircraft's flight characteristics. In particular, it was hoped they would explain the reduction in longitudinal stability at high-altitudes and mach numbers (M = 0.8). Afterwards the V3 was to be transferred to Rechlin.[409]

The aircraft first reappears in Preuschen's notebook on 14 October 1944 with the comment "trim." Dusk flights were carried out from 20 to 23 October, and on the 25th the aircraft was cleared to fly to Rechlin. There the aircraft was test flown on 8 November to evaluate two different types of flame damper. The V3 subsequently crashed on landing as a result of pilot error.[410] Ju 388 WNr. 340 098 was supposed to serve as its replacement at Dessau. Curiously, this aircraft is always referred to as "K0-098" in all BMW records. It is puzzling why a K-0 should suddenly appear in a series of L-1s.

Junkers Ju 388 K-1/V4

This aircraft, which was assigned the *Werknummer* 500 004 and the manufacturer's code PE+ID, was built at Dessau. Construction was based on the fuselage and wings of a Ju 188 E-1. Construction supervision had already criticized the poor workmanship on several converted fuselages, and it found the situation no different when it examined the V4 fuselage, which was completed on 21 March 1944. Damaged frames, defective riveting, and damaged stringers

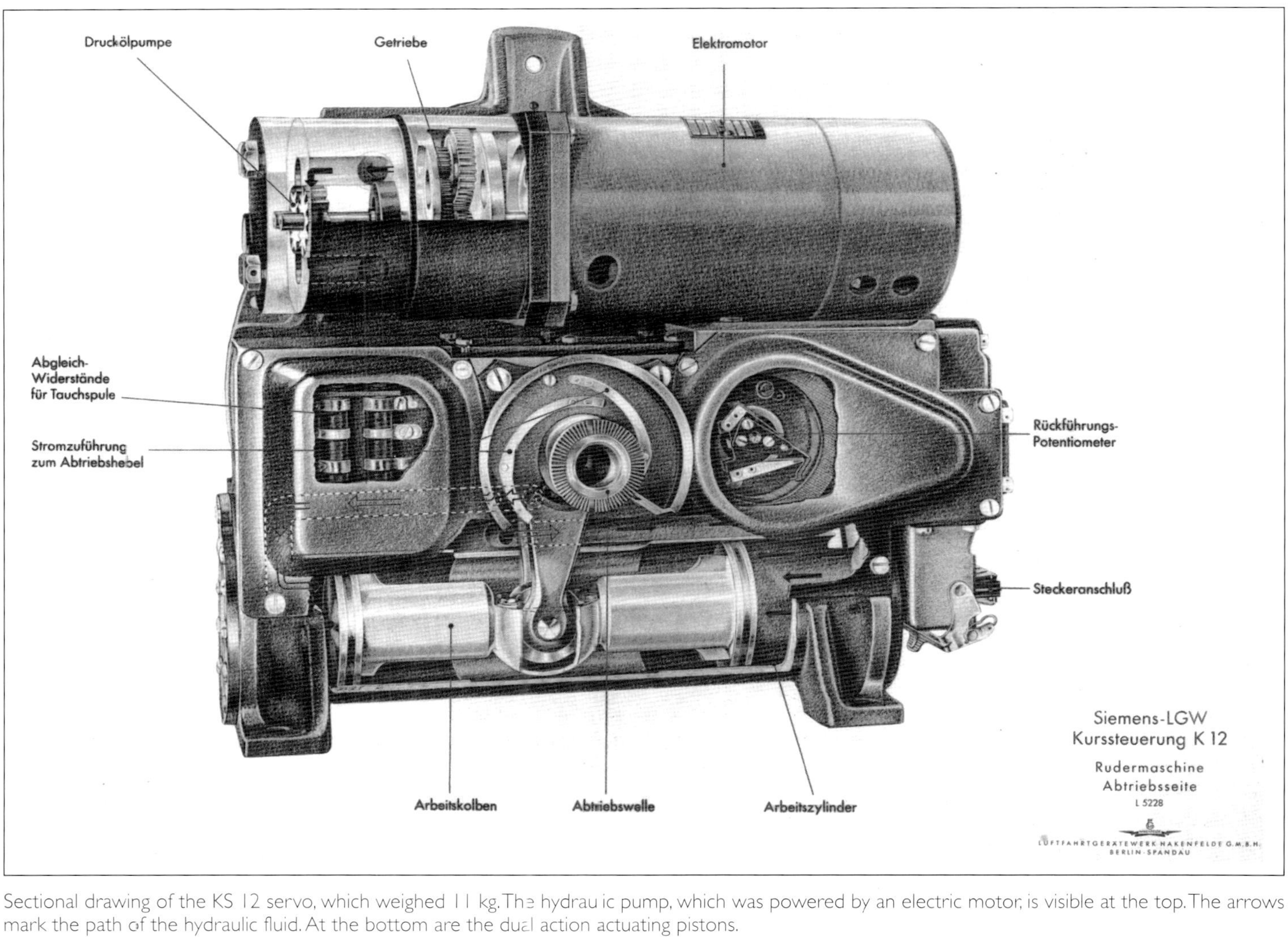

Sectional drawing of the KS 12 servo, which weighed 11 kg. The hydraulic pump, which was powered by an electric motor, is visible at the top. The arrows mark the path of the hydraulic fluid. At the bottom are the dual action actuating pistons.

and outer skin were common, repetitive errors. As well, the rear of the Ju 188 fuselage on which the conversion was based was misaligned.[411] These errors are a clear indication of who was working in aircraft production at that time. Direct reference is made elsewhere to the fact that conversion work on the fuselages of Ju 388s was being "carried out in an irresponsible fashion by (Russian) prisoners" because of a shortage of skilled workers.[412]

Among other things, the Ju 388 V4 was to be used to test the Patin two-axis autopilot envisaged for the Ju 388 K-0/K-1. The two-axis autopilot was a development of the autopilots already in widespread use that held an aircraft on course by automatically moving the rudder. The two-axis system operated both rudder and ailerons, and provided control about the vertical and longitudinal axes to maintain a selected course. According to a summary of modifications to be carried out to the prototypes, the two-axis autopilot was at first envisaged only for the V4.[413] A second damping gyro regulator (127-367 B) had to be installed after Patin modified the system.[414] As well, a toggle switch was to be installed in all aircraft to shut down the rotary current gyro when the autopilot was not switched on, thus avoiding unnecessary wear.[415]

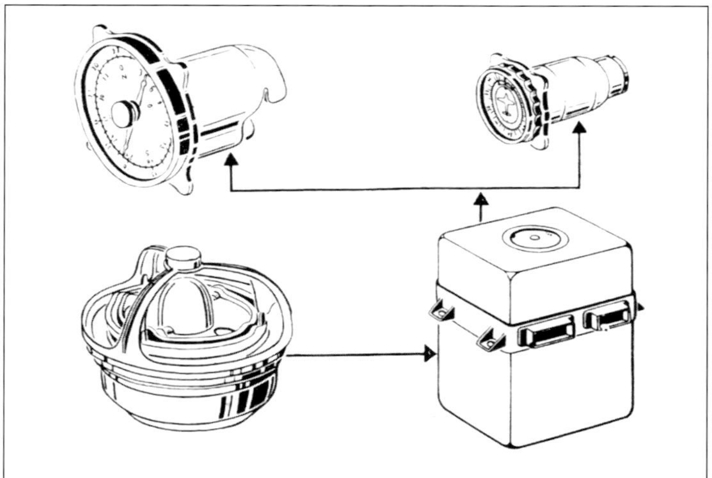

The Patin directional gyro (bottom right) was an electrically operated gyroscope that was controlled by the master compass. Not only did it steer the indicator roses in the repeater compasses, it also served as signal transmitter for the Patin, or the Siemens K 12 autopilot.

Cockpit of a Ju 388 at the beginning of September 1944. To the right of the control column is the central panel, with all the instruments needed for instrument flying. Beneath the combined artificial horizon and turn-and-bank indicator is the master compass indicator, which was fed by the Patin directional gyro. To the right of the instrument panel are the mounting rack and controls for the FuG 217 radar scope. The scope itself has yet to be installed. Griehl Collection

Visible in this photo, which was taken from the entry hatch, are the blind flying panel, the FuG 217 display, and an American radio set to the right of the DF repeater compass. Vernaleken Collection

Director Thiedemann congratulates Ing. Ernst Seibert and Dipl.-Ing. Kurt Heintz (2nd from right) in front of the record setting Ju 88V5. On 19 March 1939 the crew set a new world record by covering the 1000 km course to the Zugspitze and back with a 2 tonne payload at an average speed of 517 kph. Vernaleken Collection

The Ju 388 V4 was supposed to begin flying on 13 June 1944. It was fitted with the new bomb doors, modified hydraulic fluid pump, and bombsight, but lacked the latest modifications for automatic aileron control. Subsequent flight trials were to be used to test the Lotfe 7 H bombsight. At the conclusion of the tests it was planned to fit the aircraft with a new exhaust fairing, the cooler for the climate control system, and new individual valves instead of the rotary slide valve. At the same time the automatic aileron control was to be "put in order" in preparation for flight trials. The most important modifications, or so called "A-Modifications," were to be carried out at the same time.[397] These included the replacement of leaky astro panels with ones that had been checked and certified.[398] The aircraft final report dated 15 June 1944 reveals that when accepted on 14 June the aircraft was fitted with the BMW 801 TJ-0 engines 385 212 (port) and 385 129 (starboard). Because of poor quality control during production, the 10 mm gap between the bomb doors called for in the plans was not achieved. Consequently, the aircraft's maximum speed was initially restricted to 350 kph at a yaw angle of 6°. As well, the FHL 131 Z/1 A tail turret was not installed in the V4, nor was the FA 15. It may therefore be assumed that while the Ju 388 V4 had the rear fuselage of the Ju 388 K-1, the cutout for the tail turret was probably covered over with sheet metal. The climate control system was found to be unsatisfactory, both in terms of design and installation. As well, because of missing parts (shrouds, nozzle exit tubes), the propeller de-ice system was only completely installed and ready for use as far as the cooling fan.[416]

The aircraft completed one maintenance test flight between 16 and 18 June 1944.

Flugingenieur Heintz, who had obviously flown the machine, noted several deficiencies in the V4 on 19 June. The elevator forces were excessive, the hydraulics too sluggish, and the ailerons were 5° off center. As well, the landing flaps vibrated and there were problems with the entry hatch. Heintz also criticized the missing propeller spinners and a malfunctioning oil temperature indicator. He ordered the aircraft counterbalanced, and the window screws fitted with rubber seats like the V6.[417]

According to Preuschen, the aircraft was flown again on 21 June, apparently to test the autopilot. A high-altitude flight took place on 24 June, possibly to gather information on use of the bombsight at high-altitude. The port engine was replaced in the period from 25 to 30 June. The aircraft subsequently made a high-altitude flight to 11700 meters, and finally the aircraft was made available to Patin for testing of the two-axis autopilot.[363]

Preuschen noted a "simulated bombing run" on 2 July, and according to his writings the aircraft was also flyable on 5 and 6 July. On 12 June the V4 entered Hangar 203 for conversion. The next day it underwent a thorough examination by inspector Wähler. The outstanding modifications, along with a few complaints (such as damaged right wheel cover, rusted compressed air lines on the access hatch) were noted in a three page memo. In addition to minor changes to the controls, the installation of auxiliary oil tanks was also envisaged. Work on the port wing was required for installation of the FuG 101, and the much criticized climate control system was finally to be modified, with the rotary slide valve replaced by compensating and butterfly valves. Furthermore, the automatic aileron control was to be hooked up and a function test carried out. These jobs should, in fact, have been completed long before. Also seemingly worthy of mention are planned modifications to the fuel tank heads, installation of the outboard fuel tank in the port wing, installation of reinforced tailwheel doors, removal of the tailwheel fender, and modification of the exhaust fairings as per the Ju 388 K-1/V3. On verbal direction from the design bureau, the bottom part of the extended nacelles was to be altered. Probably the most important measure was the replacement of the bomb doors with the new version with widened trailing edge. This allowed the flight restrictions imposed by the Statics Design Bureau to finally be lifted. The aircraft obviously had a relatively complete offensive armament system, for the report mentions damage to the bombsight cap fairing, and

"heat tests with BC 1000 in the presence of inspectors" were ordered under the heading "SO-Modifications."[418] Originally, the program of modifications was to have included connection of the wing de-icing system; however, the Series Production Design Bureau subsequently pointed out that this must not happen until BMW had installed the so called Y-branch pipes in the engines; otherwise the de-ice air would be too hot (approx. 280°C). As well, it was clarified that bleed air for the climate control system was to be taken from behind the intercooler.[419]

The work obviously went on for a while, for on 24 July Preuschen noted that "the V4 is to arrive today." Two days later he was considering scheduling the aircraft for performance testing.

The work was completed on or about 29 July. The V4 now had a complete climate control system, an auxiliary fuel tank (425 l) in the port wing, and new exhaust fairings. New, better fitting bomb doors had also been fitted, and the gap with the doors closed was now just 8 mm.[420]

Flugkapitän Rupprecht Wendel took the V4 up on 31 July 1944, and in a flight that lasted from 1554 to 1710 hours reached an altitude of 10800 m. He also flew the aircraft from Dessau on 1 August from 1844 to 1851 hours. Minor repairs or modifications were obviously required, for on 31 July Preuschen noted that the V4 would probably be ready by the coming Saturday or Sunday.

While there are no official documents concerning testing by the V4 in July and August 1944, Flugkapitän Wendel's logbook reveals that other test flights were carried out from Dessau:

04 Aug. 1944:	1053 – 1124 hours
05 Aug. 1944:	1753 – 1830 hours
08 Aug. 1944:	1605 – 1640 hours
11 Aug. 1944:	1719 – 1747 hours
11 Aug. 1944:	1918 – 2021 hours
14 Aug. 1944:	1030 – 1139 hours, 11000 meters achieved.

According to Preuschen, on 5 August the aircraft developed engine trouble. This was obviously not serious, as the aircraft flew again three days later at the latest (see above). The flights on 10 and 11 August were used to measure performance, and one day later it was discovered that the port engine was leaking oil. The high-altitude flight on 14 August was used to determine maximum speed. It is not certain, however, that the aircraft reached its calculated maximum speed of 580 kph (see Chapter IV) on this day.

It was originally planned that the Ju 388 V4 would be assigned to Rechlin until 19 August 1944 in place of the damaged V5. Obviously the aircraft's power plants (or those earmarked for it) were damaged by enemy action prior to that date.[421]

On 29 August 1944 385 129, which had begun its career as the starboard engine in the V7, was removed from the Ju 388 V4 at Dessau. The sparkplugs were constantly fouling, the supercharger was leaking oil, and even though the cooler and forward cowling had been replaced, the old problem with high oil temperatures still existed. The high-altitude supercharger lost boost pressure, and as replacement of the boost pressure regulator brought no improvement, the cause was suspected to be slippage in the high-altitude supercharger coupling. The coupling was therefore to be checked for slippage prior to disassembly; furthermore, supercharger sealing was to be closely inspected. At the beginning of September 1944 the V4 was fitted with engines 385 196 (11.5 hrs flying time) and 321 226 (7 hrs),[422] and at 0900 hours on 7 September the aircraft was ready to fly again. In a flight lasting about one hour, the V4 climbed to 10 km at maximum allowable continuous power, and then to 10.9 km at combat power before beginning a power off descent. At an altitude of 9 km the port engine began leaking oil from the vent, and the flight was subsequently aborted at 10.9 km. On this occasion the starboard engine was fitted with the new Fimag intercooler. Although the flight was aborted, it was determined that with the new intercooler the supercharger air was 30°C colder than with the standard cooler at combat power at 10500 meters. It was as yet uncertain whether the oil loss was coming from the tank or the engine, but the subsequent oil consumption measurements, 50 and 35 liters per hour, were not surprising. The V7, on the other hand, lost no oil, even under the most adverse conditions. On 10-11 September 1944 the Ju 388 V4 was fitted with a new engine, and the landing flaps from the Ju 188 V11, which was obviously grounded the same day. This engine change is confirmed by a BMW field service report, according to which the aircraft was equipped with engines 321 219 (5 hrs) and 321 226 (13 hrs) on 29 September 1944, both with Fimag intercoolers. The aircraft thus flew just six hours in September.[423] The V4 was serviceable again on 15 September and flew one day later. Another flight is known to have taken place on 23 September.

At the end of September the aircraft was magnetically compensated. The master compass was installed between Frames 24 and 25, the most favorable location magnetically. Nevertheless a deviation of 1.5° was observed when the U11 DF transformer was switched on. Compared to other Ju 388s, only very minor deviations were observed from the OK 42 emergency compass, which was mounted on the connecting rod in front of the control column.[434]

At the beginning of October it was planned to use the machine to determine maximum speed with the new intercoolers. It was originally envisaged that the aircraft would be handed over to Rechlin with a functioning two axis autopilot, but apparently this did not happen.[425]

Records concerning flight testing of the machine in October 1944 are rather sparse. Preuschen mentions a serviceable V4 on 14 October, and the aircraft was flown twice on the 15th, a Sunday. During the second flight there was a drop in boost pressure from the starboard engine. The problem apparently lay in the boost pressure control. One of the V4's engines underwent modification work of some type between 21 and 23 October. The aircraft was flown again on 25 October and 3 November. The cooling fan developed a problem on or about 12 November, and the unit was replaced on the 13th. A routine 25-hour check was carried out at the same time. On 18 November both engines were removed from the V4, and one of them, WNr. 321 219, was installed in the V7. The V4 was supposed to receive new engines with improved supercharger sealing prior to assignment to the Dessau flight department to serve as an experimental aircraft for performance measurements with the new Fimag intercoolers.[426] On 9 December the apparently brand new engines 387 837 and 387 776 (no flying hours) were installed. In addition to improved supercharger sealing, they were also equipped with a new exhaust system. The external supercharger air intakes had just been installed. Flights to measure cooling performance were supposed to begin after installation of the new intercoolers.[427] Measurements were taken on 10 December 1944, and the aircraft was supposed to fly the next day.

As there are two drawings of unknown origin showing what is claimed to be a Ju 388 J-1 V4 with a faired-over tail turret and Morgenstern nose cone, the authors wish to state emphatically that the surviving original documents from December 1944 still refer to the Ju 388 V4 as a K-1.

While there are no clear indications of any flight testing during the rest of December, the V4 is known to have flown twice on 9 January 1945. Why the aircraft was in Merseburg on 13 January 1945 is as much a mystery as the story behind the notation "intercooler to Oranienbaum" (referring to the V4) found in Preuschen's notes on 25-26 January 1945.

According to Preuschen, the V4 had to wait at Dessau for parts longer than expected—until mid-February—1945, and it was finally supposed to be flyable again on Monday, 19 February. By then, however, the Armaments Staff had ordered the cancellation of the Ju 388, and on 20 February, after removal of the intercoolers, the V4 was grounded, together with the V7 and WNr. 230 151 (KS-TA).

Junkers Ju 388 L-1/V5

Built at Dessau and assigned the *Werknummer* 500 005 and manufacturer's code PE-IE, the Ju 388 V5 was the first aircraft of the L-1 series. Like the V4, construction was based on the fuselage and wings of a Ju 188 E-1. According to the aircraft final report of 26 July 1944, whose wording and contents were almost identical to that on the V4, the aircraft was equipped with BMW 801 TJ-0 engines with the serial numbers 385 189 (port) and 385 195 (starboard). It had strengthened camera doors, but no bomb door actuators. In the closed position the fit of the camera doors, which did not follow the fuselage contours, left something to be desired. The V5's cameras were installed in an experimental oblique arrangement. Drive motors for the powered frame advance were not available, however.

The V5 also lacked both the FHL 131 Z/1 A tail turret and the FA 15 system. Cutouts for the periscope were, however, present in the upper and lower fuselage shells. The notched Ju 388 rudder and the characteristic thickening of the L-1's aft fuselage are, however, clearly visible in the several poor quality surviving photographs of the aircraft. It appears that a metal fairing was fitted over the opening for the tail turret.

The Ju 388 V5 also ran into fit problems when the pressurized cockpit was attached to the fuselage. Washers had to be used to compensate for gaps of as much as 15 mm. The engine exhaust fairings were produced from a pattern drawing dated 5 May and fit the airframe. A 425-liter rubber fuel tank was installed in the port wing in place of the unprotected tank. The engine firewalls and tailwheel doors were still the earlier versions without additional strengthening. The wing de-icing system was not hooked up because of the problem of excessively high de-ice air temperatures; instead, the air was vented directly into the open.[428] The electrical system underwent a system check by the BAL on 22 April 1944.[429]

Several outstanding complaints and minor design changes were supposed to be addressed before the aircraft was handed over to the flight department at the beginning of April 1944. In addition to changes in the climate control system's rotary slide valves and locking of the jettisonable canopy section, it was anticipated that a function check of the pressurized cockpit and climate control system would be carried out. Also to be determined was whether the aircraft could be flown with the fuel tank in the forward fuselage bay. As well, talks were to be held with the design bureau to determine whether the external thermometer and identification light could remain on the entry hatch, or had to be moved.[430]

Work on the exhaust pipe of the Ju 388 V5. This photo was taken in July 1944 after an extension was added to improve flow over the wing. Note the so called "little boat."

On 4 May 1944 the port engine of the Ju 388 L-1/V5 began leaking oil heavily, because something was wrong with the turbine oil venting. *Werknummer* 300 004 had encountered the same problem on 12 April 1944. A few days later, on 11 May, the throttle valve refused to close, even with the throttle lever fully closed. The engine had to be shut down on account of this problem with the control system. In addition, on several aircraft an abnormal rise in oil and supercharger air temperatures had been observed above maximum boost altitude. Furthermore, turbine boost pressure fluctuations of 0.1 atm were observed in new engines.[431]

Installation of cameras took place in mid-May. The antenna mast, which vibrated in flight, was supposed to be modified at the same time. As well, the drives for the rudder and aileron trim tabs, which froze up at high-altitude, were reassembled with cold proof grease. A reminder was also sent to adjust the pressure tensioners in the cockpit glazing, and to order the ultimate propellers for the performance flights to be conducted at Rechlin.[432]

In mid-July complaints about the V5 were to be addressed, and the most urgent modifications carried out. The installation of new exhaust fairings was also envisaged for this aircraft, and the climate control system's rotary slide valves were to be replaced with separate valves. Once performance flights had been completed, the V5 was to be delivered.[397]

At the beginning of July, after a number of modifications had been completed, the V5 resumed flying activities. Time constraints and procurement difficulties meant that several complaints had gone unaddressed, and several planned installations had not been carried out. A mockup was installed instead of the actual cooler for the climate control system. The pressurized cockpit leaked, and operating pressure dropped within just 10 minutes. The principle causes for this were the opening for the life raft release cable, the emergency exit, and the non-return flow valves in the wing. The fuel tank (8-4318 C) installed in the first fuselage bay was a makeshift solution, for its use during flight operations was restricted(!). Dives and turns were forbidden when the tank was used. The camera drive motors were also still absent, as was the DF repeater compass.[433]

An initial "lighting session" was held on 6 May, resulting in a number of criticisms.[434] On 14 July another cockpit lighting test was carried out with the V5, which was later to conduct night trials at Rechlin, in the presence of Flugkapitän Matthies, *Herr* Schlader of Junkers, and Inspector Knebel of the BAL. With the exception of the undercarriage and flap indicators, illumination was found to be adequate with an electrical system working voltage of 22 V. Matthies planned a night flight for a final check of the instrument lighting before "finalization of the lighting" by the design bureau.[435]

At the end of July the Ju 388 V5 was supposed to be transferred to Rechlin for performance flights. Before this, however, a certain amount of work remained to be done at the request of development control under Prof. Hertel. It was planned to bring the exhaust system up to the latest standard and install propeller spinners; if necessary, the spinners could be taken from the Ju 388 V1. The V1's outside air thermometer was also to be installed on the V5. BMW planned improvements in the oil cooler sealing, and both engines were to be fitted with sensors with which to measure supercharger air temperature.[436] The aircraft was also to be fitted with the sensor probe from Ju 188 WNr. 0051, plus two engine tachometers. The corresponding instrument leads were to be run to the cockpit.[437]

On 1 August 1944, however, the Ju 388 V5 was damaged while taking off on a measuring track flight. The port undercarriage leg suddenly retracted at an indicated airspeed of 150 to 170 kph.[438] As it had been raining heavily, the aircraft slid along the ground on its port undercarriage gondola and wingtip "without serious resistance," until it came to a stop at a right angle to the runway. Pilot Wendel reacted quickly and shut down the engines. Consequently, the only damage to the port engine was the propeller, with no damage to the propeller shaft. Fortunately, the other damage was so minor that it was initially hoped that the aircraft could be made flyable again within 10 days, provided replacement parts were procured in time. The next day, after a night outside in heavy rain, isolation was measured at just 10,000 ohms. Based on this, a short circuit caused by a jolt during taxiing was seen as the only possible cause of the uncommanded retraction.

Unfortunately, this rather poor quality photograph of the Ju 388 V5 is preserved only on microfilm. The notch at the base of the rudder is clearly visible, however. Vernaleken Collection

Following repairs, at the end of August 1944 the aircraft was handed over to *Erprobungskommando* 388 at Rechlin. On 28 August 1944 Heinz Lehmann flew the aircraft from Dessau to Lärz.

Junkers Ju 388 L-1/V6

The Ju 388 L-1/V6, WNr. 500 006, had the manufacturer's code PE+IF, and was the second prototype for the L-1 series. It was also built using modified major components from a Ju 188 E-1 (fuselage cell, wings, and tail surfaces). The aircraft was powered by two BMW 801 TJ-0 engines, with WNr. 319 058 installed in the port wing and 305 213 in the starboard. The pressurized cockpit to fuselage joint was not a precise fit, and rounded washers and the adjustable cockpit struts were used to compensate for this when the two components were joined. The clevises on the struts did require reworking, however. The undercarriage could not be fitted with the standard struts because of missing parts. The wing de-icing system was not hooked up, because at 280°C exhaust temperatures were too high. The exhaust gases were instead vented into the open from

The instrumented probe on the starboard wing is a typical feature of an aircraft undergoing flight testing. At the end of the fuselage one can just about see the fairing covering the opening for the FHL 131 Z turret, which was not installed on the V5. Vernaleken Collection

The canopy has already been covered with a tarpaulin. From this angle it is possible to see the bent propeller tips. Curiously, the company photographer gave the date of the accident as 31 July 1944. Vernaleken Collection

Another view of the crash landed V5. The notched rudder and bent propeller blade tips are also visible in this view. The fate of the original photographs is not known. Vernaleken Collection

Undercarriage and firewall of the Ju 388 V6, which was still under construction at the beginning of June 1944. The prototype number (V number) and serial number are both marked on the undercarriage leg. Griehl Collection

The cockpit of the Ju 388 V6 during construction in late May 1944. Some instruments and the control yoke are still missing, however, the port side console with throttle and turbine control levers is already complete. Griehl Collection

the intended attachment point at the exhaust collector. Plans called for the installation of a horizontal stabilizer with a light metal leading edge; however, only stabilizers with steel leading edges were available. Also different than envisaged, the aircraft was fitted with a 425-liter unprotected fuel tank in each wing for improved safety during takeoff.[439] From the beginning the aircraft was equipped with the air cooler for the climate control system and the new exhaust fairing.[397]

A series of urgent modifications were supposed to be carried out on the V6 in mid-June. These included the installation of camera heating, and modification of the bomb bay side doors. As well, the hand crank and the universal-joint shaft for operation of the camera doors had to be reinforced.

Because the V6 was one of the first machines to be fitted with a periscope, FA 15, and production tail turret, the "Priority A" modifications concentrated on these installations. Provision had already been made for the FA 15 control box to fold out of the way; now a periscope fairing based on a design by Meister Teichmann was installed. The mount for the FA 15 and the periscope were still to be modified, as was the FA 15 shaft layout.

The V6 was also one of the first aircraft to be fitted with four separate valves for the climate control system, instead of the rotary slide valve. Two butterfly valves—also based on a Teichmann design—were installed, together with two Type 19-3416 A high-speed

camera drive motors were not available.[441] It is not surprising, therefore, that the Ju 388 V6 was not delivered to the testing station in Tarnewitz until 22 August 1944. Prior to this, some test flights were carried out in the course of factory testing. Rupprecht Wendel is known to have flown the aircraft on 12 August (1119 to 1146 hours), 14 August (2037 to 2057 hours), and 15 August (1740 to 1806 hours). Wendel made his last two flights in the aircraft on 20 August, just prior to delivery: one in the morning (0836 to 0922 hours), and one at noon (1215 to 1245 hours).

The aircraft's subsequent history will be described in later chapters dealing with testing at Tarnewitz and aircraft captured by the Allies.

The V1 and V2 Night fighter Prototypes

The prototypes of the high-altitude Zerstörer and night fighter variant were the Ju 388 V1 (WNr. 500 001, PE+IA) and V2 (WNr. 500 002, PE+IB). Other prototypes were at least at an advanced stage of construction. We have dedicated a separate chapter to the J prototypes because of the clear design differences between them and the K and L series, even though there were overlapping areas concerning testing of engines, the tail turret (not present on the Ju 388 V1), and pressurized cockpit.

On 18 April 1944 an attempt was made to compensate the Ju 388 V1's emergency compass, but no suitable location for it could be found. It was therefore decided to demagnetize the Ju 388 V2's forward and center canopy, the armor plates, and the pilot's seat to prevent a recurrence of the problem. In the event that Junkers was not equipped to carry out the demagnetization, it was planned that Opel should carry out the procedure.[442]

Firing tests were carried out to determine the effects on the cockpit and its equipment from the vibration and shock waves produced when the guns were fired. The weapons housed beneath the fuselage were fired past the bottom of the pressurized cockpit. In the first series of tests, a total of 600 rounds of 20-mm and 83 rounds of 30-mm ammunition was fired past the airframe shell, which was under just 0.2 atm of overpressure. In March 1944 another series of tests was planned to determine the behavior of the airframe under pressure (0.2 atm) with ventilation valves and entry hatch in place. The armament again consisted of two MG 151/20 and two MK 103 cannon, with the flash suppressors set horizontally. The time required for pressure to fall from 0.2 to 0.1 atm was measured before and after firing, once with the pressurized cockpit completely sealed, and once with pressure reduction through the valves. The procedure began with the firing of one MG 151/20, then both; the process was then repeated with the two MK 103s. Finally, all four weapons were fired simultaneously. Altogether 2,100 rounds of MG 151/20 and 1,100 rounds of MK 103 ammunition were fired, including 1,200 and 800 rounds simultaneously. The experiments were complicated by problems with the MK 103s, which frequently jammed or broke down. It was found that the airframe and entry hatch were incapable of standing up to the stresses imposed by firing the guns. An abrupt drop in pressure was observed when the MK 103s were fired, and finally the cockpit began to leak. While rivets and spot welds in the pressurized cockpit held, several profiles at the bottom of the fuselage structure failed. Structural reinforcement was definitely required there. The entry hatch locking mechanism was also found to be unsatisfactory. The entry hatch sprang outward so far that drilling chips and drilled rivet heads

The only known photo to provide an overall view of the Ju 388 J was taken in autumn 1944, and depicts the Ju 388 V2, PE+IB, which was retrofitted with external ram air intakes for the turbosuperchargers. Haubner/Misna Collection

reached the seal. With no internal pressure, the hatch simply sprang open after fifty rounds had been fired from each MK 103. Complaints about the pressurized cockpit included loose pressure tensioners, broken moisture traps, and a floor window completely covered in powder residue.[443]

After the cockpit structure had been reinforced, in April 1944 the components most sensitive to vibration (autopilot, valves, climate control system) were tested again, but still without the periscope and the blast shields for the MK 103s. The armament again consisted of two MG 151/20 and two MK 103 cannon with horizontally aligned muzzle brakes. All four weapons were fired simultaneously in short bursts of up to 30 rounds, and this time a total of 200 rounds per MG 151 and 150 rounds per MK 103 was fired. No new damage could be found on the airframe structure; however, the entry hatch was still incapable of standing up to the imposed stresses. It still sprang outwards when the guns were fired, causing the pressurized cockpit to become unsealed. As well, both fairing strips were torn, and the spot welds of the added profiles failed, causing them to break. Furthermore, the window pressure tensioners did not hold, and six simply dropped off during firing of the guns. Consequently, the method used to attach the windows had to be changed. And last of all, the left MK 103 frequently stopped firing, and the ventilation valve group was found to be leaking.[444]

Obviously as a result of these tests, it was subsequently decided not to equip the Ju 388 J-1 with the MK 103 at first. The first fifty production aircraft and the V1 were to be armed with just four MG 151/20 cannon (Waffensatz 388 W11). The installation of Waffensatz 388 W12 (two MG 151/20 and two MK 108) was to begin with the 51st production aircraft and the Ju 388 V2.[445a]

In May 1944, therefore, the Ju 388 J-1 V1 was used to conduct firing tests with Waffensatz 388 W11, the alternate armament variation. The objective was to investigate the behavior of the pressur-

Ju 388 J-1 pressurized cockpit under construction at the end of March-beginning of April 1944. This cockpit was probably used for the V1 and the tests to determine the cockpit's ability to maintain pressure when the cannon were fired. Griehl Collection

ized cockpit (0.2 atm) with glazing and all accessories (except periscope) with the guns suspended beneath the fuselage. The tests followed basic firing trials with the weapons. The number of rounds fired was kept relatively low, on the one hand because only the tightness of the cockpit with glazing and entry hatch was to be tested (behavior of the airframe as such was determined in parallel tests), and on the other because the aircraft was to begin flight testing as soon as possible. During the tests the gunner and observer were in the cockpit, which was under 0.2 atm of pressure, and short bursts of about 20 to 30 rounds were fired from all four guns simultaneously. A sudden drop in pressure of about 0.01 atm was observed during firing, but they were not quite sure whether this was caused by the unseating of the windows, or the access hatch. No damage at all could be found on the airframe after a total of 1,600 rounds had been fired (4 x 400), the only effect being the loss of some putty. The cockpit's pressure integrity had even improved somewhat, probably because the shaking received during the firing tests had caused the windows and seals to settle somewhat. The glazing was at risk, however. The window pressure tensioners loosened when the cockpit was under pressure, and in some cases fell out when the guns were fired. The instruments were not tested after the firing tests. Based on the earlier tests no damage was expected, and a thorough check of the aircraft before the start of flight testing was standard practice anyway.[445] After the entry hatch was reinforced, the tests were continued from 10 July to 14 August 1944. As the type of periscope to be used in the Ju 388 was not available, a dorsal periscope from a Ju 288 was installed instead. The periscope fairing was lengthened to match the later type by adding an extension between the glazed cupola and the periscope. The periscope was then covered with a teardrop-shaped fairing made of sheet metal (1.5 mm). Before the periscope was installed, the seal around the entry hatch was checked. As the hatch locking mechanism had come open during firing (just two rounds from the MK 103s), the hatch handle was secured with 0.8-mm wire. This simply sheared off after 40 rounds had been fired from each MK 103. In the next step the outer handle was secured with a Lyra spring, the inner handle by means of a counterweight, and the locking mechanism's lever arm with a tension spring. Success was not achieved until a 200-gram weight was used on the inner lever arm, and after 50 rounds from each MK 103 all was in order. The fairing did, however, prove to be too weak; two rivets were torn from a deformed part, and the metal fairing cracked in two places.

The periscope fairing—intended to protect the periscope from being struck by stones during takeoff and landing—was still not equal to the loads imposed by the MK 103, being literally torn up when the guns were fired. It was removed, and tests with blast shields for the MK 103s continued until the forward mounts were torn from the airframe. The tests had finally revealed that either the periscope fairing or cannon blast shields could be installed, but not both. The design would have to be changed once again.[446]

On 18 May 1944 the flight department compiled a report entitled "Ju 388 J-1/V-1 acceptance flight complaints to be addressed by the design bureau." Like all Ju 388s the aircraft vibrated, especially in high-speed descents. According to the pilot, visibility for landing was worse than that of the Ju 88, and he asked for the installation of the promised bad weather window. It was felt that padding on the left corner of the armor plate was an absolute necessity. As well, he asked that armrests be provided, and the gunsight mount moved in the event that "the curious control yoke is retained for the J-series aircraft." The yoke collided with the trigger for the lower weapons, and also blocked the instrument panel. Also criticized were the incorrectly wired FuG 101 and the missing emergency compass.[447]

At a meeting on 6-7 June 1944 it was decided to use the Ju 388 V1 for performance flights. On 6 June Preuschen also mentioned that the V2 had been damaged. This was obviously the result of enemy action, for in a memo dated 30 June 1944 Prof. Hertel complained about the completion target date, which had been extended to the end of July because of enemy action. In Hertel's view this was unacceptable from a development standpoint; therefore, he asked his associates to "also strive to have the V-2 ready to fly as soon as possible."[448] In any event, work on the Ju 388 V2 was still in full swing in mid-June 1944. Frames 10 and 11 had to be reinforced on both the V1 and V2, probably as a result of the firing tests. As well, the box spars and fittings for the ammunition boxes were supposed to be changed. The planned strengthening of the Ju 388 J-1's armor plate from 1.5 to 2 mm could be omitted for the time being.[413] The previously mentioned file memo "Ju 388 Prototypes: Work Program" dated 11 June 1944 said of the V2: "Complete fuselage with armament, so that fixed armament and tail turret can be tested."[397] The FA 15 control system installed in the V2 was already at the latest standard, and the periscope fairing based on the Teichmann design was fitted as in the V6.[398]

The work program envisaged the installation of the Ju 88 control yoke, after which the aircraft was to be test flown. Also anticipated was the firing of 1,000 rounds per gun. Further changes were planned afterwards.[397]

- Installation of climate control system cooler in wing,
- Installation of the new engine exhaust fairing,
- Modification of the gunsight installation, and
- Implementation of A-Modifications.

Delivery would finally take place once performance measurements had been completed.

In the second half of the month the V1 carried out autopilot test flights; however, the autopilot was initially unserviceable because of the missing gyro. The V1 was also used for exhaust system tests. Between 25 and 30 June 1944, after the fitting of deflectors (probably to shield the engine compartment as on the V6), test flights—or at least static tests—were carried out (according to Preuschen, the V1 was supposed to fly on the evening of 25-26

The V2 was photographed extensively at the beginning of November 1944. From this angle the ventral weapons tray with two MG 151/20 and two MK 108 cannon is plainly visible. EADS

June). These revealed that flow on the wing upper surface behind the exhaust system was unsatisfactory without the added bulge and the "little boat" fairing. Wool tufts had been glued to the surface to observe flow behavior. When the flaps were retracted, the tufts on the rear two-thirds suddenly leapt forward 180°. As a result of these tests, it was decided to modify the V1's exhaust system based on experience with the Ju 88 V93 and V94, which were equipped with TK 11 superchargers.[363] Parts from the V93's exhaust fairings were installed on the Ju 388 V1. The aircraft was flown with the modified exhaust system between 2 and 8 July 1944, and no problems were encountered with flow over the wing upper surface. The main advantage of this "TK 11 solution" was probably that the flow of exhaust gases was diverted away from the wing, and thus was unable to interfere with the boundary layer flow.[449] On 2 July 1944 Preuschen noted that at high-altitudes the aircraft was "in order with the TK 11," with no malfunctions. On 6 July the aircraft was obviously used to measure maximum speed with antennas in place, and three days later modifications were carried out. In a memo dated 12 July, Prof. Hertel characterized the report on the speed measurements with the new exhaust system and SN-2 antennas (which, unfortunately, we do not have) as "good news." While the aerodynamically optimized antennas had resulted in a speed decrease of just one to two kilometers per hour, Hertel nevertheless encouraged the flight department to continue working on the SN-2 array.[450] The V1, V3, and V7 were all test-flown on 11 and 12 July 1944. On the following day the V1 was ready to fly at 1000 hours, and later was flown to 9000 meters. On 14 July the port engine developed a problem with cylinder No. 10.

The Ju 388 V1 required much modification and repair work in July 1944. This was carried out in Hangar 201 beginning on the 15th. The upper surfaces of both wings had been damaged by enemy action, as well as the port wingtip and the starboard landing flap. The tail section had also sustained significant damage. The upper surfaces of both horizontal stabilizers were damaged, and on the port side there was also damage to the bearers and stringers. Both elevators, including trim tabs, were dented, and the trailing edges were deformed, as was that of the rudder. The fuselage end cap was badly dented, and there was damage to the fuselage be-

Close up of the nose of the Ju 388 V2 with SN2 antennas and the air intake for the cockpit's forced air ventilation system. Haubner/Misna Collection

tween Frames 21 and 22. Overall the aircraft's technical state was not good. The port relief tank was blistered, the mainwheel doors had been holed, and the inner skinning of the tailwheel doors had separated in places. All of these components had to be replaced. As well, there were loose control cables in the fuselage and vertical stabilizer. One of the SN-2 antenna masts was defective and due for replacement; the antenna was to be permanently connected at the same time.

Along with the repair work, the V1 was supposed to be brought up to the latest technical standard. First, streamlined wooden fairings were to be fitted to the antenna mounting posts for improved aerodynamics. Also planned was the installation of the new climate control system, now to be connected behind the intercooler. Modifications to the engine exhaust fairings and engine ventilation were also envisaged. The first aid box in the fuselage was replaced by a first aid bag in the cockpit, and the necessary fittings had to be installed. As with the other prototypes, the oxygen valves and hydraulic hand pump were to be relocated. The electrical fittings on the control yoke were missing, as was the reflector sight with mount.[451,452] At the end of July 1944, the type supervisor requested the demagnetization of the canopy and steel parts of the Ju 388 V1 prior to delivery. Junkers initially limited itself to a promise to determine if this was even possible without significantly postponing the delivery date.[453]

Firing range tests took place at the beginning of August, and on the 9th the V1 fired 500 rounds. Two days later a maximum speed of 405 kph was measured with SN-2 antennas—not exactly a sparkling performance for a heavy fighter. Flugkapitän Wendel flew the V1 on 10 August 1944, taking off from Dessau at 0854 hours and landing again at 0936. He also flew the aircraft on 14 August from 1516 to 1710 hours. On 11 August the V1 was at Merseburg, probably for work on the radio or oxygen systems. On 15 August the aircraft took part in a demonstration together with the V6, and it was flown again the next day.

In mid-July it became apparent that the V2 had to be fitted with a heating system for the guns. The necessary documentation was delivered from Prague at the beginning of August. If it proved impossible to install the heating system by the aircraft's "hangar deadline"—meaning the planned completion date—the components were at least to be prepared so that installation could take place between flights, or during the night shift. As well, the aircraft's antenna mounting posts were to be fitted with streamlined wooden fairings like those of the V1. Dzulko also felt it necessary to state explicitly that the V2's canopy and armor plate must be demagnetized.[454] The V2 went to the firing range at the end of August, and tests continued until mid-September. On 24 August the aircraft fired 500 rounds from its MK 108 cannon. At the beginning of September the V2 was at Dessau without engines, as there was only one

The production standard periscope fairing based on results of cannon firing trials, seen here on the Ju 388 V2. It was intended to protect the periscope from being damaged by debris during takeoff, but it also had to be able to withstand the vibration created when the guns were fired. Locating the fairing between the barrels provided a measure of protection against the shock waves created by the cannon. EADS

The periscope installation atop the fuselage of the V2 was similar to that of the Ju 388 L. Beyond the antenna mast another variant of the exhaust system may be seen on the upper surface of the wing, the "little boat" solution with flame damping system (FlaV). Note the raindrops on the misted canopy. EADS

Remains of the compass swing platform graduations in front of the former hangars in Dessau could still be seen in 2000. It was here that efforts to compensate the emergency compass probably took place. Handig Collection

spare engine there.[455] In mid-August the V2's engines had been installed in the V3. It must therefore be assumed that the firing tests were conducted without engines. Junkers obviously made the best of the situation, conducting tests for which engines were not a major requirement.

Sometime at the beginning of September work began to install the Ju 88 control yoke in the Ju 388 V1 and V2. Numerous difficulties were encountered. When turned to the left, the yoke collided with the open bad weather window and its silicagel moisture trap, and also with the thumb switch on the starboard throttle lever, no matter what position it was in. When moved forward, the yoke struck the forward armor plates before reaching maximum elevator deflection. The window was subsequently modified, and the thumb switch was deleted. The armor plates were moved forward 30 mm, and the control yoke's grips shortened by the same amount. As well, the control column's neutral position was reset 1° to the rear.

There was too little room between the rudder pedals and the armor plates, making it impossible for a pilot wearing boots to operate the rudder. The V1's armor plates were therefore reworked by about 30 mm, while the bottom edge of those in the V2 were shortened by a similar amount. There were also some wiring errors in the electrical system. If the Kärcher heater was turned on while the flaps were being lowered from cruise to landing position, the cruise position indicator lamp remained on until the Kärcher heater was also switched offÖ.[456]

Finally, at the end of September, it was decided to install in the Ju 388 V2 the engines removed from the Ju 388 L-0 WNr. 300 006 (PG+YF) at Köthen.[457] On 29 September Inspector Wähler inspected the Ju 388 V2 in Assembly Hangar 203, and determined that extensive modifications had to be carried out before the Ju 388 V2 could be released. As well as fitting the strengthened undercarriage and a fuel tank (1 050 l) in the second fuselage bay, these included fitting the latest climate control system air intake in the leading edge of the wing. It was planned to equip the engines with external supercharger air intakes, construct new extended nacelles for the aircraft, and modify the upper part of the engine exhaust fairing, including the mount and exhaust stack fastener. The fire protection sheeting on the upper wing surface, and the hydraulic cylinder and drive shaft for the cooling air flap were removed at the same time. The cooling air flap was blocked in a certain position. After the removal of the hydraulic fluid cooler, the associated opening in the ventral bath was to be covered, and the hydraulic system modified accordingly. A fairing was to be fitted over the tail turret opening. Wähler also clarified that the compressed air for the pressurized cockpit was to be bled by means of a steel tube behind the intercooler, and that both prototypes needed a ladder for climbing onto the wing, plus control locks for all control surfaces.[458] During servicing of the hydraulic filter located on the upper right side of Frame 9, hydraulic fluid leaked onto both the fitter and nearby electrical equipment. Instructions were subsequently issued to replace the filter.[459]

At the beginning of 1944, the flight department's plans called for the Ju 388 V2 to return to the firing range for further tests after all work in Hangar 203 had been completed. 100 rounds were to be fired with each MK 108 with the cockpit pressurized. The gunsight for the fixed armament had to be aligned. After the firing tests the machine was to complete its acceptance flight program. This included autopilot and radio flights, plus flight testing of the FA 15 with dead-man switch and 60-atm overpressure valve in the hydraulic system with no cooler (19-6205). As well, performance measurements were to be made to determine the maximum and climbing speeds at Zerstörer weight. Afterwards the V2, equipped with external supercharger air intakes and, if possible, flame dampers, was to be handed over to Werneuchen for testing of the SN 2 D.[460]

In any event, the V2 resumed its armament tests at Dessau on 13 October 1944. A test run was carried out on or about 20 October, and in the days that followed the V2 was fitted with new propellers. A performance flight followed on 25 October 1944.

As before, compensation of the emergency compass caused problems, the location specified by the *E-Stelle* Rechlin also proving unsatisfactory. At the end of October 1944 the type supervisor, Hptm. Stahl, decided to transfer the Ju 388 V2 to Werneuchen in the coming days "for urgent installation of radio equipment and performance flights"—probably testing of the SN 2 D. These investigations thus had to be broken off, to be resumed when the Ju 388 V1, which was back at Dessau, had been fitted with production-type armor plates.[461,462]

At the end of October 1944 Hptm. Dr. Stahl and Hptm. Bonow, commander of the trials unit, demanded that the Ju 388 J-1's propeller pitch indicators be moved from the engine nacelles to the cockpit, as they could not be seen at night. This proved a simple matter, for all that was required was replacement of the mechanical position indicator with an electric transmitter.[463]

After the V1 had been sent to Tarnewitz, on 12 October 1944 *Erprobungskommando* 388 took charge of the aircraft. It had apparently done little flying at Tarnewitz because of a shortage of fuel.[605] The aircraft made one brief flight of about an hour on each of the next two days, with deputy type supervisor Alfred Kuhn at the controls. The first flight, which lasted 1 hour and 7 minutes, produced no complaints apart from a loose connection in the oil temperature gauge. The second flight (1 hr, 3 min) was plagued by engine trouble, about which nothing more is known. On 15 October the port engine's cooling fan failed, and had to be replaced; consequently, the next flights did not take place until 17-18 October. The first lasted 2 hours and 3 minutes, the second 39 minutes. There were apparently no significant problems. The V1 contributed significantly to the high attrition in cooling fans, for the next day another fan disintegrated while the engines were being run up. Finally, on 21 October 1944 the V1 was apparently ferried direct to Dessau in a flight that lasted 27 minutes. There the aircraft was to undergo modifications that would continue beyond the end of the month.[464] Those at Dessau were already waiting for the aircraft, Preuschen having noted its return in his notebook on 24 October. As the Ju 388 V1 was to be used for performance measurements, priority was given to modifications that would influence performance. As well, the machine was to undergo a refit program to bring it up to the latest standard. According to "Ju 388 V1 Work Program,"[466] dated 17 October 1944, external supercharger air intakes were to be installed, the gap around the engine oil cooler sealed, and a gap in the area of the ailerons covered. Additional skinning on the wings and flaps installed to protect against heat from the exhausts could be removed. After completion of flame damper tests using the Ju 388 V3 and V7, the configuration planned for production was to be removed from the affected aircraft and installed on the V1. Seams were to be filled and puttied in preparation for the planned performance measurements. Finally, it was also planned that the SN-2 antenna should be replaced by a Morgenstern antenna, with a wooden nose cap, provided that the necessary design bureau documentation was delivered in early November as planned. Otherwise the conversion would be postponed.

Before anything was done, however, on 2 November 1944 the aircraft underwent a thorough acceptance check by Inspector Wähler in Assembly Hangar 203.[466] Among other things, he discovered that the exhaust collector on the starboard engine (WNr. 385 169) had been destroyed by a fire after just 43 hours of operation. Pieces of the burned collector lay in the turbine, which would have to be checked for damage by BMW. The fire had also affected the metal skin of the fixed undercarriage fairing. In addition to installing a new collector and repairing the visible damage, all lines in the area of the fire had to be removed, checked, and repaired. The defective collector was to be sent to BMW. Several lines on the engine, which interestingly had flown in the Ju 388 V7 in spring 1944, showed signs of chafing, and numerous clamps and screws were loose. The port engine, WNr. 385 127, had begun its career as the V3's port engine, and had been replaced in May 1944 because of oil leaks. Now the outer upper intercooler segment was defective, and the inner sides of the exhausts stacks were torn. The pipes that ejected hot air into the open (a makeshift solution after the exhaust gases were found to be too hot to allow the wing de-icing system to be hooked up) were burned. With the installation of the external supercharger air intakes on both engines, which did not yet have the 40-tonne attachment points, the rods that operated the cooling air flaps could be removed.

There was also airframe damage. Numerous rivets had popped on the lower edge of the canopy near Frame 5, and there were cracks in the skin, and several cross members. As a result, the bottom of the canopy was to be strengthened. In addition two canopy windows were loose, the fairing over the ventral weapons was wrecked, and a circular access panel on the port tailplane had been torn off. The trailing edge of the outer starboard aileron was destroyed, and the starboard mainwheel was so worn that the fabric was visible. The port collector tank had to be replaced because of damage in the area of the tank head and contents transmitter. Technically the ma-

The rear of the V2's cockpit. Apart from the absence of armor plate for the radio operator, it was identical to that of the Ju 388 L-1. Behind the tracking controls for the FA 15 is the antenna tuning device. EADS

chine was not up to the latest standards, for the list of modifications that had not been carried out contained no less than 39 points, from missing auxiliary oil tanks, to modified retraction struts on the undercarriage, the relocation of the outside thermometer from fuselage to the wing, and new placards for the fuel system, to oil ventilation. Before the aircraft could be handed over to the flight section, the undercarriage—brakes, oleo pressure, tire pressure, and toe-in—had to be checked and, if necessary, adjusted. The hydraulic system was to be checked for possible leaks, and the cockpit sealing was to be tested with and without the climate control system ("depressurized").

The design bureau put off some of these modifications, however, probably to get the machine into flight testing as soon as possible. Among those postponed were the auxiliary oil tanks, reinforcements on the elevators and wings, and improved intercoolers, as the V1 did not yet have them. Other modification work was tied to the availability of materials (retraction struts) or, as in the case of relocating the hydraulic fluid filter, the timely completion of other work.[467] Preuschen next mentions the aircraft on 17-18 December 1944. At that time the aircraft was in Assembly Hangar 203. The V1 was serviceable again at the beginning of January, and on the 13th, after problems with a missing switch panel had been addressed, it was flown back to Rechlin.

There was much commonality between the Ju 388 J and the Ju 88 G. On 14 November 1944, therefore, discussions were held with Hptm. Dr. Stahl, the type supervisor, concerning the installation of the ammunition boxes from the Ju 88 V104 (WNr. 710 612) in the Ju 388 V2, which was then at Werneuchen. The old ammunition boxes would be reinstalled in the V104.[468] Moving the Ju 388 J-1's ammunition boxes and feeds from the first fuselage bay would make room for the 475-liter fuel tank used in some Ju 388 L-1s, resulting in increased range.[469] Plans were therefore made to install the tank in the V2's forward fuselage bay, along with the associated fuel jettisoning system.[470] More detailed plans were revealed on 9 December. The Ju 388 V2 was to be fitted with the ammunition boxes and feeds, plus the 475-liter fuel tank and fittings, from the V104, which was expected to be at the firing range until 11 December. The parts were to be removed from the V104 after the end of the firing program, and forwarded to the DLH Test Center at Werneuchen. After these parts and an oblique armament[471] had been installed in the V2, it was to be transferred to the *E-Stelle Tarnewitz* for weapons testing.[472] According to Preuschen, on 10 December the V2 was initially transferred to Rechlin, which might explain the delay.

The Ju 388 J's radio and radar systems were also installed with the type supervisor on 14 November 1944. Dr. Stahl advised that the Naxos ZD was now to be used instead of the Naxos ZC. Only about half of the aircraft would be equipped with this system, however, as the rest were to be fitted with "Flensburg" at a modification facility.[474]

Testing of the Ju 388 J at Dessau obviously came to an end with the release of the V1 and V2. Testing of the Ju 88 G continued, however. Even after a heavy air raid at the beginning of March 1945, Preuschen's last entry contained a reference to the Ju 88 G-7. Apparently temperature gauges with a greater measuring range were to be procured for specialized anti-Mosquito and standard production aircraft.

Externally there were obvious parallels between the Ju 388 J and Ju 88 G. The armament and radar equipment carried by the two types were almost identical. Haubner/Misna Collection

Pre-Production Aircraft as Prototypes

For many German aircraft developed late in the war, it became standard practice for prototypes to be gradually replaced by pattern aircraft from the production line. There were a variety of reasons for this. For one, it was significantly cheaper to simply replace worn out or destroyed prototypes with production aircraft; for another, the numerous improvements incorporated into production aircraft often meant that further testing with much modified and retrofitted "old" prototypes was no longer worthwhile, nor produced reliable results.

Testing and mass production of the Ju 388 began almost simultaneously, making it the most striking example of this phenomenon. The first prototype to be completed (the V7) was also the first pre-production machine; the "true" prototypes followed later. While aircraft of the L-0 series were involved in basic testing of the pressurized cockpit and power plants—some with prototype (V) numbers, some without—the K-0 series was almost completely converted to other purposes after the high-altitude bomber version was cancelled in principle. Some of the aircraft were converted or built as the "Ju 388 L-0 Nacht" version. The rest were either planned or built as prototypes, and some were used as engine test-beds, or built as prototypes for the J-series. In October 1944 the fixed rearward-firing armament was deleted from all such prototypes and test-beds.[475]

As with the first prototypes, the original plans were modified. Originally, the Ju 388 V11 to V15 were to have been taken from the L-0 or K-0 series to be used as test-beds for the Jumo 213 E.[476] Ultimately, the V11 to V15 were test-beds for the Jumo 222.

Most of the subsequently described prototypes were never completed, though some reached an advanced stage of construction. The Ju 388 V32 (WNr. 300 295) has been intentionally omitted from the description of prototypes with the Jumo 213 E, as its testing, done exclusively by the *Luftwaffe*, will be described in detail in Chapter VII.

Prototypes with the Jumo 213 E

As related in Chapter IV, at Zindel's urging five Ju 388 K-0 aircraft were supposed to serve as test-beds for the Jumo 213 E. In the beginning these aircraft were designated the V11 to V15, and their construction was to be kept as simple as possible. The fuselage was to have no bulged bomb bay, and was to be similar to that of the L-0, but without the camera installation behind Frame 15. The wiring

When production problems with the BMW 801 TJ became apparent, Director Zindel, seen here with a model of the Ju 52, which he designed, advocated equipping the Ju 388 with the Jumo 213.

for the cameras and the tail turret was to be omitted. The only armament envisaged was the MG 131 in the dorsal position. The radio equipment was to be similar to that of the Ju 388 K-0, but without the FuG 217. The main differences from the L-0 series resulted from the use of Jumo 213 E engines. Fire protection for the landing flaps was not needed, and the lines for the hydraulically operated nose ring needed to cool the BMW 801 could also be abandoned, as cooling of the Jumo 213 was controlled by a temperature regulator.[477] Two fuel tanks were to be installed in the fuselage: a forward tank with a capacity of 1220 liters, and a rear tank holding 580 liters. Two fuel tanks were also envisaged in each wing, with capacities of 405 l (inner) and 425 l (outer), and no fuel jettisoning capability.[478] Initially, the Kärcher heater for horizontal stabilizer de-icing was to be dropped. Like the Jumo powered versions of the Ju 188, de-icing of the horizontal stabilizer was to be achieved by warm air from the engines.[479]

As a result of a requirement issued by the KdE, in June 1944 the last three aircraft, at least, were to be equipped so that "an element could be added" if required, which meant a camera installation similar to that of the Ju 388 L-0.[480,481] As the ultimate design documentation for conversion to the Jumo 213 E could not be expected from Prague until 15 August 1944, this meant another delay in delivery of the aircraft, which was supposed to take place from August to October (already a month later than originally).[482] It was subsequently decided to build only the last three aircraft using the design drawings from Prague.[481] For reasons of time, however, the other two machines were improvised. Both were to be equipped with a Kärcher heater, for example.[483]

Apparently, at the beginning of July the five Jumo powered machines were given the amended designations V21 to V25.[484]

The data contained in the construction directive of May, and the first amendment from July, were modified, and stated more precisely in another amendment dated 12 August 1944. Based on decisions made by the RLM and the type supervisor, in addition to serving as test-beds for the 9-8213 D (Jumo 213 E) engine, the Ju 388 V21 to V25 were now also to serve in part as prototypes for the Ju 388 J-3.[485]

Ju 388 V21, WNr. 230 157

The first Ju 388 to be equipped with the Jumo 213 E engine was the Ju 388 V21. A letter from the design bureau (series production section) to the Merseburg factory reveals that it initially planned to equip the V21 with the fifth and sixth Type 8-213 D power plants (Jumo 213 E) delivered by Jumo. As this would have delayed completion of the aircraft until the end of August 1944, the V21 was to be equipped with Jumo 213 A engines as an interim measure, as comparative data with the BMW 801 powered Ju 388 was urgently needed. Through the intervention of Professor Hertel, who

This Jumo 213 installation exhibits clear differences compared to that of Jumo 213 engines installed in the Ju 88 G-6 and Ju 188 A/D. The exhaust pipe cover strip and vertical ventilation louvers were not seen on any other production engine. The photo is from a series taken in October 1944, bearing the caption "Ju 388/20153." Was this another error by the photographer, or did he actually mean the V21, WNr. 230 153? Deutsches Museum

stressed the urgency attached to the Ju 388, it was subsequently decided that the first two Jumo 213 E engines, which had been promised by 29 July at the latest, would be earmarked for the Ju 388 V21. Installation of the Jumo 213 A, which would have taken about 14 days, was made unnecessary. At the beginning of August, it was expected that the aircraft would be weighed and cleared to fly by the 12th.[487] For the most part, it proved possible to hold to this schedule, for on 17 August 1944 Heinz Lehmann took the V21 up on its maiden flight from Merseburg. The next day Lehmann also carried out the radio and autopilot check flight, before the aircraft was flown from Merseburg to Dessau on 21 August.[487a] The first two engines to be installed (numbers 104 161 0031 and 104 161 0038) thus logged about two flying hours before they had to be removed between 21 and 28 August after failing an inspection by the BAL. The crankshafts were not properly chamfered at the main bearing, creating fears of seized bearings. According to Preuschen, a static test by the Ju 388 V21 was planned for 31 August 1944. The aircraft was supposed to fly at noon on 8 September, and two days later an engine was changed. On 10 September engine 104 151 0125 also had to be removed because of low oil pressure, heavy vibration, and brass shavings in the oil filter. The other engine (105 161 0041) also had to be removed, as the supercharger control had become locked in second gear during the flight. Pieces of the laminated retainer had fallen into the engine housing. The engine had only a flown a total of 33 minutes.

On 16 September an engine again had to be replaced. Number 104 161 0085 had to be removed without ever having flown, because at temperatures over 100°C oil pressure could not be maintained within allowable limits. With newly upgraded engines, the aircraft was finally supposed to carry out a static test run on 19 September.[488]

On 23 September 1944 company pilot Gottschalk flew the aircraft and observed its stalling behavior. At 2,700 rpm and 1.40 atm of boost pressure, flow separation occurred between 195 and 205 kph in cruise configuration at an altitude of 2500 meters. If the engines were throttled back to 1,250 rpm, separation occurred between 205 and 210 kph. At the same engine speed, with the flaps and undercarriage down airspeed could be reduced to 165-170 kph before flow separation occurred. The stall itself was preceded by wing shudder; thus, behavior approaching the stall was the same as that of a normal Ju 88.[489]

On 30 September a cylinder head had to be replaced on the Jumo 213 A with the number 104 161 0117 (D-engine No. 5) after six flying hours, because the adaptors for sparkplugs M1 and M2 on the first cylinder were loose. On the same day, after another 3 hours, 45 minutes, another dummy plug on the mechanical transmission began leaking. On 10 October D-engine No. 5, installed in the V21, exhibited the previously mentioned cooling problem. After 10 hours flying time, a 30-mm crack was discovered in the base of the left upper radiator segment's water compartment beneath the take-off connection. Two days later, after 11 hours flying time, cracks also appeared in the right upper and right lower segments. Nevertheless, the aircraft was serviceable again by 1100 hours the next morning. The V21 was flown on the morning of 14 October; then, the left upper radiator segment of the other engine (104 161 0090, D-engine No. 10) was struck by the same problem. As a precautionary measure, three reinforced radiator segments were immediately installed, and the aircraft was supposed to be ready to fly by the next day. As the result of an installation error, on 19 October a fine filter in the oil circulation system was crushed, and a guide tube broke. On 22 October one of the reinforced segments was found to be leaking. Because of this damage the aircraft was unable to fly for at least a day, and according to Preuschen, it did not take to the air again until the 24th. Finally, on 26 October D-engine No. 5 developed radiator trouble again: the bottom right segment was leaking at the clamping point.[490]

At the beginning of November *Herr* Kersiek of the flight department complained that propeller pitch drives had had to be changed eight times during 18 hours of flight testing with the Jumo 213 E. He demanded much improved operating reliability, especially in the face of an engine failure.[491]

The Ju 388 V21 appears in Preuschen's notes again on 6 November 1944. He made reference to "minor defects," and on the following day noted problems with an engine control system. The aircraft was at work on 10 November. On 11 November, after 16 flying hours, the threaded sparkplug mount on No. 1 cylinder of D-engine No. 10 was loose at both sparkplugs; furthermore, the exhaust valve shank was broken. A test run was planned for Tuesday, 14 November 1944. Work was not limited to the engines, however. Installation of the (new?) tail section was also to take place on that day. Then, on 14-15 November the cooling system was removed. A second test run was carried out at 1100 hours on 17 November, and the aircraft was supposed to be ready to fly on the 19th. On that day there were again problems with the cooling system, but these could not be fixed. There was no flying on the 21st, but on the following day the V21 returned to the air. One day later, after 17 hours, 30 minutes flying time, the aircraft's port engine (D-engine No. 5) was found to be operating far below rated power. No. 9 cylinder had no compression because of problems with the intake valve.[492] The next significant notes by Preuschen appear on 10 December, when the supercharger pipe had to be sealed. The aircraft was previously noted as serviceable on 29 Nov., 3 Dec., and 7 Dec. Modifications to the V21 were noted on 13 December, a maximum performance flight was carried out on 21 or 22 December, and loose parts had to be tightened on 5 January 1945. On 8 January minor problems associated with the wings had to be fixed before the aircraft could fly. An engine change took place on 20 January. For several days Preuschen made reference to the machine with the note "coming tomorrow," but in fact it did not resume flying until 4 February 1945. The aircraft continued to be flown even after work on the Ju 388 was halted, probably because it was equipped with Jumo 213 E engines. After problems with the electrical system, the V21 was cleared to fly again on 20 February 1945.[382a] Preuschen's last reference to the Ju 388 V21 appears on 1 March 1945—six days before the big air raid on Dessau—when he noted that the aircraft would be ready on Saturday, 3 March. The aircraft's subsequent fate is not known.

Ju 388 V22, WNr. 230 691

The V22 was envisaged as a test-bed for the Jumo 213 E with no cameras, and unlike the V21 and V23, it was to be equipped with a "special material system" for methanol-water injection.[493] The aircraft's fuselage was to be delivered to final assembly in Merseburg by 12 August 1944 at the latest; at that time, however, there was no firm date for delivery of the power plants, and as yet no completion date had been announced at Merseburg.[494] As the second prototype destined for the Otto-Mader-Werke, equipment was to include circuitry for engine sensors, and a MW 50 system in the second fuselage bay (*Rüstsatz* B7). The *Rüstsatz* itself was ordered from ATG, with delivery to take place at the end of August.[495] The technical documentation for the installation of the methanol-water system in the Ju 388 J-3 was not, however, ordered until 23 September 1944. It was to be created by the Prague design bureau. The injection system was very similar to the one used in the Ju 88 G-7. Low pressure delivery of the methanol-water mixture was again achieved by means of supercharger air, which was bled from a point behind the second stage and the intercooler. One of the main problems when the Ju 388 J-3 was equipped with the methanol-water system was the supercharger air bleed for ventilation of the pressurized cockpit, for poisonous methanol absolutely had to be kept out of the cockpit. Therefore, for combat and emergency power—power settings that might involve active methanol-water injection—a point behind the first supercharger stage, which was in front of the methanol-water injection site, was selected for the supercharger air bleed. At power settings below combat power, supercharger air was to be bled from a location behind the second supercharger stage's intercooler.[496]

Ju 388 V23, WNr. 230 697

This aircraft was envisaged as an engine test-bed for the Jumo 213 E, and was to have been fitted with a Dachlaurin fire extinguishing system in the fuselage and wings.[497] It was completely destroyed in an air raid on Merseburg on 25 November 1944. The aircraft's fuselage was equipped for the installation of cameras.[493] The following overview, like all those that follow, is based on a damage report dated 28 November 1944 that is held in the EADS Unternehmensarchiv in Ottobrunn.

Ju 388 V30, WNr. 300 293

This aircraft was originally built as a standard Ju 388 L-0 pre-production aircraft with BMW 801 TJ-0 engines at Merseburg in July 1944. After several flights one of the engines (385 269) suffered a cooling fan failure, and pieces of the disintegrating unit pierced the cockpit, injuring the pilot. This incident grounded the aircraft for only a brief period, however. Pilot Lehmann and test engineer Lange carried out a speed course flight on 28 July 1944, recording a maximum speed of 429 kph at low level; the day before they had completed a high-altitude flight in the aircraft. On 11 August 1944 the machine was temporarily grounded because of a shortage of BMW 801 TJ-0 engines.

The aircraft was later fitted with Jumo 213 E engines, and on 10 October 1944 the crew of Lehmann and Gödecke test flew the aircraft with its new engines at Merseburg. After two high performance flights by the same crew on 12 and 13 October, on the 15th an engine failed after a burst oil tank resulted in a seized piston. The Ju 388 V30 was flown again on 20 October, however, this time by the crew of Lehmann and Kerksiek. In November the aircraft

The test beds equipped with Jumo 213 E engines probably possessed a marked similarity to the Ju 388V31 shown here and, though some were converted K-0s, were built without the bulged bomb bay. Note the small diameter of the annular radiators. Ott Collection

Staatsgeheimnis!
Geheimhaltungsverpflichtung beachten!

8-388 J mit H-Stand (N) mit Fla V mit Schrägbewaffnung mit Morgensternkanzel.
Bewaffnung: 2 MG 151 + 2 x MK 108 Schrägbew.: 2 MG 151 H-Stand: FHL 131 Z

Zustand:		II.m.Kuto Nase	III.o.Kuto Nase				Weiterentwicklung
Motore:		801 TJ-0 außenl. Ansaugschacht	801TJ-1 außenl. Ansaugsch. Fimag	213 E	213 E 1900 PS Basis	603 E/F	222 E/F
Startleistung	PS	1590+11 kg	1615+11 kg	1750+103 kg	1890+108 kg	2000+ 85 kg	2420+137 kg
Steig- u.Kampfleistung in H = 0	"	1450+ 8 "	1472+ 8 "	1580+ 85 "	1680+ 90 "	1610+ 70 "	2220+126 "
" " (in H km)	"	1370+43 "(12,5)	1430+62 "(12,3)	1270+140 "(10,2)	1430+137 "(8,8)	1550+102 "(7,1)	1720+198 "(10,5)
Sondernotl.+ MW 50 (in H km)	"	–	–	1630+188 "(9,1)	1810+193 "(8,2)	1930+116 "(5,7)	2360+238 "(8,5)
Einsatz:		o.MW 50 o.AB	o.MW 50 o.AB	m.MW 50 o.AB	m.MW 50 o.AB	m.MW 50 o.AB	m.MW 50 o.AB
Rüstgewicht	kg	10485	10485	10410	10410	10510	12000
Besatzung	"	400	400	400	400	400	400
Kraftstoff	"	2490	2490	2410	2410	2410	3130
MW 50	"	–	–	280	280	280	280
Schmierstoff	"	260	260	260	260	260	260
Munition	"	340	340	340	340	340	440
Abfluggewicht	kg	13975	13975	14100	14100	14200	16510
Flächenbelastung	kg/m^2	250	250	252	252	254	295
mittl.Fluggewicht	kg	12820	12820	12750	12750	12850	14945
Kraftstoffvorrat		3280	3280	3280	3280	3280	4220
v_{max} in H = 0 km	km/h	385	402	439	440	432	482
v_{max} (in H km)	"	543 (12,5)	583 (12,3)	576 (10,2)	588(8,8	557 (7,1)	650 (10,5)
v_{Not} + MW 50 in H = 0 km	"	–	–	480	492	486	524
v_{Not} + MW 50 (in H km)	"	–	–	626 (9,1)	633 (8,2	575 (5,7)	678 (8,5)
v_{max} bez.Not + MW 50 ohne Fla V	"	558 (12,5)	598 (12,3)	647 (9,1)	654 (8,2	594 (5,7)	692 (8,5)
Flugzeit (in H km)	h	3,15(11) 3,28 (8)	3,02(11) 3,03 (8)	3,0 (9,2)		3,21 (6)	2,75 (8)
bei $v_{R\ max}$	km/h	495 476	538 520	520 (		508	548
Flugzeit (in H km)	h	3,25(11) 4,45 (8)	3,69(11) 4,45 (8)	3,65 (9,2)		3,75 (6)	3,50 (8)
bei $v_{R\ gedr.}$	km/h	481 404	480 405	434		444	443
w_{st} 2-mot. in.H = 0 bei G_A	m/s	5,84	6,3	7,85	8,61	7,9	9,8
Steigzeit (bis H km)	min	43 (11) 24 (8)	34,5(11) 21 (8)	27,7(9,2)		13 (6)	17,0 (8)
H_D 2-mot. bei G_A	m	11250	12500	11100	10900	9800	11250
H_D 2-mot. " G_m	m	12900	13000	11800	11500	10450	11800
H_D 1-mot. " G_m	m	–	6100	2300	4000	3850	2900
w_{st} 1-mot. " $G_{1-mot.}$ in H = 0	m/s	0,5	1,2	1,4	1,83	1,39	1,95
Roll-/Startstrecke		[illegible]/940	675/940	670/920		630/885	675/950

Junkers Flugzeug- und -Motorenwerke AG. Flugzeugbau Stammwerk Dessau
Bearbeitung Kobü-Entwurf-Leistungsdaten Da./Fr.
Datum 1.12.44
Kennzeichnung Nr.: 388/615 Bl.1
6. Ex.

This performance comparison of the Ju 388 J with various power plants reveals clear performance advantages at low altitudes for aircraft with the Jumo 213 E.

was handed over to the Rechlin test station, and it was still there on 7 February 1945.

Ju 388 V31, WNr. 300 294

This aircraft was also originally built with BMW 801 TJ-0 engines, and was first flown by Lehmann on 25 July 1944; this flight also served as the radio and autopilot check flight. The next day, the aircraft was test flown by the crew of Lehmann and Lange. Like its sister machine, in August 1944 it was grounded at Merseburg without engines.

The aircraft was initially re-engined with Jumo 213 A-1s. On 9 October 1944, during preparations for the first flight with the new power plants, one engine was destroyed during a ground run-up. The crew of Lehmann and Gödecke flew the re-engined aircraft for the first time on 13 October, the flight again serving as the radio and autopilot check flight. Serious problems developed during a high-altitude flight on the following day: one of the engine mount struts had been installed incorrectly, resulting in severe airframe vibration in flight. The aircraft was supposed to be refitted with Jumo 213 Es at the end of October, as soon as the engines were delivered from the *E-Stelle* Rechlin. It is likely that the re-engined aircraft was supposed to be tested there. Photographs confirm that the aircraft was in fact equipped with Jumo 213 E engines at war's end.

Aircraft:	**Owner**	**Description:**	**Damage:**
8-388/230 697 (V23)	RLM	Fuselage with pressurized cockpit, port and starboard wings, tail section completely, undercarriage completely.	100%

As the damage report makes no reference to power plants, it seems quite certain that the aircraft was never flown before it was completely destroyed.

Prototypes with the Jumo 222

The Junkers factory in Merseburg was given the double task of producing engine test-beds for the Jumo 222, and building prototypes of the Ju 388 J-2. Therefore, on 6 October 1944 Dipl.-Ing. Dzulko of the series production design bureau in Dessau turned to factory manager Bräuer to decide on the weapons suite to be installed in these prototypes. Instead of the previously envisaged weapons suite W10 (2 x MG 151, 2 x MK 103), W12 (MK 108s instead of MK 103s) would have to be used, as firing trials with weapons suite W10 and the pressurized cockpit had yet to be successfully completed. In any event, the weapon mounts and ammunition boxes for the MG 151s from W10 could still be used.[498] Other references to the construction of J model aircraft at Merseburg are found in a message from Kovermi Merseburg to the local management dated 26 October 1944. The message contained instructions to immediately halt work on the originally envisaged oblique armament in the forward fuselage bay, and replace it with a similar armament behind Frame 15. The necessary drawings were to be produced by the Prague design bureau. Furthermore, heating was now to be provided for the guns in the ventral tray. The strengthened Type 188.20 or 188.21 undercarriage envisaged for aircraft powered by the Jumo 222 was obviously not available, and an interim solution had to be found. Additional Kärcher heaters had to be installed in the wings of Jumo 222 powered aircraft for the wing de-icing system, with air inlet scoops in the leading edge similar to those for the climate control system.[499]

Ju 388 V9, WNr. 230 695

This aircraft was originally to have been built as a K-0, but then it was selected as the first of six engine test-beds with the Jumo 222. The aircraft was to be delivered to the KdE in January 1945.[500] While camera fittings were present in the fuselage, which was probably similar to the L-0 version without the bulged bomb bay, the installation of cameras was not planned.[493]

An early reference to the Ju 388 V9, with no indication of serial number of variant, appears in Preuschen's writings. On 3 October 1944 he noted problems with the aircraft's engine cowlings.

During a visit to Merseburg on 9 November, Preuschen was surprised to find that so far only the fuselage was being worked on, as the strengthened wings for the aircraft had only arrived on the day of his visit. Also, a strengthened undercarriage still had to be procured. Preuschen promised to provide data on the necessary ballast for the rudder as soon as possible. These had to be cast from concrete, which took considerable time.[501]

On 25 November 1944 the aircraft, located in Hangar II, was damaged in a heavy air raid. Though it sustained not inconsiderable damage, the V9 was apparently repaired, for according to Preuschen it began static tests on 18 December 1944. Static tests by a K-0 would certainly not have seemed worthy of mention to Preuschen if it had been an aircraft with BMW engines.

In fact, in a message to Prof. Hertel on 22 December 1944, Schilling complained about BMW's failure to deliver a short intake tube for the Jumo 222 in the Ju 388 V9. A drawing of the part had to be quickly made up, for "the aircraft's finish date must not depend on this piece of tube." At the same time, Schilling made reference to Merseburg's allegation that missing documentation had affected completion dates. While it is true that the switch for the

Aircraft:	Owner	Description:	Damage:
8-388/230 695	RLM	Fuselage, left and right sides badly dented.	
		Outer skin pierced in several places. Left tailwheel door deformed.	5%
		Pressurized holed in several places	20%
		Wing skin holed in port leading edge, upper surface at bearer I	5%
		Port undercarriage fairing holed in several places	30%
		Starboard aileron pierced in several places	25%
		Fin	25%
		Rudder	30%
		Port and starboard elevators	30%

Kärcher heaters for cockpit and wing heating (the Jumo 222 had no heating) had been mistakenly left out of the electrical diagrams, the necessary drawings were delivered one day after a query to the Merseburg "construction site." Aside from that, according to Schilling the aircraft could have "made its first flights without Kärcher heaters."[502]

The V9 was still at Merseburg on 9 January 1945. Preuschen's references to the aircraft ended on 14-15 January with the comment, "did not fly at Merseburg." The V9 was probably also the Ju 388 at Merseburg, in which a Jumo 222 B-3 converted from a B-2 (No. 48) exhibited excessively low oil pressure. During static test runs to determine the cause, the gearing was completely destroyed due to faulty assembly. The engine was completely reassembled, and was ready for collection on 5 February.[503] With the cancellation of the Ju 388 program, the aircraft was grounded on 20 February 1945.[382a]

Ju 388 V10, WNr. 230 721

This machine was envisaged as the first of four Zerstörer with Jumo 222 engines. Except for the engines, they were to be largely similar to the Ju 388 V2, and were also to be fitted with a tail turret.[504] In June 1944 existing plans called for the cockpit and radio equipment to be the same as that of the V2. While wing de-icing was to be identical to that of the Ju 188 L-0, details of the engine de-icing and control systems were still open.[505]

After the type supervisor's decision to have just one J-2 prototype (the V15) built instead of four, the V10 was one of the three aircraft that was now to receive the BMW 801 TJ. On 5 December 1944 Junkers Merseburg reported to Rechlin that, "a J-0 with production standard equipment, the V10, is available for emergency compass installation trials." These trials were of particular importance, because with production starting, no location had yet been selected for the OK 42 emergency compass. On examination of the aircraft, however, Hauptnautiker Vogel of the *E-Stelle* Rechlin discovered that the windscreen's port armor-glass panel with steel frame had not been installed, nor had the head and back armor. As the corresponding parts were missing, it was not possible to install them at Rechlin. Vogel was thus only able to make a rough assessment of the aircraft's magnetic state, and issued a reminder that all steel and iron parts in the cockpit should be carefully demagnetized. As soon as the V10 or another aircraft was completed, he intended to repeat the measurements and investigate the influence of heated flight suits and other electrical equipment.[506] The BAL inspected the weapons installation on 15 December 1944, also at Merseburg. The aircraft had a fixed armament of two MG 151 and two MK 108 cannon in a tray beneath the fuselage. As the report mentions no other weapons, it may be assumed that the V10, like all K-0s and most L-0s, was not equipped with a tail turret, hence the designation J-0. There were a number of complaints about the aircraft. The weapons tray was not a good fit to the fuselage, the diagrams for inserting ammu-

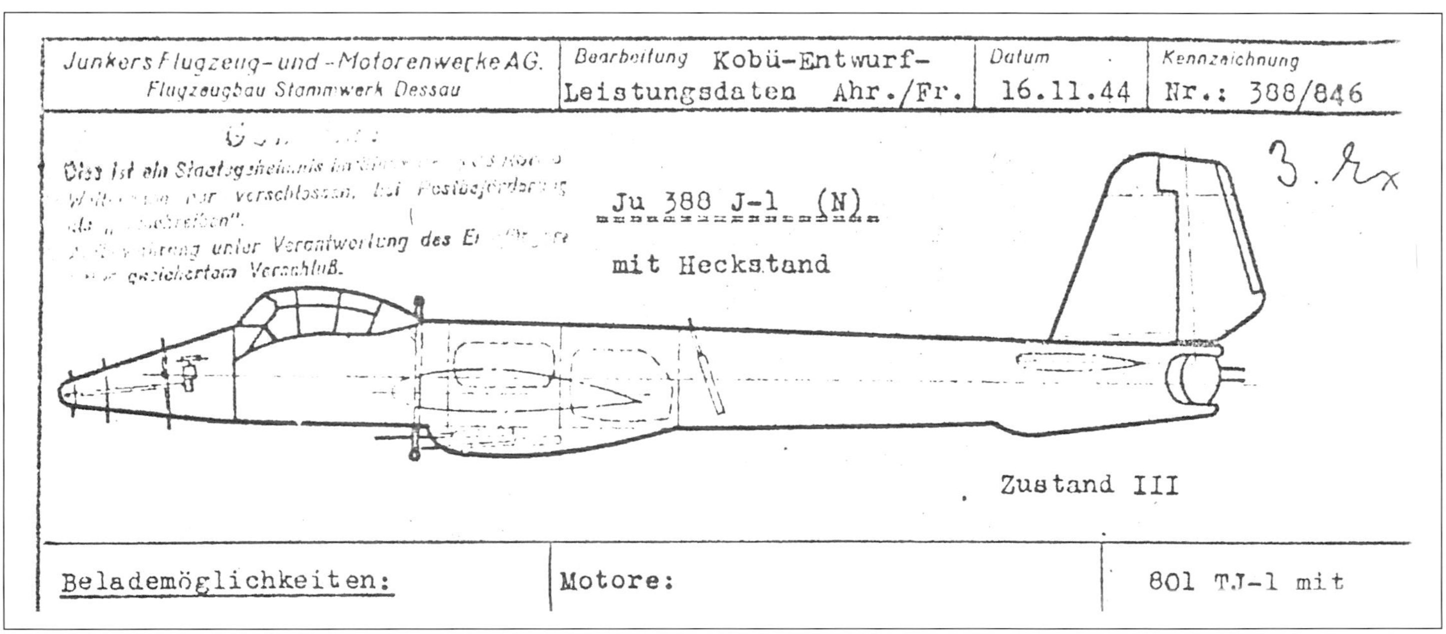

Drawing of the Ju 388 J with wooden nose cone for Morgenstern antennas.

nition into the MK 108s were in the wrong place, and the wrong compressed air system had been installed for the weapons (DHAG 4 instead of DHAG 5). The gunsight installation was also criticized, and was to be moved to within 10 cm of the armor-glass windscreen.

On the inspection date the Ju 388 V10 was also demonstrated on the firing range. During test firing there were numerous misfires (ignition current interruptions) and breakdowns, especially of the starboard MK 108. The firing demonstration was subsequently cancelled. Initially, a specialist was supposed to be requested from Rheinmetall-Borsig to determine if the interruptions were due to mistakes in the manufacturing or installation of the guns. It was also planned to provide the firing range personnel with instruction on the MK 108 and procure spare parts. The BAL would then be advised, and schedule another live fire session to check weapons function and alignment. After discussion with the type supervisor, however, it was decided to deliver the aircraft to the *E-Stelle* Werneuchen as soon as possible, and drop the requirement for a serviceable weapons system. The complaints about the weapons system were only to be addressed as required for firing before and after magnetic compensation. While the weapons were to remain in the aircraft, the armament electrical system was to be disconnected at the circuit breakers.[507]

Pilot Lehmann subsequently made the first flight of the aircraft on 3 January 1945, this also serving as the radio and autopilot check flight. The machine was flown again the next day.[507a] At present, it is not known if the aircraft was transferred to Werneuchen.

Ju 388 V11, WNr. 230 727

The aircraft was envisaged as a test-bed for the Jumo 222, and had fittings in the fuselage for camera equipment. The installation of cameras was not anticipated, however.[493] Construction of the aircraft was apparently at an advanced state when Merseburg was attacked on 25 November 1944. The damage suffered by the aircraft is revealed by the following summary:

Aircraft:	Owner	Description:	Damage:
8-388/230 727 (V11)	RLM	Fuselage, Frame 9 to tail	60%
		Pressurized cockpit, glazing damaged, canopy deformed and airframe holed in several places.	30%
		Port wing, bottom skin and spar in Area I holed, upper skin in Area III at Spar I holed. Bottom skin and leading edge in area of Cross-Members III and V holed.	25%
		Starboard wing, leading edge dented. Upper skin from Area V to end rib holed in several places. Bottom skin in Area III behind Spar II holed.	30%
		Rudder	100%
		Fin	100%
		Tailplane, port and starboard	25%
		Tailwheel doors, port and starboard	25%
		Elevators, port and starboard	20%

Because of the considerable damage, the aircraft was probably not completed.

Ju 388 V12, WNr. 230 733
Like the V10, this aircraft was envisaged as a prototype of the Zerstörer with Jumo 222 engines. The aircraft was under construction at Merseburg in November 1944, and was damaged in Hangar II when it was bombed on the 25th.

In view of the not insignificant damage, it is unlikely that the aircraft was ever completed. The presence of BMW 801 TJ-0 engines confirms that plans to equip the machine with Jumo 222s had obviously been cancelled. Subsequently, the aircraft was one of three Ju 388 J-2s that were to be completed as J-1s with BMW 801 TJ-0 engines at the request of the type supervisor.

Aircraft:	Owner	Description:	Damage:
8-388/230 733	RLM	Fuselage, Frame 15 to tail destroyed	40%
		Port wing	75%
		Starboard wing	30%
		Port and starboard tires	100%
		Port and starboard undercarriage fairings	75%
		Entire undercarriage	100%
BMW 801 TJ-0 387 628 329 011	RLM	Engines removed for thorough check. Engine cowlings destroyed. Shrapnel holes in exhaust collectors, rocker arm covers holed.	

Ju 388 V13, WNr. 230 737
Like the V12, this aircraft, envisaged as a test-bed for the Jumo 222, was under construction at Merseburg, and was damaged in Hangar II in the same raid in November 1944.

The aircraft had fittings in the fuselage for the installation of camera equipment; however, there were no plans to install cameras.[493] Nothing is known of the aircraft's subsequent fate.

Aircraft:	Owner	Description:	Damage:
8-388/230 737 (V13)	RLM	Fuselage 5225, side from Frame 12 – 26 badly crushed. Skin holed in several places.	30%

Ju 388 V14 – V14/2, WNr. 230 901
This aircraft was originally one of the batch of 50 K-0s ordered, but was then changed to an engine test-bed for the Jumo 222. It was supposed to be delivered initially as a *Zerstörer*, like the V10, and was largely similar to the V2.

Following another change in plans, on orders from the type supervisor the first three J-2 prototypes were to be completed as J-1s with BMW 801 TJ engines. While the first two aircraft were obviously to be sister aircraft of the V2, the V14 was apparently given a special role. In mid-November 1944 an MW 50 tank was ordered for the aircraft,[508] followed soon afterward by a wooden nose cone for the Morgenstern radar antenna.[509] It is striking that the construction directives with which these parts were ordered are in reference to the six Ju 388 J-3s planned for the anti-Mosquito role. It is also noteworthy that the V14 was now referred to as the V14/2 with no change in *Werknummer*, not as a Mosquito hunter, but under another contract number allocated to the "Construction of 3 8-388 with BMW 801 TJ engines." As the use of MW 50 with the BMW 801 TJ-0 was not envisaged, one might conclude that the machine must have been conceived with BMW 801 TJ-1 engines. In fact, however, the aircraft was equipped with Jumo 213 E engines.[382a] The new designation V14/2 was probably based on this complete rededication of roles.

The aircraft was still under construction at Merseburg in February 1945. In a telephone conversation between *Herr* Löbnitz (Merseburg) and *Herr* Dzulko of the Series Production Design Bureau in Dessau, it was decided that the aircraft would not be fitted with the wooden nose cone for the Morgenstern antenna after all. It would instead be fitted with conventional "antler" antennas.[510] A corresponding amendment to the construction directive was issued on 2 February 1945. From this one can see that the aircraft was to be given a night camouflage finish as per drawing S 18800-6005. Another indication that the aircraft was to be used as a night fighter is the directive to install upwards firing armament at an angle of 80° behind Frame 15. Due to unavailability, both the flame damping and fire extinguishing systems had to be abandoned. The envisaged propellers were three-blade VS 9s with 9-30372.211 blades (3.6 m diameter, B/D 12.2%). The planned radio equipment consisted of a FuG 10, Peil G 6 (DF), FuG 16 ZY with equipment set, FuB L 2 F, FuG 101, FuG 25a, and FuG 220 D with Rü antenna.[511]

The aircraft was grounded on 20 February 1945.[382a]

Ju 388 V15, WNr. 230 903

This aircraft's *Werknummer* indicates that it was also one of the batch of fifty Ju 388 K-0s originally planned. It was soon decided, however, that the V15 was to be one of four prototypes of the Ju 388 J-2 with Jumo 222 A/B engines. After the other three prototypes were assigned to other roles, the V15 was the only aircraft left resulting from the "construction directive for the building of a prototype of (the) 8-388 J-2 with 9-8222 A/B engines (Jumo 222)" dated 30 October 1944.[512] The second amendment, dated 29 November, confirmed that the V15 was to be the prototype for the Ju 388 J-2, and provided the first details. Unlike the Ju 388 V2, the V15 was to have the Ju 88 control yoke, a bad weather window for the pressurized cockpit, and a Dachlaurin fire extinguishing system. Of course, several modifications to the firewall and engine controls were required to accommodate the Jumo 222, and Kärcher heaters had to be installed for the wing de-icing system. The engines were to be fitted with flame dampers, and four-blade variable-pitch propellers with a diameter of 4 meters. As a crew of four was envisaged, an additional oxygen system had to be installed. Installation of a FuG 220 D/R with a scope in the observer's position was planned, in addition to the usual radio equipment (FuG 10, Peil G 6, FuG 16 ZY, FuBl 2 F, FuG 101, and FuG 25a). Defensive armament would consist of a FHL 131 Z tail turret. There were to be no ammunition boxes—the entire ammunition supply of 300 rounds per gun would be contained in the belt feeds. The forward firing offensive armament was to consist of two MG 151/20s (mechanical firing) and two MK 108s. Compared to those of the V2, the guns had modified ammunition boxes and feeds made possible by the installation of a fuel tank. The gunsight installation was also to be modified. Also planned was an upwards firing armament of two mechanically fired MG 151/20 cannon. The gunsight for these weapons was to be mounted in the fuselage.[513]

Ju 388 V16

This aircraft was envisaged as one of the prototypes for the Ju 388 *Zerstörer* powered by Jumo 222 engines. According to a message dated 16 November 1944, the V12, V14, and V16 were envisaged as additional J-series prototypes at that time. At the request of Dr. Stahl they were to be equipped with infrared lighting.[514] Nothing is known of the aircraft's *Werknummer*, or its state of construction at the end of the war.

Ju 388 V18, V19, and V20 (WNr. 230 908 – 910)

These aircraft were supposed to be built as pure test-beds for the Jumo 222, with no special equipment whatsoever.[493] It is highly unlikely that they were ever completed.

List of Ju 388 Prototypes

	WNr.		Man. Code	Remarks/Location
V1	500 001	PE+IA		J-1 EKdo 388 and Tarnewitz
V2	500 002	PE+IB		J-1 Dessau, Werneuchen
V3	500 003	PE+IC		K-1 Dessau, Rechlin
V4	500 004	PE+ID		K-1 Dessau
V5	500 005	PE+IE		L-1 EKdo 388
V6	500 006	PE+IF		K-1 Tarnewitz, captured by British
V7	300 001	PG+YA	L-0	First prototype built
V8	300 002	PG+YB	L-0	Merseburg, Rechlin
V9	230 695	KS+TO	K-0	Jumo 222 A/B test-bed
V10	230 721			K-0 J-0 with 2 x MG 151 and 2 x MK 108
V11	230 727			K-0 Jumo 222 A/B test-bed
V12	230 733			K-0 also prototype for Ju 388 J-2
V13	230 737			K-0 Under construction in Merseburg
V14	230 901			J-1 Completion not confirmed
V14/2	230 901			J-3 Prototype with Jumo 213 E
V15	230 903			K-0 J-2 operational trials aircraft
V16				Ju 388 J prototype
V17				*No existing documentation*
V18	230 908			K-0 Exclusively engine test-bed[493]
V19	230 909			K-0 with Jumo 222 A/B
V20	230 910			K-0 Construction not confirmed
V21	230 157	KS+TG	K-0	1st aircraft with Jumo 213 E
V22	230 691	KS+TK	K-0	Test-bed for 9-9213 D engine
V23	230 697	KS+TQ	K-0	(Jumo 213 E) also prototype for Ju 388 J-3, construction begun, otherwise uncertain
V24	230 725			K-0 Like V23 and V24, but cancelled
V25	230 729			K-0 End of October 1944[515]
V26				*No existing documentation*
V27	230 902			K-0 Conversion to DB 603 U test-bed
V28	230 906			K-0 by DLH Staaken planned
V29				No existing documentation
V30	300 293	DW+ZA	L-0	From 10/1944 with Jumo 213 E at Merseburg
V31	300 294	DW+ZB	L-0	Converted to 9-8123 D engines
V32	300 295	DW+ZC	L-0	Flew as T9+DL with Jumo 213 E
V33	300 296	DW+ZD	L-0	VfH, Oranienburg
V34	300 298	DW+ZF	L-0	Jumo 213 E?

List of known prototypes of the Ju 388 based on Amendment IV to Construction Directive 7 388 26 10/000 of 12 August 1944 (V10 to V14 and V21 to V25, V27 and V28), Kobü-Rh memo 3767 of 16 November 1944 (V16) and offer no. 3766 of 28 November 1944 (V9, V15, V18 to V20) and damage report from Merseburg dated 28 November 1944.

Handling Characteristics with the BMW 801 TJ and Jumo 213 E

The stall characteristics of Ju 388s powered by the BMW 801 TJ, and equipped with the late September 1944 variant of the exhaust system, differed greatly from aircraft to aircraft. Flugkapitän Matthies suspected that the main reason for this was the joint between the engine nacelle and airframe.

Stall speeds of the aircraft with the BMW 801 TJ at 2000 meters were as follows:

Cruise at maximum allowable continuous power, 12 tonnes	V4	indicated airspeed 215 kph
	V3	indicated airspeed 195 kph
	V7	indicated airspeed 235 kph
Landing, undercarriage down, engines at idle	V4	indicated airspeed 190 kph
	V3	indicated airspeed 175 kph
	V7	indicated airspeed 190 kph

While aircraft with the BMW 801 TJ no longer experienced tail vibration, there was an aperiodic oscillation about the lateral axis, which became stronger as altitude increased. Matthies blamed this on the exhaust stream striking the horizontal tail.

Instead of the Jumo 222, the V14 was ultimately to be equipped with BMW 801 TJ power plants. Here is a rear view of an early example. MTU

Below: These performance graphs show the Jumo 222 would have enabled the Ju 388 to achieve an impressive performance.

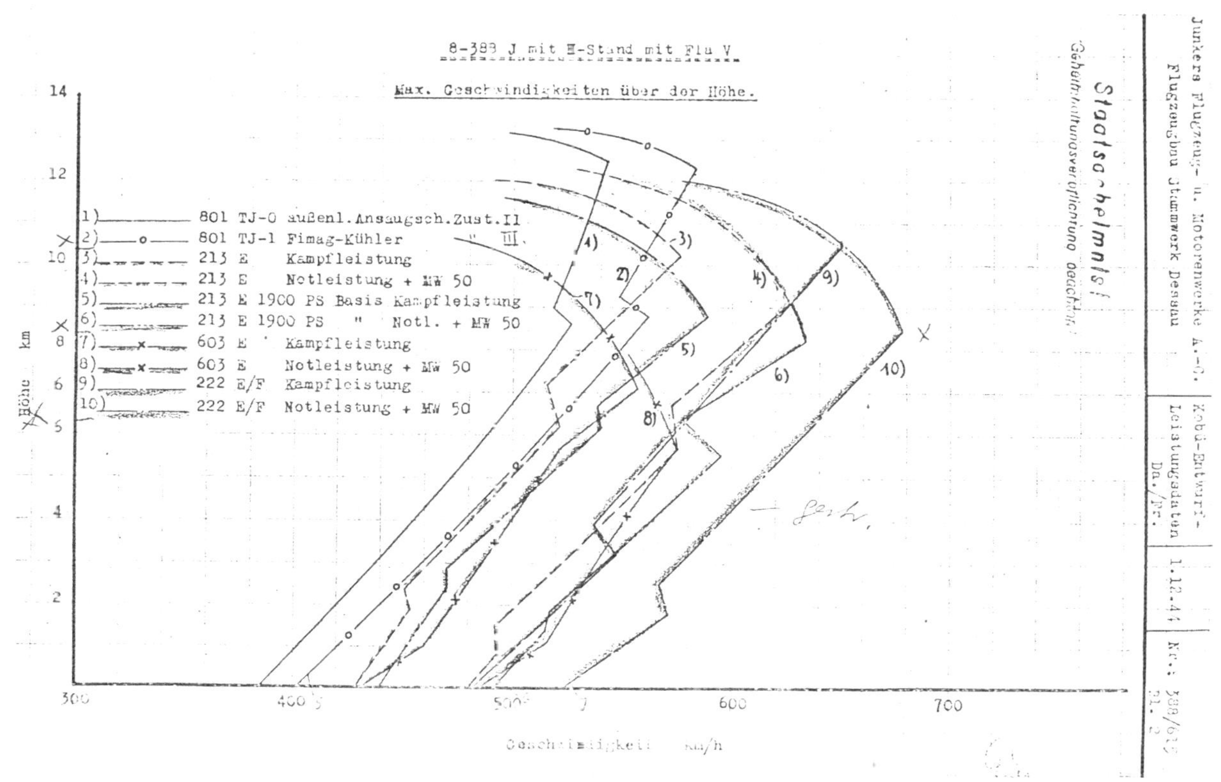

In a wings level descent with engines throttled back, the aircraft vibrated badly at speeds in excess of 400 kph, which Matthies suspected was due to engine shake, and not the exhaust system. Tests with engines shut down were to settle the matter. Concerning the Ju 388 with BMW 801 TJ engines, he reasoned:

"Aerodynamically, there are still problems with the Ju 388 with BMW 801 TJ engines, as the early stall and the oscillation about the lateral axis prove."

With Jumo 213 E engines the flight characteristics of the Ju 388 were significantly better. At a takeoff weight of 11 tonnes, the Ju 388 V21's stall speed was 185 kph, while in landing configuration it was 168 kph. The aircraft, which by then had achieved altitudes to 8000 meters, was "better aerodynamically, and because of the vibration-free engines was also steady in a high-speed descent with no oscillation about the lateral axis." Takeoff distance also appeared to be shorter.[516]

Factory Testing Using Production Aircraft

As previously described, pre-production were converted into prototypes for factory testing; however, they were not the only aircraft "without V numbers" used in this role.

After the Ju 388 V8 was released, Ju 388 K-0 WNr. 230 151 (KS+TA) was supposed to replace it in the test program. At the beginning of April 1944, therefore, the aircraft was to be equipped with the same measuring points as the earlier machine. Eighteen different temperatures and 16 different pressures were measured. In addition to indicators on an additional instrument panel, sixfold recorder and two pressure recorders in the cockpit, two additional pressure recorders were envisaged in the bomb bay to record altitude, speed, boost pressure, etc. As well, there were installations for measuring performance.[517]

The aircraft was finally test flown at Merseburg on 29 May 1944 by company pilot Lehmann and flight engineer Lange. After several more flights at Merseburg, the main purposes of which were to tune the radio equipment and adjust the autopilot, at 6:10 in the morning on 8 July 1944 Lehmann flew WNr. 230 151 to Dessau, where it was to be used by the flight department as a test aircraft. Some factory testing apparently took place in Prague, where Junkers maintained its own flight department. Flight tests were to be undertaken there with KS+TA, mainly to test the bombing equipment. An M20 equipment set had to be made available for the aircraft. The following points were to be investigated during flight testing:[518, 519]

1. Lotfe (bomb sight) tests, especially at altitude
2. Performance measurements with old fuselage spine
3. Modifications
    - a Strengthening of fuselage spine
    - b Exhaust fairings
    - c Camera installation for fall of bombs
    - d Installation of bomb racks, etc.
4. Performance measurements with strengthened spine
5. Bomb dropping
    - a Function tests
    - b Bomb sight tests from high-altitude with Zeiss

The aircraft was in fact in Prague some of the time, for on 11 August 1944 WNr. 230 151's port BMW 801 TJ-0 engine (WNr. 385 185, 4 flying hours) was removed. The reason was the failure of the main rod on cylinder No. 9, which resulted in the piercing of cylinder No. 1. There was also a supercharger gearing failure. The maximum boost altitude of 385 172, which was also installed in KS+TA, proved to be too low, and not until the turbosupercharger was replaced was pressure behind the turbosupercharger high enough. The removed turbosupercharger was supposed to go to Munich with engine 385 185.[520]

The flight departments, at the beginning of October 1944, envisaged WNr. 230 151 being used mainly for further testing of the pressurized cockpit. Aircraft powered by the Jumo 213 E used air compressors (instead of supercharger air) to pressurize the cockpit. Tests were to be carried out to determine how to avoid misting of the canopy with the reduced volume of air. As well, measurements were to be made of the mass volume, and the air flow rate during forced air ventilation of the cockpit. The engineers wanted to come up with some ideas for overcoming the loud noise produced by the forced air ventilation system. Also to be examined was regulation of the cockpit air cooler for summer and winter operations. And finally, tests were to be made of engine ventilation with a stationary aircraft, a new cockpit window attachment method using tensioning screws, cockpit pressure integrity, canopy rainproofing, moisture traps, and finally, of the oxygen supply with increased range.

Also planned were aerodynamic tests for a better understanding of the aircraft's high-altitude behavior. Wool tufts were to be attached at the critical points—cockpit and bulged bomb bay—to better observe the air flow. Pressure distribution was also to be measured, and the influence of the much-modified exhaust system on the wing's aerodynamics investigated. In the latter instance, c_a measurements were to be made for comparison with the Ju 188.[521]

At the end of November 1944, KS+TA was definitely back at Dessau as a flight department prototype. It had been retrofitted with external aircraft air intakes, and it was now to be used for comparative measurements with WNr. 230 156. Thermal measurement of the turbosupercharger was also planned, and maximum boost altitude was to be determined with and without the honeycomb filter in front of the external supercharger air intakes.[522]

Climbing flights were made on 19 and 26 November to determine the leading edge temperature with two variants of the wing de-icing system. In both versions, pipes that followed the path of the former internal supercharger air intake delivered fresh air to the Y-pipe, through which hot air flowed from the engines into the leading edge of the wing (see drawing on following page). Initially a 38-mm pipe was used for this purpose, but results were unsatisfactory, as the limiting curve for 160°C was exceeded somewhat on average. A larger diameter fresh air pipe was subsequently used, and the entrance was rounded, changing another parameter. The limiting value was now clearly exceeded, and experimentation obviously continued in December or January. With the proposed ultimate solution (38-mm diameter, rounded mouth), the maximum allowable temperature was only exceeded above 10 km, and it was felt that the threat posed to the leading edge by overheating no longer existed.[523]

Another aircraft that was used for factory testing for a time was Ju 388 K-0 WNr. 230 156, KS+TF. It was converted into a night reconnaissance aircraft (Ju 388 L-0 Nacht), and sent to the Ergänzungs-Nachtaufklärerstaffel (night reconnaissance replacement training Staffel) at Brieg. Beginning on 23 September 1944, ten to twelve Ju 388 L-0s and K-0s that had already been delivered were to be returned to the main facility at Dessau to be retrofitted with equipment important for their operational roles, such as external supercharger air intakes. WNr. 230 156 was supposed to be the first such aircraft.[524] On 23 September Flugkapitän Wendel flew the machine from Brieg to Dessau, taking off at 1554 hours and landing at Dessau at 1704. KS+TF appears in Preuschen's notebook on

The Ju 388 K-0 WNr. 230 151 had the standard Ju 188 tail surfaces. Note the heavy staining on the wing behind the exhaust fairing. The mount for the MG 131 at the rear of the canopy has been sealed. EADS

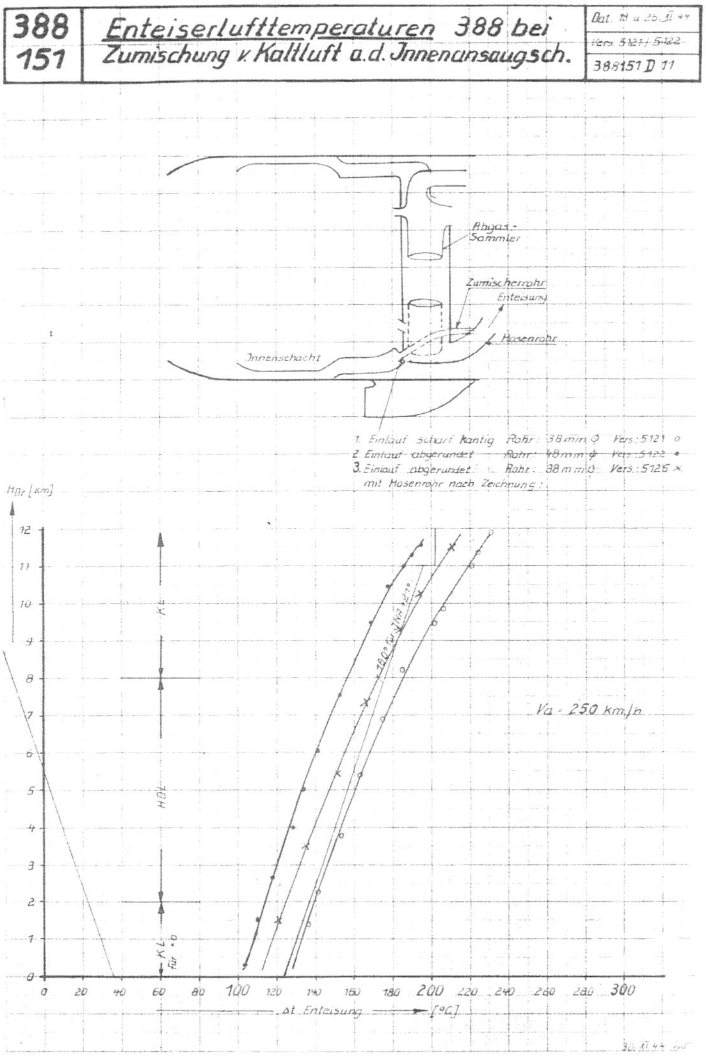

Sketch of the cold air mixing, and a graph showing deicing air temperatures.

24 October, with the comment "aircraft to be finished." Unfortunately, it is not quite clear whether the aircraft was still in Dessau or already back again. The aircraft was obviously in Dessau for a long time in November and December 1944. During the installation of the wing de-icing system and external supercharger air intakes in November 1944, Junkers was able to gain valuable insights into necessary design changes. It was discovered, for example, that the hot air Y-pipes had to be dented during installation to prevent a conflict with the undercarriage swing arm! Skilled improvisation was also required during fitting of the external air intakes, because parts did not match the drawings. As well, the designers had forgotten to include water drainage holes in the supercharger air intake scoops. As much as five liters of water had collected in the scoops of several prototypes.[525] WNr. 230 156 was also one of the very few aircraft to actually have a weapon installed in the cockpit dorsal position. Several design flaws were discovered in the process. For example, the wiring for the ignition current cutout as depicted in the drawings was unusable.[526] On 19 November 1944 KS+TF was in Assembly Hangar 203 for the installation of an auxiliary oil tank. On 26-27 November Preuschen noted "BMW checks," which probably means that BMW technicians carried out checks on the power plants. The aircraft was flyable on 29 November, however, the oxygen system had obviously not been filled.[527] There are several notes for the days that followed, indicating that WNr. 230 156 was serviceable or could be flown. No flights are verified, however, until 5 December 1944, when Wendel achieved an altitude of 12000 meters during a test flight that lasted from 1442 to 1551 hours. According to Preuschen, the aircraft did not fly two days later. Then, on the morning of 10 December, Wendel took the aircraft up to 12700 meters. Cockpit sealing was obviously not satisfactory.

Right: As well as the Ju 388, company pilots also tested the final variants of the Ju 88, here a Ju 88 S-3 with Jumo 213 E engines. Griehl Collection

Taken on 22 March 1945, this photo shows no activity at Dessau after the last heavy raid of 7 March 1945; however, numerous wrecked machines may be seen in front of the large hangars and on the right side of the airfield. Luftbilddatenbank Würzburg

Testing by BMW

In addition to testing by the manufacturer and the air force, which will be described in a subsequent chapter, trials were also carried out by the engine manufacturer.

In the summer of 1944 the BMW Flight Department at Fürstenfeldbruck used two Ju 388s—WNr. 300 003 (PG+YC) and WNr. 300 010 (PG+YJ)—as engine test beds. BMW was supposed to have received a third Ju 388 for high-altitude and vibration tests after two Ju 88 test beds were "lost."[528]

At the end of November BMW conducted a rigidity test of the BMW 801 TJ-0 engines in a dive. The main purpose was to prove the rigidity of the engine cowling. The aircraft used was Ju 388 L-0 WNr. 300 299 (DW+ZG), with BMW 801 TJ-0 engines 385 279 on the port wing and 385 281 on the starboard. The aircraft had been modified to the latest standard,[529] and the engines were fitted with the improved exhaust system and external supercharger air intakes.

Eight dives were made in a total of eight flights. Structural loads were steadily increased until, during the eighth dive, an indicated airspeed of 684 kph was achieved, equivalent to a true airspeed of 774 kph. The aircraft dove from an altitude of 5.83 km to 2.31 km in just over a minute. The maximum loading just before pullout was 3.18 g, or slightly more than three times the force of gravity. In spite of these loads there were no significant problems

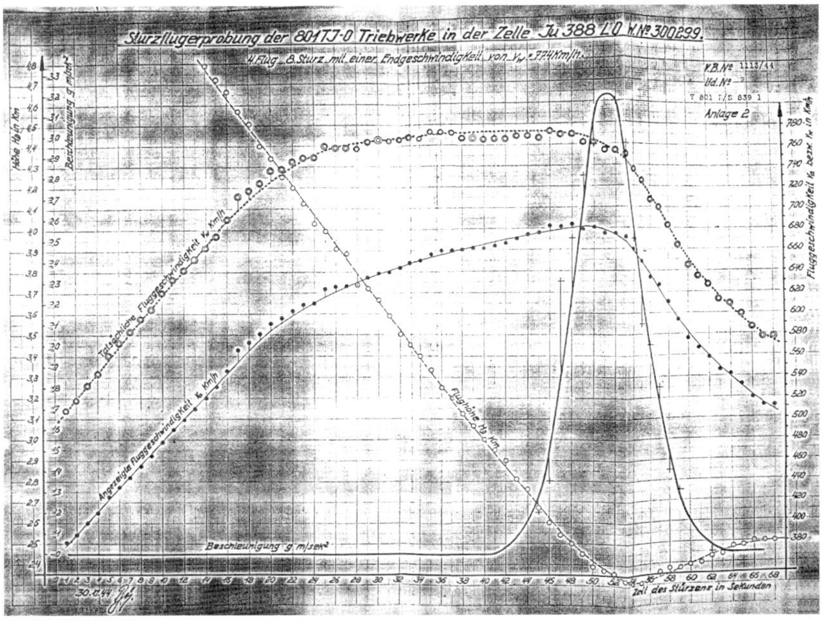

Record of altitude, airspeed (indicated and true), and g-forces during the 8th diving flight on 30 Nov. 1944.

This Ju 388 L-0, WNr. 300 299, manufacturer's code DW+ZG, was used as an engine test bed by BMW. At war's end it was found by the Allies at Fürstenfeldbruck. Note the external ram air intakes on the BMW 801 TJs. Crow Collection

with the engine mounts or cowlings. As for the airframe, at high indicated airspeeds a strong clockwise rotation about the longitudinal axis was observed; however, this was attenuated by retrimming. Between 630 and 650 kph (indicated airspeed) the aircraft began to oscillate about the vertical axis. The control forces encountered during the pullout from the 3-g dive were so great that the pilot could not have achieved this feat alone.

A thorough inspection after the flight revealed some minor damage to the power plants. Some of the cooling gills, which were fixed at 15°, and their mounts were bent, and the flange for the attachment of the hot air Y-pipe (wing de-icing) had rubbed through the intake funnel leading to the turbosupercharger. Strangely, much of the hydraulic fluid had also been lost, although no leaks were found when the undercarriage was later cycled with the aircraft raised up on jacks.[529]

There is much to suggest that this is also DW+ZG. On the right another Ju 388 L-0 with covered engines and no propellers. Its manufacturer's code also begins with "DW." Crow Collection

Chapter 6: Manufacturing Facilities, Production Planning, and Production Numbers

Introduction

A characteristic of the mass production of German aircraft in the Second World War was thCe system of "replica firms." Thus, most of the classic Junkers aircraft types were traditionally built not just by the parent company, but under license, as well. For example the legendary Ju 52 was built by the Allgemeinen Transportanlagen-gesellschaft mbH (ATG) in Leipzig, and the Ju 87 for the most part by Weserflug (WFG).

With the Ju 88 program and the "central control" created for its implementation, the Junkers Flugzeug und Motorenwerk (JFM) attained a special place among the "development works," the aircraft firms that developed new aircraft themselves. Endowed with numerous privileges and far reaching authority, JFM controlled and oversaw mass production of the Ju 88 in its own branch plants, and also in four licensed manufacturing concerns.

For mass production of the Ju 88, the department titled Head Office Aircraft Construction-Technical (FTH) assumed the role of a central coordinating and control point between the Junkers branch plants and other licensees. The FTH handled the organization of the entire program delivery to the RLM on the scale of 10,000 completed aircraft. However, the final assembly of JFM production aircraft was also carried out in the Bernburg aircraft branch plant (FZH). The major subassemblies for the Ju 88s completed there came mainly from the Junkers branch plants in Aschersleben (FZA; fuselages) and Halberstadt (FZH; wings); tail sections were delivered from Leopoldshall (FZL); and individual parts and jigs from Schönebeck (FZS). As a rule, the Jumo engines used in the aircraft came from the Jumo factories in Köthen[531] and Magdeburg, the Mitteldeutschen Motorenwerke Taucha (MIMO), and the Pommerschen Motorenwerke Stettin (POMO).[532,533]

Production of the Ju 388 should be viewed in close context to production of the Ju 188 and its predecessor, the Ju 88, as significant subassemblies were used either unchanged, or after modification. At Dessau in November 1943 exchange lists were compiled

While none of the four large assembly halls has survived, the Junkers administration building of the FSD master aircraft works in Dessau still stands at the edge of the former company airfield, right next to the remains of the wind tunnel. Handig Collection

for the Ju 388 J-1, K-1, and L-1 with parts from ongoing Ju 88/Ju 188 production that required modification. The lists were extremely careful to differentiate between prototypes, test beds, and mass production.[534] All of the factories and subcontractors involved in the production of the Ju 388 were later directed that all parts originally built for the Ju 88/188 and now to be used in the Ju 388 (such as wing control surfaces) were to be re-greased with the cold proof "new instrument grease."[535]

As recounted in the previous chapters, the production story of the Ju 388 began with new construction and conversions in the Junkers "parent factories" of Dessau (FSD) and Merseburg (FSD-M). Neither of these two production sites had been envisaged for series production (manufacturing of major components and final assembly), however, though Merseburg had handled a number of small production series and conversion programs. Small scale production of the Ju 188 had begun in 1943, and the type was built by Junkers, ATG, and Siebel.[536] Furthermore, the Henschel Flugzeug-Werke AG (HFW) and the Siebel Flugzeugwerke Halle GmbH (SFW) in Halle and Schkeuditz produced the Ju 88 and Ju 188 in addition to their own products. From the start of planning it was envisaged that these typical Junkers licensees would also be intimately involved in production of the Ju 388. ATG and WFG in particular played a significant role in final assembly. But Henschel and Siebel were also involved, as they had been with production of the Ju 88/188.

Produced by the RLM's Technical Office, the "Aircraft Program" fixed aircraft production delivery plans, and conveyed a good impression of the planned scope of Ju 388 production. The following account concentrates on Delivery Plans 226 to 228, as these were created shortly before or during series production of the Ju 388.

The first issue of Delivery Plan 226 was dated 15 May 1944, and like all the others discussed here, it established production figures until the end of March 1946. It called for 1,914 Ju 388 J night fighters and

Zerstörer, 1,888 Ju 388 K bombers, 1,302 Ju 388 L reconnaissance aircraft, and 655 Ju 388 M torpedo bombers. The bomber and reconnaissance aircraft were to be powered by the BMW 801 TJ initially, and later by the BMW 801 TQ. The Jumo 213 E was also envisaged for the night fighter. The Ju 388 J-2 Zerstörer was to have been powered by the Jumo 222. Several other companies were to be included in the production scheme besides ATG, Henschel, Siebel, and Weserflug: the Böhmisch-Mährischen Maschinenfabriken (BMM, 342 Ju 388 L-1), a firm with the abbreviation "MNH" (perhaps the Metallwerke Niedersachsen Brinckmann und Margell in Harburg, 768 Ju 388 K-1); and the Volkswagenwerk (as sole producer of the Ju 388 M-1).[537]

The optical effects of dispersed component production are visible on the vertical tail surfaces of this ATG-built machine. Although the camouflage is quite primitive, the schemes on the fin and rudder are completely different. Crow Collection

The sole surviving Ar 234 B-2 (bomber version) in the colors of KG 76. The importance of the Ar 234 rose with the cancellation of the Ju 388. Handig Collection

The second issue of the delivery plan, dated 15 July 1944, was a reduced Fighter Staff program, and as such there was a clear shift in emphasis. While the total number of Ju 388s ordered was virtually unchanged at 5,638, the emphasis now lay on the Ju 388 J, with 2,684 of this variant planned (including 525 Ju 388 J-2 Zerstörer). This increase came at the expense of the Ju 388 L, whose numbers were almost halved to 764. BMM was now to build 342 Ju 388 J-1s, which reduced the number of bombers to 1,535—not as serious a reduction as that of the reconnaissance version.

In addition to the change in favor of the night fighter version and the first appearances of the L-3 and K-3 variants, there was also a change in the manufacturing firms[538]: with 182 Ju 388 K-1s and 470 K-3s, Blohm & Voss obviously took the place of MNH, while Pirth, in Hungary, was added to the program (115 Ju 388 J-3s).

In the months that followed, the second issue of Delivery Plan 226 underwent further changes. The next version dated 28 August saw the elimination of the planned 525 Ju 388 J-2s from Henschel-Flugzeugwerke, reducing total production of the Ju 388 J to 2,159. Otherwise the production numbers remained unchanged.

In the next issue, dated 15 September 1944, the Ju 388 no longer appeared among bomber aircraft, as that role had been taken over by the Ar 234 and Me 262. The production capacity freed up as a result of this decision was largely used to increase night fighter numbers to 906 Ju 388 J-1s and 2,349 Ju 388 J-3s. On the other hand, production of the reconnaissance version was reduced to 506 aircraft. Production of the Ju 388 was now to be restricted to the "traditional" licensees—ATG, Henschel, Siebel, and Weserflug.

Together with the Ar 234, the interim bomber version of the Me 262 A, seen here with two SC 250 bombs, was to replace conventional bombers. Handig Collection

Just two weeks later, however, a new issue of the production plan dated 30 September cancelled the increase in production of the J series. The numbers for night fighters and reconnaissance aircraft were now exactly the same as they had been on 28 August, while the bomber and torpedo-bomber versions were cancelled without replacements.

The first issue of Delivery Plan 227, released in mid-December, reflected the diminishing significance of the Ju 388. The B-29 had not been deployed in Europe, and the new construction program envisaged 1,465 examples of the Ju 88 G-6/G-7 night fighter, but just 207 Ju 388 J-1s (for the first half of 1945) and 788 J-3s (beginning April 1945, with 80 aircraft per month later). Night fighters from other manufacturers included 210 He 219 A-7s (and five D-1s), 1,620 Do 335s, and 580 Ar 234 C-7s. Production of long range reconnaissance aircraft was to consist of another 84 Ju 188 D-2s, 330 Ar 234 C-4s, and just 426 examples of the Ju 388 L-1. Of the latter, 33 were noted as delivered by the end of November. The reconnaissance versions were supposed to begin leaving the factory sheds in December 1944, with an initial production rate of 35 per month, and later 20 per month. Beginning with the 116th aircraft, the

Luftwaffe wanted the aircraft delivered with the capability of being converted for the night photo reconnaissance role. The air force obviously anticipated that the aircraft's performance would no longer be sufficient for the daylight reconnaissance role by mid-1945. Also retained in this delivery plan were 153 Kalkert 8-430 transport gliders, for which the Ju 388 was envisaged as a glider tug (see Chapter III).

The Ju 388 also remained in the conversion program. By the end of November, seven K-0 pre-production machines and 15 L-0s had already been converted into

"high-altitude, high-speed reconnaissance aircraft with 3-man pressurized cockpit." Twenty more Ju 388 L-0 night reconnaissance aircraft were supposed to be delivered between December 1944 and March 1945; these were the remaining Ju 388 K-0s, which were to be completed as L-0

Nacht reconnaissance aircraft.[539]

Following the cancellation of the Ju 388, the Ju 88 G-6/G-7 was the only Junkers type left in the last delivery plan (228), which was issued on 16 March 1945. Not a single Ju 388 remained.[540]

Planning and Implementation of Ju 188/388 Production by Junkers

Junkers (Technical Head Office Aircraft Construction - Overall Planning) devised production plans from the RLM delivery plans. The delivery plans differentiated between deliveries of prototypes and conversions on the one hand, and actual series production on the other. The first step in the administrative-technical implementation of development and production contracts by the RLM was usually the sending of preliminary agreements to the corresponding manufacturers.[541] At Junkers, the next formal stage was the issuing of a construction directive, which was the responsibility of the depart-

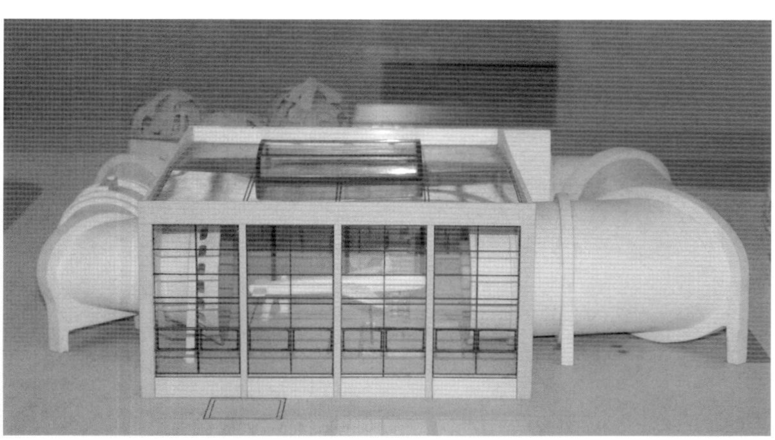

In 1934 Junkers built a wind tunnel, one of the first prestressed concrete structures. Seventy percent destroyed during the war, today the remains are protected as a historic site. The model shows the original appearance of the installation. Handig Collection

The hangar erected for the Baade 152, an East German airliner, now houses the "Technikmuseum Hugo Junkers." It stands on the spot formerly occupied by the prominent Hangars 201 and 202 in front of the wind tunnel at the master works in Dessau. Handig Collection

ment titled Aircraft Construction Technical Sales. Construction directives usually also served as "production directives." The construction directive was given a contract number/cost center, with which the completed delivery costs could be settled with the buyer. Construction directives were not necessarily closely linked in time to the actual work, however. An example of this is the construction directive for the shipping of major components of Ju 388 K-0s WNr. 230 152 and 230 154. The corresponding amendment to construction directive 7 388 26 10/000 bears the date 12 August 1944; however, by that time the Resistance had already done its work in Toulouse (see Chapter I). And the V21, which is also mentioned in this construction directive, could hardly have been completed by the end of August if work did not officially begin until 12 August.

Dessau was the "nerve center" of production by Junkers. Construction plans and jigs for mass production were both produced there. The "Technical Head Office Aircraft Construction" (FTH) in Dessau controlled mass production. Production planning was derived from the RLM delivery plans, which were broken down into aircraft and component delivery plans. Technical control of production in branch plants, in the sense of program control, also lay in the hands of the FTH. This included requirements planning for materials, and the coordination of all modifications to the type.[542] Every airframe or equipment modification to aircraft already in production, or planned new production aircraft, no matter how minor, had to be presented to the RLM for approval.[543] There exist several hundred modification notifications and directives, which contain minor and major manufacturing changes to types planned or already in production, from the period 1943 to 1945 pertaining to the Ju 388 and the Ju 188 G/H. These changes were either results of simplification measures (use of non-strategic materials and production rationalization), or fallout from testing.

In mid-June 1944, therefore, Junkers was working feverishly to overhaul the technical drawings for the Ju 388 in preparation for the start of production. Prague obviously played a major role in the design—the drawings were overhauled, and the production version of the climate control system designed there. In mid-June 1944 experimental version drawings were still being used at Merseburg, and these were also provided to ATG. On 12 June 1944 prototype pipes were ready at Merseburg for pickup by ATG. Another change still being worked on was the relocation of the O_2 valves. Already delivered were changes to the hydraulic fluid pump, radio armor, FA 15 shafts, FA 15 control box, and drawings for the relocation of the EZ 6.[544] The wide-ranging consequences of such modification messages and preliminary decisions are illustrated by a table compiled by Prof. Hertel of the "ATG-Maschinenbau GmbH" following a Ju 388 meeting in Berlin on 5 January 1945:

1944	**Alteration Messages**	**Alteration Points**	**Alteration Advance Notices**
October	145	1106	206
November	155	1079	169
December	301	1991	191

The term "Alteration Point" indicates the number of drawings and parts lists ultimately modified.[545] According to Prof. Hertel, the extent of these alterations was "extraordinarily high," and he asked Schilling to check them over, and give his opinion as to whether the many adjustments were "normal corrections" to the aircraft, or if they too were based on new requirements "from the authorities."[546] A discussion ensued as to whether the figures given by ATG were not too high. Of the 566 preliminary alteration decisions named, ATG was only able to produce 480; 57 of these concerned types not built by ATG. In an internal listing, on the other hand, the design bureau only came up with 250 preliminary alteration decisions, of which the overwhelming majority (188) resulted from requirements from the OKL. Internal preliminary alteration decisions had not been counted, however.[547]

The ongoing changes were a major problem. While under interrogation by the Allies after the war, the leadership of the Allgemeinen Transportanlagengesellschaft mbH (ATG), one of the most important builders of Junkers designed aircraft, stated that the constant changes in planning had resulted in a production loss roughly equal to that caused by Allied bombing!

How much truth is there in this statement? If one looks at the minutes of the meetings held by the Chief of Aircraft Procurement and Supply, the Fighter Staff, and later the Armaments Staff, one is immediately struck by the enormous amount of discussion that took place, while the decisions made were few. After ten years of a dictatorship based on blind obedience, the democratic culture of debate had probably faded so far into the past that the men in these positions were no longer capable of reaching majority decisions, and justifying them to superiors like Göring.

Chronology of Series Production

...based on official meetings

In July 1944 the RLM began dealing with the startup of Ju 388 production. It is obvious that consideration was originally given to also producing the Ju 388 outside Germany. During a Fighter Staff conference on 3 July 1944, Rowehl asked if any thought had been given to possible production of the Ju 388 in Hungary. Saur replied that he would prefer to

"partially shift production to France and build considerable numbers there." It is also interesting how the monthly production figures for the Ju 388 were determined: they were derived by adding the former numbers for the Ju 88, 188, and 288. Saur also addressed an alarming report on the Me 262, which contained a series of previously unmentioned disadvantages of the highly-praised jet aircraft:

"...for example, that the speed is 710 [kph], in some cases just 680, and that it has a tactical radius of just 170 km, or things that are simply out of the question, an optimum flying time of 45 minutes, for pure flying 27 minutes in practical terms."[548]

Unfortunately, it was not mentioned whether the report concerned the high-speed bomber or fighter version; the Fighter Staff thus had good reason for concern about a complete switch to jet powered aircraft.

The official production startup meeting for the Ju 388 was held on 5 July 1944.[549] It brought together representatives from industry, and the offices of the RLM and the Fighter Staff. Dr. Lübke was named startup commissioner for the Junkers Ju 388. The initial situ-

Cancellation of the Ta 154 made available additional production capacity, which benefited production of the Ju 388. Crow Collection

ation was as follows: that the Ju 288 was to be cancelled had been common knowledge since 1 July. The Ju 88 and Ju 188 were to be gradually replaced by the Ju 388, which initially would be built with different engines, because they could not and would not wait for the airframe with the ultimate engines. Before HDL Saur turned the startup meeting over to Frydag, he praised the surprisingly quick resumption of Ju 88 night fighter production as an outstanding accomplishment, and expressed his appreciation to Junkers and its director Thiedemann. He went on to add that he hoped that Junkers would also be able to accomplish the significant increase in production of the Ju 388.

The development question was discussed first. Obviously, there were no fears that the aircraft might not reach production for development reasons. The potential benefits from the cancellation of the Ju 288 had gone unexploited.[550] On the other hand, cancellation of the Ta 154 and Ta 254 had made more capacity available for production of the Ju 388 than originally anticipated. Frydag again brought up for discussion the subject of building the Ju 388 in Hungary—proposing that Pirth produce 20 fuselages per month, instead of the Ju 52s previously built there. Saur was opposed to the idea, however, for in most cases, production in Hungary rarely reached even a third of the target figure. Thiedemann agreed with him.

"I can underscore that. It was easier to put through 50 complete aircraft in Paris with nine German engineers and otherwise just Frenchmen than five in Hungary, an allied nation."

The following production breakdown was adopted "for reasons of security," according to Frydag:

"The fuselage is being built by ATG with an offshoot at Fokker, by Menibum, Henschel underground facility, Opel. Pirth and then again by Junkers, or in six places.

The wings are being built by Siebel, the Volkswagen factory, Lettow, Weser, BMM and later by Junkers, so that the entire machine is relatively well divided up. [...] The wings are being built in six places."

The meaning of the "offshoot" at Fokker is that 75 fuselages were to be built by ATG, plus 80 sets of tail surfaces by Fokker in Holland.[551]

Startup commissioner Lübke praised the dispersal efforts by Junkers, which had begun systematically moving its production facilities to outlying areas. He was concerned, however, about ATG and Siebel, for attacks could be expected as soon as the enemy

DW+YY was one of the first Ju 388 L-0s completed at Merseburg, and the most photographed. Because four-blade propellers were difficult to obtain, the first pre-production aircraft were equipped with standard three-blade propellers. EADS

identified the firms involved in production of the Ju 388. At that time ATG had suffered damage from air raids, but not SFW. No dispersal had yet been decided on for Siebel, while ATG had been assigned an underground factory. The U.S. Bombing Survey's report on ATG suggests that the dispersal had not progressed very far, however.[552]

The talks naturally focused on the construction of jigs, an area in which resources were especially limited. As a result, suitable labor forces had to be brought in from everywhere, a time consuming process. Jig building capacity from the Me 262 sector was to be made available for the Ju 388, and tools and dies were to be "accommodated" in Italy. It was envisaged that additional help would later come from the "A 4" (the V2 program).[549] Inevitably, however, with this shifting of labor forces here and there oversight was lost, and decisions were ultimately made based on personal discretion. Not four days later, desperately needed jig building capacity would be removed from the Junkers complex, as the following extract from the Fighter Staff meeting of 9 July 1944 proves:

Krome: I ask that the entire Siebel jig building operation be transferred to the Me 262.
Saur: Can we do that?
Frydag: I actually had the 388 in mind. But we can draw off water (?). Come see me tomorrow morning. I'll get everyone together then.[553]

ATG desperately needed 25 weak current electricians for production of the Ju 388. The insertion of specialists from subsidiary companies could serve as a stopgap measure for three months, and during that time ATG planned to provide other personnel with the necessary training. As well, from July to October Heinkel Oranienburg was to assist by loaning concentration camp inmates. In July alone there was a shortfall of 2,008 workers.[554]

In the summer of 1944 the Fighter Staff made increasing use of concentration camp inmates. At that time, however, the SS was suffering from a shortage of guard personnel. It therefore asked that the "small concentration camp groups of 50 or 100 men be abandoned as quickly as possible," because otherwise it would be unable to get by with the guard personnel available. Junkers, which had four times requested 500 men, was therefore requested to "concentrate them in a 2,000-man camp."[555]

Junkers was supposed to receive 2,000 concentration camp prisoners from Heinkel-Oranienburg for ATG, Siebel, Halberstadt, and Aschersleben. While housing was available, guard personnel were not. On the other hand, labor forces were also needed for the planned Do 335 production in Oranienburg. Industry requested a total of 70,000 concentration camp inmates, which caused Wolf to remark: "We simply can't bring in as many as are needed."[556] The engine manufacturing sector alone had urgently requested the allocation of "20,000 female concentration camp inmates and Hungarian Jewesses."[557]

Decisions concerning production of the Ju 388 were made here in the State Aviation Ministry (RLM). Today the building houses the Federal Finance Ministry. Handig Collection

This thoughtless handling of many thousands of individual lives was marked by ice cold rationality, with no trace of humanity. The Fighter Staff occasionally discussed social aspects, but only insofar as this might lead to a rise in productivity. It is in line with this shocking callousness that ideological fanaticism was apparently held within bounds, for belief in ultimate victory was not absolutely compulsory within the Fighter Staff. In the course of a discussion as to whether at least part of the production capacity in the "Mittelbau" (A4, or V2, construction site) could be given up in favor of engine production, Wegener declared:

"Without engines we'll be paralyzed. The war will be over within two months."[558]

While most of his countrymen would have been sent to a concentration camp, or immediately turned over to the hangman for such a statement, in the Fighter Staff it went unchallenged.[559]

Because of the initial difficulties with the BMW 801 TJ, which according to type supervisor Stahl "had just begun testing," just a week before the startup meeting consideration was obviously given to equipping the first aircraft "with the old engine" (meaning the BMW 801 D), resulting in a machine with the flight characteristics of the Ju 188. According to Prof. Hertel, however, it was decided "that the 801 TJ would remain in the startup phase, and be given sufficient priority to make it work." The "other concern, finding 801s" was thus negated.

The huge interruption in production of the BMW 801 had resulted in serious problems. Relief was therefore sought in the area of the Ju 88G night fighter. Half of the August production (according to Thiedemann's estimate about 125 aircraft) was to be converted to the Jumo 213 E, making about 250 urgently needed BMW 801 D engines available for the Fw 190. As the BMW 801 TJ-0 was based on the BMW 801 D, it is not surprising that series production of the Ju 388 in the late autumn of 1944 was hindered mainly by a shortage of power plants. VDM promised that complete four-blade propellers would be available from August, and that they would help out with repaired propellers until then.

At the Fighter Staff meeting on 6 July 1944. Generalmajor Diesing,[560] head of the Planning Office, asked HDL Saur to approve priorities for "significant aircraft types." Saur approved the request without delay:

Saur: So, the 109, 110, 163, 262, 410, 152, 190, 88, 188, 287, 388, 335, 96, 234, 396, 219, 181, 204, Fi 103. That is the first priority. Did I leave any out?"

After the Fighter Staff meeting on 6 July 1944. Saur met with representatives of the defense districts, and discussed the new aircraft program with them. He characterized the Arado 234 as one of the most important types of all, and praised the aircraft, which was to become the long sought after "Blitzbomber," as "the greatest technical advance ever made." The minutes provide details of the planned transition from the Ju 88/188 to the Ju 388. Plans called for production of the Ju 88 night fighter to rise to 300 aircraft per month, before being reduced to 145 per month in 1946, and finally replaced completely by the Ju 388. Replacement of the Ju 88 bomber was to take place in mid-1945. There was also to be a complete halt in production of the Ju 188. Prior to this, however, there was to be an

With the forced switch to fighter production for defense of the Reich, the Ju 88 G and the Mistel became the last Junkers types in production. The Ju 88 G night fighter was built in large numbers by JFM until 1945. Here a Ju 88 G-6 with FuG 218 antenna array and Jumo 213 A bomber engines is seen in the hands of the victors. Haubner/Misna Collection

increase in production, especially of the long range reconnaissance version, before output could be reduced to 50 per month, and finally eliminated in favor of the Ju 388. Production of the Ju 287 for the bomber role was to begin during this time, with an initial output of 21 aircraft per month in mid-1945, and 50 per month by the end of the year. By that time preparations would have been made for the production of 100 aircraft per month.

Production of the Ju 388 night fighter was to begin in January 1945, with an output of 60 aircraft per month by June. Production targets were 270 aircraft per month by the end of the year, and 300 per month in 1946. It was also envisaged that 60 aircraft per month would be built as heavy fighters (Zerstörer), and beginning in August 1944 a consistent output of 150 bombers per month would be achieved after 5, 21, 29, and 65 aircraft. It is a small irony of fate that, according to the delivery plan, the Ju 388 L-1 was to have the shortest career of any variant—production was to begin in August 1944 with at first 1, 4, 5, and finally just 80 aircraft per month by mid-1945. It was, however, planned to scale back production to 40 aircraft per month "when the other aircraft come," probably referring to the Ju 488.[561]

Difficulties in preparations for series production are confirmed by a comment made by Saur at a meeting on 9 July concerning the so called "Führer Program." There he mentioned the "considerable damage at Erla, ATG, Oschersleben, Bernburg, Wimo, and Siebel-Halle." He subsequently ordered the jig building operation transferred to other factories—in the case of ATG and SFW, however, this almost surely never happened. This confirms that the two most important licensees were engaged in the building of jigs in preparation for quantity production. With respect to production by Weser at the dispersed site in Rabstein, in June the BAL requested that Junkers ensure the jig building operation be up and running on time, so as not to threaten production startup.[562]

During a discussion concerning priority levels (0 or 1) for the individual aircraft types at an armament staff meeting on 2 August 1944, Stoffregen (responsible for raw materials) proposed assigning the aircraft Priority 0. Frydag responded by characterizing the

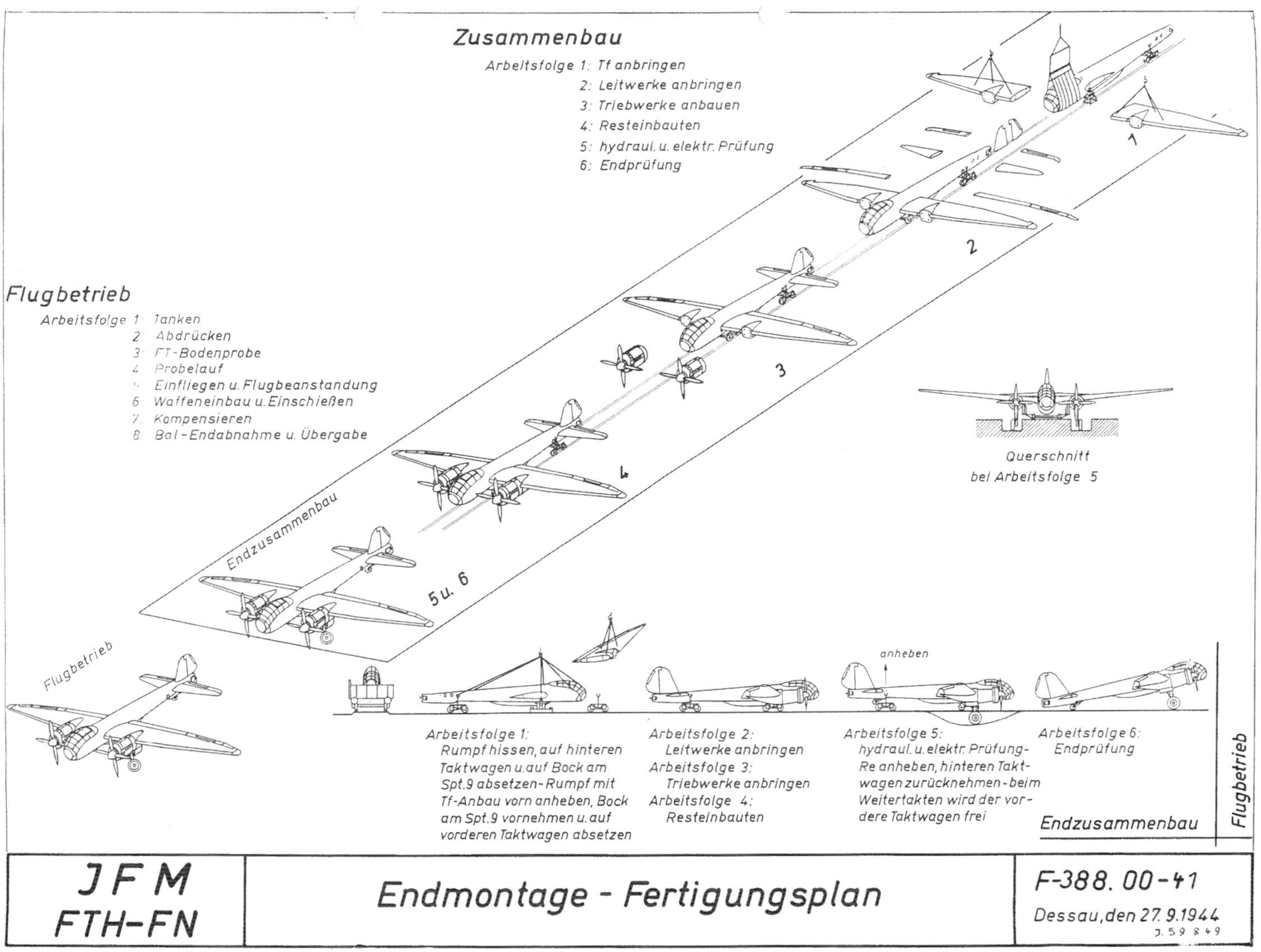

The final assembly process with the individual production stations was planned in detail by Junkers.

Ju 388 as a "most uninteresting machine." In contrast to the He 219 and Do 335, the Ju 388, along with the Ju 287, Ju 88, Ju 188, Me 110, and training aircraft, had been assigned the lower Priority Level 1. It is not surprising that *Generaldirektor* Karl Frydag, general manager of the Heinkel concern and chairman of the "Aircraft Construction" steering committee, should offer a biased view of the Ju 388. It is also interesting that, in spite of the catastrophic fuel shortage, fuel was still being allocated to *Lufthansa* by order of the OKW—the aircraft were not to be grounded "for reasons of prestige."[563]

The end of the Ju 388 came rather stealthily. Through bad luck, those prototypes envisaged for performance measurements were the ones lost in crashes, seriously delaying the determination of the actual performance of production machines (see Chapter VII). In December 1944 the K.d.E. finally decided that the performance of the Ju 388 with BMW 801 TJ-0 engines was insufficient, and demanded energetic measures.[564]

In a situation report dated 21 December 1944, under Point 12 the specialist department Fl-E 2 made the following succinct comments about the Ju 388:

8-388

Performance measurements at Rechlin have produced unsatisfactory results. At present, tests are under way to achieve the projected performance figures. If these are unsuccessful, it is to be decided if the aircraft is justified.

Initially, however, this had no serious consequences, for on Christmas Eve 1944 HDL Saur decreed that, as high performance aircraft, the Ta 152 and Ju 388 were to be treated with the appropriate priority. This of course also applied to all shipments and deliveries from subcontractors.[565] A precise delivery plan was also ordered, so as to be able to achieve the complete delivery of repair, retrofit, and conversion stocks to high performance aircraft within four weeks.[566] A clear indication that the Ju 388's star was waning

Large caliber weapons remained a great hope for dealing with enemy bombers, although better results were achieved with rockets. Griehl Collection

came on 10 January 1945, when Saur suggested that production be focused on the Me 262 and Ta 152, and rejected a request that labor forces earmarked for the Ju 388 be shielded from the "SE-6 Action," which allowed them to be inducted into the armed forces. On the same day, however, it was also decided that the possible installation of the 50-mm rapid firing cannon in the Ju 388 was to be thoroughly investigated by 18 January.[567] Thus, the Ju 388 had not yet been written off entirely.

Increasingly, fuel was becoming a critical resource in aircraft construction. Denial of fuel for test flying was therefore usually the first, but still reversible step in the cancellation of an aircraft type, as the following extract from the Armaments Staff general report of 9 February 1945 shows:

"6-9 Feb. (3457) Proposal for cancellation of various aircraft types was initially approved by the Reichsmarschall only with respect to test flying, expanded final decision to follow after consultation with HDL Saur."[568]

Saur did not want to get involved in this matter, however. Göring alone was to make the decision on cancellation of certain aircraft types

"as a consequence of the fuel—and in the final analysis—materials situation."[569] The fuel shortage finally became so serious that an order was issued at the end of February 1945 that Ju 88 night fighters could only be built if, in addition to engines, sufficient fuel was available for acceptance test flights.[570]

The official end of Ju 388 production can be traced to two high level meetings concerning required changes to Delivery Plan 227 held on 8 and 10 February 1945. On 8 February, in the presence of Generalmajor Diesing, the Chief of Aircraft Procurement and Supply (TLR), *Oberst* Petersen, commander of test stations, and *Oberst* Knemeyer, head of development, a preliminary decision was made on the aircraft to remain in production based on an assessment of the raw materials situation.[571] Ultimately, the only types left in the program were types 8-262, 8-152, and 8-162, with 1,500 aircraft per month, and the 8-234, with 200 examples per month. At the same time, as the 8-219 and 8-335 were dropped, testing and development of types 8-388, 8-155, 8-163, 8-263, and the engines 9-8222, 411, 803, and 605 was temporarily suspended "pending a decision by the Reichsmarschall." Also among the cancelled aircraft was the Ju 88; however, production of the Ju 88 G-7 was allowed to continue at the present rate (50 aircraft per month).

This became even clearer in another meeting with Saur on 10 February 1945. According to program master plan No. 22, orders were issued for production to cease, and no more raw materials were to be taken from stockpiles for construction of these types.[572] Concerning the Ju 388, it said:

"Production is to be halted at the end of February. Labor forces thus made available are to be reported to HAF/A for reassignment."

Production of the Ju 88 was also to continue with, "consumption of materials and maximum of 50 aircraft." There were to be "no new orders for raw materials," and all pending orders were to be cancelled immediately.

Aside from the reduction in the means of production caused by the war, and the effects of the bombing campaign against the transportation infrastructure, without which decentralized production was impossible, one must bear in mind the Allied air raids on the airframe factories[573] in central Germany, where Junkers had dispersed its production sites. Planning for the raids began in December 1943, and their execution began in early 1944. One of the first culminations of these operations occurred in late February 1944 during the so called "Big Week." Allied bombers struck 21 aircraft manufacturing and repair facilities, including Leipzig, as well as several Junkers operations, especially Bernburg, where production of the Ju 88/188 briefly fell by 100%. The first American attack on the main facility in Dessau followed on 28 May, however, only 12 of 164 bombers hit the target. The aero-engine works, factory airfield, airframe halls, and test stands all suffered damage. Another attack by 268 bombers in five waves followed on 30 May, and it was more successful. Several manufacturing and assembly halls suffered 30 to 70 percent damage. The Otto-Mader-Flugmotorenwerke was almost completely destroyed, resulting in the loss of 80% of Jumo 222 production. Although the buildings were severely damaged, the dispersal of production facilities, which was already in progress, prevented any serious drop in output. Further attacks on the main facility at the north end of the airfield took place on 20 July and 16 August 1944. The latter raid, in particular, caused severe damage to the final assembly area, the acceptance flight hangar, Hangars 225 and 226, and the wind tunnel. Production of Jumo engines fell by

75-80%; consequently, the production of aero-engines was halved in the following quarter. In spite of all the damage, the actual target of the Allied raids—the production of jet engines—was not hit. On 28 September there was a massive raid by 1,000 bombers on Magdeburg, Leuna, and Merseburg. Bombs also fell on Leipzig and Dessau. The heaviest attack on Dessau did not come until 7 March 1945, however, when the city itself was about 80% destroyed. The Junkers aircraft and engine factories also suffered heavy damage. On 10 April the Volkssturm was called up in Dessau, and on the 21st American troops entered the grounds of the Junkers factories and the city. They stayed until 1 July, then the Soviets moved in.

...from the monthly reports

The June report by Junkers' aircraft operation placed fresh emphasis on the Ju 388. A new preliminary decision was received for the total of 23 Ju 388 test beds with Jumo 222 (some with lengthened fuselages) and Jumo 213 E engines. However, the omens for further development of the Ju 388 and Ju 488 were anything but good, given the massive cuts everywhere else.[574] The fact that the jigs for the Ju 188 G-2 built for Opel by ELBAG had to be accounted for suggested that the program was about to be cancelled, or at least that company's part in it. An order for production equipment for construction of the Ju 388 in the amount of 2.9 million Reichsmark was mentioned. Delivery Plan 225 was in effect at that time, however, it was soon to be replaced by Delivery Plan 226. Of the ten L-0 aircraft ordered, to date only three had officially been delivered.

Remarkably, a backlog of nine Ju 188 G-2 fuselages was reported in the August report. At that time the production of Fw 190 wings by Junkers was proceeding apace. The monthly report stated that 109 of a total of 273 sets of wings had already been delivered. The order overview of 30 August 1944 brought considerable changes to production aircraft. In addition to various major components for the Ju 88 A-4/S-3, and small numbers of the Ju 188 A-2/D-2 (one each), Ju 288 C-1 (2 examples), and Ju 352 A, the following table shows the types to be built by Junkers (see below).

A report entitled "Ifa Forest Camp," compiled by the aircraft division's accounting department at the at the end of October, and based on accounts received by the end of September, reported the cancellation of several projects of significance to the company.[575] By order of the Armaments Staff, work had to be suspended on the Ju 188, Ju 287, Ju 288, Ju 290, Ju 390, and the Ju 488. Production of the Ju 88 was limited to the G version, while only the J and L versions of the Ju 388 were to continue. At that time Issue J8 of Delivery Plan 226 was in effect, limiting the aircraft types built by Junkers to the Ju 88 G-6/G-7 and the Ju 388 J-3. Beyond that, all the company was building was wings for the Fw 190 and Ta 152. It was also involved in the delivery of Mistel composites. Junkers took over development of the Me 163 C. Initially designated "Type 142," it was later renamed the Ju 248.

Altogether, the order summary for 30 September 1944 shows 1,435 Ju 88 G-6s, 980 G-7s, and 1,375 Ju 388 J-1s, plus 127 L-1

Junkers Type	Price	Number Ordered	Remarks
Ju 88 G-1	130,000 RM	669	
Ju 88 G-2	130,000 RM	2,106	In fact no G-2s were built.
Ju 88 G-6	130,000 RM	899	
Ju 88 G-7	130,000 RM	570	
Ju 188 G-2 (fuselages by Opel)	60,000 RM	71	Interestingly just the fuselages were ordered, all other components were to come from standard Ju 188 production.
Ju 287	(illegible)	433 (154 by FSD)	
Ju 388 J-3	200,000 RM	140	At least the J-3 with Jumo 213 E engines was to (also) be built by JFM.
Ju 388 J-1 cockpits	50,000 RM	194	Obviously the majority of cockpits for aircraft built by licensees were to be built by JFM.
Ju 388 L-3 cockpits	50,000 RM	180	
Ju 388 J-3 cockpits	50,000 RM	329	

The Ju 88V58 was the prototype for the Ju 88 G-1. The similarity between the ventral weapons trays of the Ju 88 G and Ju 388 J is striking. Griehl Collection

cockpits and 229 J-3 cockpits (plus one Ju 188 G-2). It also includes orders for 115 Mistel composites.

Also mentioned was cooperation with Japan on the Ju 388 and Ju 390, as both types were expected to be produced under license there in the future.

Overall, in the aircraft division's monthly reports the Ju 388 played a rather subordinate role to other aircraft types under development and in production between mid-1943 and mid-1944. Given that production of the type was almost completely in the hands of other manufacturers, this is not surprising. The reports instead concentrate on the advanced pressurized cockpits, the Ju 88 G-6, which was built by Junkers itself, and the G-7, which was just beginning flight trials.

...on the basis of construction directives

The following list of construction directives provide a precise chronological overview of the orders received for Ju 388 prototypes and pre-production aircraft beginning in 1942:

21 Oct. 1942, Dessau Construction Directive 7 188 0010/025/000	Construction directive for the development and representation in mockup form of the new 3-man cockpit for the night-fighter and high-speed bomber versions of the Ju 188; "six fixed weapons (MG 151 or MK 108) are to be installed for the night-fighter role."
24 May 1943, Dessau Construction Directive 7 188 26 10/001	Production of 10 Ju 188 L-0; *as per contract RLM GL/C-E 2, Ju 188 aircraft Werknummern 300 001 to 300 010 are to be produced, the first aircraft to be converted in Dessau, all others in Merseburg.*
26 May 1943, Dessau Amendment I to Construction Directive 7 188 00 10/037/001	By order of the RLM GL/C-E 2, Ju 188 J-1 night-fighters are to be produced by conversion, the aircraft will have the *Werknummern* 500 001/V4 and 500 002/V5.
31 May 1943, Dessau Amendment I to Construction Directive 7 188 00 10/038/001	By order of the RLM GL/C-E 2, Ju 188 K-1 bombers are to be produced by conversion; the aircraft will have the *Werknummern* 500 003/V6 and 500 004/V7.
6 Aug. 1943, Dessau Construction Directive 7 388 26 10/001/000	Production of 10 Ju 188 L-0 high-speed reconnaissance aircraft; *as per RLM GL/C-E 2 contract, a further 10 Ju 188 aircraft are to be built using major components [...] for the Ju 188 L-0 [...]. The number of prototypes ordered is thus increased to 20.*
4 May 1944, Dessau Amendment I to Construction Directive 7 388 26 10/000	Production of 20 K-0 bombers (formerly 188 K-0), "*as per the RLM GL/C-E 2 contract, an additional 30 aircraft are to be converted in Merseburg using major components for the Ju 188 L-0, raising the total number of bombers to 50.*" Two sets of components are to be transported to Toulouse for the special conversion.
11 May 1944 Construction Directive 7 388 00 10/005/000	Construction of 8 test-beds from the Ju 388 with 9-8222 A/B (Jumo 222) engines as per advance order SS 4940-5101-0576/43 of 14 March 1944
11 May 1944 Construction Directive 7 388 00 10/006/000	Construction of 4 operational aircraft from Ju 388 with 9-8222 A/B (Jumo 222) engines; as per SS 4940-5101-0576/43 of 14 March 1944 the Ju 388 J-1 *Zerstörer* is to be developed.
19 June 1944 Amendment II to Construction Directive 7 388 26 10/000	Construction of 50 Ju 388 K-0 (formerly 188 K-0) bombers; the aircraft are to be constructed or retrofitted with auxiliary fuel tanks in the port wing and two auxiliary oil tanks in both wings.
18 July 1944 Amendment I to Construction Directive 7 388 0010/007/001	Construction of 5 test-beds from Ju 388 with Jumo 213 E engines as per construction directive of 23 May 1944. Two are to be built as engine test-beds and three as L-0 model operational aircraft
18 July 1944 Amendment III to Construction Directive 7 388 26 10/000	The 50 Ju 388 K-0 bombers are to be built with Equipment Set M 14, but only on the starboard wing. It is anticipated that no provision will be made for jettisoning this equipment set.

12 Aug. 1944 Amendment IV to Construction Directive 7 388 26 10/000 K.A Numbers Ad 1: 7 188 00 10/401 Ad 2: 7 188 10 10/005 for engine test-beds; 388 00 10/006 for four J-2 prototypes Ad 3. 7 388 00 10/007	Major components are to be taken from the batch of 50 Ju 388 K-0 bomber aircraft for the following purposes: 1. WNr. 230 152 and 230 154 for the Ju 188 special version flying mock-up (V401/V402) 2. 7 sets of major components 230 695 (V9), 230 721 (V10), 230 727 (V11), 230 733 (V12), 230 737 (V13), 230 901 (V14), 230 903 (V15) as test-beds for the Jumo 222 A/B engine or prototypes for the Ju 388 J-2. 3. 5 sets of major components 230 157 (V21), 230 691 (V22), 230 697 (V23), 230 723 (V24), 230 729 (V25) as test-beds for the 9-8213 D (Jumo 213) engine or prototypes for the Ju 388 J-3. 4. 2 sets of major components 230 902 (V27) and 230 906 (V28) for the conversion of DB 603 U test-beds.
12 Aug. 1944 Amendment II to Construction Directive 7 388 00 10/005/002	Construction of 5 test-beds from Ju 388 with Jumo 213 E. The second Ju 388 test-bed with 9-8213 D-0 engines, destined for the OMW flight department, is to be equipped with wiring for sensors and MW 50 system (*Rüstsatz* B7) in the second fuselage bay. The equipment set has been ordered from ATG and is to be delivered at the end of August.
12 Aug. 1944 Amendment I to Construction Directive 7 388 00 10/005/001	Amendment of contract for construction of Ju 388 test-beds with 9-8222 A/B (Jumo 222) engines, reducing number of aircraft from eight to just three, as per a decision by the type supervisor.
12 Aug. 1944 Construction Directive 7 388 00 10/008/000	Development and mock-up of an extended-range Ju 388 plus construction of three prototypes.
8 Sept. 1944 Construction Directive 7 388 00 10/011	Development and prototype installation L 10 in Ju 388 M-1; development, design and prototype installation of Equipment Set L 10 beneath the fuselage and modification of the PVC intermediate bearer and fairing for ETC 2000.
17 Oct. 1944 Amendment III to Construction Directive 7 388 00 10/007/003	Construction of 5 Ju 388 test-beds with Jumo 213 E engines; the V23 WNr. 230 697 is also to be equipped with a Dachlaurin fire-extinguishing system.
23 Oct. 1944 Amendment V to Construction Directive 7 388 26 10/000	After the deductions from the 50 Ju 388 K-0 bombers already ordered, there remains a "remnant small series" of 21. WNr. 230 698 has been earmarked for high-altitude ballistic tests at Udetfeld. To be equipped with *Rüstsatz* M 20. The remaining 20 K-0s will be designated L-0 Nacht and as per construction directive and amendments I-IV will be built with the following changes from the K-0: • no fire protection on wings and landing flaps • flame-dampers (if not available, delivery to follow later) • *Rüstsatz* M 20 (2 Mount II and Rack 8 in forward fuselage bay • APZ A 6 deleted from Peil G 6 • Reinforced undercarriage with steel version of lower retraction strut All aircraft are to be modified to the latest standard and in particular are to be fitted with external ram-air intakes. As a result of the cancellation of the Ju 388 V24 and V25, the conversion order for K-0 aircraft is increased from 33 to 35.
26 Oct. 1944 Amendment I to Construction Directive 7 388 26 10/001	Construction of 5 test-beds from Ju 388 L-0 with Jumo 213 A and E engines. After delivery of engines by the *E-Stelle* Rechlin, the V31, WNr. 300 294, is to be equipped with 9-8231 D power plants with Jumo 213 E engines by FSD-M (Merseburg).

30 Oct. 1944 Amendment I to Construction Directive 7 388 0010/006/000	As per a decision by the type supervisor, the Ju 388 J-2 program is to be reduced from four operational *Zerstörer* to just one, the V15, WNr. 230 903.
1 Nov. 1944 Construction Directive 7 388 00 10/013/000	As per a decision by the development steering committee and the type supervisor, 6 Ju 388 J-3 prototype aircraft will be built for the anti-Mosquito role. The aircraft are to incorporate the aerodynamic improvements of the Ju 88 G-7 and will be equipped with MW 50. Tanks for the MW 50 system are being ordered for these six J-3 prototypes, the Ju 388 V14/2 (WNr. 230 901), and seven Ju 88 G-6s or G-7s.
13 Nov. 1944 Amendment II to Construction Directive 7 388 00 10/005/002	Subsequent to the reduction in the contract for construction of 8 test-beds based on the Ju 388 with 9-8222 A/B (Jumo 222) engines to just 3 test-beds, as the result of a decision by the type supervisor 6 aircraft are now to be built from the K-0 series. Dorsal position with MG 131, day camouflage, 4-blade VS 19 propellers with 4 m diameter, 900-l external tank, strengthened undercarriage; no camera installation.
20 Nov. 1944 Amendment I to Construction Directive 7 388 00 10/013/001	Creation of 6 Ju 388 J-3 prototypes for anti-Mosquito role, to be equipped with wooden nose for Morgenstern antenna. In addition to the six nose cones for Ju 388 Mosquito hunter, two reserve nose cones and one nose cone for the Ju 388 V14/2 are also to be procured. Contract number for V14/2: 7 388 00 10/12/001.
29 Nov. 1944 Amendment II to Construction Directive 7 388 00 10/006/002	Creation of a prototype for the 8-388 J-2 with 9-8222 A/B (Jumo 222) engines; differences between the Ju 388 V15, WNr. 230 903, and the Ju 388 V2 are listed.
5 Dec. 1944 Amendment VI to Construction Directive 7 388 26 10/000	The 20 L-0 *Nacht* under construction are to be equipped with night cameras, infrared lighting and operational dorsal position.
2 Feb. 1945 Amendment III to Construction Directive 7 388 00 10/012/004	Creation of three Ju 388 J-1s with BMW 801 TJ engines. V14/2 is not to be fitted with wooden nose for Morgenstern antenna. Flame dampers also deleted, as it cannot be delivered. Oblique armament behind Frame 15 at an angle of 80°.

Analysis of the Actual Distribution of Production

The following list of companies or factories engaged in production of the Ju 388 is based on a production distribution overview compiled for the Allies:

	Fuselage	**Wings**	**Tail Section (Rf)**	**Tail Section (Tfl)**	**Final Assembly**
Junkers	planned	planned	planned	planned	planned
ATG	X		X		X
Siebel		X			X
Volkswagenwerk		X	X	X	planned
Weser		X			X
AEG			X	X	
Württ. Metallw-Fabrik			X		
Neue Kühlerfabrik				X	
Metallwerke Niedersachsen	X				planned
Henschel	planned				planned
Letov		planned			
Böhmisch-Mährische Maschinen-Fabrik		planned			planned

The original table shows that production by those firms that actually built the aircraft ended in February 1945. Planned participation is indicated by the remark "Startup planned, not implemented."

The big surprise in this list is the Württembergische Metallwarenfabrik, which today is known worldwide as a maker of housewares under the abbreviation "WMF." It had already taken part in production of tail sections for the Ju 188. On 4 January 1945 the company, which was based in Geislingen, was asked "to ensure that the tail sections for the 8-388 are sent as soon as possible."[577]

This overview also confirms that, apart from prototypes and pre-production aircraft, production of the Ju 388 was to be carried out exclusively by companies other than Junkers. At first glance its may seem odd that the Junkers Ju 388, the company's most technologically advanced design next to the Ju 287, was not assembled in the Bernburg plant, with the usual delivery of subassemblies from the other branch plants, even though by mid-1944 Junkers-Bernburg, like the other Junkers branch plants involved in production of the Ju 88/188, had suffered only moderate damage from Allied bombing. FZB's full capacity was never used, not even during the period of highest output in 1944. Why were no Ju 388s assembled there, even though Junkers records show that damage to the buildings was within limits, and that production could have been resumed quickly on at least one production line at any time?

From our point of view, the reason for the outside production of the Ju 388 may be found in Delivery Plan 226, 1st Edition, 15 May 1944.[578] Of the total 5,759 Ju 388s of all versions planned until the end of March 1946 (the end of the planning period), just 75 Ju 388 J-3s were to be built by Junkers itself, and while Henschel and Siebel were supposed to begin deliveries in mid-1945, production by Junkers was not to begin until December. The reason for this late start was that all of the 5,205 Ju 88 Gs envisaged in the delivery plan were to be built by Junkers itself. 1,644 of a total of 1,921 Ju 88 G-1s remained to be delivered, plus 2,050 Ju 88 G-2s with Jumo 213 E engines, and 1,511 Ju 88 G-6s with Jumo 213 As. Nearly 300 Ju 88 Gs per month would have to be built for this ambitious program to succeed, and the only factory that could probably have come close to that was Bernburg where, in spite of Allied bombing, six to seven aircraft were being completed per day in 1944 (after the introduction of a three shift operation).

In addition to mass production of various sub-types of the Ju 88 G, Junkers was supposed to begin production of the Ju 287 in February 1945, and achieve a monthly output of 100 machines by January 1946. With this, Junkers would probably have reached the

A Ju 88 G-1 at Grove in 1945. This type was built in large numbers by Junkers itself; consequently, the Ju 388 had to be built by licensed companies. Griehl Collection

limits of its production capacity, making it necessary for the Ju 388 to be built by other companies.

The fact that Siebel and ATG were envisaged as playing a leading role, along with Weserflug, Henschel, and several others, can probably be explained by their participation in final assembly of the Ju 188. While the last Ju 188 was supposed to leave the Junkers production line at the end of July, production by ATG was not scheduled to terminate until the end of the year, while Siebel was to keep building the type until March 1945. Production of the Ju 388 was to begin slowly, keeping pace with this schedule.

In spite of the numerous changes undergone by Delivery Plan 226, even in one of the latest versions of Edition 2, dated 15 September 1944, production startup of the Ju 388 J-3 by Junkers was not envisaged before April 1945, even though the number of aircraft to be built there had risen to 1,375.

Other factors that ultimately hindered the startup of production of the Ju 388 J-3 by Junkers in early 1945 were the constantly deteriorating production conditions. Most significant were the increasing relocation (dispersal of factories and use of underground facilities) because of enemy bombing, and the simultaneous, progressive destruction of the transportation system. Another reason was the concentration on a few fighter types. Thus, in 1944 most former "bomber factories" had to begin the tedious process of converting to the production of fighter components.[579] In the end Junkers was left with just the Ju 88 G and the Mistel, plus other companies' aircraft, such as the Fw 190/Ta 152, and especially the He 162.[580] In mid-1944 plans called for a total of 12,000 examples of the latter to be built by Heinkel and Junkers.[581] The licensees, especially SFW and ATG, were also increasingly brought into the production of fighter components.

Even though events quickly turned the plans contained in Delivery Plan 226 into so much waste paper, its decisions regarding the basic distribution of production—the Ju 88 G by Junkers, and the Ju 188/388 family under license—were instrumental in the actual production of the Ju 388 in autumn 1944.

Production by Junkers (JFM)

Dessau and Merseburg

The production of prototypes and pre-production aircraft began in parallel at Merseburg and Dessau. As the building of prototypes has already been dealt with in connection with factory testing, this chapter will concentrate on the construction of the pre-production machines.

The two surviving hangars at Merseburg in 2002 after decades of use by the Soviet armed forces. The intact hangar is used as "Luftpark Merseburg," a private aviation museum. Handig Collection

The second hangar fell victim to fire in 2002, and its remains were removed. The second Ju 388 (the V8) was built in one of these four Junkers Hangars. Handig Collection

Taken on 23 March 1945, this aerial photo provides an excellent view of Merseburg airfield and the production buildings. Several Ju 388s may have been among the aircraft in the blast pens and parked on the right side of the airfield. Luftbilddatenbank Würzburg

Production of the *Werknummer* block 300 002 to 300 010 began at Merseburg at the end of 1944. There exists a report by Dessau Production Inspection concerning 300 002/V8, the only prototype in this block, by Inspector Hasse dated 25 February 1944, with comments by "Kovermi" on 6 March:[582]

A document on the stocks of and requirements for BMW 801 TJ-0 engines for the Ju 388 in Dessau and Merseburg dated 24 January 1944 gives information about in which sequence, and in what location, the next prototypes were to be built. By that time nine of the rare turbosupercharged power plants had arrived at Dessau; however, four of these had been given to *Lufthansa*. A total of three engines had been used up for the Ju 388 V7, as one had already been replaced following a failure on 20 January 1944. The two remaining engines were supposed to be installed in the Ju 388 V3. A total of five engines were on hand in Merseburg (FSD-M), two each for WNr. 300 002/V8 and WNr. 300 003, plus one for WNr. 300 004. It is impossible to say for certain if these engines were in fact installed in the specified aircraft, for another document contains differing information on the delivery of engines for the aircraft completed at Merseburg.

Two engines were required for each of the subsequent prototypes at Dessau. The delivery dates, coordinated with the completion of airframes, were as follows:

- **Ju 388/V5:** 20 January 1944
- **Ju 388/V1:** 10 February 1944
- **Ju 388/V4:** 20 February 1944
- **Ju 388/V6:** 10 March 1944
- **Ju 388/V2:** 20 March 1944

One of the first pre-production machines built in Merseburg at an advanced stage of construction in early March 1944. The engines have already been installed. It is not known if the "No 5" on the cockpit refers just to the cockpit, or to the whole aircraft. Griehl Collection

As well, delivery of one spare engine for each aircraft was to take place on 15 January, and on 10 and 20 February. Three more engines should therefore have arrived at Dessau by the end of January.

The fuselages used in the construction of the prototypes at Dessau were delivered by FZA. These exhibited serious shortcomings in quality, such as poor riveting and dents in the skin. After the delivery of the first two fuselages, the chief inspector from Aschersleben was ordered to Dessau. This resulted in a short lived improvement in quality; however, the 6th and 7th fuselages were somewhat worse. But there were also errors in the conversion of Frames 21 to 33 carried out in Dessau, as the company had been forced to use Russian prisoners to do the work.

The Ju 388 L-0 WNr. 300 004 was completed at Merseburg at about the same time as the V8. It was fitted with engines 385 187 and 385 183, which had been delivered on 28 February 1944. Static tests were completed at the beginning of March, and only new cooling fans remained to be installed. The aircraft was still not serviceable on 11 March, however, because of work on the pressurized cockpit. By that time the engines had already been installed in the next aircraft "388 L0/003 and 388 L0/005."[585] WNr. 300 003 and 300 004 had been earmarked for testing at Rechlin and BMW.[586] WNr. 300 004 was later tested extensively at Rechlin, and on 13 May 1944 300 003 was flown by Lehmann from Merseburg to Fürstenfeldbruck. The aircraft subsequently served BMW as a test bed. Like the other aircraft, WNr. 300 004's reduction gearing was changed as part of a BMW modification program (15 March 1944). Lehmann flew the aircraft for first time on 22 March; however, the flight had to be aborted after the starboard engine began leaking oil from the crankcase air vent. Six days later during a test flight (crew Lehmann/Lange) the starboard engine developed the same problem, and both engines lost oil from the reduction gearing. On 12 April, during a radio check flight, the oil pressure in the port engine was only 7 atm, and the aircraft subsequently had to land on one engine. Three days later the same engine caused problems during a high-altitude flight by Lehmann and Lange. It leaked so much oil that 55 liters were consumed in just two hours; the flight had to be aborted. There were also frequent engine problems with the other L-0 machines. On 20 April 1944 Lehmann and Lange completed a high-altitude flight in WNr. 300 003, having taken the aircraft up on its maiden flight two days earlier. The port engine's boost pressure fluctuated and the engine vibrated. Engineer Schabert of the BMW field service was able to form his own opinion of the state of power plant testing on 24 April 1944, when he accompanied pilot Lehmann during a test flight in WNr. 300 003.

Cockpit of a Ju 388 under construction about mid-April 1944. The control yoke, airspeed indicator, and FuG 217 radar scope have yet to be installed. The view through the canopy suggests that the cockpit is sitting on the ground. Crow Collection

Because of delays in delivery of four-blade VDM 9-12188 propellers, both WNr. 300 003 and 300 004 initially had to be equipped and delivered with three-blade VDM 9-12152 propellers. This procedure was coordinated with the *E-Stelle* Rechlin. A message from the Dessau design bureau's development section dated 13 April contained provisional pitch settings, as the three-blade propellers had yet to be tested in flight.[586] This raises the question, however, as to whether WNr. 300 004 made its maiden flight with "borrowed" propellers, or if Dessau was not quite up to the latest standard.

During static tests on 5 May 1944, WNr. 300 005's starboard engine only developed 1.45 atm of boost pressure. Not until take-off did it rise to 1.52 atm. On 10 May a static test was carried out prior to a high-altitude flight. The starboard engine's turbine was found to be running rough. As there was no replacement turbine available, the entire starboard power plant had to be removed. There was little better luck with the replacement engine. On 23 May a high-altitude flight was aborted because of control problems with the starboard engine. Three inlet valves had broken, and the engine had to be changed again. Again on 31 May a high-altitude flight had to be aborted because of engine problems. This time the source of trouble was the port engine's boost pressure regulator. The starboard engine also had to be changed again, on account of a defective turbine bearing.

WNr. 300 007 also had problems with boost pressure. On 20 May 1944 a high-altitude flight had to be aborted because of fluctuating boost pressure (up to 0.1 atm), and there was a similar occurrence four days later. Another high-altitude test flight had to be aborted on 31 May because of surging boost pressure. The fluctuations only appeared when the turbines were engaged. The reason was unknown.[383]

These engine related problems were summarized in a letter from

Herr Roeder of the flight control section at the "Merseburg construction site" to the head of development at Dessau dated 1 June 1944, entitled "Faults encountered to date with BMW 801 TJ-0 engines during static tests and flight operations." It gave telling insights into the progress of flight testing at Merseburg in the spring of 1944.[383]

According to Roeder, the main problem was the BMW 801 TJ-0 engine. There were repeated engine failures, and their cause—at least in Merseburg—was a mystery. It was believed that all the criticisms of the airframe to date had been rectified, or at least identified:

"...several remaining [problems] such as: cooling of the air in the pressurized cockpit, butterfly valves in the climate control system, airframe vibration at higher airspeeds, and a pressure-tight mount for the dorsal weapon, have been identified and the responsible offices are working to resolve them. The necessary test flights can be carried out by a single aircraft. The situation in which we find ourselves is that Merseburg cannot adhere to its delivery program because the engines are still suffering teething troubles, and that because of this we will have to continue making test flights for the foreseeable future that are necessary only because of the engines."

Merseburg may have underestimated the airframe related complaints somewhat, perhaps because—in contrast to the engine problems—they believed they knew the causes, and could draw upon a vast store of experience in dealing with such problems. However, the airframe vibration at high speeds mentioned almost in passing concealed a serious aerodynamic problem: the original version of the central exhaust pipe on the BMW 801 TJ-0 significantly disrupted the airflow over the upper surface of the wing, causing flow separation, and thus leading to vibration.

In any case, the amount of engine testing Merseburg was forced to carry out was clearly too great, causing the overall test program to become inefficient:

"The BMW engine company is represented by *Herr* Schabert, a field service engineer, and a party of fitters under the leadership of Senior Fitter Ritt. The way it works is that *Herr* Schabert reports problems that arise here to BMW Munich by telephone, by letter or in person. In each case, valuable days are lost until new directions are received from Munich, even when *Herr* Schabert travels to Munich to give a verbal report. One must now also bear in mind that the telephone lines between here or Dessau and Munich are often down for days at a time, resulting in further unnecessary lost time."

At Merseburg, however, they had a very clear and relatively reasonable idea as to how the situation might be remedied:

"We therefore propose that, before any thought is given to delivering Ju 388 aircraft to anyone else, all aircraft—provided the airframes are in order—be delivered to Munich so that BMW might ensure that the engines are fit for handover. Along with the previously-mentioned time savings, the most important argument for this proposal is that every office that is or will be involved with the BMW 801 TJ-0, such as the development design bureau, bench testing, etc., is concentrated in Munich. The basis for modification proposals, evaluation of experience, etc., is thus vastly broader than if we as the airframe manufacturer—whether here or in Dessau—were to carry out these flights. It would, however, then be advisable for Junkers to send someone familiar with the Ju 388, *Herr* Lange of the flight department probably being the best choice, to Munich as an observer. It would also be advantageous if the specialist in Rechlin responsible for the Ju 388, Hauptmann Dr. Stahl, also sent a responsible officer to Munich. He could ensure that the tests by BMW are carried out quickly and do not end up in a purely scientific channel."

BMW apparently gained the upper hand on the engine problems, however, for there is no record of any aircraft being delivered to BMW.[587]

The aircraft themselves were not above criticism, however. A communication by Preuschen on 5 May 1944 reveals that the rear-view sight on the Ju 388 K-0 and L-0 was supposed to be fitted with an improved click stop mount.[588] A few days later type supervisor Hauptmann Dr. Stahl visited the parent factory in Dessau. The central theme of his visit was the rapid interchangeability of

This Ju 388 L-0, probably WNr. 300 292, in the final assembly stage at Merseburg, in roughly early July 1944. Another machine is visible in the background, while a fuselage and wings for another aircraft lie at the left end of the hangar. The hangar, which was not designed for large scale production, could not accommodate more than two aircraft. The last three numbers of the engine serial numbers are painted on the cowlings of the engines, which still lack cooling fans. Assuming that it does not refer exclusively to the pressurized cockpit production batch, the marking "L 12/23" is a further indication that the aircraft is WNr. 300 292, the 12th L-0. The jettisonable canopy section has not yet been fitted. EADS

The panels on the underside of the wing missing in the previous photo could not be fitted until the internal equipment, such as oxygen bottles, had been installed. Deutsches Museum

the periscope, and the use of the ZFR 4 A telescopic sight for the Ju 388 (only in conjunction with the MK 103).[589]

Very early on, Merseburg had to struggle with the problem of missing parts. In many cases, such as the 19-6205 hydraulic fluid cooler, which was only required for aircraft with a tail turret, or bomb rack mounts, which should have been ordered by the *Luftwaffe* itself, it was soon found that a misunderstanding was responsible for the absence of the parts. Some switches (for example, the Fl 32 346-3 toggle switch) were in fact unavailable, however, and had to be replaced with others.[590]

At the beginning of June 1944 the Merseburg factory documented further production and delivery difficulties involving the Ju 388 K-0 and L-0. Initially, several points still remained to be cleared up, such as the installation of the FuG 216, the click stop turbine control switch in the cockpit, and the DF loop installation, even though they had already been listed in "Kobü-Rh Minutes 3155" of 6 May 1944. This stated, for example, that no DF loop would initially be installed for the FuG 16 ZY, as its position could not be determined until tests on the Ju 88 V60 (the aircraft with the bulged bomb bay of the Ju 388 K/L) had been completed at Rechlin. Permission was also given to deliver the first six L-0 aircraft with three-blade propellers in the absence of four-blade units.[591]

Design drawings existed for the cooling of air in the pressurized cockpit, but no documentation was available. As before, access to the FuG 217 was unsatisfactory, and the dorsal gun position was still unacceptable, as air escaped through the gun mount padding. The aircraft would therefore have to be delivered without functioning dorsal gun positions. Merseburg also doubted the necessity of a new exhaust fairing, because to date flight testing of the V7 had failed to produce vibrations of the same magnitude as those described by the Dessau flight department. It questioned whether the cost of changing the fairing was justified. The completion dates for Ju 388 L-0s WNr. 300 009 and 300 010 were delayed because of the unavailability of "Weguper" adhesive, which was required for the canopies. The BMW engines continued to cause problems:

"To make a determination based on the condition of L-0s 005/007, which flew 10 and 7 hours respectively and encountered repeated engine problems, [in our opinion] on-time delivery cannot be expected in the foreseeable future. Particularly bothersome is the poor communication with the engine builder, which is quite apparent.

Furthermore, little improvement is to be expected through the installation of production engines, as the improvements made to this engine are unrelated to the problem sources identified here."[592]

Ultimately, a total of eight pre-production aircraft were delivered in June 1944. Two of these went to Rechlin. WNr. 300 007 went to Rechlin between 19 and 24 June. Already there at that time were WNr. 300 005 (PG+YE) and WNr. 300 004 (PG+YD), which had obviously been delivered at the end of April.

WNr. 300 006 (PG+YF) was delivered to the Experimental Air Signals Regiment in Köthen in the same period, however, it was not flown until the engines were replaced in mid-September. WNr. 300 008 and 300 009 also completed their BAL acceptance flights on 28 and 30 June 1944, "in preparation for delivery to Rechlin." They were probably among the next three aircraft ready for collection at the beginning of July, of which two were earmarked for Rechlin and one for BMW, where it was to be used as an engine test bed. The first two Ju 388 K-0s were also "industrially complete," apart from test flights and adjustment of the two axis autopilot by Patin.[593,594,595]

Taken a short time later, this photo may show the same machine during depressurization, the process of testing the integrity of the pressurized cockpit using compressed air (from the tank in the foreground). The air inlets for the climate control system's cockpit air cooler were installed after the wings had been painted. EADS

Type/WNr.	Date	Altitude/Temp.	Max. Cruise Speed	Max. Speed
L-0/292	26.7.1944	250 m/20 °C	388 km/h	430 km/h
L-0/293	28.7.1944	200 m/22 °C	388 km/h	429 km/h
K-0/158	28.7.1944	200 m/19 °C	377 km/h	405 km/h

At Merseburg, the low level performances of several aircraft were checked after completion. At the end of July WNr. 300 292, WNr. 300 293 (L-0), and WNr. 230 158 (K-0) carried out performance measuring flights equipped with broad-blade cooling fans and four-blade propellers.[596] In most cases the aircraft were crewed by Lehmann and Lange.

Production progress until 4 August 1944 can be traced on the basis of a document from Merseburg's production inspection branch.[597] For aircraft already delivered, a number of minor changes, such as the labeling of the fuel transfer system, were to be carried out by the users themselves:

Ju 388 L-0: WNr. 300 002 to 292
Ju 388 K-0: WNr. 230 151, 153, 155, 156

Construction of the second batch of Ju 388 L-0s (WNr. 300 291 - 300 00) had thus already begun, and the first ten Ju 388 K-0s had almost all been delivered. Next in line was the completion of aircraft WNr. 300 293 (L-0) and WNr. 230 158 (K-0), which were to be modified at the factory.

In addition to an overview of the aircraft already delivered, the records of a meeting held at Merseburg on 11 August 1944 also provide an insight into subsequent production planning at Merseburg. The first eleven L-0 machines had already been delivered. The status of the remaining machines was as follows:

L 13	WNr. 300 293 (DW+ZA)	Set aside without engines.
L 14	WNr. 300 294 (DW+ZB)	Set aside without engines.
L 15	WNr. 300 295 (DW+ZC)	Sent to Oranienburg by rail by 12 Aug. Remaining complaints were resolved either at FSD/M or Oranienburg.
L 16	WNr. 300 296 (DW+ZD)	Set aside without engines on or about 14 Aug.
L 17	WNr. 300 297 (DW+ZE)	Set aside on 20 Aug. at latest.
L 18	WNr. 300 298 (DW+ZF)	Set aside by 30 Aug. at latest.

The first five Ju 388 K-0s had also been delivered. Plans for the rest were as follows:

K 8	WNr. 230 158 (KS+TH)	Set aside without engines. A new experiment with attachment frame for modified dorsal gun position was carried out.
K 9	WNr. 230 159 (KS+TI)	Set aside without engines.
K 10	WNr. 230 160	Set aside without engines.
K 12	WNr. 230 692	Pressurized cockpit delivered to final assembly on 11 Aug. Aircraft set aside by 23 Aug. at latest.
K 13	WNr. 230 693	Completion delayed, as Weguper (special glue) for canopy construction not available.

KS+TH had already been flown, apparently with "borrowed" engines.

KS+TH had already flown, obviously with "borrowed" engines!

Delivery of the two Ju 388 K-0s—WNr. 230 153 (KS+TC) and 230 155 (KS+TE)—is confirmed by inspection reports filed by the Inspection Group at *E-Stelle* Rechlin dated 29 July[600] and 2 August 1944.[601] The reports, which contain 51[599] and 66 points, respectively, were compiled after an initial inspection (complaints and work to be completed) by aircraft inspectors Buttler and Rödiger, engine inspectors Hardt and Dahle, and equipment inspectors Opkamp and Desch. A similar complaint report exists for Ju 388 L-0 WNr. 300 009. Dated 10 July 1944, it is in the form of a telex from "junkers rechlin" to the parent factory at Dessau, and bears the comment "extremely urgent." It was sent after one of the ball and socket joints used to attach the wings to the fuselage was found to have been installed incorrectly.

The initial inspections of WNr. 300 005 (inspection report Ze 355/44) and 300 009, both delivered from Merseburg, "revealed serious faults, which are unacceptable for flight safety." A communication from the production inspectorate in Merseburg spoke of "errors and shortcomings, some of vital importance."[602] On 17 July a telephone call was made from the parent factory in Dessau to Merseburg to complain about the state of WNr. 300 008 after completion. These grievances resulted in a lively exchange of correspondence between the Junkers technical field service office in Rechlin, the parent factory, and "Construction Site Merseburg," where the aircraft had been assembled. It turned out that hoses fraying on the wing root fairing flanges and the canopy jettison system were the only real design shortcomings on the aircraft. All of the other complaints could be traced to production errors. Steps were taken immediately, and directions were issued for all main attachment points for the cockpit, power plants, undercarriage, wings, and tail section to be closely scrutinized during final acceptance checks.

Copies or edited versions of inspection reports from acceptance inspectors and complaint reports were distributed internally by Junkers to all responsible offices, including the "Series Production" design bureau (Kobü Rh). The latter initiated measures to address the complaints. When the problem was a design fault or production defect, the design bureau usually issued advance modification bulletins and modification directives, which were incorporated into planned or existing series production. If relevant advance modification bulletins already existed, the production branch was reminded explicitly to adhere to them.

The first eight Ju 388 L-1s (WNr. 340 081 (RT+KA) and WNr. 340 083-89 (RT+KC to RT+KI)) were also assembled at Merseburg. The fuselages, at least, are known to have been built by ATG. On 31 July 1944 ATG delivered the fuselage of a Ju 388 K-1 to Junkers Prague for structural load tests; this could explain the error in accounting for WNr. 340 082. As mentioned earlier, the new construction fuselage for WNr. 340 083 also arrived from ATG on 12 August 1944, and initially was supposed to be used in an extended range experiment. Thus, it could very well have been the Ju 388 L-1 fuselage which, according to the ATG production summary, was delivered to the design bureau on 12 August.

Ju 388 L-1, WNr. 340 084, RT+KD. Although its *Werknummer* is part of an ATG block, the aircraft was assembled in Merseburg, and was one of the first aircraft delivered to the Ju 388 trials unit in Rechlin. EADS

Ju 388 fuselages with tail turret under construction in about mid-October 1944. They must have been aircraft of the 340 *Werknummer* block. Griehl Collection

These eight aircraft must be assigned to ATG production, for the six Ju 388 L-1s delivered by ATG in September 1944 can only be explained by assuming that they included aircraft whose final assembly took place at the "Merseburg Construction Site"; likewise, the total of 54 aircraft delivered by ATG by the end of January 1945.[604] The reason for assembling the first eight ATG aircraft at Merseburg, which already had experience with the type, surely lies in the Ju 388 L-1's significance to testing at Rechlin, which wanted to obtain production aircraft as quickly as possible.

From 1 August to 19 September 1944, six Ju 388s at Merseburg completed 24 flights, logging a total of 23 flying hours. One was a Ju 388 L-0, the rest were Ju 388 L-1s. All of the aircraft had BMW 801 TJ engines, and one Ju 388 L-1, WNr. 340 083, had already been equipped with external ram air intakes. None of the aircraft fitted with internal intakes reached the high-altitude supercharger's maximum boost altitude of 11500 meters. Many times bright red flames appeared from the exhausts at intervals of two seconds, caused by defective sparkplugs and fuel injectors.[605]

The initial flight tests at Merseburg were not without incident, however. On 27 September 1944 Ju 388 L-1 WNr. 340 088 took off on a test flight that was to include checks of the radio equipment and autopilot. The crew consisted of pilot Lehmann, radio acceptance inspector Schiedhering, and Patin acceptance inspector Neubauer. Gross weight was about 12.2 tonnes. During takeoff the port engine reached 2,800 rpm, which was clearly too high. When the two engines were throttled back to sustained power (1.28 atm), engine speed was still too high at 2,550 rpm. An attempt to manually set the propeller speed failed, and the port engine began shaking badly. Lehmann therefore decided to land immediately with undercarriage and flaps lowered. He reduced power on the port engine to idle, and advanced the throttle for the starboard engine to

An Allied reconnaissance photo of Merseburg taken on 19 September 1944 shows four aircraft below the last hangar by the firing range. They must have been Ju 188s or Ju 388 pre-production aircraft, especially since several are parked without engines. Luftbilddatenbank Würzburg

combat power. What happened next was described in a report by the flight department:

"According to the pilot's statement, his speed during the approach was very low because of the low altitude, and he was forced to touch down 100 meters from the airfield boundary. After rolling 100 meters (first on the port mainwheel, then on the starboard), the aircraft struck the road embankment. This appears to have caused the port undercarriage retraction strut to break off. The aircraft then hopped over the road, the ditch, the airfield fence and the section of track on the airfield perimeter. When it touched down again the port undercarriage leg collapsed. The aircraft slid about 100 meters on the port nacelle and the starboard undercarriage before swinging 170 degrees to the left."

The crew survived the accident with nothing worse than a fright. One undercarriage leg and an engine with propeller had to be completely written off, however.[606] The aircraft was repaired, and at the end of October 1944 it was delivered to *Erprobungskommando* Ju 388. Of course, it was not possible to meet the aircraft's handover deadline, especially since a (another?) problem with the propeller pitch control could not be fixed, even after replacing the pitch control device (three times), rpm governor, propeller, and engine. According to BMW records the problem was probably caused by low oil pressure.

By 17 October the number of flights at Merseburg by Ju 388s with BMW engines had climbed to 35 since 1 August, with a total flying time of 35 hours. All eight of the Ju 388 L-s completed there had now been flown.

A fuel line in the port engine (315 320) of WNr. 340 085 broke as a result of a poorly brazed joint. All of the BMW 801 TJ-0 engines leaked fuel from the engine vents at combat power above 7000 meters, resulting in oil consumptions of 20 to 30 liters per hour. As well, the propeller pitch control devices on three machines had failed because of seized slide valves.

Two Ju 388 L-0s powered by Jumo 213 A and Jumo 213 E engines had also begun flying. The two aircraft had completed nine flights, logging seven hours flying time. On 9 October a Jumo 213 A (WNr. 102 152 27281) was destroyed during a static test by metal shavings in the oil filter, probably while installed in WNr. 300 294 (DW+ZB). During a flight on 14 October the port engine vibrated so badly that the flight had to be aborted. An engine strut had not been connected, the result of human error.

During a flight by WNr. 300 293 (DW+ZA) on 15 October 1944, Jumo 213 E WNr. 104 161 0092 failed after an oil tank ruptured, causing a seized piston in cylinder No. 8.[607]

The cockpit of a pre-production machine, perhaps WNr. 300 297, being fitted to the fuselage in September 1944. An undercarriage leg is also ready for installation. What appears to be damage to the canopy is actually a flaw in the negative. EADS

This enlargement from the front view of RT+KD shows another Ju 388 parked in the open without engines. EADS

This description of flight operations at Merseburg brings something interesting to light. With one exception, only Ju 388 L-0s with Jumo 213 engines were flown, and not a single Ju 388 K-0. Completion of the pre-production aircraft had obviously been put on hold in favor of the first eight Ju 388 L-1s. The completed airframes were either put in storage, or shipped out by rail as major components (like WNr. 300 295). It is likely that so few BMW 801 TJ-0 engines were delivered that priority was given to installing them in the Ju 388 L-1s, which were of vital importance to the test program. No further flights by K-0s at Merseburg can be verified until the beginning of November (Schreiber logbook).

Astonishing information about production at Merseburg is found in Priority Test Report No. 47/44 compiled by the *E-Stelle* Tarnewitz, and dated 25 November 1944. It states that a representative of the test station was supposed to go to Merseburg the following week to participate in firing trials by the first of three 8-388 J-1 aircraft. The first three J-2 night fighter prototypes were ultimately to be completed as J-1s, probably because of delivery problems with the Jumo 222. On 15 December 1944 the construction supervision specialist attached to Junkers Dessau reported from Merseburg about an inspection of the weapons installation in Ju 388 J-0 V10, consisting of two MK 108s and two MG 151/20s in the ventral tray.

Definitive cockpits for the Ju 388 J were also built at Merseburg. There are reports by the construction inspector dated 4 and 9 December 1944 concerning two Ju 388 J-1 pressurized cockpits with the serial numbers 1001 and 1002, both destined for ATG. The reports contain information on sealing, faults that had been found, or modifications that had not been carried out (as per modification directives), and missing parts. Among the latter were the obliquely mounted gunsight, lower periscope fairing, FA 15 control links, and the wooden nose cone for the Morgenstern antenna (8.388.109).

As a supplement to the preceding accounts of individual aircraft, there follows a complete list of known Merseburg production batches by block (see *Werknummer* summary in Appendix 1):

1. Nine Ju 388 L-0s, WNr. 300 002 (PG+YB) to 300 010 (PG+YJ), completed by August 1944, and delivered to various test stations, including one prototype, WNr. 300 002 (V8).
2. Ten more Ju 388 L-0s, WNr. 300 291 (DW+YY) to 300 000 (DW+ZH). Factory photographs confirmed that WNr. 300 291 had been completed by the beginning of April 1944. Just one more aircraft from this series was completed by August, while the rest followed in the late summer and autumn of 1944.
3. The first ten Ju 388 K-0s, WNr. 230 151 (KS+TA) to 230 160 (KS+TJ), were used for a wide variety of purposes. KS+TA was completed at the end of June-beginning of July 1944, and was used for factory testing at Dessau and Prague. KS+TB and KS+TD were supposed to become Ju 488s. KS+TC and KS+TE were in Rechlin from the end of July 1944, where KS+TE was destroyed on 25 August. KS+TF was the first Ju 388 L-0 Nacht. WNr. 230 157 became the V21, and in mid-August WNr. 230 158 to 160 were parked in the open without engines.
4. Other Ju 388 K-0s (WNr. 230 691 et al) were in the final assembly stage as of August, and a number of aircraft are known to have flown in November. The exact number of machines completed is not known.
5. Ju 388 K-0s WNr. 230 721 et al and WNr. 230 901 et al. Some are known to have been in the final assembly stage as proto types in November, and several were heavily damaged in an air raid.
6. Eight Ju 388 L-1s (WNr. 340 081, 340 083 to 340 089).

Where did the major components that were used by Junkers to build the prototypes and pre-production aircraft come from? The answer to this question is found in Delivery Plan 226.2, 2nd Edition, in the version of 15 September 1944. In the first half of the year ATG had already delivered 32 of the 50 Ju 188 A-2s it was to deliver, Siebel 81 of 100. The picture was similar for the Ju 188 D-2. At ATG 26 of 101 aircraft had left the assembly halls, while at Siebel, which was to build 196 aircraft, the figure was just 13. Behind these numbers, however, is a clue that some aircraft had been envisaged for "conversion Ju 388 L-0/K-0 in major subassemblies" by Junkers Merseburg, as follows:

	ATG	SFW
Ju 188 A-2:	8	18
Ju 188 D-2:	12	7

As approximately 40 pre-production machines had been produced at Merseburg by mid-August 1944 alone, or were in an advanced state of construction, they probably consisted of major subassemblies delivered previously, although the list does not make this clear. The numbers given for ATG coincide exactly with the factory data. Eight Ju 188 A-2s and 12 Ju 188 D-2s had been delivered to "JFM Merseburg as fuselage components" by 31 December 1944.[604] Some of the serial numbers assigned to the small batch of K-0s built at Merseburg fit seamlessly into those of Ju 188s produced by ATG. At the beginning of July 1944, for example, a Ju 188 with the *Werknummer* 230 147 appears in the Köpp logbook.

Other Branch Plants

According to our documents and statements by contemporary witnesses, no Ju 388s were completed at Bernburg-Strenzfeld (FZB), Junkers' most important final assembly facility.[609] In the period from 1943 to 1945, FZB mass produced such types as the Ju 52, Ju 88 C-6, R-2, G-1/G-6/G-10, S-1, and the Ju 188 E-1, F-1, A-2, and D-2.[610] Production of the Ju 88 G by FZB began in December 1944. Final assembly had already largely been transferred from Bernburg to Fritzlar and Langensalza ("Langenwerke AG," located on the airbase there). In the final phase the facility also produced Type 3a, 3b, and 3c Mistel combinations, and finally the He 162, 29 examples of which were completed by FZB.

As production of the Ju 388 J-3 by Junkers was not to commence until December 1945, the company's own efficient production network and modern four cycle final assembly,[611] which was even praised by the Allies, were apparently not used for the Ju 388. A few references to various Junkers branch plants and Bernburg in connection with production of the Ju 388 are found in the many surviving modification reports.

Generally, the RLM's order to disperse final production was a consequence of the attacks on the air armaments factories.[612] Factories were restricted to assembling 75, 100, or 150 aircraft per month, regardless of production capacity, in order to limit production losses in the event of attack.

How far the startup of Ju 388 production had progressed in other Junkers branch plants is uncertain at this time. The many surviving documents from German sources provide no clue to any such beginnings in the branch plants. A different picture is painted in

Numerous Ju 388s were obviously brought to Merseburg toward the end of the war. The reason for this is not known at the present time. Seen here is WNr. 560 044, a Weser built machine, which was still relatively intact at the end of the war. Petrick Collection

several reports by American intelligence teams. These technically capable units visited Dessau immediately after the cessation of hostilities, interrogating Dir. Thiedemann, Prof. Zindel, and Prof. Baade, along with several factories in the Dessau region that had been identified as targets. Independently of one another, the following documents all contain references to production measures relevant to the Ju 388, or the discovery of aircraft components:

1. CIOS Evaluation Report No. 149 of 27 June 1945,
2. CIOS Evaluation Report No. 69 Junkers Werke Dessau (Jumo and JFA Plants; as supplement to No. 149) of 8 June 1945,
3. CIOS Final Report XXX-99 of July 1945; main points of interest work cycles, spare parts questions, use of replacement materials/plastic, jigs and maintenance[613]
4. CIOS File No. XXXI-36 Junkers Aircraft and Engines Facilities (U.S. Technical Industrial Intelligence Committee; main point of interest production of jet engines) undated, but probably written in May 1945,
5. U.S. Strategic Bombing Survey Airframes Plant Report No. 1 – Junkers Aircraft and Aero-Engine Works Dessau, Germany of 4-8 May 1945,[614]
6. United States Strategic Bombing Survey Aircraft Division Airframes Plant Report No. 3 – ATG Maschinenbau GmbH, Leipzig, Germany, of 16-29 May 1945 (IWM).

As well, interrogation of plant managers revealed that their factories were in the process of converting to the He 162 and Ju 388. As the interrogations and inspections took place immediately after the plants were occupied, the statements can be judged to be largely accurate.[615] The descriptions by the teams which, for example, reported finding "50 Ju 288 night fighter fuselages" at Leopoldshall should, however, be treated with the necessary caution. While this could be the result of a simple typing error, it could also be a true error in identifying the components or aircraft they found. Finally, one must bear in mind that the only knowledge the team members had about the aircraft they were evaluating—if any at all—came from various reports and drawings made by the intelligence services, and that they were seeing them "in natura" for the first time. Correct identification was also hindered by the fact that they were often dealing with half completed subassemblies. The following information relevant to the Ju 388 is taken from the six reports named above:[616]

A He 162 A, WNr. 310027, built at the JFM branch plant in Bernburg. It was photographed in front of one of the former *Luftwaffe* hangars at Bernburg, in which final assembly of the type took place. Crow Collection

Plant	Information from Reports #1 to #5	Remarks
FZB Bernburg and ATG Plant 18 Bernburg	#1: Several buildings totally destroyed, others intact, such as, for example, the *"flight shed where He 262 was assembled."* Of the 70 of these delivered by the Tarthun underground factory, 15 were found in final assembly, 5 more destroyed on the airfield. It was also discovered that, in addition to Ju 88/Fw 190 combinations, the *"Ju 88 G-7 (long fuselage) with two Jumo 213 E engines, this being the night-fighter version"* was built there. #4: Ju 88, Ju 188, and Misteln (total of about 50) were found there, some disassembled, some apparently in flyable condition. #6 concerning ATG Plant 18: *"above-ground fuselage assembly plant at Bernburg."*	What was meant was probably the He 162 and Ju 388 G-10. The small intact hangar was a former *Luftwaffe* hangar. Plant Manager Kürner could not be found.
FZA Aschersleben	#1 factory heavily damaged, but main assembly shed intact, *capable of turning out 70-80 Ju 288s per month."* #4: about 20 Ju 188 fuselages loaded on rail cars, and another 30 Ju 188 fuselages in an assembly shed. According to manager Schönborn, the factory was in the process of converting from the Ju 388 to the He 162. Eight He 162 fuselages were discovered. Completed fuselages were delivered to Bernburg and Langensalza.	What was meant was probably the Ju 188. Here too there was a JFM dispersed operation in the "Zornitz Mineshaft."
FZH Halberstadt (and Klusberge) (Gross Schierstedt)	#4: "*The plant was apparently producing night-fighters of the Ju 388 category.*" Salt mine (Kalisalz) as dispersed operation of FZH.	JFM facility specializing in production of wings, therefore a switch to the Ju 388 would have been a simple matter, but possibly they mistook Ju 188 wings. Company airfield for JFM and Focke-Wulf from 1944, underground production in the subterranean plants MAKRELE I & II,[618] however, just the mechanical workshops had been transferred there, MALACHIT as well.
FZB Schönebeck	#1: main supplier of pressed parts for JFM and licensed production.	FZS was responsible for machining and shaping.[619] Moved to Tarthun (mineshaft).

FZL Leopoldshall (actually in Stassfurt)	#4: *Approximately **50 Ju 288 night fighters** were found in the main production building, and it was apparent that the plant was concerned largely with the production and modification of various conventional Junkers types, as well as the pilot production of a small number of He 162s."*	FZL specialized in the production of tail sections, and had several underground plants in salt and potash mineshafts (KALAG), which extended as far as Unseburg-Tarthun.
Klein Schierstedt (west of Aschersleben)	#4: Partly "normal" production and partly Ju 88 fuselage construction underground in potash mine.	
Raguhn		The intact dispersed plants No. 1-3 for "metal sheet production," as well as an evacuated drawings office.
Tarthun (Unseburg) (salt mine)	"Machine shop and fuselage manufacturing and assembly [...] He 262 main fuselage and undercarriage [...] Production commenced since August 1944 [...] some 7 delivered (to Bernburg for final assembly) some 50 in production."	Located here was the underground production in Shaft IV of FZA's "Anhalt'schen Salzwerke Westerregeln," as well as dispersed Heinkel and BMW operations.
Siebel Halle	#1: Damage not serious, however, one assembly shed completely destroyed. Capacity of 235 Ju 88/188/388 wing sets per month. "Fw 152" production followed "JU388" production with an output rate of 1.5 hours per wing set. Productivity was highly praised. #3: Interest was shown in the Ju 388's plexiglass windows. Also discovered was the "*plastic lining for the cockpit*" and "*silica gel.*"	To #4: Mentioned as suppliers of cockpit parts were Venditor, Troisdorf and Rommler, and Spremberg.
Siebel Schkeuditz	#1: about 300 workers were able to produce 60-65 Fw 152s" or 30-35 Ju 388s. *"The assembly of the Ju 388 night-fighter version had commenced, but was stopped in March 1945, in preference to the Fw 152. Eight fuselages were available nearly completed, but minus armament."* #3: *"Interrogation of works manager Dr. Seitz [...] Built Ju 88, 188, 152 and 388 [...] and wood wings for 204"*. Was described as a badly-damaged factory, tools, and equipment were destroyed or removed.	It is to be assumed that the fuselages found there probably came from ATG or from Merseburg.

ATG in (or in the region around) Leipzig, Altenburg, Kölleda, Prague	#1: Technical director Herbert Schade and Gerhard Schmitz, head of production planning, were among those questioned. The degree of destruction sustained by the 15 plants, 6 of them in the Leipzig area, differed greatly. Plant 1, for example, had suffered little damage, and had been cleared of aircraft production to allow repairs to be made. Because ATG was purely a production works, the Allied intelligence team had no particular interest in it, except that "*Instructions have been given to ATI to dispatch 3 pressurized cabins (Ju 388) to the U.K.*" The average output was given as 85 fuselages for the 88/188/388 per month, of which on average about 55 went to the company's own final assembly, while the rest were delivered to JFM.	
ATG Plant 15 Solvay salt mines near Bernburg	#1: Underground production in the Solvay salt mines and Solvay Hall. #3: ATG dispersed operation. According to ATG records jigs and tools were moved to a dispersed factory near Bernburg. ATG records reveal that the RLM even wanted to disperse the complete fuselage and pressurized cockpit production processes, but then changed its plans again.	It is safe to assume that this was Tarthun, as "mixed" JFM-ATG production there cannot be ruled out.

Mistel combinations like these were produced at Bernburg until the end of the war. The Ju 88 G-10/Fw 190 combinations were referred to internally as "Nero." Together with the Ju 88 G variants, they were the last Junkers designed types built by the company. The Misteln were frequently taken into the OKW Reserve and dispersed at other airfields. Haubner/Misna Collection

Interestingly, the CIOS reports deal throughout with Ju 388 night fighter finds. That Ju 388s could actually have been found at Bernburg cannot be discounted; the aircraft could have been diverted to Bernburg because of enemy action. The report by the Allied military government confirms that the Ju 388 was not being produced at Bernburg. No Ju 388 appears in the document, which meticulously lists the plant's production from 1943 to 1945.

The countless modification directives relating to the Ju 388 also include various Junkers branch plants among their many addressees. The modification directives for the Ju 388 J-3 received by JFM-FZH in Halberstadt provide a particularly interesting example in relation to the previous accounts. In March 1945, however, the Vorka Production Bureau pointed out that a modification directive 1504/45 for the Ju 388 J-3 at FZH "is no longer being worked on, as the Ju 388 J-3 is not being produced by FZH."

License Production

ATG

Overview of Junkers License Production by ATG

ATG was closely associated with Junkers even before the war through the license production of its aircraft, as the production numbers to 31 December 1944 show:[620]

An aerial photo of Leipzig-Mockau taken on 29 May 1944 after a heavy raid. Many large craters can be seen, but only a few aircraft—probably Ju 188s, as final assembly of the Ju 388 did not begin until October-November. Luftbilddatenbank Würzburg

Type	W33	W34 hi	W34 hau	Ju 52	He 111 E & H	Ju 86 A/D/G	Ju 88 A-1	Ju 88 A-5	Ju 88 A-4	Ju 188 A-2	Ju 188 D-2	Ju 388 K	Ju 388 L-1
Number Built	21	95	104	1058	326	162	130	160	1177	72	133	1	46

ATG played an important role in the building of jigs during preparations for production. At the beginning of June 1944 Dipl.-Ing. Lange, member of the Fighter Staff responsible for machinery matters, and the head of the airframes steering committee, *Generaldirektor* Frydag, reported to the Fighter Staff that 500 ATG jig builders were currently engaged in production of jigs for the Ju 388. This work was being hindered by a succession of modification directives from JFM.[621]

Production of the Ju 388 really got under way in autumn 1944, but slowed in December when the completion of aircraft was delayed by a critical level of missing parts. Deliveries were also delayed by the need to retrofit or modify aircraft in response to the steady stream of modification directives, a result of production and testing proceeding in parallel. These circumstances made it difficult for the builders to make "regular" deliveries to the air force, aside from individual examples. Notwithstanding this, the BAL continued accepting aircraft until the "production halt" ordered by the armament staff, in order to meet the specified targets. The BAL handed the aircraft over to the *Luftwaffe* by delivering them from ATG via "Autobedarf AG,"[622] a fictional company, to ferry units, operational units, or test stations. The Fighter Staff, and later the Armaments Staff, pushed the manufacturers hard to meet production targets, and the elimination of shortfalls was closely monitored.[623] The same uncompromising attitude was shown when the production halt was ordered and immediately implemented.

In 1944 license production for Junkers by ATG was organized in the following plants:[624]

The first Ju 388 delivered by licensee ATG was a structural test airframe, like the Ta 154 shown here. Handig Collection

Plant	Ju 388 Components	Remarks
Plant 1 Leipzig-Kleinzschocher	Fuselage production	Ju 88 fuselages from half shells
Plant 2 Leipzig-Eutritzsch	Fuselages, canopies, cockpits, ammunition belt feeds and boxes, and oblique weapons into February 1945.	Ju 52 fuselages.
Plant 3[625] Leipzig airfield Mockau/Seehausen Herr Kuckertz	Fuselages, assembly, final assembly, acceptance flights were later moved to Plants 15 and 17, cockpits.	Final assembly of Ju 52, Ju 88. Testing of the competing type Do 217 P was also carried out at the airfield[626].
Plant 4 Böhlitz-Ehrenberg	Cockpits also for J-1, Ju 388 tail fins[627].	Former piano factory, in which tails for Ju 52s and Ju 88s and finally Fw 190 tail sections were built.
Plant 5 Leipzig-Grosszschocher	Jig construction for ATG, Henschel, Weserflug, also Siebel and Menibum.	
Plant 6 Leipzig Fairgrounds, Halle	Fuselage production until bombing forced dispersal.	Fuselage equipment, seats, electrics, hydraulic, weapons. In February 1945 used to store cockpits and tail fins for 188/388 no longer needed, as well as Ta 152 fuselage parts.
Plant 7 Leipzig-Plagwitz Opposite Plant 1	Electrics, wiring bundles.	
Plant 8 Leipzig-Plagwitz	Installation of electrical equipment in J-1 fuselages[629] began there in January 1945	Former cotton mill, various assemblies and operation management offices.
Plant 15 Altenburg	Ju 188/388 test flying operation.	
Plant 16 Solvay and Solvay Hall salt mines near Bernburg	Preparations for underground production, originally entire fuselage production was to be transferred, this plan was not realized.	
Plant 17 Kölleda (modification facility at local air armaments depot)	Modification facility for 188/388.	In addition to the ATG modification facility used for Junkers aircraft, in Kölleda there was a Luftwaffe air armaments depot for the storage and repair of camera equipment. It also installed and adjusted cameras in aircraft[630].
Plant 18 Bernburg	Possible production of fuselages.	

Subcontracting was done by Siebel/Halle (wings), the Mitteldeutschen Motorenwerke (in Taucha-Portitz, Jumo 211), and Rheinmetall (Sömmerda, MG 131). As well, propellers were provided by JFM. At the end of 1943 production was switched to the successor type. Between January 1944 and January 1945, 72 Ju 188 A-2s and 139 Ju 188 D-2s left the ATG assembly halls; however, discounting the aircraft destroyed in air raids and those delivered to Junkers as major subassemblies, the number of aircraft that went directly to the

Luftwaffe must only have been 38 Ju 188 A-2s and 109 Ju 188 D-2s. Production of the Ju 388 began in the late summer of 1944.

In addition to building jigs for itself, JFM, Menibum, Weser, and Henschel, plus to a lesser extent for Siebel and BNM (or BMN?), early on ATG played a part in production of the Ju 388 by delivering the previously mentioned eight Ju 188 A-2 fuselages and twelve Ju 188 D-2s to the Junkers plant in Merseburg for the Ju 388 K-0/L-0 pre-production series. The delivery of a K-1 fuselage to the JFM bureau in Prague on 31 July 1944 for static tests marked the beginning of fuselage production. The final assembly of complete aircraft did not begin until September-October. Prior to this, as described earlier, the aircraft of the first production batch (WNr. 340 081 and others) were obviously assembled by FSD-M under contract.

From the end of February, instead of the cancelled Ju 388, ATG was reduced to building major components for the Ta 152. The implementation of Delivery Plan 228 by the ATG production planning of the TH-P[632] shows that, beginning in March-April 1944, ATG was to take over production of the fuselage, rear fuselage with empennage, and the final assembly of the Ta 152 H-1/R11 and C-1/R11 series in close cooperation with SFW. Preparations for final assembly were in the initial stages when the war ended, as confirmed by the CIOS and Air Intelligence reports.

In December 1943 ATG had become a target of the Allied air forces.[633] The RAF struck for the first time on 4 December 1943, putting the airframe construction plant out of action for several weeks. 96 fuselages and 18 complete Ju 88s were destroyed at ATG

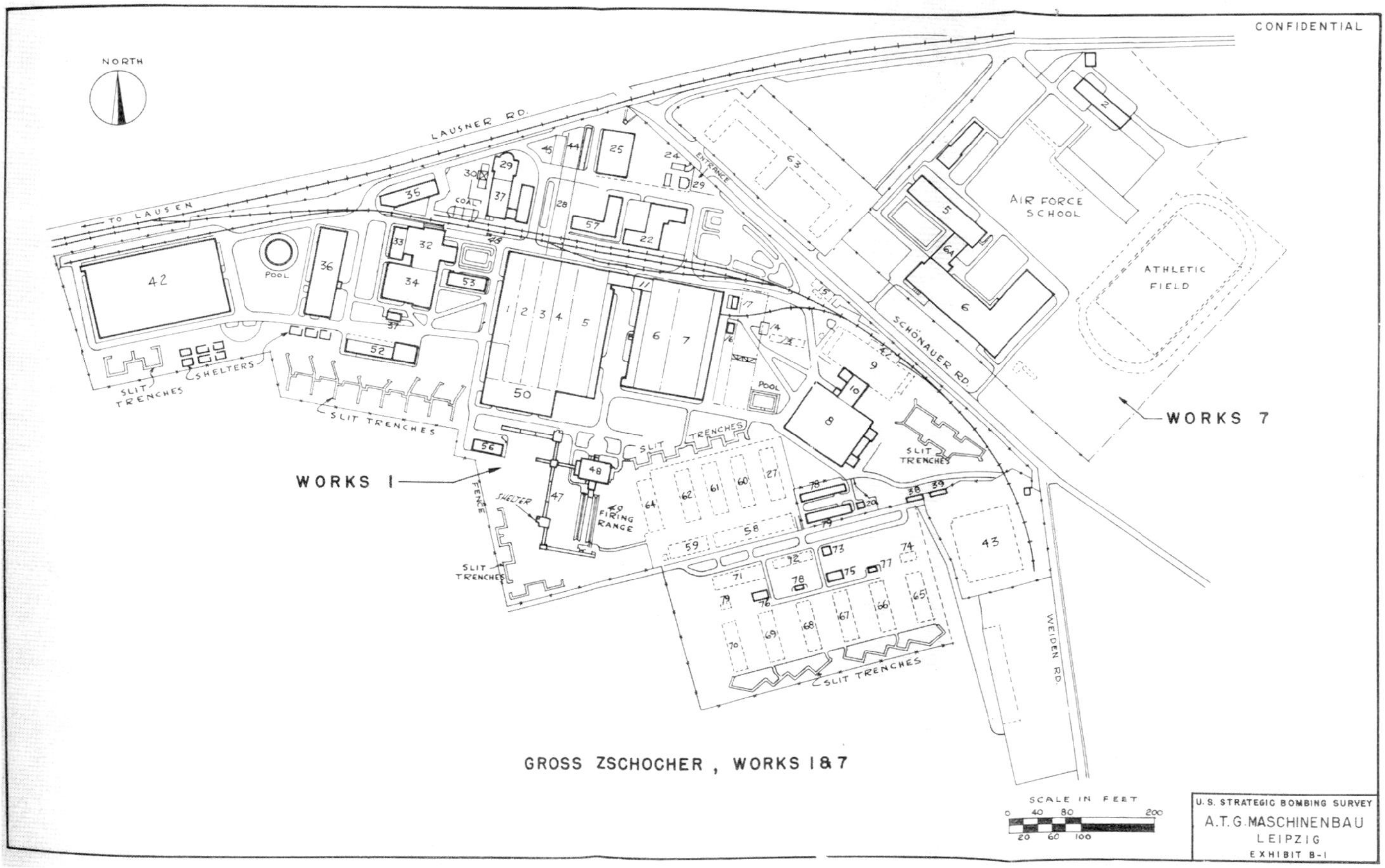

Allied map of ATG Plant 1 (Leipzig-Kleinzschocher) and Plant 7 directly opposite (Leipzig-Plagwitz). IWM

alone.[634] The next attack, by the RAF and the "Mighty Eighth," came on 20 February 1944. It caused the loss of another 65 fuselages and 25 aircraft. Further attacks by the 8th Air Force followed on 29 May and 7 July 1944; only Erla was hit, but it sustained 80% damage. ATG's own records show significantly lower totals, reporting that only 46 Ju 88 A-4s, 11 Ju 188 A-2s, and 12 Ju 188 D-2s were destroyed, along with 15 Ju 188 A-2 fuselages. A final heavy raid on 27 February 1944 destroyed another 44 fuselages, and probably inflicted the most damage on the Ju 388 final assembly operation, which had since come to a standstill. The most lasting destruction affected the fuselage production area in the big halls of Plant 6 on the fairgrounds, and Plant 3, with the three final assembly hangar halls on Mockau airfield. At the beginning of April orders were given to destroy the runways, but these were not carried out. In mid-April groups of Allied fighters strafed the hangars and wrecked buildings—the Bombing Survey gives the date as 10 April. American Sentinel and Piper Cub communications aircraft landed immediately afterwards, probably delivering the first ATI teams. A few

This is how the Allied intelligence teams found Plant 2. They described the damage in a comprehensive report on the effects of bomb damage. The fuselages produced here were outfitted for the FHL 131 Z turret. IWM

This view reveals the extent of the damage to the hall. Obviously only part of the building was used, as ATG had already dispersed production. IWM

days later C-47s landed at Leipzig airport, which was occupied by American troops from 1 May to 1 June. The remaining ATG hangars were completely disassembled by the Soviets in 1946.[635]

In terms of the number of aircraft actually produced, ATG was the most important production company for the Ju 388. By a fortunate coincidence, the meticulous internal correspondence for the period from December 1944 to February 1945, when production of the Ju 388 was in full swing, has survived in the Federal Archives. Regrettably, similar documentation does not exist for JFM, Siebel, or Weserflug. These original documents make it possible to reconstruct in detail the most important chapter in Ju 388 production.

The Internal View of Production

Engineer Fritz Schabert of the BMW main department EZA/2 visited Leipzig on 15 September 1944, and there discussed the state of work on the Ju 388 with plant manager Kuckertz and plant engineer Hempel. Construction work on the first aircraft to be assembled and flown by ATG, WNr. 340 090, had reached the point where the BMW 801 TJ-0 engines (WNr. 321 227 and 315 317, already fitted with reinforced cooling gills) could be installed. This proved more difficult than planned, however. In addition to minor assembly errors, ATG criticized the absence of cooling fans, supercharger air hoses (for the climate control system), and Y-branch pipes (for the wing de-icing system). That the latter were not to be installed at first due to temperature problems had not yet filtered through from Junkers to ATG. Instead, the hot air had to be vented using locally designed connection pieces. The target date for commencement of static tests was set as 18 September. The schedule was obviously held too, for the aircraft is known to have made its first flight on 26 September 1944 with Flugkapitän Baltabol at the controls.

During Schabert's visit it was also clarified that filters did not need to be checked and cleaned after stationary engine runs and test flights. This could now form part of the final check, as was being done by the "JFM construction site Merseburg". In view of the growing fuel shortage, consideration was also given—subject to the approval of the BAL—to eliminating the climb "to maxi-

An unidentified Ju 388, probably an ATG built machine. At least one other Ju 388 is sitting in the background. A comparison with the photographs of WNr. 340 084 suggests that this may be Merseburg. Griehl Collection

mum altitude."[636] On average, 110 liters of gasoline was needed for each test run by the Jumo 213 powered Ju 188, also being built by ATG. According to ATG, this quantity would probably not suffice for the BMW 801 TJ-0. At the conclusion of the meeting it was also decided that ATG would initially have to build the supports for mounting the exhaust port extensions itself, because they did not yet come with the engines. The modified exhaust fairing also made it necessary to change the engine vents (upwards exit from the engine). Until the arrival of the new vent pipes and associated drawings, however, these were to be produced as before, while in Merseburg the new venting arrangement was being installed on the production line.[637]

After a tour of Plant 3 on 21 September, on the occasion of the first delivery of a Ju 388, the technical management group complained in a letter to *Herr* Heller of the SO[638] plant management that "the tempo of ATG production leaves something to be desired in comparison to production at Merseburg."[639]

Preuschen visited Plant 3 at the end of October to learn of possible complaints that had arisen, and to brief the test pilots on the experiences of their opposite numbers at Dessau and Merseburg. At ATG seven aircraft were either in the final assembly stage or flight preparation, and three aircraft had been ferried to Altenburg, one of which (WNr. 340 090) had already flown to Rechlin. Only one aircraft had even attempted a climb to high-altitude, as the 400 liters of fuel per aircraft allocated by the "fuel commissar" was insufficient for flight preparation and a high-altitude flight, even under the most favorable conditions. Plant manager Kuckertz had therefore reached an agreement with type supervisor Hptm. Stahl that all aircraft destined for the *E-Stelle* Rechlin would only have to carry out one low level flight for systems checks and a "cockpit pressurization check" at an altitude of 3000 meters, and then be considered accepted. Consequently, under favorable conditions the ferry flight from Leipzig-Mockau to Altenburg served as both initial test and acceptance flight. To carry out these flights chief test pilot Baltabol had two other test pilots, both former military pilots. As they only flew the Ju 388 at low altitude, they considered it a step backwards compared to the Ju 188. Disappointed, Preuschen noted this meant that no further information could be gathered about the aircraft's performance during these low level flights.

Final assembly and flight preparation bemoaned the inadequate scrutiny given major subassemblies prior to assembly. This made it more difficult to find the problems after assembly, or during static tests. There were not even simple electrical block diagrams, which made the search for faults in the radio system and electro-hydraulics very time consuming. Important parts of the engines (housing ventilation, exhaust bearer arm, etc.) still had to be modified. Apart from the frequent failure of the Siemens generators, no significant faults had appeared in the six engines flown to date. During the high-altitude flight to 12 km oil leaked heavily from the engine housing vent; however, this had not been modified to the latest standard.[640]

On 15 November 1944 Ing. Reichel of BMW met with Director Keller and Plant Manager Kuckertz of ATG. There he learned that, beginning in December 1944, the entire ATG program was to be divided among the two assembly plants in Leipzig-Seehausen and Kölleda, with each taking over half the program by February-March. Test flying for both facilities would be done in Altenburg by the Erich Fritzsche company.

Total production for November at Leipzig-Seehausen suffered mainly because of a shortage of engines. Five completed aircraft were sitting in the open without engines. The biggest problems were caused by the constant shortage of cooling fans, Y-branch pipes (with heat hose), and the supercharger air hoses for the climate control system. While Reichel was able to assure plant manager Kuckertz that since 7 November all engines had been leaving the BMW factory with cooling fans installed, no remedy was in sight for the supercharger air hoses that supplied the pressurized cockpit. It was therefore decided to fly the completed aircraft to Welzow without the hoses. At the modification facility there, the engines were supposed to be brought up to the latest standard and fitted with external supercharger air intakes. Because of the modification program at Welzow, ATG cancelled plans to convert the engines to external supercharger air intakes, especially since approximately 50 hours work per engine was required for an interim solution to the oil ventilation problem, and another five hours for the installation of the so called "fish tails." Installation of the "fish tails" was necessary because the fittings for the hot-air deicing of the wing leading edge were missing, and the hot air had to be vented.

ATG-Leipzig was therefore able to distribute the forty external supercharger air intake conversion sets it had in storage. Ten of these were to go to Welzow, 12 to Dessau for conversion of prototypes, and the remaining 18 to BMW at Eisenach, or again Welzow.[641]

On 1-2 December Ing. Reichel of the Fritzsche company paid a visit. He learned that the planned completion of seven aircraft in October 1944 had been achieved, but that ATG had been unable to meet its November target. It had only been possible to complete eight of the planned 14 aircraft because of a lack of engines. On 1 December 1944 two of the missing six examples were still unserviceable at Altenburg with engine trouble.[642] There were no significant engine problems, as the aircraft made only two or three brief flights at Altenburg (maintenance test, radio, and acceptance flights) before they were ferried to the modification facility at Welzow. However, the BMW 801 TJ-0 engines proved hard to start at temperatures around 0°C, requiring large amounts of increasingly scarce starter fuel. There were obviously disagreements as to who was responsible for the procurement of spare parts for the engines. BMW stressed that this was a matter for the airframe company, and not the field service. Finally, BAL representatives Köpp and Reichel were able to convince plant manager Fritzsche to order sufficient spare parts in the future. In his trip report, Reichel asked that steps be taken to ensure delivery of the replacement parts ordered by ATG weeks earlier, otherwise, there would be difficulties in airframe production for which BMW would be blamed. The BAL asked to be provided with assembly directives on a regular basis, so as to be able to keep track of engine deliveries to each manufacturer. The distribution of work between ATG and BMW had also not yet been decided upon—as purely a mass producer [of aircraft], ATG insisted that it receive ready-to-install engines, and that all work on the engines be handled by BMW fitters, which repeatedly caused tensions. Efforts were being made, however, to clarify the situation.

The first ATG-built machines delivered to *Erprobungskommando* 388 came under considerable criticism. It is not surprising, therefore, that a *Herr* Döring of ATG inspection management paid a visit (or was perhaps ordered) to the *E-Stelle* Rechlin on 4-5 December 1944. A series of complaints were discussed in a meeting with deputy type supervisor Fliegerstabsingenieur Kuhn, Oblt. Elbing (technical officer of EKdo 388), Fl.Stabsing. Weidmüller (group leader E 5 VII, hydraulics), and Prüfmeister Zimmermann of Ekdo 388. These ranged from landing flap hydraulics, through difficulties jettisoning the canopy, excessive concentrations of fuel in the oil, and oil cooler rings opening independently in flight, to the frequent failure of altimeters and airspeed indicators.[643]

The first serious difficulties in the delivery of components appeared at the beginning of December, and remained throughout the life of the production program. On 5 December 1944 the management of Plant 2 compiled a report on "Difficulties in production of the Ju 388 L and Ju 388 J caused by missing parts." Fifteen Ju 388 L-1s had been delivered with no fuel pumps in the fuselage tank, and no tail turret. These parts were also missing from another 11 aircraft in the "fuselage set up."[644]

As for the Ju 388 J-1, the situation was even more dramatic. Ammunition boxes and belt feeds for the armament in the fuselage tray were still being developed, as was the entire oblique armament installation behind Frame 15. As a result, it was not possible to install the fuel tank in the first fuselage bay directly above the weapons tray. Consequently, five aircraft were delivered without a tail turret or offensive armament. Five more J-2s in Plant 3 faced similar problems. In response, ATG sent specialists to Dessau to make patterns using one of the prototypes there. Retrofit work was thus

Among the Ju 388 built by ATG in December 1944 were WNr. 340 118 (foreground) and WNr. 340 284. The tall grass in places suggests that the airfield was not used for some time. Crow Collection

being carried out in ATG Plant 2, and by WFG and SFW on the scale of approximately 140 to 340 hours per Ju 388 L-1/J-1.[645] In addition to documenting the hindrance to production caused by missing parts, the message also confirms that, in addition to L-1 reconnaissance aircraft, several Ju 388 Js must also have been in an advanced state of construction.

Another memo dated 11 December 1944 confirms the obvious production crisis, caused mainly by the hours spent retrofitting and modifying production aircraft. It was envisaged that "the retrofitting of fixtures for the Ju 388 J armament in the first aircraft" would take place "in a modification facility." As the completed L-1 aircraft already had to go through ATG's own facility, it was anticipated that another modification facility would be needed for the J-1.

Prior to this, the installation of the J cockpit was to be carried out in Plant 3, followed by a function test of the cockpit and the entire fuselage. The installation of weapons in the aircraft from Plant 2, however, was to be carried out subsequently in Plant 6. The cockpits for the first five machines were again to come from Merseburg.[647] This information is in line with the acceptance of the first J-1 pressurized cockpits at Merseburg (see above).

A message about a tour of Factory 3 dated 10 December provides a precise overview of the true production situation.[646] The delivery target was 16 aircraft, however, only 11 sets of wings were on hand. The following table traces the progress of individual aircraft in final assembly to first flight in the second half of December:

WNr	13.	14.	15.	16.	17.	18.	19.	20.	21.	22.	23.	24.	27.	28.	29.	30.	31.
114	5	6															
115	3	5	6														
116	3	3	3	3	3	3	4	5	6								
117	2	2	2	2	2	3	3	4	5	6							
118	1	1	2	2	2	2	3	3	4	5	6						
119	2	2	2	2	2	2	3	3	4	5	6						
281	2	2	2	2	2	2	2	3	3	4	5	6					
284	1	2	2	2	2	2	2	2	3	3	4	5	6				
286	2	2	2	2	2	2	2	2	3	3	4	5	6				
287	1	1	2	2	2	2	2	2	2	3	3	4	5	6			
288	1	1	2	2	2	2	2	2	2	3	3	4	5	6			
290				1	1	2	2	2	2	2	3	3	4	5	6		
296					1	1	2	2	2	2	2	3	3	4	5	6	

Legend: 1 = engine static test; 2 = radio (check?); 3 = aileron gap, sealed run-through, periscope and tail turret, entry hatch modifications; 4 = firing range; 5 = cockpit pressurization test and takeoff checks; 6 = flight to Kölleda (modification facility).

On 20 December the SO plant management reported the delivery of the following machines in December 1944:

1. Ju 188: "Planning target 15, of which three have been delivered, three in Kölleda ready for Altenburg, four more ready by the end of December, giving a probable shortfall of five machines."
2. Ju 388: 340114 and 115 from the November program are ready to fly at Altenburg, but have not flown due to weather; December target: 16 aircraft, all of which are to pass through the ATG modification facility, 10-11 to go to Kölleda. In the event that the weather is (too) bad, "serviceable" aircraft from Leipzig are to reported to BAL as "delivered, because modification work is being carried out in Plant 3 [instead of in Kölleda]."

The production situation was obviously coming to a head under pressure from the Armaments Staff for deliveries according to schedule. As well, weather was disrupting the regionally decentralized production. All efforts were obviously made to maintain output, probably at the expense of ongoing production.

A memo by the management of Plant 2 about the "Cessation of Production of Ju 388 L-1 Canopies" dated 21 December 1944 shows the progress of the crisis.[648] Although the 113th L-1 and 5th J-1 cockpits were supposed to be delivered according to the current TH production plan, there were only enough special canopy windows from Prague and Eilenburg for 74 L-1s and 2 J-1s.[649]

At the beginning of January the technical head office created a new internal plant delivery plan (40.398) with a series allocation for Plant 3. According to this, the final assembly of L-1 fuselage deliveries would no longer take place in blocks of fifty (766.7101), but instead in eleven blocks of twenty. Production of the first batch was to commence on 4 January 1945, beginning with fuselage number 766.7101, and number 766.7301 by 1 September 1945.[651]

Efforts by ATG to build its own J-series cockpits had obviously not progressed very far. The installation of other military equipment in the J-1 was laid down in such a way by the Technical Head Office-Production Preparation that the sixth aircraft was to be the first to receive the first ATG cockpit, together with the fixed weapons, oblique armament, and FuG 220 D/R. The fuselage was to be prepared for the FuG 220 in Plant 2, but "overhaul of the system first in a modification facility to be determined, documentation for the rearwards-facing antenna still not received." A prototype installation of the fixed armament was to be made in the 25th J-1 fuselage. The production version of the fixed weaponry would then be retrofitted, beginning with the sixth fuselage.[652] Nevertheless, 14 fuselages had already been delivered to SFW and five to Plant 3. The design of the oblique weapons installation was still not finished, nor was that of the rearwards looking radar antenna. The modification facility, about which no decisions had yet been made, was supposed to retrofit the fixed armament into the first five fuselages. Finally, thought was already being given as to whether the J-1 could be delivered "to the modification facility without having

After acceptance by the BAL, several Ju 388 L-1s were apparently stored in Altenburg, some covered with tarps. There they were discovered intact by American troops. Crow Collection

been flown and with no final inspection, in order to save time and fuel."

The main problem at the beginning of January, however, was the shortfall from the previous month, and the state of work in the Kölleda modification facility. On 2 January the technical head office reported that 19 Ju 388 L-1s had arrived at Kölleda, and were being worked on by two BMW specialists who had just three incomplete conversion sets. Other modification work consisted of installing a modified DF loop, cooling gill modifications beginning with the 72nd set of L-1 wings (SFW) and the 113th set of ATG wings, and the 6th set of J-1 wings (SFW) and 26th ATG set. A modified camera heating system was also needed, as tests indicated that the Kärcher heaters were no longer adequate at the increased altitudes made possible by the improved engines. As well, the Production Special Committee was urging that the necessary changes be made in Plant 3, beginning with the 48th set of SFW wings.[653]

Plant 3 described the delivery situation on or up to 11 January 1945 as follows:

"To date, two aircraft have been delivered to the modification facility since 1 Jan. 1945. All major subassemblies and engines for five more machines are presently on site here. If possible, these are to be flown to Kölleda by the 20th of the month. Fuselages and 'Besa' (cockpits) are on hand for another 16 aircraft. Missing for these are 16 power plants, 12 sets of wings, 15 tail fins and 15 elevators, [...]. Assuming that the next engines arrive here in modified condition, the 8th of January machine will be the first aircraft from Leipzig to go straight into flight operations. [...] 1.5 machines are being completed on the final assembly line daily."

The situation was also not the best when it came to the delivery of major subassemblies to other manufacturers:[655]

"Subsequent deliveries of missing parts to WFG: Parts delivered from Plants 2 and 4 are repeatedly rejected by receiving inspection, to date five major subassemblies per month to WFG."

This confirms that the ten Ju 388s assembled by Weser/Liegnitz used ATG parts; further proof is provided by a manufacturer's plate in the fuselage interior of Weser aircraft WNr. 560 049.

Two days later Plant 2's production shortfalls were listed:[656]

Component	Authorized	Actual
L-series canopy	119	83
J-series canopy		
Glazed	12	2
Canopy frame without glazing	15	8
J-series ammunition box	50	2
Ammunition belt feed	50	2
Oblique weapons	2	2

Ju 388 WNr. 340 118, photographed in Altenburg after the war. On 27 January 1945 this aircraft was taken up on a high-altitude test flight by test pilot Karstens and BAL Inspector Köpp. Williams Collection

ATG was at least far advanced in the construction of the missing components for two J-series machines.

On 16 January 1945 consideration was given to striking a deal between Director Herbert Schade of ATG and SFW-Schkeuditz (*Herr* Petereit) for the purpose of cooperation in the test flying of new aircraft.[657] However, chief pilot Baltabol of the ATG test flying section advised Director Schade that Altenburg could take care of test flying aircraft that had passed through the Kölleda modification facility itself, and that, if needed, support from SFW crews had been guaranteed in a previous agreement. At that time the Armaments Staff was energetically demanding improvement in the organization of industrial test flying, which in Saur's view frequently caused "production delays."[658]

The following day Plant 3 reported the cockpit delivery situation[659] as per its internal "Delivery Plan 40-358" of 28 December 1944, which brought a significant increase in the planned target compared to the previous plan 40-358: previous December target 23, now 28; previous January target 27, now 44; and February previously 35, now 50. In fact, by 16 January the plant was already behind with 19 cockpits.

On 18 January the SO plant complained about the inadequate production of engines. There were 27 fuselages without engines in Plant 3, and 14 at SFW. Eight fuselages were ready for SFW in Plant 2. Altogether, there were now 49 fuselages without engines in final assembly. This does not include fuselages without engines at Liegnitz, or still needed by JFM. The necessary fuselages and cockpits for Plant 3 and WFG (for J and L) were supposed to be completed on time, however, this only resulted in the "piling up" of major subassemblies in final assembly. The output of major subassemblies and fuselage was supposed to reach its highest level of 95 by April, and then fall to 50 from April to July. Consideration was therefore given to delaying peak output in order to eliminate the backlog. As well, as of the 15th all necessary installations in the J-1 were to be carried out by ATG itself.

The next day brought a new total overview of the aircraft under construction, and a report on storm damage in Stations 8 and 15 in Plant 3. The repairs had to be makeshift in nature because of a shortage of building materials (roofing felt). The state of production was also reported to the management of ATG:[660]

"1.) L-1

As of today there are 17 L-1 machines in the Kölleda modification facility. The 18th machine will arrive at Kölleda on 20 January 1945, the 19th on 22nd January. Horizontal stabilizer assemblies and wings are on hand for another ten fuselages. These are aircraft numbers 340 283, 340 289, 340 292, 340 294, 340 295, 340 301, 340 302,

After production was halted, aircraft under construction from the 340 4XX *Werknummer* block (or their major components) were stored in this hangar in Leipzig-Mockau. The type of destruction seen here is more indicative of intentional rather than bomb damage. The night fighter fuselage with the severed tail in the right foreground is noteworthy; the oblique armament clearly identifies it as a night fighter. Lying about in front of WNr. 340 410 with the burned out cockpit and unusual supporting wheels are Ju 188 canopy frames. Lutz Collection via Rosch

340 304, 340 306 and 340 307. Missing for these machines are power plants, fuel tank pumps [...], canopy windows, [...] and wiring [...].

Forward fuselages or cockpits are on hand for another five aircraft and also wings for two aircraft. These are the aircraft (*Werknummer* 340 299, 340 303, 340 305, 340 308 and 340 310). Furthermore, we have ten fuselages without cockpit. This applies to aircraft 340 402, 340 405, 340 407, 340 408, 340 411, 340 412, 340 414, 340 415 and two aircraft whose serial numbers are not known. Plant 4 has advised that the next delivery dates for cockpits are: one each on 20 Jan., [...] 23 Jan., 24. Jan. and 25 Jan.. [...]

2.) J-1

On site are 5 fuselage ends, 3 cockpits, 2 sets of wings, 1 set of horizontal tail surfaces, arrangements for the completion of these machines will not be made until we receive your final decision, which was promised to us for 20 January 1945."

The plant complained about the inefficient use of labor forces—which were on a 67.5-hour work week—resulting from the sluggish arrival of major subassemblies. The absence of parts meant that many things had to be done twice, and the amount of shunting work was unusually high.[661] The document cited above obviously describes the conclusion of production efforts, for some of these aircraft may be seen in half-completed condition in the well known USAAF photos taken in a hangar at Leipzig-Mockau. These photographs show that, while progress had been made toward completion of the batches (340 300 and 340 400) since 19 January, the aircraft were obviously not completed. On the one hand, the report illustrates the production advances, while it also shows the problems of dislocated production at the time of highest tension.[602] These were exacerbated by the war's impact on the transportation system. A report on a tour of Plant 3 on 21 January clarifies this picture:

"Station 8	6 fuselages with cockpit, 2 of these with engines 5 fuselages without cockpit, 1 of them with no tail assembly
Station 3	fuselages (L) with cockpit, 2 fuselages without cockpit, 2 J fuselages with cockpit
Station 1	1 aircraft with cockpit without tail section; 6 fuselages set aside
Station 2	3 aircraft
Airfield 4	4 aircraft, some without tails; 1 Ju 188

Thus, there are 32 Ju 388s in a more or less advanced state of construction.[663]"

This enlargement shows a Ju 388 J fuselage with mounting tubes for FuG 220 radar antennas behind a collapsed roof support. Lutz Collection via Rosch

The slump in production at the end of January meant that the planned 10,000 work hours in Plants 2 and 4 (Plexiglas windows, power plants, retroactive modifications), and the planned output could not be achieved. The excess of labor forces was viewed as temporary, for a capacity bottleneck was expected in March.[664]

On 29 January the SO plant asked Plant 8 how it should deal with the bills received from Merseburg for the work done on *Werknummer* 340 097 to 099.[665] Junkers in Merseburg had taken over these aircraft in flyable condition, and in November had put them through flight safety checks. A number of faults, some dangerous, were discovered, such as missing stops on the rudder pedals of WNr. 340 098. The bearing block for the rudder and aileron levers on the rear face of the control block under the pilot's seat had been destroyed as a result. The bomb doors did not close properly on any of the three aircraft, although this problem could have been alleviated by properly adjusting the supports. Other modifications that could have been carried out at minimal cost, such as covering or removing the hydraulic fluid cooler, were also not undertaken. As well, the lubricators installed in all the aircraft were too short. It is likely, therefore, that Junkers billed ATG for correcting obvious production errors.[666]

On 1 February the management of the SO plant reported that, while 126 pressurized cockpits had been delivered by the end of January, it was still 34 in arrears because of delivery difficulties with the pressure tight canopy openings. Plant 1 had delivered 131 seat and control mounts (for the L-1) and 16 for the J-1 to Plant 4. Of the former, however, 11 had been sent to the Meinl company, and they were now in default of delivery of 20 cockpits.[667] Just before production was halted by the Armaments Staff on 6 February 1945, *Herr* Kuckertz of the management of Plant 3 prepared an overview of the Altenburg, Kölleda, and Leipzig facilities for Director Schade in Leipzig:

This view of the other side of the hangar (with WNr. 340 405 in the left background) shows that not much of the roof remained. Apart from its burned tires, WNr. 340 307 appears undamaged. Lutz Collection via Rosch

"Altenburg: 1 modification facility aircraft set aside, 1 aircraft in final inspection, 4 aircraft in flight operations but no fuel for flying as of 6 Feb.

Kölleda: 21 aircraft (of which the above-named) 6 went to Altenburg, the remaining 15 are being worked on by ATG/BMW, with materials for 12 aircraft on site.

Leipzig: on hand for 26 machines are 26 L-1 fuselages, 6 J-1, 20 sets of L wings and 5 sets of J wings, however all aircraft lack engines, the airframes of 7 machines are complete, another airframe is completed each day, therefore the delivery of fuselages from Plant 2 was stopped.[668]"

The production of fuselages was also beginning to break down, however, for the next day Plant 2 reported to Director Schade that Plant 1 was in default of delivery of 12 L-1 and 15 J-1 fuselages.[669]

Plant 4 in Böhlitz-Ehrenberg also reported that it was behind in the production of tail surfaces, the main reason being the amount of modification work. February's target of 57, including 20 shortfalls from the previous month, would not be achieved again until March.[670]

Production at ATG was also not spared the effects of war. On 22 February Kölleda reported an air raid alarm and damage to *Werknummer* 340 290, both of whose wings were holed, and *Werknummer* 340 291, whose fuselage was holed.[671]

In mid-February production of the Ju 388 was stopped by the Armaments Staff, and instructions were issued for the release of 1,077 "follower members" from Plants 1 to 4.[672]

Consequently, on 28 February the ATG jig bureau compiled a "status report"[673]; the "operating resources"—which probably meant the jigs and tools needed for quantity production—for the Ju 388 J

This bomb crater was most likely created before the Ju 388s were pushed together in the hangar, for otherwise WNr. 340 410, seen in the foreground, would almost certainly have been damaged by flying debris and fragments. Lutz Collection via Rosch

and L were completely halted as of 8 February and, in some cases, 16 February. In keeping with a JFM requirement, one set of "operating resources" for the production of 30 aircraft per month in a one shift operation was set aside. All the remaining jigs were scrapped, as were those for the Ju 188.

Thus ended the story of ATG's association with the Ju 88 family of aircraft, which in the case of the Ju 388 had begun in December 1943 with Delivery Plan 225/1. The final version of the ATG program called for the construction of 447 L-1s (166 of these with the BMW 801 TJ, the rest with the TQ). Nineteen aircraft were to have been completed with Menibum fuselages, and 25 fuselages removed for K-0/L-0 conversions. Also envisaged were 260 J machines, 80 of which were to have been delivered with BMW 801 TJ engines, but none of these were completed.

In conclusion, an overview of actual production by ATG to the turn of the year 1944-45:

Ju 388 L-1 Production by ATG as per Aircraft Production Records[604]

1944						1945
July	August	September	October	November	December	January
1*	1**	6	5	15	19	8

Remarks: * K-1 to Prague to be tested to destruction, ** L-1 as design bureau mock-up.

Not included in these 53 L-1s delivered are 20 Ju 188 fuselages delivered to FSD-M for use in the Ju 388 pre-production series.

The state of production at Leipzig-Mockau is documented by USAAF photographs. They show aircraft in an advanced state of completion, albeit often without engines. Damage to the machines is only partly due to bombing, and it appears that efforts were intentionally made to set the parked aircraft on fire.

The fact that no production lines, parts waiting for installation, or tools can be seen confirms the cessation of Ju 388 production. The remaining subassemblies were apparently all pushed together in an already damaged hangar.

Another impression of ATG production is gained from the U.S. Strategic Bombing Survey (Aircraft Division Airframes Plant Report No. 3, ATG Maschinen Bau GmbH). The assessments by the ATG managers are very enlightening. They stated that the constantly changing production requirements (with the associated construction of jigs, etc.) resulted in a production loss equal to that caused by enemy bombing. The Ju 388 was explicitly cited as an example. According to ATG records, the program began in October 1944, and began getting up to speed by the end of the year, when orders were received to switch to production of Fw 190 tail sections. In early 1945 this was changed to delivery of Ta 152 components, and work was almost complete when American troops arrived. Compared to 1943, 1944 production losses in fuselages were placed at 36%; furthermore, 55% fewer completely assembled machines were delivered in 1944, although this could be traced to the startup of Ju 388 production. The report also points out that the plant's production capacity was significantly higher than actual output. The American analysts could therefore not conclude that the production losses were the result of bombing alone. Only in the case of the attack in February 1945, which led to the loss of 44 fuselages (two weeks' production), can it be assumed that, in addition to Fw 190 components, the remaining stocks of Ju 388s, production of which had just been halted, were also hit. It is noteworthy that not a single completed machine was listed as lost. It can therefore be assumed that the completed aircraft were in fact all delivered or moved to other airfields (Altenburg, Kölleda, etc.). Altogether, evaluation of the damage was rather sobering for the Americans. Damage inflicted upon feeder sources (raw materials and minor components) was rather insignificant, and was quickly made good from stockpiles. The damaged buildings themselves were usually not repaired; instead, production was moved to smaller buildings. ATG named destruction of the transportation system as the greatest hindrance [to production]. Overall, damage inflicted on the means of production (jigs, presses, and tools), some of which were well protected, was rated as minor.

Weserflug (WFG)

The Weser Flugzeugbau GmbH (Weserflug) had plants in Bremen, Lemwerder, Nordenham, and Einswarden. These facilities on the lower Weser were subsequently joined by a large plant in Berlin-Tempelhof, and later another in Rabstein, near Böhmisch-Kamnitz, which served as the Berlin operation's dispersed facility.[674] Previously the largest manufacturer of the Ju 87 (with more than 3,000 Ju 87 Ds alone), Weserflug was another of the traditional Junkers licensees. When production of the Ju 87 ended in 1944 it was only natural that the company begin building another Junkers design. Preparations for participation in the Ju 388 had begun long before. The Tempelhof plant planned its personnel requirements for 1944-45 for its anticipated role in production. Calculations were made of the exact number of personnel required by each plant associated with the project, and their involvement, ranging from "production of minor components" to "completion and flight testing." For example, in March 1944 production of wings was to commence in Plant 8, wingtip extensions by Kahrmann, end boxes by Bücker, and cut parts in Rabstein.[675]

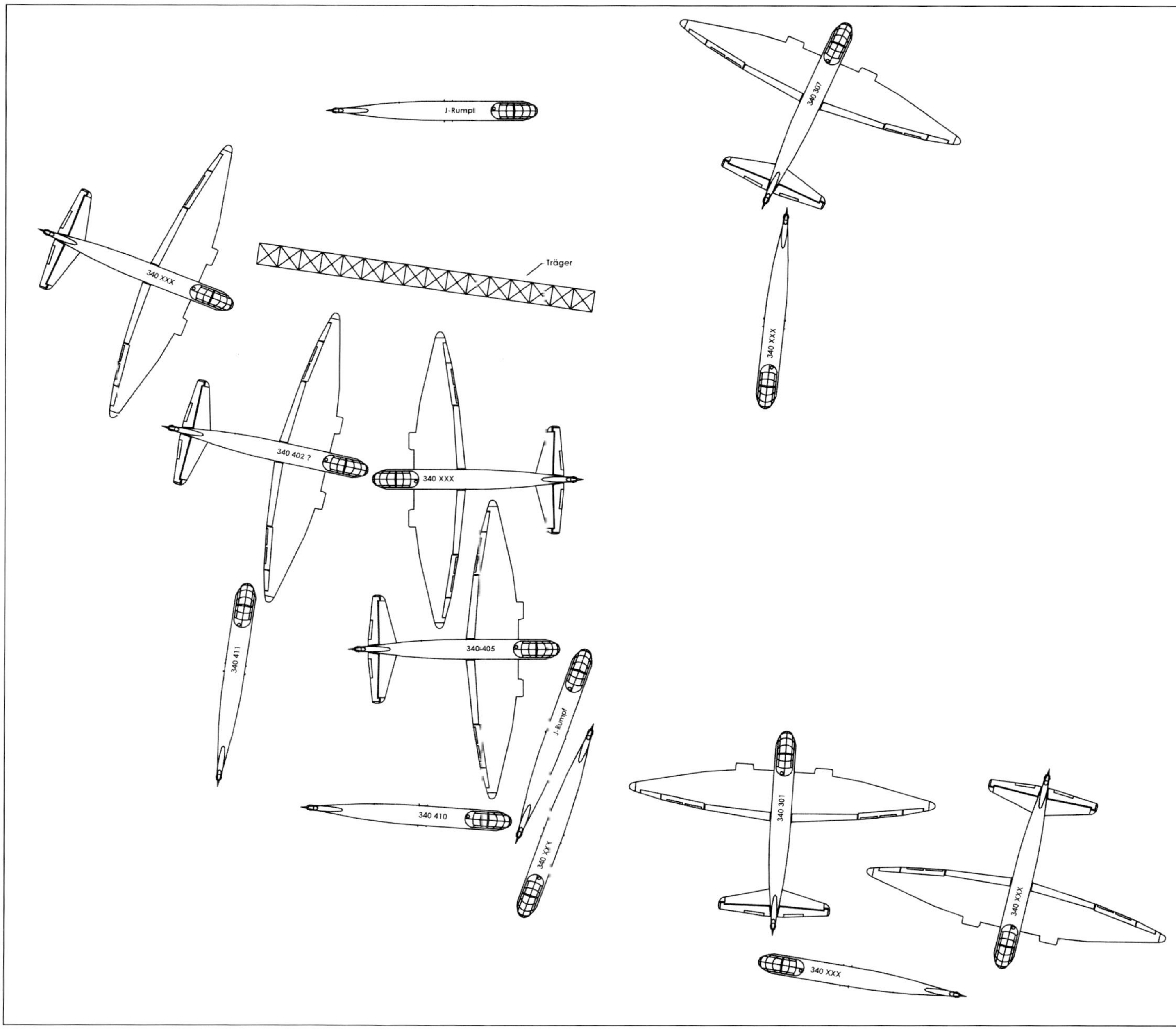

Overhead view of the Ju 388s in the ATG hangar reconstructed from the previous photographs. Vernaleken

In reality, production startup occurred somewhat later. On 22 June 1944 the BAL at Junkers' parent facility in Dessau decided that wing jigs and tools from Dessau for production of the Ju 388 by WFG were missing or unsatisfactory. A handwritten note stated that the "start of Ju 388 production in Rab[en]stein is jeopardized."[676]

According to JFM production planning in May 1944, WFG was initially to take part in production of the Ju 388 J-1. As the result of a correction to "FTH Plan 44.216, Page 2," the "44 J-1 fuselages scheduled for HFW" were instead "envisaged for WFG."[677] With the adoption of GL/C Aircraft Program 226, 1st Edition of 15 May 1944, WFG received an order for 914 Ju 388 J-1 night fighters powered by BMW 801 TJ/TQ engines. Production was to begin in January 1945, and reach 90 aircraft per month by September.[678]

The second edition of Delivery Plan 226, which appeared on 15 July 1944, brought about a significant change. WFG was now to produce 127 Ju 388 J-1s from January to August 1945, plus a series of 700 Ju 388 J-3s, production of which was to begin in the Tempelhof plant (WFG'T[679]) in April 1945. As of October, 90 aircraft were supposed to leave the factory monthly, with production falling to 75 per month in January 1946. As well, a small series of 22 K-1s was planned from September 1944 to March 1945, also at Tempelhof.[680]

Another new version of the delivery plan issued on 15 September 1944 included a reduction of the J-3 series to 497 aircraft, and deliveries were now to begin in March 1945 (with the maximum production rate of 50 aircraft per month by August). At the same time, there was to be a significant increase in production of the Ju 388 L-1, which was to be built instead of the K-1. Initially, the production of 33 aircraft with BMW 801 TJ engines was envisaged between September 1944 and April 1945. After this, an additional 300 L-1s with improved BMW 801 TQ engines were to be built at Tempelhof (30 aircraft per month). Concerning the 33 L-1s, it was noted that because of the construction progress made in the interim, it was likely that a conversion from K-1 to L-1 would have to take place.

In fact, production of the Ju 388 K-1/L-1 did not take place in Tempelhof, but in Liegnitz, Silesia, in a repair facility that Weserflug operated on the air base there.

The previously mentioned BMW trip report on Schaberts' visit to ATG in Leipzig also describes a subsequent visit to the Weserflug facility in Liegnitz on 21 September 1944. There Schabert met with plant director Dorn, technical director Tiarks, production inspector Kienbaum, and *Herr* Koch and *Herr* Jäschke of customer service and work preparation. The first aircraft, a Ju 388 K-1 (WNr. 560 041), was already 80% complete, and the BMW 801 TJ-0 engines that had been delivered (WNr. 385 277 and 315 346) were to be installed the following day. Before it could be flown, the aircraft still had to be converted from a Ju 388 K-1 to an L-1. At Liegnitz Schabert discovered that Junkers had also failed to advise Weser about the temporary deactivation of the wing deicing system. As well, procurement of the supercharger air hose from the intercooler to the firewall had not been thought of. Junkers Merseburg had, however, already delivered the parts for the modification of the exhaust system, while Weser, like ATG, built the machine with the

Assembled by Weser in Liegnitz, WNr. 560 044 was at Merseburg when the war ended. The separation line between the upper and lower camouflage is unusually low on the fuselage, while the mottling on the fin and rudder is similar to that of WNr. 560 049. No other aircraft is known to have worn a camouflage scheme like this. The stones in the foreground may have come from a wrecked blast pen. Petrick Collection

Ju 388 L-1 WNr. 560 049, the only survivor, was also built in Liegnitz, and is seen here at Merseburg after its capture by the Americans. Several American soldiers are resting in the shade beneath its wings. The camouflage scheme is similar to that of ATG aircraft, however, the fin and rudder were clearly painted with more care. Rosch Collection

old style engine venting. The list of complaints about the engines was somewhat longer than at Leipzig, however, especially for 315 346. Probably for this reason, Schabert came to the conclusion that the static test scheduled for 26 September 1944 would not be carried out. At the conclusion of his report, BMW engineer *Herr* Koch submitted a "Replacement Parts List for 50 801 TJ Engines in Flight Service with the Aircraft Manufacturers," with the request that they be ordered soon from Special Committee T3.

In the course of his tour of the plants building the Ju 388, in late October Preuschen arrived at the Weser repair facility in Liegnitz. Two aircraft were already ready to fly, and a third was about to take part in static tests. Not a single flight had been made by 31 October, however, and as at ATG, a safety flight at low level was all that was envisaged. Thanks to effective support from Dessau no special problems had appeared, only the exhaust fairings ("Little Boats") sat too far back on the first three aircraft.

A few things had gone wrong in construction of the first aircraft, however. A production inspector from Dessau had ordered the completion of 59 urgent changes. The aircraft had already been disassembled when the Armaments Staff and the Dessau head office ordered the aircraft reassembled in its original condition, even though the modifications could have been completed in a maxi-

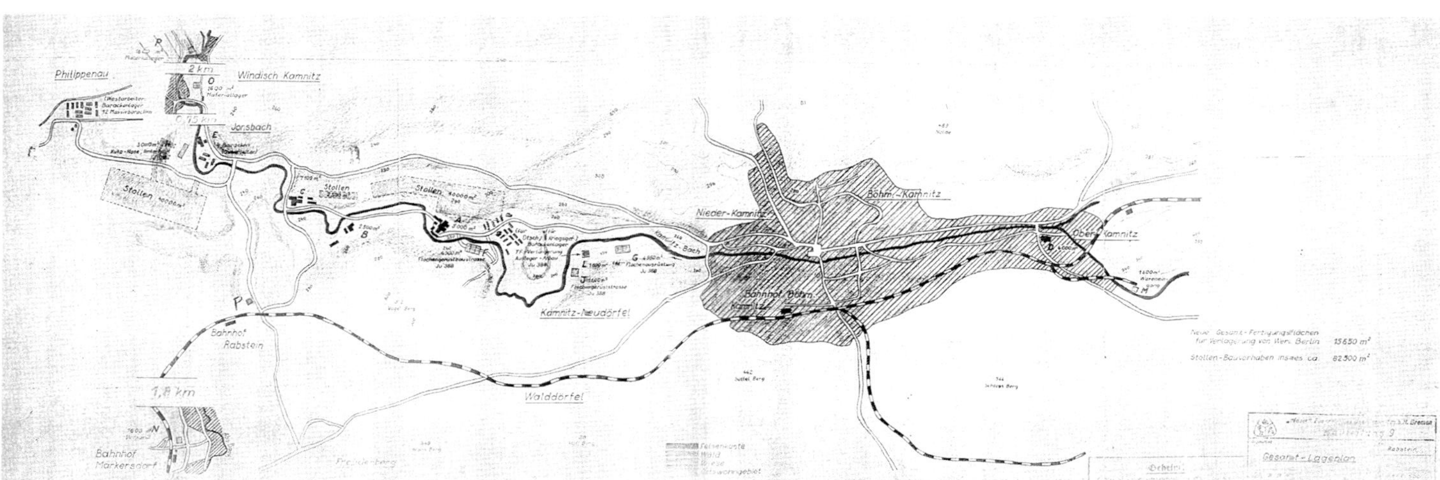

Site map of the largely underground dispersed production facility at Rabstein, near Böhmisch-Kamnitz. Forced laborers and POWs were to have built wings for the Ju 388 here.

mum of 14 days. The aircraft, which was now supposed to go via Rechlin to Welzow, would not only have been flyable, but Weserflug would have become familiar with the modifications, which would later have to be applied to the production line.

The absence of wiring diagrams was also very troublesome for Weser. In one case, it took two days and one night to isolate a problem with the switch controlling the flaps.

Initial test flying at Liegnitz brought to light a not insignificant problem: while taking off from rain soaked airfields, mud thrown up by the wheels entered through the gap between the bomb doors and the fuselage, completely covering the glass panels in the bomb doors and the lenses of both cameras. Dust was expected to cause similar problems. The side screens installed in the camera bay provided inadequate protection. Another concern was that mud might reach the aft fuselage through the lightening holes in the frame in front of the aft entry hatch, resulting in moisture damage to the equipment there.

The RLM construction inspectors responsible for Weser-Flugzeugbau, which had meanwhile transferred its headquarters to the Rabstein plant, came to the conclusion that the Ju 388 L high-altitude reconnaissance aircraft would often become "operationally unserviceable" during takeoff, and therefore felt that in its current configuration it was only conditionally suitable for front line use. Therefore, on 11 December 1944 they proposed that a bellows be fitted between the glass panel and the camera lenses, or that two protective sheaths be installed on the glass panel and the camera lenses, which would interlock tightly when the bomb doors were closed. It was proposed that the lightening holes be sealed to protect the equipment in the rear fuselage.[681]

Just before Christmas 1944 four aircraft of the December program were still at Liegnitz, apparently waiting for installation of the external supercharger air intakes. The approximately five aircraft planned in January were to undergo the same modification.[682]

Production of the first batch was at an advanced stage in January 1945. Of the eight machines completed, just one was still at Liegnitz.[683]

As a result of the cancellation of the Ju 388 in February 1945, nine or ten Ju 388 L-1s of the first production batch (WNr. 560 041 to 049/50) were probably still at Liegnitz, especially as the city fell to Soviet troops in early February as part of their winter offensive.[684] In addition to the production of complete aircraft, it appears that several dozen sets of wings were also built there.[685]

Siebel Flugzeugwerke (SFW)

The Siebel Flugzeugwerke GmbH in Halle an der Saale (SFW) was one of the most important producers of the Ju 88, the others being JFM-Bernburg and ATG. 1,633 complete aircraft left Plant 3 in Schkeuditz between 1941 and 1945. Plant 1 in Halle produced Ju 188 wings and tail sections for Siebel's own final assembly, and that of ATG. Coming off the assembly line at a rate of one every six hours, the Ju 188 wings differed little from those of the Ju 388.[686] Beginning in February 1944 they went to Plant III in Schkeuditz, where they joined ATG fuselages and Jumo engines on the final assembly line. In 1944 a total of 353 aircraft left the SFW halls in Schkeuditz.[687] It seems certain that the overwhelming majority of Ju 188 wings came from Siebel.

Apart from the documents mentioned in the section on ATG, there are few references in the surviving records from German

This aircraft, which obviously sustained severe damage when its wing tanks were set on fire, is allegedly WNr. 560 047. Based on the camouflage scheme on the fin and rudder, it is highly probable that this machine was built by Weserflug. Hafsten Collection

sources about production of the Ju 388 or major components by Siebel. A letter from Dzulko (Series Production Design Bureau) dated 17 October 1944 states that Junkers was also making Ju 388 exhaust fairings for Siebel at that time. In response to an inquiry from a *Herr* Kepper of Siebel, Dzulko advised that, unlike the nacelle extension, Siebel would not be building the exhaust fairing, as the Ju 388 J-1 would be receiving "the flame dampers one-hundred percent," as would the Ju 388 L-1s built for the night turbosupercharger role. Provided the flame dampers did not result in performance losses compared to the normal exhaust ports, they intended to immediately equip all Ju 388 L-1s with them.[688]

Another clue to Siebel's participation in production of the Ju 388 can be found by closely examining a trip report on a visit to Welzow by the "Retrofit Detachment 388 L-1." The following appears among the 19 complaints and observations brought forward by Lantzsch and Brachvogel of ATG:

"On wing truss QV 6 (SFW), the bolt installation locking tabs are not folded over."

This is a more than clear indication that Siebel was delivering wings for series production of the Ju 388 by ATG from the beginning, for at that time only ATG machines from the first two blocks of ten (WNr. 340 095, 096, 100, 101, and 340 104) were at Welzow.[689]

At Siebel in Schkeuditz preparations, at least, were underway for final assembly of the Ju 388. A pressurized cockpit for the Ju 388 J-1 (serial no. 1003) that was accepted on 2 December 1944 was supposed to be delivered to SFW.[690] A letter from the BAL at Siebel to the BAL at Junkers dated 12 January 1945, concerning "Startup of 8-388 J-1 at Siebel," not only bemoans the absence of parts, but also Junkers' failure to react to corresponding complaints. It was feared that output in January would be very low.[691] Another reference is found in a report on the BMW 801 TJ-0 modification

An oblique aerial view of the Siebel Flugzeugwerke in Schkeuditz taken in June 1946. There is little to suggest that aircraft were produced here. Luftbilddatenbank Würzburg

program in February 1945. In addition to the other locations (especially the modification facilities at Welzow and Kölleda), where complete aircraft were waiting for the necessary engine modifications, the modification program was also to take into account five engines at Merseburg and Schkeuditz.[692] As well, tail fins for the Ju 388 were being built or modified in Schkeuditz, as stated above.

Finally, by April 1945 Siebel made preparations for, or began production of wings for the Ta 152 C. This is confirmed by CIOS Evaluation Report 149 Target No. 19/79.[693] It reported on the Siebel company's production sites in Halle, and described its hangars and buildings as not seriously damaged. It also stated that this plant had the capacity to produce 235 wings for the Ju 88 family per month. The report also had some interesting things to say about Schkeuditz. The Plant 3 hangars used for assembly and repair work had been heavily damaged, especially by the USAAF raids on 7 July and 16 August 1944.[694] Production capacity was given as 60 to 65 Ta 152s or 30 to 35 Ju 188s, which were produced by about 300 workers. Two heavily damaged Ta 152s without engines had been found. Of central interest to the Allies, however, was production of the Ju 388 night fighter. This was said to have taken place until March 1945. This statement was backed up by the presence of eight "almost complete" aircraft, none of which carried armament, however. These could have been the original 14 fuselages delivered by ATG that were mentioned earlier. That these were in fact night fighter versions is confirmed by the report, which describes in detail a tail barbette controlled by means of a periscope, and an armament of two MK 108s and two MG 151s. The presence of an actuating cylinder on the Ju 388's entry hatch (for emergency opening using compressed air) led to the incorrect conclusion that it could be used as a speed brake. This was an unusual error, considering that the team had identified the presence of a pressurized cockpit. There were even plans to ship a Ju 388 fuselage and a set of wings to England. These were cancelled when it was found that the Allies already had intact examples for the U.K. and USA.

Henschel Flugzeugwerke (HFW)

As an important producer of the Ju 88 under license, including the Ju 88 S, *Oberst* Rowehl proposed that Henschel be included in the

Photographed while serving with KG 3, Ju 88 A-4, WNr. 3554, was built under license by Henschel. Planned production of the Ju 388 never began, however. Vernaleken Collection

Ju 388 production program, especially since the Henschel Hs 130 high-altitude aircraft had been cancelled in favor of the Ju 388, and spare capacity was available following the cancellation of the Me 410. A mood of crisis was thus spreading through HFW in 1943. Its own designs, such as the Hs 129 and Hs 130, were falling victim to the RLM's "rationalization measures," and license production of the Ju 88 A-4, A-14, and S-3[695] was no guarantee of relief in the future; indeed, it was obvious that the design was approaching obsolescence. The company faced the threat of "transfer of labor forces" to other plants producing more advanced types. In principle, it was a development that faced all bomber manufacturers, as their types were dropped from the delivery program in favor of fighters. *Generaldirektor* Karl Frydag, the company's senior representative on the Fighter Staff, tried to secure sufficient orders for his former firm—for major subassemblies, for example. In September 1944 Frydag already assessed "the Ju 388's chances in the long run as not favorable," and advised the company's management to strive instead for the Do 335/435, Me 262, and Ar 234. When this did not succeed, the company secured production of the Hs 117 anti-aircraft rocket, and began development of its own jet powered close support aircraft, the Hs 132.

Henschel was also supposed to become involved in producing the Jumo 222, and because of the existing links with JFM and HFW, it seemed logical that HFW should build the Ju 388 J-2.[696] HFW's monthly report to the RLM for January 1944 reveals that, in addition to converting Ju 88 S-3s into T-3 reconnaissance aircraft, preparations were being made to produce the Ju 388 J-2 with Jumo 222 engines.

At the same time, on 31 January 1944 the GL/C E-2 advised Henschel by letter of the final cancellation of the company's own high-altitude type, the Hs 130 E. The letter also contained an order to,

"...undertake development of the Jumo 222 standard power plant in cooperation with Jumo, and begin the design work required to convert the Ju 388 to the Jumo 222."

With the adoption of GL/C Aircraft Program 226/1st Edition on 15 May 1944, HFW received a contract for production of the "Ju 388 J-2," which was classified as a Zerstörer, and the rather modest total of 160 J-3 night fighters with Jumo 213 E engines, to be delivered between the second quarter of 1945 and the first quarter of 1946. The first batch was also to include 20 J-2 prototypes from the GL/C "V-Series Program." Planned delivery dates for these J-2-V machines lay between April and July 1945. The plans changed again in July: in addition to 565 J-2s, HFW was now to build 175 J-3s and 22 K-1s.

The J-2 disappeared from the 15 September 1944 version of the delivery plan, and on 30 September even the twenty prototypes were officially "withdrawn." On the other hand, production of the J-3 was now to comprise 477 examples, to begin in March 1945 and end in March 1946.

The Jumo "monthly report" for the first quarter of 1944 confirms that the advanced Jumo 222 engine was only included in the RLM GL/C-B-3's Delivery Plan 325a retroactively at the urging of Junkers, even though six of the promising Ju 388 E/F were already under test. There is also a reference to the transfer of design work on the production engine to Henschel with "technical support from Jumo."[697] At the beginning of July 1944 Saur decreed: "Henschel's main plant in Schönefeld envisaged as primary development facility!"[698]

As per the delivery plan valid in July 1944, HFW projected limited production of the Jumo 222 engine (August: 4 examples, September: 6, then alternating 8 and 12 examples per month until February 1945, with 20 per month beginning in March[699]). The actual delivery situation at that time is not known.

Advances in the planned production startup remained modest. The September report merely contains the statement that the 8-388 J-3 series was in the startup process with increased pressure. An attempt was being made to complete the first aircraft in March instead of April 1945. The requirement for the Jumo 222 was rated as questionable because of the cancellation of the J-2 series. It is surprising how little progress was made, as the capacity made available by the cancellation of the Hs 293, Hs 129, and Ju 88 S-3 grew steadily, and led to the increasing "loaning out" of non-permanent workers to other aircraft manufacturers, a situation that might become permanent without follow-up orders.

A conference on the installation of exhaust-driven turbines and the apportioning of saved Jumo 222s decided that Henschel was to assemble Jumo 222 E/F engines into power plants for "RLM Program Requirements," and deliver them to "Ifa" (Junkers Flugzeugwerke AG) for installation in the Ju 388. At that time 26

Jumo 222 A/B and 14 E/F engines were available, having been earmarked for a wide range of test purposes (Hü 211, "Fw 152," Ju 388, Ju 488, He 219). While the first A/B-III power plants for the Ju 388 were to come from Muldenstein in August 1944,[700] delivery of all others by Henschel between September and December 1944 was envisaged as follows:

	Serial Numbers	Number per version
September	42-45, 49, 50	6 A/B-II
October	51, 54, 56-59, 61, 62	8 A/B-I
November	55, 60, 67-68; 15, 32,	4 E/F-I
	39, 40	4 E/F-II
December	45, 50, 53, 52; 36, 41	4 E/F-I
		2 E/F-II

As well, four A/B-1 engines and one A/B-II were to be built by Henschel as spares, and these were not assigned to any airframe.

The actual progress toward engine production was described as follows in a report to the Allies in October 1946:

"[...] In the engine field, toward the end of the war the Jumo 222 standard power plant was given high priority: the 222 A/B low-altitude engine and the 222 E/F high-altitude version, the latter with intercooling, were both redesigned with special consideration to production-related maintenance and spare parts issues. A prototype series of the A/B engine was complete when the plant was shut down, preparations were being made for production startup."

The report mentions other contributions by HFW to the development of the Ju 88/188 into the Ju 388—regrettably, again very general in nature:

"[...] A series of heavy weapons installations were designed and built as prototypes n preparation for conversion of the Ju 88 and 188 for the Zerstörer role (1944-45)."

A fragmentary inventory list[701] from the Henschel plant dated July 1945 reveals that spare parts and a mockup (installation) of the Jumo 222 were present there.

VW and Opel

As mentioned earlier, the subcontracting of components was not restricted to JFM and the "usual" licensees. Production of pressurized cockpits was dispersed in the same way as other components. Typical subcontractors, such as Opel and the Volkswagenwerk, were also tasked with the delivery of complete aircraft.

Well known for the production of motor vehicles, in 1940 the Adam Opel AG in Rüsselheim began building undercarriages, cockpits, fairings, and other components, as well as producing Junkers aero-engines under license. The company was also contracted to assemble the wiring (complete wiring harness, including instrumentation) for the Ju 88, this continuing until early 1945.[702] The envisaged production of fuselages for the Ju 188 does not appear to have come to fruition, however (see above). An overview of the production breakdown made for the Allies in May 1945 confirms Opel's participation in installing engines in the Ju 88, and building fuselages for the Ju 188, although this ended in July 1944.

Available production capacity led the Volkswagen factory—then the most modern automobile plant in the world—in the "City of the Strength through Joy Car" (later Wolfsburg) to become increasingly interested in involvement in the field of air armaments. Mass production of the "Beetle," which was supposed to begin in the late summer of 1939, was halted by the war, as was civilian automobile production at Opel.[703] Koppenberg suggested that the resulting free capacity be used to build parts for the Ju 88.[704] Göring gave the job of examining the possibilities of building the Ju 88 at VW to Wilhelm Heinrich Evers, the designer of the Ju 88 trained by Fokker in the USA. Evers determined that the company was capable of producing 1,000 to 1,500 machines. The project failed, however, because of resistance by Porsche, whose efforts to defend his production of automobiles were initially successful.[705] Nevertheless, in September 1939 VW Fallersleben began producing drop tanks, bombs (SC 250), and large pressed parts for the Ju 88. In the years that followed, the aircraft industry began coveting all available production areas, like VW's branch plant in Neudek, northwest of Karlsbad, a former wool combing works. Beginning in August 1944, 736 complete examples of the Ju 188 G and H were to be built there, plus 655 Ju 388s, in total one quarter of planned production by JFM![706] In the Junkers production plans of March 1944, the Volkswagenwerk (VWW) was envisaged as the manufacturer of 425 examples of the Ju 188 H-1 (reconnaissance), and 687 of the G-3 (torpedo-bomber), with production running from autumn 1944 to March 1946.[707]

In GL/C Aircraft Program 226[708] of 15 May 1944, in addition to 71 examples of the "Ju 188 G-2 with Jumo 213 A," VWW was ordered to produce the "Ju 388 M cuto," classified a torpedo-bomber, as the sole manufacturer of the type. A total of 655 examples of this version, powered by "BMW 801 TG" engines, were to leave the assembly line between January 1945 and February 1946. These plans were scrapped in the 15 September version of the delivery plan.[709]

Aircraft repair (Ju 88/188) and part production were ultimately transferred completely to Neudek,[710] where they clung to the idea of building complete Ju 188s to the end.

Constant changes in planning, however, meant that most of the work exhausted itself in production preparations. At Neudek it progressed no farther than the repair of Ju 88s and Ju 188s, and initial steps toward a wing building operation.[711]

Hints of the Volkswagenwerk's actual participation in production of the Ju 388 are found in the previously mentioned production breakdown. According to it, the Volkswagenwerk delivered wings and tail components until February 1945.

Production of Components by Subcontractors

The "E.A. Naether Kinderwagen GmbH," which obviously served as a subcontractor for the JFM plant in Merseburg,[712] began producing cockpits for the Ju 388 in January 1944. Completion of the first "Ju 388 high-altitude cockpit" was reported on 24 January. That same day the cockpit was inspected by representatives of the

Lower shell of a Ju 388 pressurized cockpit under construction. None of the internal equipment has been fitted. Most production of cockpits obviously took place in dispersed operations. Griehl Collection

BAL assigned to JFM and JFM production inspectors. They were extremely dissatisfied with the workmanship. The interior of the glazing had been scratched during sealing, and some of the riveting work was faulty. The plant management promised improved work on subsequent cockpits, but as a precaution JFM Dessau sent trained specialists to Zeitz to improve the quality of the riveting.[713]

At the end of February the quality of the work by Naether was still unsatisfactory, however. Pressurized cockpit 13/4, which was delivered on 16 February 1944, was selected as an example. It was reported complete, although much of the work still remained to be completed. Because of time constraints this work had to be done at Dessau. Dessau production inspection subsequently sent a reminder to Naether to conduct more rigorous inspections.[714]

An order synopsis of the "8-388 HK J-1 series" by FSD-M[715] dated 28 January 1945 lists the following locations and quantities for the production of pressurized cockpits:

1. Pressurized cockpit shell by the Naether company, Zeitz (60 examples), and the Fleischer company[716] in Eilenburg (100 examples),
2. Pressurized cockpit canopy by the Naether company, Zeitz (110 examples), and Fleischer company in Eilenburg (50 examples),
3. Final assembly FSD-M (160 examples).

Production of pressurized cockpits (one of the above mentioned?) also took place in a Hindenburg Barracks in an unidentified location. Twenty-four pressurized cockpits assembled there were delivered by 11 August 1944. Two more were at an advanced stage of construction, and it remained to be determined "whether the 26 cockpits could be delivered unsealed."[717] In addition to cockpits for the Ju 188, the Meinl company in Grosspriesen also produced at least parts for the Ju 388's pressurized cockpit. Two side walls were completed in the last week of June, and a crew compart-

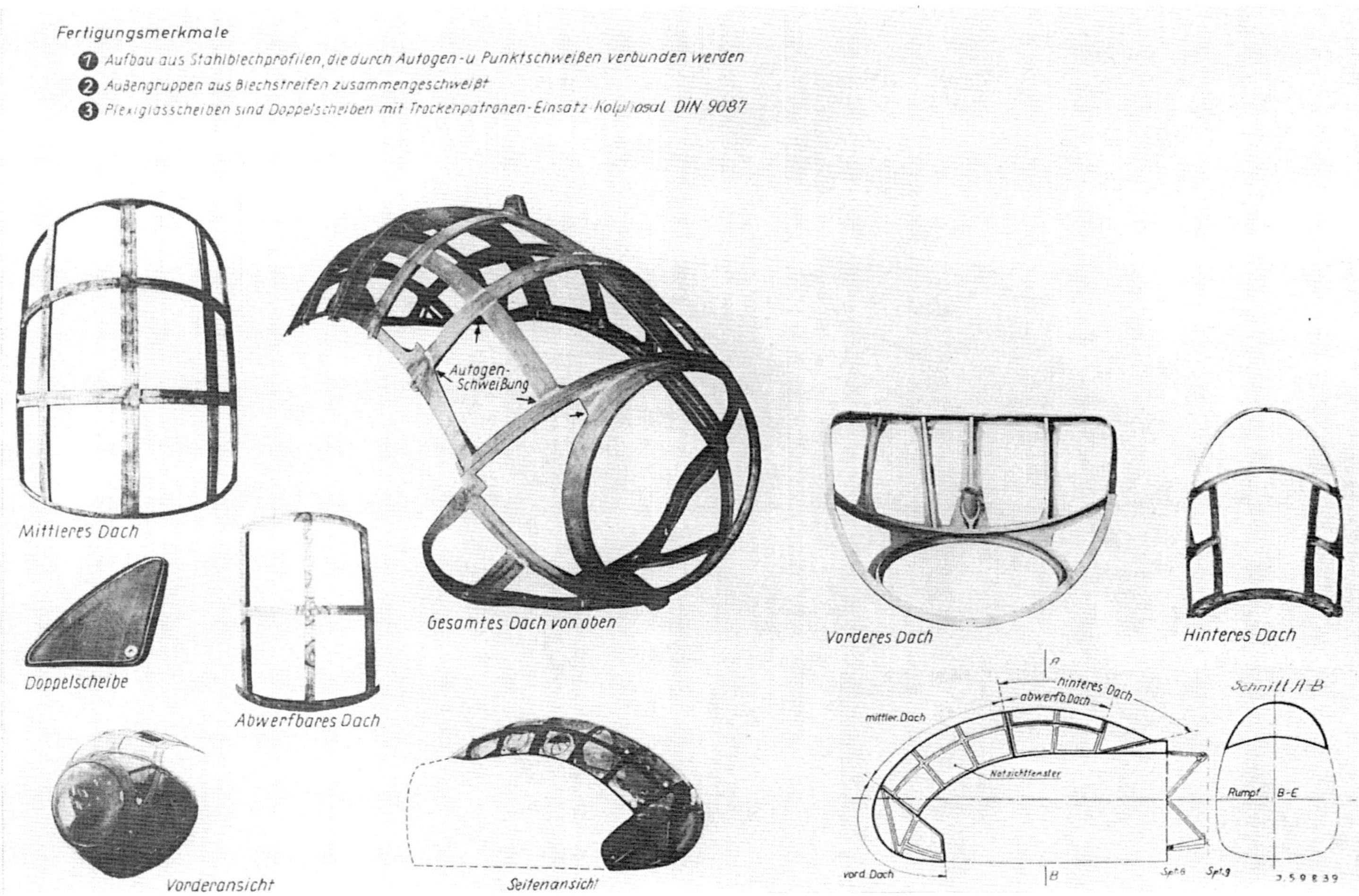

Overview of the various components and production stages of the cockpit canopy.

ment, a lower shell, two more side walls, and an upper shell were under construction.[718]

The double-glazed panels for the Ju 388 prototypes were fabricated in Dessau. Panels for production aircraft were delivered by the Röhm & Haas company of Darmstadt, and the Henze company of Eilenburg. The rectangular rubber strips (8 x 4 mm) used for sealing the panels were produced by the Wetzell company of Hildesheim. They were fixed between the panels using a special glue (KP 758). Experiments were carried out in Dessau with the objective of discovering an improved gluing method, while Röhm & Haas experimented with joining the rubber and Plexiglas through vulcanization, which offered considerable advantages compared to cold gluing. The problem of fogging encountered at Dessau, and during *Luftwaffe* testing was in part due to the use of the wrong glue, and in part to the complete absence of tightness checks by the Henze company.[719]

The main producer of fuselages for the Ju 388 was the "Metallwerke Niedersachsen Brinckmann & Mergell," which was usually abbreviated to "Menibum," because of the long company name. Production was carried out in an underground plant codenamed "Colibri"[720] (at the Hamburg-Berlin Reichsautobahn crossing).

Proof of this is found in WNr. 560 049. The manufacturer's plate on the bottom fuselage shell bears the manufacturer's code "mde," which stood for the Metallwerke Niedersachsen Brinckmann & Mergell in Hamburg-Harburg. Like most manufacturers, it too was ordered to halt production in February 1945.

On 9 June 1945 a CIOS team discovered eight Ju 388 fuselages in the yard of the bomb ravaged Harburg plant.[721]

Acceptance and Modification

After final assembly and an inspection by plant inspectors, the aircraft came to the *Luftwaffe* Construction Inspection (BAL), which was responsible for accepting aircraft for the air force. The completed aircraft were put through a final inspection, after which an acceptance flight took place. At least one further flight was usually required to eliminate minor faults.[722] The aircraft was then accepted by the BAL, and delivered to the RLM. With acceptance by the BAL, the aircraft became the property of the air force, and could be considered "delivered." The manufacturer could then submit the bill to the RLM.

Completed machines were subjected to extensive tests by the manufacturer. At JFM (FSD) this was done by Production Inspection (Feprü/Fp). This organizational unit filed inspection reports, examples of which have survived for the following aircraft:

1. Ju 388 V1 WNr. 500001 as of 7 September 1944 (FSD Hangar 203), dated 8 November 1944, with comments by the design bureau on 9 November.
2. Ju 388 J-1/V2 WNr. 500002 (FSD Hanagar 203). Report dated 29 September 1944.
3. Ju 388 K-1/V3 WNr. 500003 dated 10 April 1944 (FSD Hangar 203).
4. Ju 388 WNr. 230 153, response to the complaints expressed in the inspection report by the Series Production Design Bureau (*Herr* Korthaus), dated 26 September 1944.
5. Ju 388 K-0 WNr. 230 156 dated 17 November 1944; contains complaints and outstanding work.

In most cases the complaints were also passed on to the official responsible for the type, *Herr* Clement.

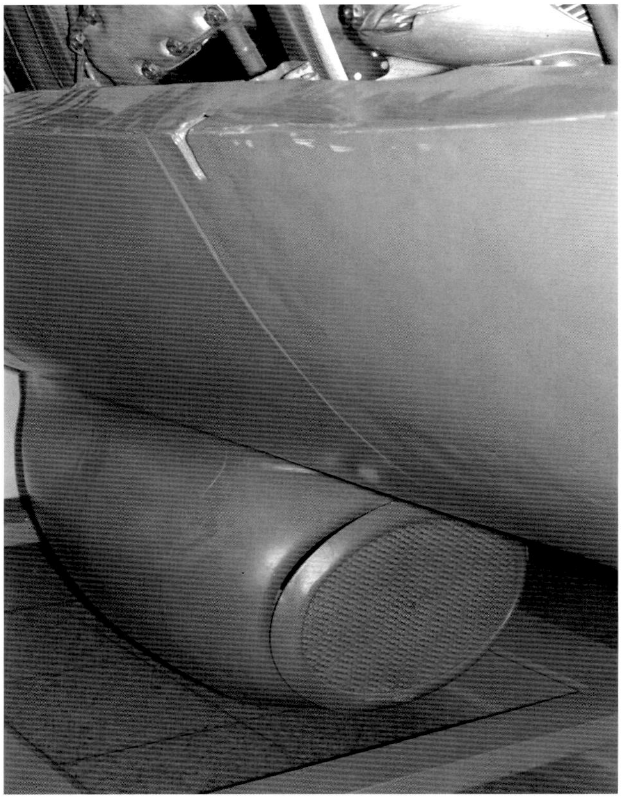

Close up of the external ram air intake on the BMW 801 TJ-1, in this case with the honeycomb FOD filter. It was one of the parts that was unavailable when the BMW 801 TJ-0 was retrofitted. Handig Collection

Amazingly, these two aircraft—probably WNr. 340 095 and 340 281—escaped being damaged by air attack, despite being parked in the open with no attempt at camouflage. WNr. 340 095 was one of the aircraft converted to external air intakes at Welzow. Crow Collection

On 22 September 1944 a meeting was held at Junkers, in which representatives of Junkers and BMW discussed retrofitting the Ju 388 with external supercharger air intakes. Hot air lines for the wing deicing and a flame damping system were discussed.[723] From January 1945 all production aircraft were to receive BMW 801 TJ-0 engines with external supercharger air intakes, and some aircraft were to come off the production line with such power plants beginning in November 1944. It was envisaged that the installation of the hot air lines needed for a serviceable wing deicing system would become standard on all aircraft in January 1945.

Those Ju 388s already delivered were to be retrofitted as soon as possible by the BMW field service with support from JFM's technical field service. The number of modification kits for the external supercharger air intake shows that they were still planning in considerable numbers at that time: BMW was to produce 80 modification kits in October, 130 in November, and 200 in December. Eight of these modification kits were already in Monfalcone, Italy, while forty more were supposed to be ready for collection as individual parts at BMW in Eisenach by the end of September. The remaining parts needed for assembly were to be made by Junkers itself as required. As well, the FOD filter was missing from all 48 kits, and had to be delivered by the BMW field service. The aircraft to be retrofitted had already been delivered to the *Luftwaffe*, and in such cases components would normally be delivered by way of the appropriate air armaments office, making delays inevitable. To avoid this, the Junkers field service was to designate an installation site, to which BMW was then supposed to deliver the modification materials direct.

Approximately 700 sets of hot air lines were required for the wing de-icing system. Because BMW no longer had the capacity to build the required number of modification kits, Junkers wanted to determine if, in view of their importance, it could possibly assume responsibility for their manufacture itself. All Ju 388s were also supposed to be fitted with flame dampers.

A large-scale modification program began in the modification facility at Welzow air base at the end of November 1944, in which the first aircraft built were to be brought up to the latest technical standard. The most important engine related modification was the installation of the external supercharger air intakes; however, several airframe modifications were also required. Seven Ju 388s had arrived at Welzow by 2 December 1944, and work on four aircraft had already reached an advanced stage. The absence of various modification and replacement parts proved a not inconsiderable hindrance, however. Originally, the first machine was supposed be ready for static tests on 6 December 1944.[724] According to Junkers records, a total of about 80 aircraft were supposed to pass through the Welzow facility by the end of January 1945. There was no shortage of personnel. BMW alone provided 37 men—mainly fitters and sheet metal workers—while the base itself contributed ten fitters and a welder. Modification materials were on hand for 30 engines, and only the Y-branch pipes for the wing de-icing were missing. Junkers was supposed to have delivered the first example by 6 December. On that day Ing. Reichel of the BMW field service visited the retrofit detachment. By that time the number of Ju 388s at Welzow had climbed to eight aircraft.[725] The following aircraft were present:

ATG	340 095	Weser-Liegnitz	560 041
	340 096		560 043
	340 100		560 045
	340 101		
	340 104		

Apart from *Werknummern* 560 041 and 043, which had come to Welzow by way of *Erprobungskommando* 388 in Rechlin, all production aircraft appear to have been delivered directly from the manufacturer.[726] All of the engines were being worked on, and half were already 70% completed. The first machine's engine modifications were now supposed to be completed on 15 December, with one more every two days thereafter. Beginning in January, one modified aircraft was to leave the facility each day. The airframe modifications lagged behind because of materials shortages; the target for the first aircraft was put back to 20 December 1944, and the entire timetable of the modification action was thrown off.

In addition to the BMW field service, representatives of the aircraft manufacturers also visited the retrofit detachment. On 5 December 1944 *Herren* Brachvogel and Lantzsch of ATG spoke with engineer Otzen and airframe inspector Wähler, who had come in from Dessau. With the exception of WNr. 340 104 and WNr. 560 045, Wähler, who was responsible for the prototype aircraft in Dessau, had already carried out initial inspections on all the aircraft. In addition to actual production errors or damage to the machines, Wähler also criticized the absence of several modifications that had already been carried out on the Dessau prototypes, but about which ATG still had no information. The two ATG representatives protested that this cast a false light on ATG's work, and asked for a separate listing of actual shortcomings and modifications not yet carried out. All eight aircraft were already being worked on, but no end to the airframe work was yet in sight. During the visit two ATG men installed the additional rivets between Frames 12 and 15 on WNr. 560 043.

At the end of December 1944 the Ju 388 L-1 was declared a "completed design." All subsequent modifications would require the approval of a modification committee in Berlin chaired by *Oberst*absinginieur Klinker. The following modifications were envisaged for 1945 production aircraft:[727]

Modification Standard 1/45
1. Fuselage sealing
2. Installation of FuG 16 (ZY)
3. Modification of turret belt feed and weapons fairing
4. Installation of fitting for aileron spring control
5. Rounding-off of nacelle extension leading edge
6. Recessing of cockpit air cooler air intake
7. Covering of slot in aileron area
8. Installation of auxiliary oil tank
9. Installation of armor for radio equipment (if armor plates not available, to be delivered later)
10. Deletion of hydraulic cooler for tail turret guns and installation of dead man button.
11. Modification of fuel system placard. This modification is also to be carried out retroactively during repairs.
12. Installation of vertical antenna (for FuG 217). As per HAF, H. Otherwise vertical antennas are said to be available; any that are delivered are to be installed on January L-1 aircraft.

Modification Standard 2/45
1. Installation of infrared lighting.
2. Installation of camera heating
3. Separate O_2 batteries for radio operator and observer. This modification is not to be carried out retroactively.

Modification Standard 3/45
1. Removal of Cuto leading edge de-ice. For speed reasons it is essential that this modification be carried out at this stage at the latest.
2. Installation of 8-88 control yoke.
3. Installation of unprotected 500-liter fuel tank in wing.
4. Installation of adjustable cockpit cooling.

Modification Standard 4/45
1. Mechanical antenna severing device (dependent on delivery of equipment)
2. Possible installation of night photography equipment.

Modification Standard 5/45
1. Installation of OK 42
2. Installation of additional turn and bank indicator. For the L-1 a modification of the mounting [meaning by the Luftwaffe itself] is being carried out, and a modification directive is being written.
3. Installation of wing load points.
4. Installation of an improved entry hatch.
5. New control run-through in pressurized frame.

To be modified	Number of Engines
From series production	
By end of October	184
November deliveries	54
December deliveries	25
Subtotal	263
Broken down	60
Repaired engines	17
Total delivered and requiring modification	220

A report by the BMW field service on 19 January 1945 provides information on the engines and power plants completed to date; it declared that the following (at left) were in need of modification:[728]

In addition to the relatively high number of unserviceable engines, the document also provides a survey of the locations where these modification actions were carried out, or were to be carried out in the future:

Welzow Air Base and Forward Modification Facility	On 14 January 12 aircraft were ready for static tests, thus 24 engines; at a rate of one aircraft per day a further 30 engines by month end, provided that the required materials arrived on time.
Kölleda ATG Modification Facility	On 13 Jan. 1945 the first aircraft was 90% modified, under favorable conditions a total of 18 aircraft and 36 engines were expected by the end of January.
WFG Liegnitz	Eight aircraft had been modified there thanks to the immediate delivery of modification materials. There were reports, however, that seven of these machines were already at Welzow.
JFM Merseburg and SFW Schkeuditz	It was envisaged that the total of eight plus five engines there would be replaced by new engines from January production.
Testing Stations	The engines there were not involved in the modification operation.

It was hoped that, as a result of these measures, all of the 220 affected engines would be brought up to standard (external supercharger air intakes) by mid-March. In fact, much of the necessary materials was still missing, or was available but in the wrong place. Finally, the document also gives an overview of the progress made at Welzow and Kölleda: work began at Welzow on 21 November 1944 with the arrival of the first aircraft. One month later, eight of the twelve aircraft being worked on were 90% modified (because of missing parts for the external supercharger air intakes), two 70%, and two 50%. The first machine commenced static tests on 30 December, and by 5 January the number rose to five. On 14 January engine related work had been completed on 14 aircraft, another 11 were being worked on, and 4 were waiting for work to begin. At Kölleda work first began on 2 January 1945. Furthermore, the document reveals that the action had been planned since the beginning of November, but in fact had begun very slowly. Originally, the modification work was only to have taken place in Welzow, or license producers were to immediately have incorporated the modifications into production, which obviously did not happen.

Beginning in 1944, role-specific modifications for reconnaissance aircraft were carried out by the "General of Reconnaissance Aviation's Modification Detachment" (see below), probably at Kölleda. There, in relatively untouched Thuringia, there was an air ordnance depot that specialized in servicing and repairing Zeiss cameras, and installing them in operational aircraft prior to their handover to front line units. Based on the Quartermaster-General's Department 6 records of "Aircraft Assigned to the Flying Units," not only the Ju 388, but also all Ju 188 D-2s, Me 410 B-3s, and Ar 234 Bs passed through this "Umrüstkommando Gen.d.Aufkl" (General of Reconnaissance Aviation's Modification Detachment), or "Modification Detachment A." As the technically demanding installation required special skills, and as only a limited number of reconnaissance aircraft were to be delivered to the *Luftwaffe*, it made sense to centralize this function in an air ordnance depot. Furthermore, beginning in August and September 1944 Ju 188s began arriving in Brieg to have wiring installed for night photography. Several also went to a "Modification Detachment Jüterbog." The daily aircraft acceptance records of the Quartermaster-General's Department 6 III C confirm that three Ju 388 L-1s were assigned to the "General of Reconnaissance Aviation's Modification Detachment" on 7 November 1944, and three more on 9 November. Two more Ju 388s were delivered in February 1945, however, these were transferred to a modification facility. At the same time the Quartermas-

Day	Number	Type	Accepted by
25.9.	1	Ju 388 L-1	OKL-Oranienburg
26.9.	1	Ju 388 L-1	Vers. Verb. OKL
29.9.	1	Ju 388 L-1	TLR
30.9.	1	Ju 388 L	TLR
5.10.	1	Ju 388 L-1	TLR
16.10.	1	Ju 388 L-1	TLR
20.10.	1	Ju 388 L-1	TLR
24.10.	1	Ju 388 L-1	TLR
28.10.	1	Ju 388 L-1	Erpr.St Rechlin
30.10.	1	Ju 388 L-1	TLR
2.11	1	Ju 388 L-1	TLR
4.11.	1	Ju 388 L-1	Erpr.St Rechlin
5.11.	1	Ju 388 L-1	Erpr.St Rechlin
7.11.	3	Ju 388 L-1	UmrüstKdo Gen. d.Aufkl.Fl.
7.11.	1	Ju 388 L-1	TLR
9.11.	3	Ju 388 L-1	UmrüstKdo Gen. d.Aufkl.Fl.
15.11.	2	Ju 388 L-1	TLR
22.11.	1	Ju 388 K-0	TLR
30.11.	1	Ju 388 L-1	TLR
5.12.	1	Ju 388 K-0	TLR
31.12.	1	Ju 388 K-0	TLR
6.1.	1	Ju 388 L-1	TLR
18.1.	1	Ju 388 L-1	TLR-Rechlin
19.1.	1	Ju 388 L-1	TLR-Rechlin
27.1.	2	Ju 388 L-1	Vers. Verb. OKL
5.2.	1	Ju 388 L-1	Gen. d. Aufkl.
14.2.	1	Ju 388 L-1	Vers. Verb. OKL

Kölleda airfield near Weimar, housed an ATG modification facility and a *Luftwaffe* ordnance depot specializing in camera equipment. Here the wrecks of several Ju 388s are seen in front of a hangar. The partial *Werknummer* 28X may be seen on the tail of the aircraft in the foreground. Crow Collection

ter-General was able to take delivery of two aircraft from modification facilities, however, their usage is once again listed as "Mod. Det. A."[729]

Of course, toward the end of the war it occasionally happened that operational units collected aircraft "at the factory." Most known examples of these "unbureaucratic occurrences" involved Fw 190/ Ta 152 aircraft in central Germany. It is therefore quite conceivable that single Ju 388s might have reached operational units by this "direct" route, or by way of a front line depot. If this did happen, the aircraft were probably just replacements for lost aircraft, for their use would have required pilots trained to fly the aircraft, and ground crews trained and equipped to service such a "sensitive high performance aircraft." As a long range reconnaissance unit, namely 4.(F)/11, is believed to have been located in Welzow from February 1945, this possibility cannot be discounted.

In closing, the confirmed acceptances by the *Luftwaffe*. According to the "daily aircraft allocations" of the Gen. Qu. 6 Abt. C III,[730] the *Luftwaffe* received a total of 33 aircraft, 30 of them Ju 388 L-1s. Of these, just six went to the abovementioned modification detachment. The *Luftwaffe* High Command's Experimental Unit (Vers. Verb. OKL) received a total of five aircraft. The *Luftwaffe* obviously did not accept a single Ju 388 J, nor even the V1 or V2.

Given these numbers, one is justified in doubting that "accurate books were kept." Aircraft are known to be missing; for example, the Ju 388 L-0 delivered to the Vers. Verb. OKL (T9+DL). Perhaps they did not differentiate between the L-0 and L-1. Furthermore, the detailed reports by *Erprobungskommando* 388 indicate a significantly higher complement of aircraft, although the constant transfers of aircraft and returns to manufacturers and front line depots make a complete listing impossible. The monthly aircraft distribution reports do not even jive with the daily allocations by the same supply department of the *Luftwaffe* High Command.[731]

From both status reports it is deducible, however, that apparently no more aircraft were accepted after the abrupt termination of production and testing in mid-February 1945.

Month	New Arrivals from Manufacturers	To Retrofit Operation	To Conversion Facility	From Modification Facility	To Modification Facility	Remarks
Sept. 44	4					1 Vers.Verb. OKL, 3 TLR
Oct 44	6					6 to *E-Stelle* Rechlin
Nov 44	23	20				3 accepted by Gen.Qu. übernommen
Dec 44	24	24				
Jan 45	10	9	1	4		2 TLR, 2 Vers.Verb. OKL
Feb 45	2	2			2	2Modification Detachment A

Chapter 7: Testing by the *Lufwaffe*

E-Stelle Rechlin and *Erprobungskommando* 388

Initial Experiences at Rechlin

Our sources do not allow us to state with certainty when the first Ju 388 arrived at the *Luftwaffe* testing station in Rechlin. On 29 April 1944 deputy type supervisor Alfred Kuhn ferried Ju 388 L-0 PG+YD, WNr. 300 004, from Rechlin to Staaken.[732] The Rechlin staff criticized the artificial horizon, which was difficult to read, and the sprung Astro-window.[733] At that early stage PG+YD, at least, was being test flown at Rechlin.

In mid-February 1944 the JFM department titled "Aircraft Construction Technical Marketing/Approvals" was already working on a type testing procedure for the Ju 388 K-0 and L-0 encompassing 30 points. In addition to departments E-2 to E-5, the process also involved JFM in Rechlin and the BAL in Dessau. The BAL was to carry out the "operating reliability flight," while the

E-Stelle Tarnewitz was assigned the task of scrutinizing the weapons and bomb dropping system.[734] In the view of JFM, however, type testing with the first ten L-0s and the first three K-0s was not feasible, because these machines had been envisaged as test aircraft.[735]

On 12 and 13 June 1944 a large demonstration of the latest developments was held at Rechlin. The aircraft that were supposed to be inspected at Lärz airfield on the old Mirow Road on the afternoon of 12 June included a Ju 388, along with examples of the Ju 88 G-1, Ju 88 S-3, He 177 B-5, Do 335, Ta 154, Ta 152, He 219, and Ar 432. The Ju 388 was not expected to take part in the flying demonstration.[736]

The final report on the type testing of the Ju 388 L-1 series (state on 31 Jan. 1945) reveals that the *E-Stelle* began receiving pre-production aircraft in June 1944, prior to completion of the first L-1 prototypes (V5 and V6). These included WNr. 300 004 (PG+YD) and WNr. 300 005 (PG+YE). At the end of June aircraft WNr. 300 008 and 300 009 also completed their acceptance flights at Merseburg prior to delivery to Rechlin.[737] To avoid further delays in delivery, the aircraft had to be accepted without dorsal armament, and with three-blade propellers instead of the planned four-blade units.

Four-blade propellers remained in short supply into August. While VDM had delivered sufficient numbers of hubs, spinners and blades remained extremely scarce. JFM was therefore initially unable to comply with the request that it send four-blade propellers for the Ju 388 L-0s already delivered, even though it very much wanted to. It did promise to send the propellers without delay once the delivery situation improved.[738]

The two Ju 388s delivered to Rechlin were initially tested by the various departments of the testing station there. WNr. 300 005, still equipped with three-blade propellers, made an engine-out landing, probably on 6 July 1944, after an oil line rubbed through on the nacelle extension during flight.[739] The amazing thing about this is that the initial inspection report, obviously written the same day, criticized that very same oil line as badly chafed.[740] As the problem was also appearing on other aircraft, the forward panel of the lower nacelle extension was immediately removed from all aircraft. Together with other minor modifications by the users, this helped improve engine accessibility.[741]

To protect against exhaust gas heat from the BMW 801 TJ-0 engine's central exhaust, a 1.4-meter wide strip of steel sheet was applied to the wings and flaps. In spite of this, in early July 1944 Ju 188 WNr. 260 308, which was equipped with BMW 801 TJ-0s, showed signs of serious damage to the landing flaps, and separation of the metal skin after just 18 flying hours. The first signs of damage (loss of paint and filler) were also observed on Ju 388s WNr. 300 004 (PG+YD) and 300 007 (PG+YG), even though they had logged significantly fewer hours. The *E-Stelle* Rechlin was convinced that the problem was caused by the exhaust fairing, which was too narrow. The jet of hot exhaust gases spread out over the wing, deflecting slightly outwards, and was thus still able to damage the wing and its metal skin. Rechlin therefore requested that

Two members of a technical unit pose in front of a Ju 383 L-0 in Rechlin in the summer of 1944. This is one of the very few photographs of an aircraft actually fitted with the rearview periscope. It may well be WNr. 300 005, as this aircraft's armament was completely installed on an experimental basis. Note the two-color spinners and overpainted markings on the engine. Regel Collection

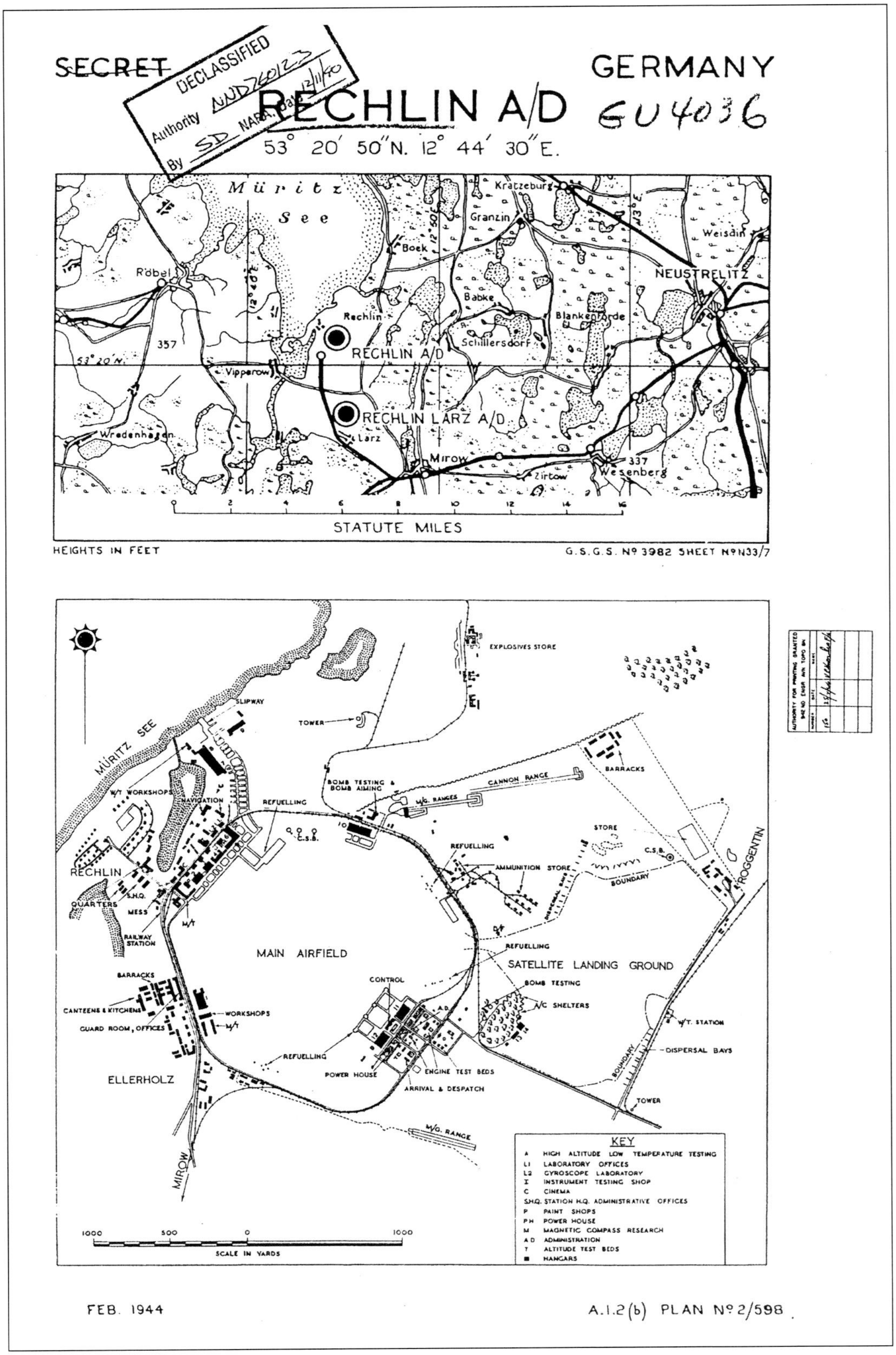

Site map of the Erprobungsstelle Rechlin from American sources. Regel Collection

the area of protective skin be extended.[742] Junkers rejected this, however, arguing that the ultimate production version of the exhaust pipe and fairing would direct the exhaust stream away from the wing in order to avoid affecting the aircraft's flight characteristics (vibration caused by flow separation). Those aircraft already delivered were to be modified as soon as testing of the production exhaust system was completed.

Damage to the wings of the existing aircraft continued, however. On 11 August 1944, the metal skin on the landing flaps of WNr. 300 007 was found to have separated from the underlying framework after 32 flying hours. The metal's tensile strength had dropped from the original 45.6 kg/mm^2 to 36.4 mm^2. After 43 flying hours, WNr. 260 308's other landing flap also began showing serious signs of deterioration.[743]

Erprobungskommando Ju 388 (Ju 388 Operational Trials Unit)

Erprobungskommando 388 was formed on 15 July 1944 with the official objective of conducting operational trials with the Ju 388 until the aircraft was ready for front line service.[744] In fact, however, the unit was more deeply involved with the reconnaissance variant, the Ju 388 L. For the first two weeks the trials unit existed only on paper. Personnel and a sufficient number of aircraft had to be procured before it could commence operations.

On 20 July 1944 (the day of the attempt on Adolf Hitler's life) type supervisor Hptm. Dr. Stahl of the *E-Stelle* Rechlin E III met with various representatives of the Junkers plant in Merseburg (FSD-M, "Construction Site Merseburg"), including *Herr* Bräuer of the plant management, to negotiate the delivery of Ju 388 L-0 and K-0 aircraft. Also present was a representative of the BAL/FSD-M by the name of Zurek. Hptm. Dr. Stahl requested that one machine be fitted with a dorsal gun position in Merseburg as quickly as possible. Sealing of the cockpit was of secondary importance to him at the outset. The Merseburg plant was also to install cockpit air coolers for the climate control system in the aircraft already delivered to Rechlin as soon as they became available. As the original version of the BMW 801 TJ-0's central exhaust had led to aerodynamic problems (flow separation, vibration), tests had been carried out with modified exhaust systems. Based on the information gleaned from these, the nacelle extensions of the Ju 388s already delivered to Rechlin would be modified, and new exhaust fairings installed. A five man modification team from Merseburg was to proceed to Rechlin to modify the nacelle extensions above the exhaust pipe; however, in August it was forced to return abruptly when the E 3 cancelled the modifications, because it feared they might result in reduction in engine cooling.[745]

Although VDM had advised Dr. Stahl that it could deliver sufficient four-blade propellers, he was also prepared to accept delivery of aircraft with three-blade propellers. He did insist, however, that the propeller deicing system be fully installed. Stahl also showed himself to be accommodating with regard to the FuG 16, even approving the delivery of aircraft with unserviceable FuG 16 systems. At that time the Lorenz company was obviously using aircraft K 3 (WNr. 230 153, KS+TC) and K 5 (WNr. 230 155, KS+TE) to test the FuG 16 system, for after completion of this work, which was expected by 20 July 1944, both aircraft were to be reported to Dr. Stahl as ready for pickup. In connection with the radio system, *Herr* Koch of Junkers requested that Rechlin determine the location of the DF loop as quickly as possible. With respect to the autopilot, Dr. Stahl declared that he was willing to accept the aircraft without complete systems, with the exception of aircraft K 1 and K 5. As well, he wanted to dispense with the FuG 217 for the engine test beds, whose serial numbers he intended to confirm in writing. Equipment set M 14 was still to be installed in aircraft K 5, and also in the remaining aircraft. The Astro window was to be installed in aircraft K 6 (WNr. 230 156), and was to become standard equipment beginning with WNr. 230 159 (K 9). Windows for the remaining aircraft would have to be delivered later. At the conclusion of the meeting Dr. Stahl raised twelve complaints from Rechlin, and asked for a report on the accident at Merseburg involving WNr. 300 293 (L-0). After several takeoffs and landings, the cooling fan on engine 385 269 had disintegrated as the aircraft taxied out for takeoff. Fragments had pierced the cockpit and injured the pilot.[746] Among the criticisms raised by Stahl were non-standard camouflage schemes on the L-0 aircraft, difficulty in opening the entry hatch from inside, and entry ladders that were too short. As well, the round counter in the cockpit had to be moved.[747]

Deliveries of the Merseburg aircraft obviously adhered quite closely to the timetable, for the *E-Stelle* Rechlin's weekly report for the period 23-29 July 1944 (filed on 29 July) records that WNr. 230 153, the test aircraft for E4 and E5, finally arrived in Rechlin on 27 July, and was subjected to a thorough initial inspection.[748] The inspection report stated that the aircraft had logged just 4.5 hours by 29 July. The list of "complaints and work to be carried out" included 51 points, a clear indication that construction of the aircraft had been rushed. Further indications of this included loose screws in the tank and equipment covers and leading edge skinning, insufficient pressure in the undercarriage oleos, and screws and pins lying about in the aft fuselage. As well, a faulty magnetic valve in the outer wing tank's fuel jettison system had been "repaired" with binding wire. The absence of the dark camouflage colors on the aircraft's upper surfaces was also criticized. While no complaints were made about the engines (385 215 starboard, 385 226 port), loose and broken lines and hoses among the engine accessories were criticized.[749] The line from the exhaust turbine to the mixture regulator had at least been in order when the aircraft underwent final inspection at Merseburg.[750]

WNr. 230 155 (KS+TE), which was the sister aircraft to 230 153, and envisaged as a test aircraft for E7, had arrived in Rechlin on 29 July, and was undergoing initial inspection.[751] Meanwhile, on the same day Ju 388 L-0 WNr. 300 009 (PG+YI) carried out a test flight that lasted from 1433 to 1714 hours.[751a]

In the final weeks of July, E3 had five Ju 388s and one Ju 188 with which to conduct endurance testing of the TJ engines. The engines proved extremely troublesome from the very start, hampering the entire test program. One broke down after 35 hours because of gearing failure, and there were numerous airframe related failures, especially of the hydraulic system. Several aircraft lost all their hydraulic fluid because of leaks in the system, and the emergency undercarriage lowering system had to be used a number of times. In addition to improved checks for cracks in the hydraulic lines, the *E-Stelle* also demanded schematics, as the Ju 388's hydraulic system was very complex, and clearly differed from that of the Ju 188. An initial evaluation of the type's performance revealed significant variations. By order of the K.d.E. the V5, which had been earmarked for testing by E5, was to remain in Dessau for at least another 14 days in order to conclude flight trials with the modified exhaust system. It was therefore expected that E2 would not be able to begin measuring the Ju 388's performance for at least three weeks.

Although WNr. 300 291 is seen in factory photos as early as the beginning of April 1944, it did not reach the trials unit in Rechlin until the end of July, with just 6 hours, 35 minutes flying time. By that time—unlike here—it was equipped with four-blade propellers. EADS

Among the pilots who conducted test flights in the Ju 388 at Rechlin was Hans-Werner Lerche, the well known Rechlin test pilot. He claims to have flown a "Ju 388 high-altitude bomber with pressurized cockpit."[752] In fact, the entry in his logbook on 29 July 1944 shows a flight in a Ju 388 lasting 1 hour, 34 minutes. The recorded manufacturer's code shows that the machine was DW+YY, WNr. 300 291. Between 1702 and 1836 hours he reached 11000 meters in the aircraft. Earlier the same day he had taken a captured Mustang to an altitude of 11500 meters.

DW+YY had just been delivered a few days earlier, and had undergone a thorough inspection on 22 July. It is amazing that the aircraft had allegedly flown just 6 hours, 35 minutes, for it appears in factory photographs probably taken at the beginning of April! The initial inspection resulted in relatively few complaints, although the flaps refused to move to the takeoff position when the corresponding button was pushed. The engines were a different story, however. The housing covers over the valve rockers on the port engine's 5th and 9th cylinders were found to be badly worn. The exhaust collector on the starboard engine was loose and not secured. Unlike in the photos, the aircraft was now equipped with four-blade propellers.[753]

Most manufacturers had their liaison offices at the testing station in Rechlin, and in the case of Junkers there was a branch station of the FTA (Aircraft Construction Technical Field Service) that was specifically responsible for the Ju 388. The reports filed by the FTA station attached to EKdo 388 in Rechlin provide a relatively clear picture of *Erprobungskommando* 388's beginnings. The detachment's first commander was Hauptmann Eduard Skrzipek, whose name first appears in the FTA station's monthly report for August 1944.[754] Skrzipek had won the Knight's Cross with Oak Leaves on 24 June 1941 while serving as Staffelkapitän of 14.(Eism.) KG 27.[755] According to the first monthly report (No. 1/44 of 1 August), an advance detachment consisting of the acting commander Oblt. Burkhard, technical officer Oblt. Ebeling, and 12 chief mechanics had arrived at the end of July. There was already a total of eight Ju 388s at Rechlin at that time, some of them having logged a considerable number of hours.[756]

300 002	27 flying hours
300 004	30 flying hours
300 005	55 flying hours
300 007	48 flying hours
300 009	28 flying hours
300 291	15 flying hours
230 153	5 flying hours
230 155	8 flying hours

By order of the K.d.E., the testing section assumed responsibility for maintaining all of the *E-Stelle*'s test aircraft in the new Junkers hangar.

These aircraft had not yet been assigned to the various test departments, however, as only E3 was using several aircraft for engine trials. The number of test beds for E3 was restricted, however, after initial plans assigned just four aircraft to that department. Three aircraft were to be "diverted" from the testing station's assets to the operational trials detachment being formed (another source places the number at two, with the third aircraft assigned to E2 for performance measurements). There were obviously also plans to send a Ju 388 to Werneuchen for the prototype installation of the FuG 204 (?) and the K12 autopilot.

The aircraft caused relatively few complaints, the majority of them found during the initial inspection, which in some cases took up to a week (probably because of the countless shortcomings). Some of the reports on the initial inspections at Rechlin have survived. Numerous metal shavings, nuts, and screws were found in the cockpit of WNr. 300 005 (PG+YE). The bad weather window (old version with plexiglass frame) stuck, the fittings were already loose, and the release button on the inner entry hatch locking lever had already broken off. Wrinkles had begun forming on the rudder, there were cracks in the metal skin near the upper hinge, and the end caps of both ailerons were loose. Loose screws were also found on the undercarriage mounts, and the undercarriage position indicator had been wired incorrectly. It is not surprising, therefore, that instructions were issued to tighten the screws on all access panels on the under surface of the wing.[757] Most of the complaints referred to chafed hoses and less serious defects. Potentially lethal defects were rare, but they were encountered. During initial inspection of WNr. 300 009 (PG+YI), it was discovered that the bottom forward nut of the threaded ball-socket connector had been secured, but not properly tightened. When the retaining wire was removed, it was found that the nut could be tightened another quarter of a turn by hand![758]

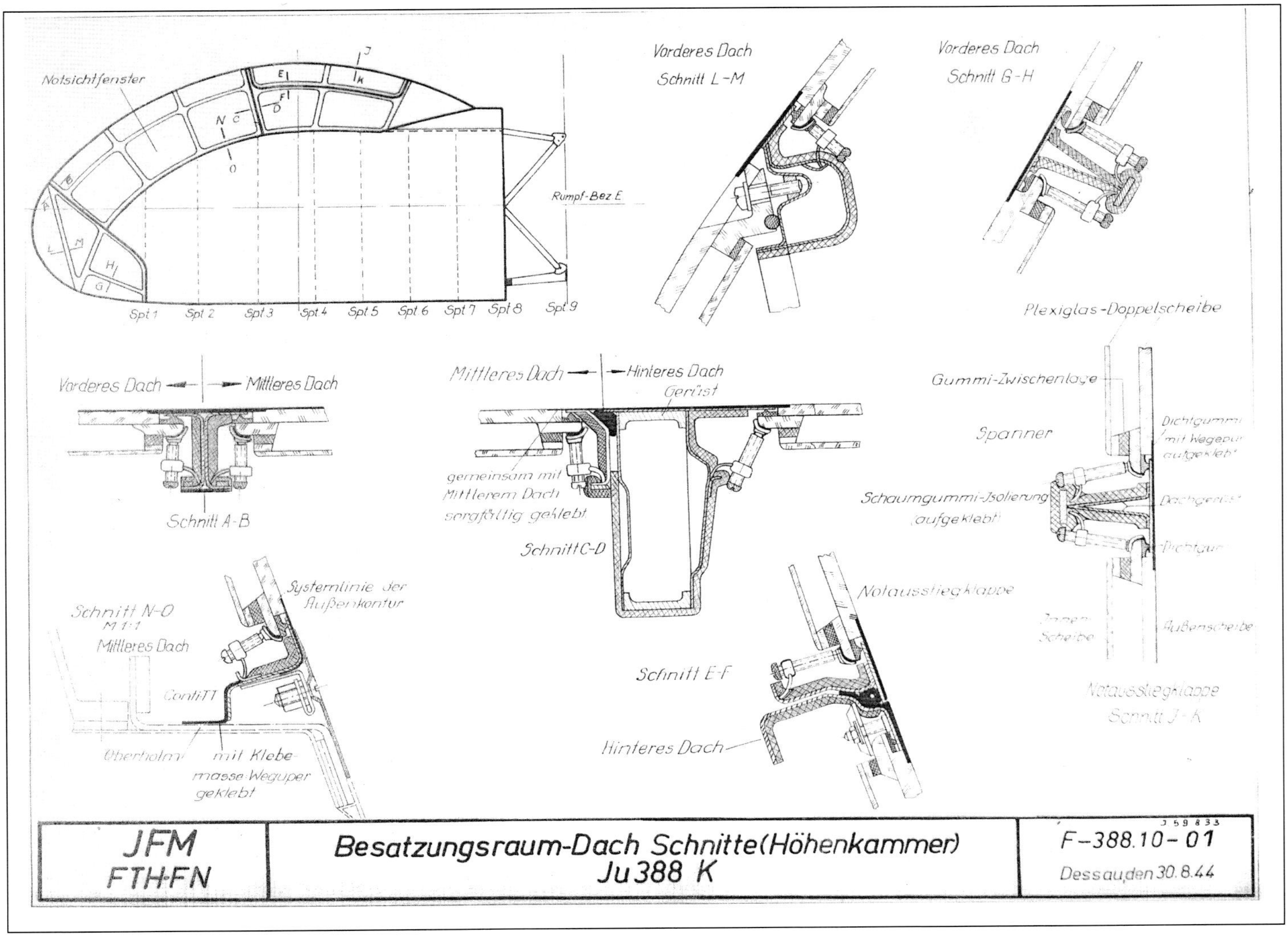

Drawing of the Ju 388 K/L pressurized cockpit with several detail drawings of the jettisonable canopy section.

In addition to these production technical problems, the first practical experience with the Ju 388 also brought to light numerous design faults. Ventilation of the pressurized cockpit proved inadequate. With an external temperature of 32°C the temperature inside the cockpit was 48°C. The metal parts in the cockpits of parked aircraft became so hot that they could not be touched with the bare hand. To improve ventilation on the ground, it was proposed that the jettisonable canopy section, designed to provide an emergency exit after a belly landing, be used as a sort of tilting roof. It was believed that a modified sealing strip "like that in the trunk of an automobile" would make it easy to open and close, while guaranteeing an effective seal. To do this, however, the folding canopy section and support would have to be built together. This had not happened to date, however, and it is not surprising that the jettisonable canopy section worked on none of the aircraft—with the exception of WNr. 300 008. As well, the mechanism to release the jettisonable section was not fully developed. Locks jammed and pull cables broke. Obviously, no one had come up with the idea of sprinkling the seals with talcum before assembly, as a result of which the jettisonable sections stuck fast.[759]

Temperatures in the cockpit during climbs were so high that the crews were forced to wear sports clothing. The crews reported temperature differences as great as 70°C compared to the outside air temperature (-40°C to -50°C) at high-altitudes; however, this was due to the absence of a cockpit air cooler, rather than the ventilation system. The supercharger air used to maintain cabin pressure reached temperatures of 50 to 100°C. In the event of an emergency exit, such a temperature difference, combined with the sudden drop in pressure, would surely have led to complications. The *E-Stelle* also criticized the inadequate forced air ventilation. Junkers subsequently calculated that, for an aircraft flying at 250 kph, the amount of air passing through an unpressurized cockpit had to be at least 250 kg/hr. Flight tests were supposed to determine whether this volume of air could be achieved with the high speed compensating valves (19-3416 A, 50 mm) used for forced air ventilation. The crews found the forced air ventilation uncomfortably loud. The intake scoop installed in the machines built at Merseburg was not yet the ultimate production version, however.[760]

After just a few hours of flight time, the wooden strip around the glazed spherical nose cap of WNr. 300 007 (PG+YG) came off. This was also suspected to be a manufacturing fault, because the problem had so far only arisen on this aircraft.

Also criticized was the non-standard, and often confusing labeling in the cockpit, apparently a problem typical of Junkers aircraft. For example, the positions of "OPEN" and "CLOSED" on the switch controlling the engine cooling gills were initially the reverse of those in the Ju 188. This was later changed. As well, the fuel transfer and jettisoning system was difficult for the crew to understand. Unlike the L-0 series aircraft, the schematics in the cockpits of K-0 machines WNr. 230 153 (KS+TC) and WNr. 230 155 (KS+TE) were marked with Roman numerals. The placard for releasing the life raft in these machines was also criticized, as was the lettering on the handle for setting the flaps prior to bail out.

Complaints were also made that it took too long to fill the fuselage tanks, especially since equipment for pressure fueling was not available. The reason for this was that the pre-production machines had been equipped with converted Ju 88/188 fuselage tanks, in which a second tank pump had been installed, and which therefore had just one filler point. The fuel tanks in production aircraft were to have two filler points (and thus the normal filling time).

The recesses in the control panel mounted on the starboard cockpit wall of the first pre-production were not deep enough. Consequently the switches stuck out too far, and could be moved by accident. While these recesses were supposed to be deepened beginning with the first production aircraft, a temporary fix was found for the pre-production machines—a folding plexiglass window over the switch recesses to prevent inadvertent operation of the landing flaps.

E3 had carried out several lengthy flights, and had found that the rudder pedals were too vertical, leading to "tired feet" among the test pilots. The pedals were subsequently modified to Ju 188 standard. Among the less serious complaints were the installation of the autopilot turn control, broken wires in the intercom sockets, and the fragility of the silicagel moisture traps.

Some of the faults that were identified were self-inflicted. Lugs and rod end bearings on the lower undercarriage retraction struts of WNr. 300 004 (PG+YD) and WNr. 300 007 (PG+YG) were found to be deformed. As it turned out, the undercarriage as designed was capable of withstanding normal loads, but it had not been designed for a landing after ten minutes of flight with almost full tanks. Unlike the Ju 88 and Ju 188, it was not possible to fill the Ju 388's hydraulic fluid reservoir through the external connection—an inattentive member of the ground crew did so anyway while servicing PG+YD, and the tank was damaged. And finally, one crew mistakenly jettisoned the entry hatch with compressed air on the ground, because he confused the red—but unsealed—emergency jettison lever with the lever used for normal opening.[758]

A design flaw in the cockpit air filter was discovered on WNr. 300 009 (PG+YI), which did not yet have the planned cockpit air cooler. 3/4 of a liter of oil had collected in a bend in a pipe beneath the filter through which the cockpit air always passed. It was no wonder, therefore, that many crews complained of oil fumes in the cockpit! A temporary solution was found, in that a 2-mm drainage tube was welded onto the lowest point on the pipe. This drained the oil, and eliminated the need for frequent visual inspections requiring the removal of 70 screws from an inspection panel. Drain screws at the lowest point were envisaged for later production aircraft.

Water drain holes were missing from the end switch caps in the undercarriage bays—one of the caps in WNr. 230 155 (KS+TE) was found to already be full of oil. The Junkers field service team member in Rechlin (Schweitzer) noted that this problem had already been identified on the Ju 88. Locking paint was no longer used on the electrical system, probably because of rationalization measures, then called "refinement." As a result, on KS+TE a loose screw on an unused connector was able to fall out. It lodged between other connectors and caused a short circuit that made it impossible to open the bomb doors. It surely took hours to locate such a problem.

A large percentage of the complaints recorded in the FTA office's monthly report for August 1944 once again concerned the cockpit. The EKdo again criticized the fuel transfer system, which was difficult to understand, especially for less experienced pilots, but at least it now had a clear idea of how it wanted to fix the problem. A standard placard with operating instructions for the fuel transfer and jettisoning system was to be installed in all aircraft. In addition to basic information about the fuel transfer procedure and the locations of the various electric fuel tank pumps, this placard was also to contain the settings to be used in single engine flight, and the capacities of the various tanks (which were of different sizes in the J, K, and L). As well, Roman or Arabic numerals on the rotary switch were to be replaced by the actual tank designations. The oxygen valves in the prototypes were located beneath the pilot's seat.[761] Pilots often jammed their fingers while trying to operate the oxygen valves, because the valves were too close together, and the cutout in the housing was too small. It is amazing that an error like this was not discovered during the mockup inspection. In addition to this problem, circuit breakers were still sticking in the main electrical panel, as it was not attached properly, and a bracket for the knob of the hydraulic hand pump had broken off because of poor riveting.[762]

WNr. DW+YY was destroyed in a bombing raid on 25 August 1944. This perspective provides a clear view of the exhaust fairing and stains on the upper wing, and the sealed MG 131 mount at the rear of the canopy. Visible beneath the fuselage are the weapons blister and trailing antenna for the FuG 10. Ott Collection

A requirement was identified to take along the ground crew chief as a fourth crew member on transfer flights. Consideration was first given to another seat with belts, but then also to a "seat on the floor" with restraining harness. Independently of this, consideration was also given to a new folding seat for the observer on the starboard cockpit wall (as in the Ju 287), in order to allow radar equipment to be installed between the pilot's and radio operator's seats of the Zerstörer version if required.[763] In the evening a lighting test was carried out in the darkened hangar using WNr. 300 009 (PG+YI). It was found that the hand lamp projected a bright beam of light even with the smallest aperture—a spot of light 4 meters in diameter was visible on the hangar ceiling. It was thought that a blue filter might help. The positioning of several lamps was not optimal; for example, the emergency compass light projected too much light onto the canopy glazing and had to be moved. The shielded UV lamp provided inadequate illumination of the flap and undercarriage switches, as a result of which three additional lamps were installed in bulges in the instrument panel. This was not entirely satisfactory, however, as the lamps projected from the panel.

Also criticized was the fact that in several aircraft the luminous markings on the switches had been overpainted with identifying colors.

While at the beginning of August a single engine reached the significant total of 63 operating hours, and another had to be replaced because of a defect in the waste gate/dump flap, further basic difficulties with the BMW 801 TJ engine were coming to light. Weak fuel mixtures resulted in engine vibration, and it was observed that the engine was difficult to start in the cold chamber—a problem that would seriously hamper testing in the winter of 1944-45. At heights above 11 km oil and cylinder head temperatures rose significantly.[764]

Department E7's installations in Ju 388 KS+TE were completed on the evening of 17 August 1944. In the days that followed, however, the weather was unsuitable for high-altitude flights. Furthermore, the aircraft's hydraulic system was unserviceable: the bomb doors were not working. Bomb sight tests were to be carried out once this problem had been fixed and the weather improved. Carrying four SC 250 bombs, KS+TE took to the air at 1658 hours on 22 August with *Fl.Oberstabsinginieur* Walter Ballerstedt at the controls. He subsequently climbed the aircraft to an altitude of 6500 meters and returned for a landing at 1745.

E2's performance measurements with Ju 388 L-0 PG+YG (WNr. 300 007) were completed by 19 August, and an interim report was being prepared. The aircraft, still equipped with three-blade propellers, achieved the following speeds at combat power at a weight of 11.7 tonnes:

Altitude		Speed
0		410 km/h
7.5	(maximum boost altitude low supercharger)	540 km/h
11.2	(maximum boost altitude high supercharger)	515 km/h

Group South's administration building at Rechlin, completely destroyed by enemy bombing on 25 August 1944. Regel Collection

Crash landed Ju 388 L-0 from the first batch of ten, allegedly photographed at Lechfeld or Fürstenfeldbruck in the late summer of 1944. Note the dark finish, apparently typical for aircraft of this batch. Photos suggest that aircraft from the second batch of ten were painted a light color overall, like DW+YY. Some later had camouflage colors applied to their upper surfaces. Unfortunately, the decisive letter of the manufacturer's code cannot be seen (PG+Y?). Griehl Collection

This aircraft, too, was still equipped with three-blade propellers. From this angle the rearview periscope is visible atop the canopy. The jettisonable canopy section is raised somewhat, an indication that this machine's emergency exit worked. The open engine access panels suggest that the search for the cause of the engine failure began right at the crash site. Griehl Collection

At a gross weight of 12.5 tonnes, service ceiling was 11.8 km at combat power.

The V5 was to have been used for performance measurements, however, shortly before it was to be delivered to the *E-Stelle* the aircraft was damaged when one of the undercarriage legs retracted during takeoff. According to Junkers records, the V4 was supposed to be handed over to Rechlin for performance tests by 19 August as a replacement for the V5. This did not happen, however, as the aircraft's engines were damaged as the result of enemy action.

In Rechlin a BMW 801 TJ had to be removed from a Ju 388 following the failure of a cooling fan after logging the considerable figure of 86 hours. By then another engine had reached 70 operating hours.[765]

On 28 August Fl.Oberingenieur Oskar Huber took off from Rechlin in the V8 (PG+YB) to conduct performance, calibration, and altitude tests. The flight, made with two crew on board, lasted from 0745 to 1015 hours, during which the aircraft flew 1050 km.

On 25 August 1944 the testing station at Rechlin was struck by a heavy air raid. Two of the trials unit's Ju 388 L-0s (WNr. 300 007, PG+YG, and WNr. 300 291, DW+YY) were destroyed by fire. Another machine (Ju 388 K-0 WNr. 230 155 (KS+TE)) was hit by flying debris, resulting in 70% damage. The engines and cockpit, however, suffered almost no damage. *Werknummer* 300 005 (PG+YE) was also slightly damaged in the attack.[766] As "the actual Rechlin airfield and all the hangars" had become "unusable," the trials unit had to move its operations "next door" to Roggentin. The attack also destroyed all the unit's records, and as a result the FTA office was unable to provide any information about takeoffs and serviceability rates in its monthly report for August 1944 (2/44).[767]

The first performance flights were obviously imminent; however, Hauptmann Skrzipek viewed the aircraft's performance with skepticism.

To make matters even worse, on 30 August 1944 WNr. 300 005 (PG+YE) made a belly landing at Stolp. After 45 minutes in the air, the recently installed port engine (WNr. 385272, 26 hours) had to be shut down after running low on oil. After another 45 minutes of single engine flight, probably at combat power, the starboard engine's (WNr. 385 260, 76 hrs) oil temperature rose sharply, then oil pressure dropped. After the loss of much oil the starboard engine failed, probably because of a broken connecting rod. After a belly landing the port oil tank was found to be completely empty, while the starboard tank contained just 15 liters.[768] Like the Ju 188, the Ju 388 carried a total of 210 liters of oil in Equipment State C.

Testing of the V5 and V6—both L-1s—could not commence until the end of August and mid-September, respectively. The V6 was sent to Tarnewitz for testing of the tail turret. Five aircraft were assigned to the especially urgent testing of the new BMW 801 TJ-0 engine. At the same time, another aircraft was used to measure the performance of the L-0.

Again in September 1944, testing was dominated by problems with the engines. The first L-1 production aircraft from Merseburg were received in early September, making it possible for performance measurements of production aircraft with four-blade propellers to begin. The aircraft still lacked the external ram-air intakes, however. The *E-Stelle* Rechlin's weekly report for the period 3-10 September 1944 states that the Ju 388 V5 with internal ram air intakes had been delivered by Junkers for performance checks. It also seems worthy of mention that the first three production air-

The Ju 388V8, seen here during factory trials with instrumented probe on the starboard wing. This machine served as an engine test bed in Rechlin, as it was fitted with the necessary measuring equipment from the beginning. EADS

craft, inaccurately referred to as K-1s, were "under test by the trials unit."[769] The planned performance tests with the V5, which the manufacturer had equipped for performance measurements, could not be carried out, because the engines could not be brought to a serviceable state.[770] Finally, after four weeks of work on the engines, the V5 had to be replaced with the first L-1 production aircraft (WNr. 340 081), which had been completed in the meantime.

At the beginning of September 1944 the Ju 388 L-0 V8, WNr. 300 002, was also at Rechlin. Assigned to the engine testing department E III b, it was equipped with engines 385 182 and 385 219, each with six hours of flying time. The port engine's cooling fan drive mount failed, and the engine was replaced with WNr. 385 225, which had 25 hours on it. There had been four such failures at Rechlin so far, three of them involving the BMW 801 TJ, and one a BMW 801 E. The engine failure involving the V8 could have happened during a 560-km flight made by Fl.Oberingenieur Oskar Huber from Roggentin on 4 September. Takeoff and landing times were 1453 and 1615 hours.

There were also problems with Ju 388 L-0 WNr. 300 004, equipped with engines 385200 (1 hr.) and 385259 (36.5 hr.). The main connecting rod in the port engine's rear bank of cylinders broke during takeoff, and the engine had to be replaced. The aircraft resumed flying by 10 September at the latest, with at least three calibration flights known to have been made by 19 September. Incidentally, at the end of the war BMW 801 TJ-0 WNr. 385 259 was captured in Oberwiesenfeld, near Munich, where it had received a general overhaul. The engine was given a thorough examination. Its subsequent fate is unknown, however, the American report praised the engine's compact design and rapid interchangeability. It recommended further examination on the spot, or shipment to Wright Field.[771]

The Ju 388 L-1 WNr. 340 083, RT+KC, was the first aircraft to leave the factory with external ram air intakes. In the background is WNr. 340 084. The photograph was taken at Merseburg in September 1944. EADS

Earmarked for exhaustive performance measurements with internal ram air intakes by E2, WNr. 340 083 was assigned to the trials unit on 6 September 1944. On 11 September the aircraft was involved in an accident at Fürth (65% damage). The four members of the crew—Ofw. Ernst Zeitter (pilot), Major Otto Behrens (observer), Fw. Ewald Barden (radio operator), and Fw. Herbert Rackow (flight engineer)—were all injured in what the official loss report called a "forced landing caused by technical defects."[772] It was subsequently decided to carry out measurements using a production aircraft to be completed later. The crash of WNr. 340 083 was blamed, at least in part, on errors by the manufacturer. The observer stated that the landing flaps refused to lower, even though he repeatedly pressed the appropriate button. This suspicion was reinforced by the fact that the magnetic slide valves (19-1408 C-1) in the landing flap hydraulics of two other trials unit aircraft were also not working because of loose contacts. The aircraft was finally transported to Dessau for closer examination. It was later discovered that the initial frequent problems with the landing flap system and the undercarriage hydraulics had been caused by maintenance errors, and the hydraulics subsequently functioned well to the aircraft's service ceiling.

The exhaust system continued to cause problems. On 2 September 1944 the heat produced by the central "exhaust scoop" burned through the top side of the landing flap and the starboard wing of WNr. 300 009. The damage could not have been too serious, however, for the aircraft logged another six hours in the air before the damage was repaired during its scheduled 75-hour check.

During work on the canopy, a heavy buildup of rust was discovered on the inside of several sheet steel frames of several aircraft. In view of the limited strength of the structure and the possibility of condensation buildup, Junkers was asked to provide corrosion protection for all steel parts of the Ju 388's canopy.[773]

The quality of the work at Merseburg was sharply criticized. The trials unit complained that it had discovered numerous faults in the aircraft recently delivered from Merseburg, problems that should actually have been identified and corrected by the manufacturer prior to delivery.

One of WNr. 340 084's fuel tanks had to be replaced after it swelled because of leaking lines. The entry hatch could only be closed with the aid of a screwdriver, because each time the locking bolt went back too far. WNr. 340 081's propeller pitch control system was incorrectly wired. The wiring for the undercarriage position indicator beneath the port exhaust were pitted, and a fuel pump in the aft tank was installed incorrectly. Because the prototype and production versions overlapped in this aircraft, an unnecessary rubber seal was installed in the engine compartment, and this was set on fire by the exhaust. These shortcomings were discussed at a meeting of the plant management held in Merseburg on 13 September.

The majority of complaints raised in September were again related to the engines. The majority of the airframe related complaints involved the electrical system. Locating and fixing problems was seriously hampered by missing and inaccurate wiring diagrams. An emergency plan for the propeller pitch control system had to be requested, but it was also found to contain errors.

A fuel injection line broke during a flight by WNr. 340 084 on 23 September. Fuel flowed into the exhaust collector and ignited, producing a long jet of flame. The forward end of the exhaust fairing, which was inserted into the collector exit, was heated to the point where it glowed. It became so seriously deformed that the exhaust had to be replaced. Indications of the same mistake were evident on other aircraft. In order to keep the affected machine flying, it was decided to replace the exhaust fairings with those from another aircraft, however, this proved impossible because the mounting holes did not match. Efforts to fit the new exhaust were still under way on 30 September. WNr. 340 085 did not arrive from Merseburg until 29 September 1944. At the end of September *Erprobungskommando* Ju 388 had the following aircraft on strength, all of which were also being used for engine durability testing[774]:

WNr. 500 005 L-1/V5
WNr. 300 009 L-0
WNr. 230 153 K-0
WNr. 340 081 L-1
WNr. 340 084 L-1
WNr. 340 085 L-1
WNr. 340 086 L-1

Rear view of WNr. 340 084. The exhaust system has clearly been modified from the original version. EADS

Monthly Summary of Ju 388 Trials Unit Aircraft (September 1944)

WNr. 500005	WNr. 300009	WNr. 230 153	WNr. 340083	WNr. 340 084	WNr. 340 081
1 Maihak pump on port engine replaced	1. — 1^{35}	1. Autopilot and radio equipment installation	1.	1.	1.
2. —	2. Top of landing flap and starboard wing burnt through. 2^{56}	2. "	2.	2.	2.
3. —	3. — 1^{20}	3. "	3.	3.	3.
4. Landing flap switched off too soon during retraction.	4. Hydraulic reservoir replaced.	4. Starboard outer fuel contents transmitter leaking.	4.	4.	4.
5. — 2^{40}	5. Starboard outer fuel contents transmitter and float leaking. 4^{23}	5. Fuel contents transmitter relief tank leaking. 0^{16}	5.	5.	5.
6. Starboard engine lost oil. 2^{56}	6. 75-hour check and wing repairs	6. —	6. Arrived from Merseburg	6.	6.
7. "	7.	7. Autopilot and radio equipment installation.	7. Acceptance checks	7. Arrived from Merseburg. 2^{51}	7.
8. " 1^{00}	8. "	8. "	8.	8. Acceptance checks	8. Arrived from Merseburg 0^{41}
9. "	9. "	9. "	9.	9.	9. Acceptance checks
10.	10. "	10. "	10. Foreign body in switch E 1004. Wiring scorched.	10.	10.
11. Sparkplug change starboard engine, turbine blade broken 0^{33}	11. "	11. "	11. Overturned at Fürth, 65% damage	11.	11.
12. Starboard engine changed	12. Accidental ground on radio transmitter plug.	12. "	12.	12.	12. Blade on starboard turbine broken
13. "	13. "	13. "	13.	13. Exhaust collector loose, left brake leaking	13.
14. "	14. "	14. "	14.	14. 12.5-hour check, landing flap too loose	14.
15. "	15. "	15. "	15.	15. — 0^{10}	15.
16. "	16. Landing flap would not retract due to defective plug on horiz. stab.	16. "	16.	16. Braking cylinders on both drums replaced 2^{48}	16. Oil temperature too high, port engine 1^{10}

WNr. 500005	WNr. 300009	WNr. 230 153	WNr. 340083	WNr. 340 084	WNr. 340 081
17. — 0^{45}	17. — 0^{18}	17.	17.	17. 0^{24}	17. —
18. Temp. transmitter on port engine not indicating, plug replaced	18. —	18.	18.	18. Forward fuselage tank not indicating, float leaking	18. Forward fuel tank not indicating because float leaking
19. 25-hour check, starboard engine 1^{25}	19. 3^{30}	19. 0^{43}	19.	19. Port oleo replaced	19. Manual pitch control both propellers not in order
20. —	20. Hydraulic magnetic valve for horiz. stab. Replaced. 133	20. 1^{45}	20.	20. 2^{11}	20.
21. Leak in hydraulic line to port oil cooler 1^{57}	21. Over land 3^{21}	21. 2^{10}	21.	21. Contents transmitter replaced 2^{15}	21.
22. —	22. 1^{35}	22. 12/5-hour check 2^{00}	22. **WNr. 340086**	22. Boost pressure interruptions 1^{14}	22.
23. 25-hour check, port engine 1^{49}	23. 2^{40}	23. Autopilot installation work 1^{10}	23. Arrived from Merseburg 0^{55}	23. Starboard exhaust glowing, injection line broken	23. Pump in 2nd fuselage tank not working 0^{54}
24. New seal in air line for emergency exit 0^{28}	24.	24.	24. Acceptance check	24. Port engine temperature transmitter, plug replaced	24. —
25. —	25.	25. Electrical fault in landing flap system	25.	25. Engine failure	25. Starboard engine nose ring damaged by cooling fan
26. —	26.	26.	26.	26.	26. Port RPM indicator jammed
27. Starter replaced port engine	27. Engine fuel pump replaced.	27. Autopilot installation work	27. Landing flap lock too loose 0^{18}	27. Exhaust pipe fairing missing	27. —
28.	28.	28.	28. Starboard engine pitch control unit replaced	28.	28. High supercharger did not engage until 9000 m 1^{12}
29. Propeller change, WNr. 340081 installed.	29. Collector ring pierced	29. Magnetic compensation	29.	29.	29. Propeller change, WNr. 500005 installed
30 Starter change	30	30 Radio checks	30 RPM fluctuations	30 Fitting exhaust fairing	30. Rubber on engine nacelle extension burned
Flying hours September: 11^{57} Total: 45^{52}	Sept.: 22^{59} Total: 107^{59}	Sept.: 8^{04} Total: 22^{45}	Sept.: 3^{14} Total: 3^{14}	Sept.: 11^{43} Total: 29^{09}	Sept.: 3^{56} Total: 14^{17}

Weapons and camera testing were also envisaged for the following month. One noteworthy aspect of the list is that obviously not all of the Ju 388s being tested at Rechlin belonged to the trials unit. Instead, it appears that some—like the V8 and WNr. 300 004—were attached to Department E3.

In September 1944 the equipment state of the power plants was still very provisional. This was clear in a meeting at Junkers on 22 September in the presence of Zindel.[775] All of the aircraft had internal ram air intakes. Even though experiments by Junkers with the Ju 388 L-0 V7 in August 1944 had demonstrated the clear superiority of external intakes, introduction of this feature on production engines was not scheduled to begin until November 1944, and become standard by January 1945. A "Priority 1" modification was envisaged for the remaining aircraft, and BMW was to produce 410 modification kits from October to December (80+130+200). Eight assembled modification kits were ready for collection in Monfalcone, Italy. As of 23 September, parts for 25 modification kits were ready at BMW Eisenach, and this number was supposed to rise to 40 two days later. A glance at the minutes from the Armaments Staff meeting of August 1944 shows who was working on the modification kits in Eisenach. With shocking casualness they mention that BMW Eisenach was to receive 300 Jewesses from the Buchenwald concentration camp. Gauleiter and "Reichsstatthalter" (Hitler's representative in a Reichsgau) Sauckel, who was also acting as "General Plenipotentiary for Labor," had decreed, however, that no Jewish females were to be employed in the "District of Thuringia." Meiwald therefore turned to Saur, and asked him to intervene on account of this "breakdown" (original emphasis), which Saur assured him he would do.

Although the existing modification kits would at least have sufficed for the experimental aircraft, testing of the external ram air intakes at Rechlin was slow. WNr. 340 083, which had left the factory with external intakes, and was envisaged for performance tests, had crashed at Fürth on 11 September. The next aircraft equipped with external intakes (WNr. 340 081), which first flew on 9 November, also crashed before analyzable performance flights could be carried out (see following). Subsequently, in mid-November, WNr. 340 085 (RT+KE) was converted to external intakes. *Werknummer* 230 153 (KS+TC) followed at the end of November. The date of 15 November 1944 for the completion of the first aircraft with external ram air intakes found in the final report on type testing is therefore rather puzzling. Perhaps WNr. 340 085 was modified at the same time as WNr. 340 081, and completed on 15 November.

At the beginning of October 1944 engine function trials were interrupted after a belly landing by WNr. 300 004 (PG+YD) was caused by an undercarriage malfunction. The aircraft was equipped with measuring devices and engines 321 221 (10 hr.) and 385 259 (47 hr.). The performance data measured in flight had so far been considerably lower than the values achieved on the test stand (by about 150 h.p.), and contained in the performance chart. The cooling fan's higher power requirement had already been factored in.

WNr. 340 085 (RT+KE) was produced in Merseburg for ATG. In November 1944 it was retrofitted with external ram air intakes for performance measurements. The aircraft is seen here in Rechlin during the winter of 1944-45. Haubner/Misna Collection

During the flight that ultimately ended in a belly landing, unsuccessful attempts had been made to transfer motor oil into the hydraulic fluid circulation system. The experiment was carried out using engine 385 259, which was fitted with measuring equipment. The result was engine failure caused by lack of oil, for after the flight aluminum and bronze shavings were found in the return filter. The engine's propeller shaft was bent 0.22 mm in the crash, and the torsion bar for the cooling fan drive was broken. The propeller shaft of the port engine (321 221) was also bent by 0.3 mm; consequently, both propeller drive units had to be replaced.

At the beginning of October engine trials involving the Ju 388 V8 had also not progressed very far on account of a variety of technical defects (failure of the cooling fan drive, generator failures, and airframe defects). Engine flight times were 26 and 7 hours (385 225 and 385 259). At the beginning of operations after a change of the port engine (see above), the generator had to be replaced three times because of electrical failures. Duration testing was based on nine aircraft at this time. Testing of the BMW 801 TJ-0 had suffered another setback when Ju 188 WNr. 260 308, which was equipped with these engines,[777] was lost in a crash at Rechlin on 17 September 1944. Attributed to pilot error, the crash cost the lives of the entire crew, which consisted of Fritz Dietz, Richard Eckert (radio operator), and Günther (flight engineer).[772]

The trials unit was also beginning to feel the effects of the growing Allied air superiority. The Ju 388 L-1 V5, WNr. 500 005, was damaged in a strafing attack on 3 October. It subsequently spent the rest of the month under repair without engines.

In October 1944 EKdo 388 underwent a change of command. Hauptmann Skrzipek, the former commanding officer, was replaced by Hauptmann Bonow, a night fighter pilot.

The General der Aufklärer sent Hauptmann Priess and Oblt. (Ing.) Bechmann to Rechlin to assess the aircraft's suitability as a reconnaissance aircraft.

In October 1944 EKdo 388 also assumed training duties. A reconnaissance unit and two other testing stations sent 25 maintenance personnel for familiarization. The men received practical and theoretical instruction.

The first three aircraft assigned to a reconnaissance unit and their crews also passed through the trials unit for training. One of the aircraft involved was WNr. 340 093.[779]

The engine's high maintenance demands are apparent in a letter from the commander of EKdo 388 to the type supervisor, Hauptmann Dr. Stahl, on 1 October 1944. It concerns the equipment change directive for the BMW 801 TJ-0 engine, and the time it allocated to each task. So far EKdo 388 had changed two engines (because of turbine failure, each 5.5 hr.), one starter (8.5 hr.), one speed governor (5 hr.), and one propeller pitch control unit. Valve play checks had also been carried out on several machines. Provided no problems were discovered, the valve play checks could be carried out in 4 1/2 hours. In addition to replacement parts, the necessary special tools and well trained personnel, carrying out of the work always required the presence of a BMW fitter. For this reason alone, the times achieved by EKdo 388, which largely mirrored those contained in the equipment change directive, could scarcely serve as a measuring stick for the air force technical service.[779]

One of the "refinement measures" proposed by the trials unit was the deletion of the additional propeller pitch switch on the throttle lever. The lightning fast operation of the propeller was only required when switching to single engine flight, but this could not be achieved using the switch on the throttle lever, as this affected both propellers. The pitch control switch, located on the starboard side of the cockpit, was completely adequate to meet requirements for a flight with manual propeller operation.[780]

During a post-flight inspection of WNr. 300 009 at the beginning of October 1944, a fist-size hole was found in the exhaust collector of engine 385 258 (88 hours total operating time). All had been in order after the previous flight. The state of the hole's edges suggested a small explosion in the exhaust collector, cause unknown.[781] On the day previous, the hydraulic line for operation of the propeller hub had come off on this engine and both engines of WNr. 300 005 due to a factory assembly error.[782]

The V8 carried out additional engine calibration flights in October 1944. At 1552 hours on 12 October, Oskar Huber took off from Roggentin, returning at 1725 hours after reaching an altitude of 9000 meters and covering 650 km. The next morning he took the aircraft up again, and in a flight lasting from 1045 to 1300 hours reached 12000 meters and flew 950 km.

Engine 385 219 (probably V8) failed on 14 October. Failure of the main connecting rod was first suspected, however, closer examination revealed that the auxiliary connecting rod on cylinder 13 had failed, caused by the seizure of the wrist pin.[783]

The final report on prototype testing of the Type 8-388 mentioned that a zero series (pre-production) machine was being tested by the deicing unit at Munich-Riem. It was Ju 388 L-0 WNr. 300 008 (PG+YH), which reached Munich by way of Rechlin.

The purpose of the test program was to answer the following questions:[784]

1. Is wing leading edge heating adequate over the entire span and at what altitude does it become inadequate?
2. Assessment of aileron characteristics in icing, which might also form in the aileron slot.
3. Is deicing of the horizontal stabilizer required?
4. Can all trim tabs still be operated in icing conditions, and do they retain their effectiveness?
5. Is the engine intake system adequate to permit satisfactory operation of the engines in icing conditions?
6. Is the absence of propeller deicing acceptable, and under what conditions?
7. Is the airspeed indicator system icing-proof?
8. How quickly does the internal fogging of canopy panels dissipate after leaving the clouds? Do some windows remain fog-free even in icing?
9. After installation of the propeller (deicing system): what quantity of deicing fluid is required for guaranteed deicing?

Deicing experiments began in Riem in mid-September, but only after turbine damage caused by blade and oil line failures had been repaired.[785] One test flight took place on 14 October, as reported by the "Junkers Testing Station Munich-Riem."[786] According to the report, PG+YH's starboard engine was equipped with an external ram air intake. On the mouth of the intake was a honeycomb filter, whose icing characteristics had been tested in the test chamber in Göttingen. During the test flight on 14 October, rime ice began forming at an altitude of 3400 meters at a temperature of -6°C. After half an hour of flying at 3500 meters the ice had become 2 cm thick. The honeycomb filter in front of the supercharger intake was also partially iced over. Interestingly, the edge of the filter was found to ice up sooner and more heavily, and air temperature in front of the supercharger was +8°C. To remain in icing conditions the aircraft climbed to 3700 m where, after another 10 minutes of flight at -7°C, the ice buildup had reached 3 to 4 cm on those parts of the aircraft with no protection against icing. The honeycomb filter was by now completely iced over, and all the air taken in by the starboard engine had to pass through a gap between the filter and the

Ju 388V8. The instrumented probe on the starboard wing was a feature of many prototype aircraft. It measured dynamic pressure at a suitable distance from the wing, so as to guarantee accurate data unaffected by possibly unknown aerodynamic disturbances. EADS

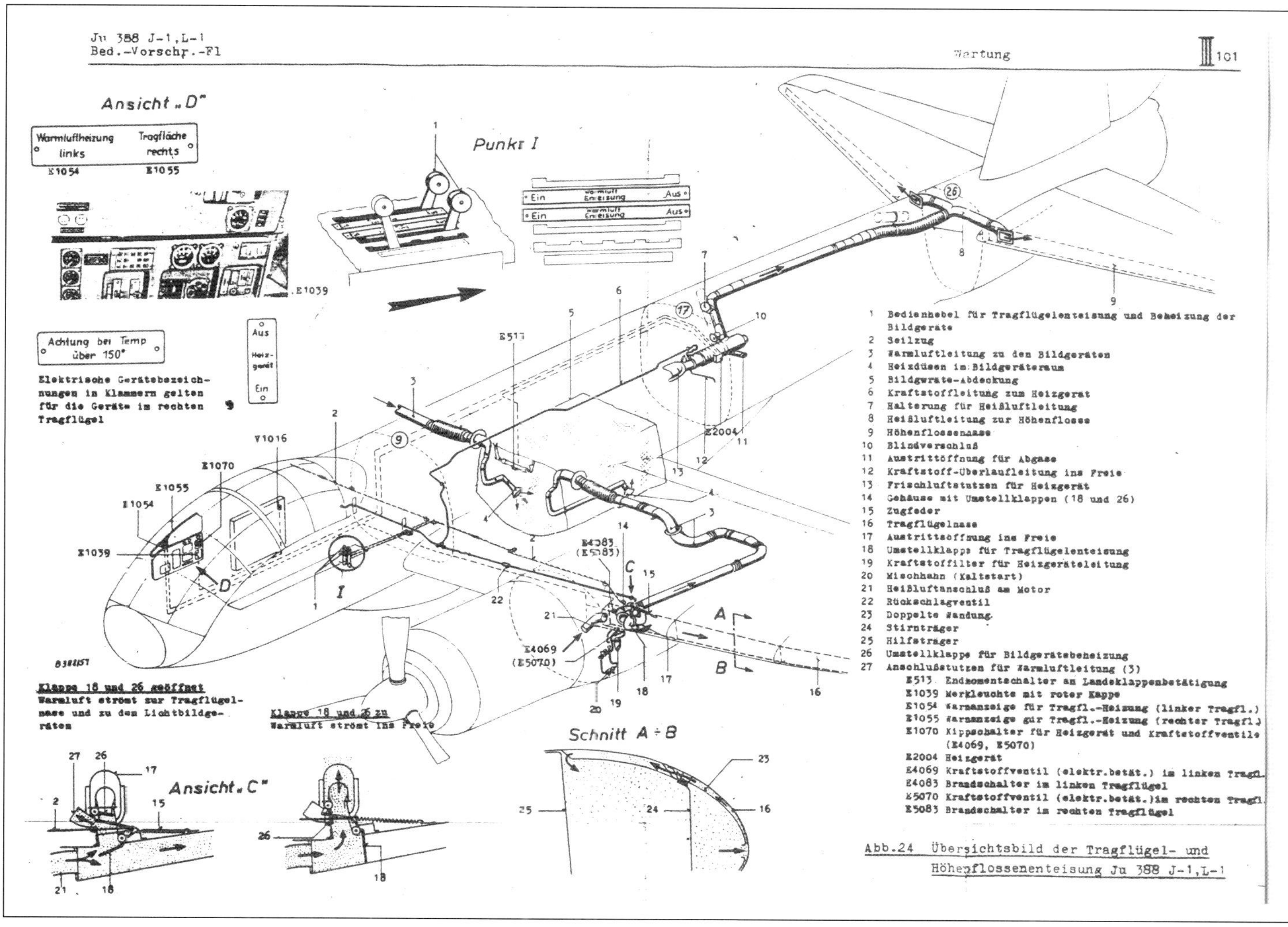

Overview of the hot air deicing system from the pilot's notes.

intake. The temperature in front of the supercharger had now climbed to +10°C. The reason for this was that, in the absence of ram air, warm air from the leading edge heating was being sucked back through the fresh air pipe, actually intended for mixing cold air (as the temperature of the heating air was too high), into the intake shaft. The fresh air intake on the port engine was also iced up. At 1.28 atm and 2,300 rpm the leading edge temperature was 124°C on the port side and 89°C on the starboard.

The top of cloud was reached at an altitude of about 3800 meters. The aircraft continued to climb, and at an altitude of 5 km with an outside air temperature of -12°C, leading edge temperature had increased to 138°C on the starboard side and 98°C on the port side. The reason for the difference in temperature between the starboard and port power plants is that the leading edge deicing system on the port side was receiving almost no fresh air because of icing, while the starboard engine was sucking in warm air.

At 5500 meters, 8 to 10 minutes after leaving the icing zone, the ice had cleared from about 25% of the filters over the ram air intakes, and supercharger air temperature had already fallen from +10°C to -5°C. It was actually planned to continue climbing with the iced-up honeycomb filter to determine if and how the honeycomb would become completely clear again, however, enemy air-

RT+KD was equipped with internal ram air intakes. In November 1944 it was used for various test purposes, including autopilot trials. Vernaleken Collection

craft were reported in the area, and the flight had to be aborted at an altitude of 5.5 km.

Back to Rechlin. *Werknummer* 300 292 came from Werneuchen on 18 October, and on the 21st returned to Dessau for modifications. On 21 October an unidentified aircraft carried out a flight to 10.5 km. The port engine's maximum boost altitude at continuous power was once again just 7.3 km (starboard more than 8 km). Achieved horizontal speeds at combat power at 10.5, 9.5, and 7.5 km were 508, 502, and 500 kph, compared to 530 kph on 21 September. It may be assumed that the flight was made by WNr. 340 081—the performance machine—and the reason for the performance differences was initially a mystery.

The Ju 388 V1 was received from Tarnewitz on 12 October 1944. It was flown on both of the following days, in each case for less than an hour. On 15 October the port engine's cooling fan failed. A two hour test flight was made on 17 October, and on the 19th another cooling fan disintegrated during engine run-up. On 21 October, after the completion of performance trials, the Ju 388 V1 was returned to Junkers. At that time the machine was still equipped with engines with internal ram air intakes and no flame dampers. In addition to the tail turret, the aircraft also lacked oblique armament and propeller spinners. The V1 was, however, already fitted with antennas for the Lichtenstein SN2 and FuG 217 radars. A brief performance check revealed a speed of 515 kph at 10 km, and 465 kph at 4 km. Company figures were 5 kph higher in each case.

On 21 October 1944 WNr. 340 090 was assigned to the trials unit from Altenburg. The acceptance check revealed that when the rudder was activated the ailerons also moved slightly. This fault was caused by excessive mechanical friction in a pressure-tight bearing on the rear cockpit wall. This was known to the head of control in Altenburg, however, the fault had not been repaired because of the amount of work required! The aircraft was declared not operational, and was subsequently sent to Dessau or Leipzig for repair.

On 15 October 1944 WNr. 300 004 made the six minute flight from Lärz to Rechlin. Two days later it took off on the one hour flight to Dessau, arriving there at 1247 hours. The aircraft had possibly been sent to Dessau for repairs or modification work.

L-1 production aircraft WNr. 340 081 was equipped with nozzles set for an air volume of 100 kg/hr, in order to check the effect of this low air flow on cockpit temperature during performance measurements. All of the crews had complained about the severe cold in the cockpit during longer flights at high-altitude. Temperatures fell too much when the cockpit air cooler was used;

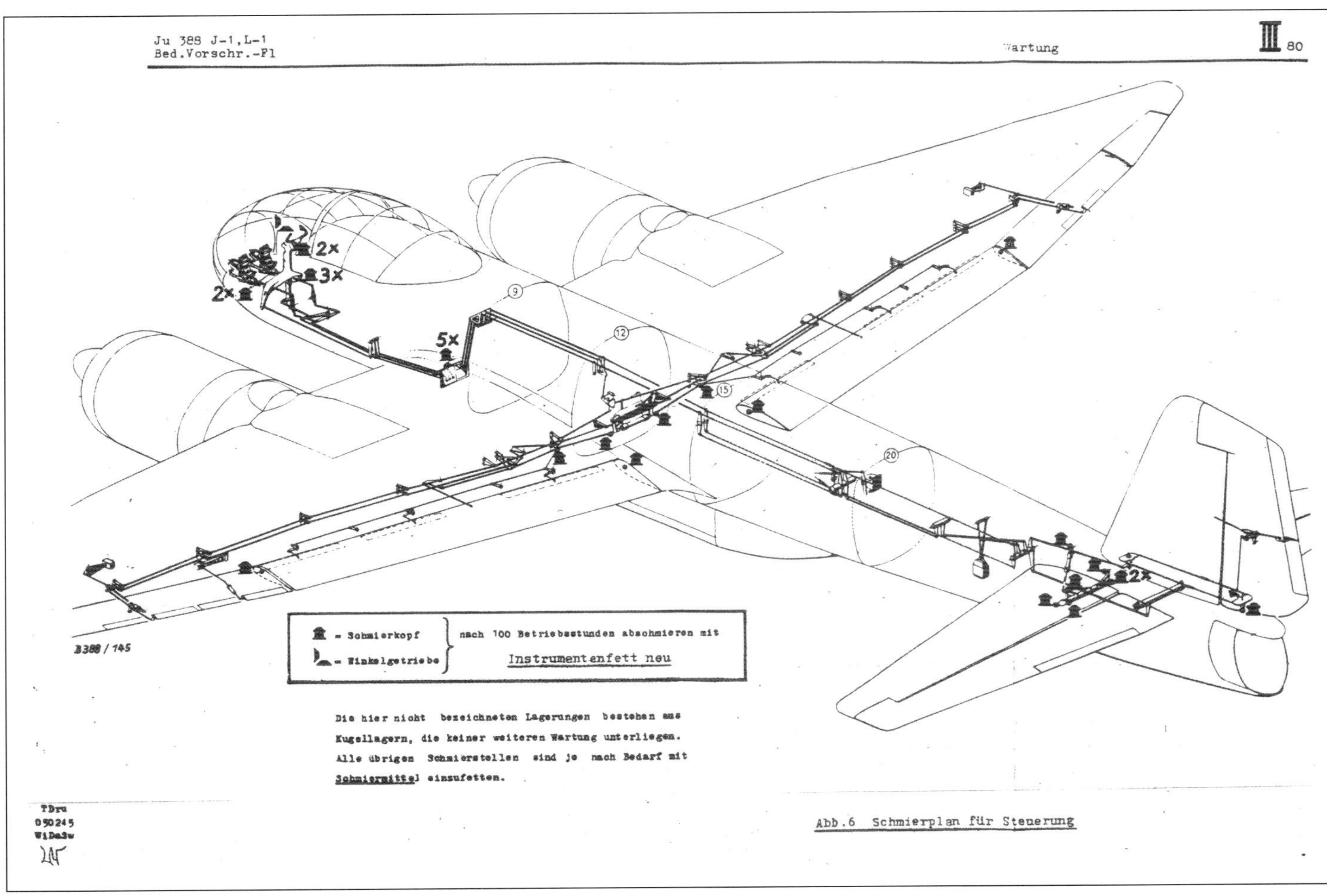

The lubrication plan from the pilot's notes provides a good overview of the arrangement of the Ju 388's elevator and aileron controls.

therefore, the tests with the cockpit air cooler covered were repeated. Performance measurements with the internal ram air intakes had obviously ended,[787] and beginning on 27 October the aircraft was to be converted to external ram air intakes by EK 388. Attached to E3, WNr. 300 004 (PG+YD) was then undergoing repairs by Junkers,[788] and during a short test flight by Oskar Huber on 26 October (1552-1607 hours) the V8 lost two cylinders, No. 9 and 10, on one of its engines. The aircraft was ready to fly again on 1 November, however, and in the afternoon it was flown from Rechlin to Brandenburg (1255-1325), and from there to Dessau (1610-1638).

The trials unit received two more aircraft towards the end of the month. WNr. 340 088 arrived from Merseburg on 27 October, and *Werknummer* 340 093 followed two days later from Altenburg, the first aircraft to be assigned to a reconnaissance Staffel. Both aircraft underwent the customary acceptance checks.

Performance measurements by WNr. 340 081 were interrupted from 29 October until 4 November 1944 because of another engine failure (main connecting rod). The three ATG production aircraft were undergoing their initial inspections, completion of which had been delayed by starter failures. WNr. 340 092 was damaged in a taxiing accident (undercarriage failure).[789]

Poor accessibility of the engine oil temperature sensor's electrical connection was generally criticized in October. In some cases it simply came off, even though it had previously been screwed tight. A securing mechanism had to be provided, for as the purpose of most flights was to test the engines, this seemingly insignificant fault almost always led to a premature landing. Also criticized was the fact that both Merseburg and ATG had ignored an experience report by FSD production inspection. The new location for the external temperature sensor in the aft fuselage also proved unsuitable, for the terminal inside the fuselage stuck out right beside the walkway, and it was broken off every time the aft fuselage was worked on. The proposed solution was a protective cover, in the event that a better location could not be found.

No testing of the FA 15 system was carried out in October because of further pending modifications. In order to nevertheless amass some initial experience, it was proposed that the system's hydraulics be activated during long flights. This would reveal any leaks in the areas of the control device, and the flange between the hydraulic fluid pump and gearing.

Tail turret of Ju 388 WNr. 340 393, which reached the trials unit at the end of October 1944. Aeroverlag Zuerl

Monthly Summary of Ju 388 Trials Unit Aircraft (October 1944)

WNr. 230153	WNr. 340081	WNr. 340084	WNr. 340085	WNr. 340086	WNr. 340087	WNr. 340089	WNr. 300009
1. FuG 16 ZY installation, magnetic compensation	1. Crack in port engine exhaust collector ring	1. Hydraulic interlock on port engine leaking	1. Acceptance check	1. Oil contents gauge starboard engine unserviceable, FA 15 leaking	1. "	1. "	1. To Merseburg for modifications
2. "	2. — 1^{04}	2. — 0^{42}	2. "	2. Propeller pitch starboard engine and oil temperature not indicating 1^{00}	2. Arrived from Merseburg, line 121 Hb broken	2. "	2. 0^{30}
3. "	3. Sparkplugs in port engine changed, exhaust was glowing 2^{33}	3. Indicator for undercarriage and flaps replaced 1^{00}	3. Hydraulic motor for tail turret sealed 1^{24}	3. Line on valve Hd 3 broken 1^{31}	3. Port engine cooler damaged by cooling fan	3. "	3. 2^{07}
4. "	4. Starboard engine seized	4. Starboard engine control device unserviceable	4. — 3^{12}	4. —	4. Acceptance check	4. Arrived from Merseburg 1^{46}	4. "
5. "	5. Starboard engine replaced	5. "	5. —	5. Port engine hydraulic pump replaced	5.	5. Hydraulic pump starboard engine replaced, FA 15 leaking	5. In Merseburg for modification work and repair of tail-wheel problem
6. "	6. "	6. — 0^{16}	6. — 1^{00}	6. Port engine from 081 removed	6.	6. Acceptance check	6. past the end of the month
7. "	7. "	7. Over land 1^{51}	7. Boost pressure fluctuations port engine	7. "	7. "	7. "	7. "
8. "	8. "	8. 2^{00}	8. Tailwheel damaged during taxiing 2^{04}	8. "	8. — 1^{58}	8. "	8. "
9. "	9. Hydraulic line 152 Hb cracked 0^{05}	9. Relief tank contents transmitter replaced 2^{30}	9. — 2^{08}	9. "	9. — 1^{53}	9. Short-circuit at switch E 4081, locking screw fallen out	9. Flying hours October: 2^{37} Total: 90^{44}
10. "	10. —	10. 1^{00}	10. Short-circuit line 1014, line lying on ground 2^{40}	10. "	10. —	10. —	10. "
11. "	11. Starboard engine seized, port contents transmitter replaced	11. Engine failure	11. Exhaust pipe broke off, engine cowling pierced	11. "	11. — 2^{52}	11. Transmitter connection for port engine oil temperature came loose	11. **WNr. 500001**
12. Electrical fault in landing flap system	12. Starboard engine replaced	12. "	12. — 4^{08}	12. Starboard generator replaced	12. Starboard engine oil cooler rubber torn off	12. — 1^{53}	12. Received from Tarnewitz

WNr. 230153	WNr. 340081	WNr. 340084	WNr. 340085	WNr. 340086	WNr. 340087	WNr. 340089	WNr. 300009
13. "	13. "	13. Starboard engine replaced	13. Oil temperature transmitter starboard engine came off 0^{45}	13. Starboard engine cooling fan and exhaust fairing replaced	13. Propeller pitch control unit and oil temperature transmitter replaced	13. Port engine propeller pitch control unit not working	13. Loose contact in oil indicator 1^{07}
14. Magnetic valve for landing flap replaced	14. Exhaust fairings cracked on both engines	14. "	14. Port engine replaced at Merseburg	14. Starboard engine replaced	14. "	14. Rivets in port engine cooling gill cylinder mount, broken	14. Engine trouble 1^{03}
15. —	15. Torsion bar on starboard engine cooling fan broken	15. "	15. "	15. "	15. "	15. "	15. Cooling fan port engine broken
16. —	16. Smaller volume control for cockpit air installed	16. Starboard oil tank replaced	16. "	16. "	16. Port engine failure, long exhaust flame	16. Port engine propeller pitch control unit replaced	16. "
17. Port engine boost pressure fluctuations	17. Contents transmitter replaced	17.	17. "	17. Short-circuit in battery	17. Boost pressure fluctuations	17. Over land 1^{00}	17. — 2^{03}
18. Landing flaps did not always extend to "Landing" position	18. Exhaust fairing burnt out and crumpled	18. Loss of RPM port engine, plugs changed	18. "	18. "	18. "	18. — 0^{36}	18. — 0^{39}
19. "	19. Exhaust pipe on Cylinder 14, port engine, broken, resulting in short-circuit, wire burned	19. Landing flaps retracted at 220 kph 0^{06}	19. "	19. Starboard engine seized 0^{04}	19. "	19. Starboard engine propeller pitch control unit problem, starboard oleo leaking 0^{39}	19. Cooling fan disintegrated during engine run-up
20. 0^{21}	20. Intake scoop for Kärcher heater cracked, turbine performance measured	20. Boost pressure fluctuations starboard engine 0^{23}	20. Rubber on starboard engine oil cooler torn off 0^{57}	20. "	20. Over land 1^{03}	20. FuG 200 installation Werneuchen	20. "
21. "	21. Port engine trouble, intercom interference 2^{20}	21. Starboard oleo replaced	21. Short-circuit on generator	21. No engine available	21. FuG 101 unserviceable, turn and bank indicator replaced 1^{05}	21. "	21. To Dessau for modifications extending past end of month 0^{27}
22. Landing flap cylinder leaking	22. Rubber strip on port engine oil cooler torn off	22. —	22. Trouble with boost pressure regulator, pitch control unit	22. "	22. Starboard engine starter replaced, short-circuit	22. "	22. Flying hours for October: 5^{19}

WNr. 230153	WNr. 340081	WNr. 340084	WNr. 340085	WNr. 340086	WNr. 340087	WNr. 340089	WNr. 300009
23. —	23. Tailwheel doors replaced	23. Boost pressure fluctuations starboard engine 1^{27}	23. Oil temperature problem, port engine 1^{35}	23. “	23. “	23. “	23. “
24. Magnetic compensation	24. Radio operator’s oxygen line broken	24. “	24. Supercharger air shaft lengthened by 10 mm, Kärcher heater air scoop torn	24. “	24. “	24. “	24. **WNr. 340090**
25. “	25. Strong interference on intercom, shield cap fitted on run-through	25. Vibration starboard engine 1^{19}	25. Intermittent long exhaust flame, antenna could not be retracted	25. “	25. Over land 0^{48}	25. “	25. Sent from Altenburg on 21 Oct., since then undergoing 1^{37}
26. “	26. “	26. Boost pressure fluctuations	26. — 3^{21}	26. “	26. “	26. “	26. acceptance check
27. — 0^{45}	27. “	27. “	27. Short-circuit in propeller pitch gauge lamp	27. “	27. “	27. “	27. Starting problems
28. —	28. Cockpit cooling air inlet covered	28. “	28. —	28. “	28. “	28. “	28. “
29. —	29. Starboard engine replaced	29. “	29. Port propeller damaged 2^{53}	29. Both engines missing	29. “	29. “	29. Fuel transfer placard written incorrectly at factory 1^{18}
30. — 0^{19}	30. “	30. “	30. Starter clutch starboard engine failed	30. “	30. “	30. “	30. Rudder inputs also move ailerons
31. —	31.	31. “	31. “	31. “	31. “	31. “	31. To Leipzig for control repairs
Flying hours October: 1^{25}	6^{36}	13^{01}	26^{01}	2^{36}	10^{39}	5^{35}	2^{55}
Flying hours total: 17^{16}	26^{06}	43^{29}	57^{59}	8^{24}	26^{47}	—	11^{55}

The Ju 388 trials unit did relatively little flying in November 1944. The following is a breakdown of tasks assigned to individual aircraft:

Radio testing:	WNr. 230 152, KS+TC
Shallow dive flights:	WNr. 340 086, RT+KF
Performance and camera testing:	WNr. 340 085, RT+KE
Hydraulic measurements:	WNr. 340 092
Flame damper tests and observation:	WNr. 340 094 and WNr. 340 088 (RT+KH)
Autopilot and electrical energy measurement:	WNr. 340 084 (RT+KD) and WNr. 340 093

Most of the radio tests with KS+TC were carried out on the ground, for the aircraft only flew nine minutes in November. WNr. 300 292 and the V1 were in Dessau for modifications during the entire month, while WNr. 340 089 was in Werneuchen for installation of the FuG 200. The V5 was still sidelined for lack of engines. The engines in WNr. 300 009 had also reached the allowable number of operating hours and were removed. From 14 November it was also parked without engines.

Installation of a MG 131 in the cockpit obviously also began in November 1944, the work being done by EKdo 388. A single MG 131 machine-gun was installed in the rear cockpit's starboard upper panel to provide defense to both sides. The weapon was to be operated by the radio operator.

Pilots complained that the aircraft could not be "held steady" while in flight at high-altitude, because minor rudder deflections also affected the ailerons and vice versa. Such behavior was of course unacceptable during photo flights. The reason lay in the excessive friction between the control rods and the pressure-tight lead-throughs. On WNr. 340 093 the control rods to the elevators and ailerons became so tightly jammed that the controls were frozen.

On 24 November 1944 this and seven other particularly important complaints were conveyed to JFM and ATG, and the construction inspectors there by telegram:[790]

1. Friction between the pressure-tight lead-through[s] and the control rods is so great that one control surface interferes with the other at altitude. On WNr. 093 the aileron control rod caused elevator control to freeze up. Check tolerances.
2. Remove wooden shims from rudder pedals. Pedal angle is too steep.
3. Pilot's oxygen hose is too short for flight without high-altitude parachute, install 1.5-meter hose.
4. Pilot's oxygen line from regulator bent while taxiing with fur-lined boots. Reroute.
5. Hydraulic system generally leaky.
6. Flettner unloading of rudder approaching over-balance, set to higher correct foot forces.
7. On WNr. 094 a wing attachment nut was not firmly attached, though secured.
8. Only use a 6 mm line for static pressure lines.

The lead-through boxes were removed and sent to Dessau for examination. Also criticized were the cover over the cockpit overpressure valve, the poor accessibility of the oil contents transmitter, the constant fogging of the oil temperature indicator (Fl 20358), the end switch on the tailwheel, and the aircraft's poor resistance to leakage. It is not surprising that the aircraft's leakage resistance

Photo of an added cockpit MG 131 installed by EK 388. It is not quite clear whether the installation shown here was in a complete aircraft, or just a pressurized cockpit. The jettisonable canopy section is missing, as is at least the panel behind Frame 9. It is likely that a separate cockpit was used to test the installation, at least initially. Aeroverlag Zuerl

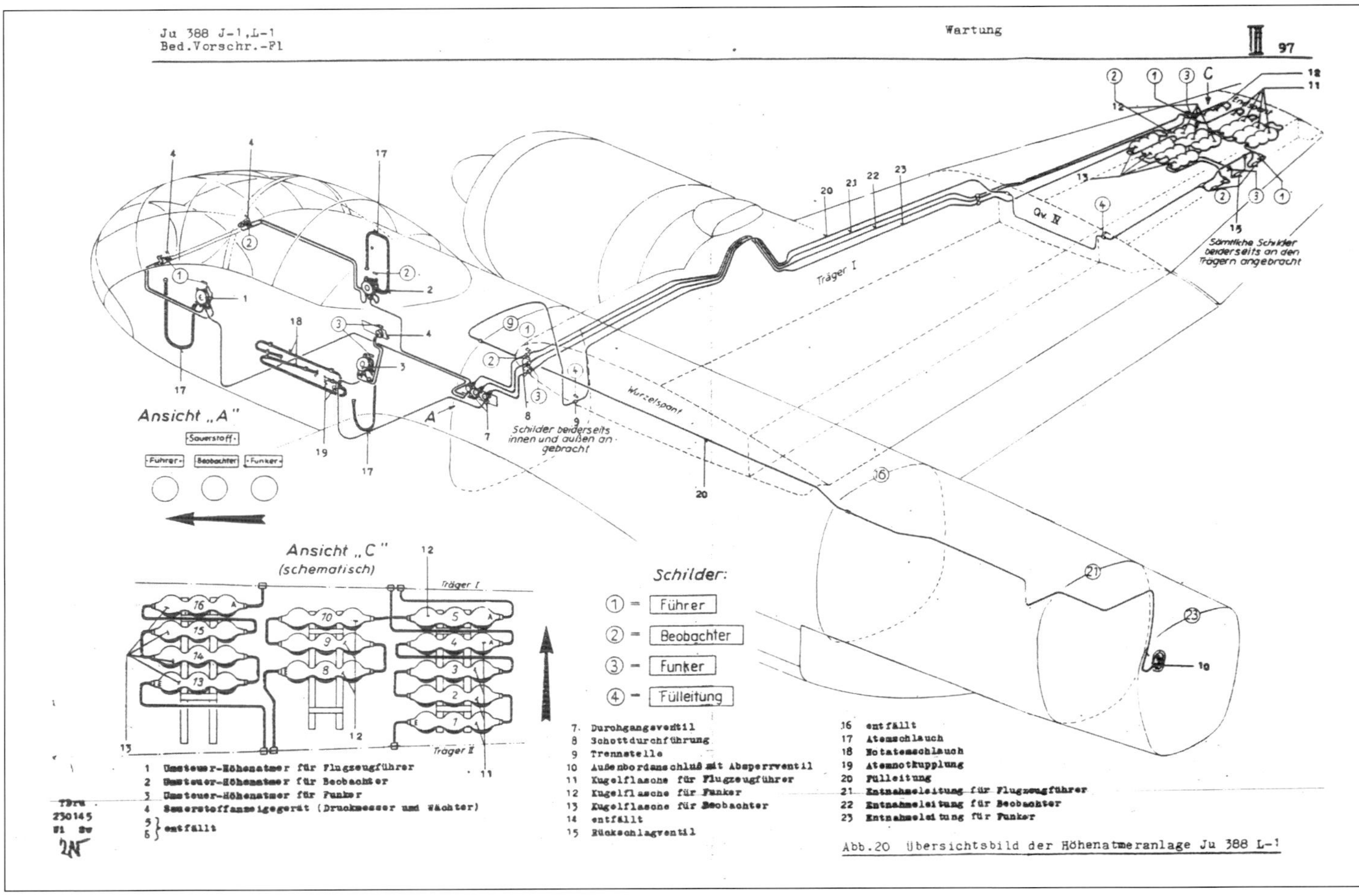

Illustration from the pilot's notes showing the oxygen system in the Ju 388 L-1.

BMW 801 TJ-0 with open access panels and daylight modification of the exhaust system on an aircraft equipped with external ram air intakes (WNr. 340 10X). Note the tarp covering the canopy. Aeroverlag Zuerl

was so low, for all aircraft were parked outside, protected only by the covers delivered with them. Water entered the fuselage everywhere during rainy weather, as well as ahead of and behind the firewall. The leakiest areas were the life raft compartment, the DF loop's flush plexiglass antenna, and the cover over the fuselage seam behind the cockpit.

Beginning in November some engines were delivered with external ram air intakes, but without the hot air lines for deicing of the wings. These were not to become standard until January 1945. By then JFM also had to construct 700 modification sets based on BMW drawings for the aircraft already delivered. As well, all Ju 388s were supposed to be equipped with flame dampers, probably because of the improvements in aerodynamics and maximum speed published in the document "Program Change."

At the end of January 1945 the testing station still did not have an aircraft with the complete deicing system. The station did, however, have intimate knowledge of the very similar systems in the Ju 188 and the Ju 88 V6, and it had reason to expect that the hot air wing deicing and propeller deicing systems would function efficiently. The wing deicing system with Blo.15 Kärcher heaters could not cope with heavy icing at temperatures below -6°C, however.[791] In spite of this, based on operational experience with the Ju 88 deicing safety was considered adequate to that degree of cold.

In November, much of the testing was again devoted to the BMW 801 TJ-0. During the course of endurance testing, complaints were made about intermittent afterburning with rich mixtures at heights up to 2000 meters. A night flight had to be aborted for this reason. The astonishing thing was that the injection and exhaust systems worked perfectly most of the time. In addition, serious thinning of the oil (up to 15% fuel) was observed on all aircraft, without the Rechlin cold start procedure having been used on any of the aircraft. On one aircraft the concentration of fuel increased from 7% to 12% within one week, probably because the cylinder temperature was too low. Furthermore, complaints about hydraulic blocking during operation of the nose ring became more frequent. Leaky hydraulic cylinders caused the nose ring to open on its own in flight. The test states of the aircraft were as seen in the table at the bottom of this page.

On 8 November 1944 WNr. 340 087's starboard engine (315 321) failed in flight. At that time the aircraft had 24.75 hours flight time. After a 45-minute single engine flight, darkness forced the pilot to make a belly landing at Lärz. It turned out that there had been a fatigue failure of the gearing clutch wheel that had caused the engine to overspeed. This had ultimately led to a seized piston.

WNr. 340 090 and 094 from ATG production began flying in the second week of November. The aircraft's performance and stall characteristics could not be verified, however, because of faulty airspeed indicator systems. The general impression of their condition on delivery was unsatisfactory. The first two production aircraft from Weser AG (WNr. 560 041 and WNr. 560 043) were delivered for inspection from Liegnitz on 8 and 9 November. Initially, however, they were not flyable because of equipment unserviceability. After acceptance checks on 27 and 29 November they were handed over to the modification facility at Welzow. Weser delivered another aircraft (WNr. 560 042) on 19 November. Numerous faults were discovered during the initial inspection of this aircraft, and the machines delivered by ATG Altenburg and Weser.

On 9 November 1944, performance aircraft RT+KA (WNr. 340 081) made its first flight following installation of external ram air intakes. This had to be aborted, however, when the airspeed indicator line froze. E2 had already encountered this problem with three Ju 388s.[792] It later turned out that this was due to an assembly error by ATG: grease had been used to help insert the branch pieces in the static pressure line into the rubber hoses, leading to a significant reduction in diameter, and ultimately to the failure of the systems.[793]

At the beginning of November the *E-Stelle* was also still waiting for new BMW 801 TJ test beds from Dessau and Merseburg. In

Werknummer	Port Engine	Starboard Engine	Remarks
300 009	385 193	385 258	94 hours flying time. Port maximum boost altitude rather too low. Engine removed as allowable number of operating hours reached and turbosupercharger checks.
340 084 (39.25 hr)	385 265 (19.5 hr.)	385 261	Port maximum boost altitude too low, waste gate/dump flap would not close completely; defective spring replaced.
340 085	387 532 (11.75 hr.)	321 213 (37.33 hr)	Starboard nose ring moved on its own. Hydraulic interlock replaced. Engine speed control on port engine stuck. Low supercharger maximum boost altitude 5.2 km. Dump flap spring broken.
340 0878	321 223	315 321	24.75 hr. flying time. Belly-landing because of starboard engine failure; see text.
340 090	317 317	315 325	13 hr. flying time. Port engine higher oil consumption (20 l/hr); nose ring hydraulic cylinders on both engines and hydraulic interlock on port engine replaced.

Close up of the waste gate of the BMW 801 TJ. At the beginning of both RT+KD and RT+KE experienced problems with this component. MTU

the meantime, the cause of the ignition interference in the Ju 388 had been discovered: an unshielded pressure tight lead-through of the ignition cables between the cockpit and bomb bay.

Testing of the flame dampers also began at Rechlin at the beginning of November 1944. On 8 November a night flight was made by the Ju 388 V3, which had been prepared by Junkers. The aircraft was equipped with two different flame damping systems. The one on the port engine was equipped with a mixing chamber, while on the starboard a cooled shroud was supposed to dampen the flames. No definitive result could be achieved, however, because the accompanying aircraft had to return early because of intercom problems and heavy afterburning from both engines. The V3 remained airborne, and from the cockpit a bright glow could be seen on the port side, while nothing could be seen to starboard. The aircraft subsequently crashed on landing as a result of pilot error. Unfortunately, the degree of damage caused by the crash landing is not known, nor is it known if the V3 was repaired. The V3's flame damping equipment was installed on WNr. 340 094 for a repetition of the night flight. First, however, fuel consumption checks were made on both engines.[794] Altogether, only three analyzable night flights could be carried out. This was not just due to the delay caused by the modification work and the ongoing engine problems. The opportunities to fly at night were also seriously limited by weather and the military situation. A night flight to test the flame dampers was scheduled in December, but could not be carried out—frost buildup restricted visibility, and no one was willing to authorize an instrument takeoff.[795]

No clear decision as to which of the two systems was better had been made when production began, and an order for 130 systems with the mixing chamber was placed on spec. Subsequent testing revealed that the system proposed by Junkers was clearly superior with respect to drag, production cost, and effectiveness. It would have taken about two months to equip the series with it. Until the flame damping system was introduced into production, aircraft were

Two views of the flame damping system. Vernaleken Collection

fitted with the exhaust fairings envisaged for aircraft assigned to daylight operations, of which 592 examples were to be built at Dessau using provisional jigs as an "aid to startup."[796]

Testing of the camera equipment was carried out simultaneously using several production aircraft. Installation of the cameras presented no problems, and an old style switch that was criticized was changed during the course of testing. During takeoffs and landings on wet airfields, however, flying water obscured the glass panel in the fuselage underside. Covering the gap between the fuselage and bomb bay doors was an interim measure. It was finally decided to remove the glass panel, and the Kärcher heaters for the camera bay were supposed to be dropped later, having proved unreliable. The electrical system, autopilot, hydraulic system, oxygen system, and equipment accessories of all aircraft engaged in flight operations were checked regularly. The pressurized cockpits of several machines were also checked regularly, as was the formation of mist in the double glazed windows.

The trials unit also complained about the BMW 801 TJ-0's high fuel consumption. As previously stated, this turned out to be due to the engines being too highly tuned, and to false information from the companies.[797]

The trials unit lost another machine on 17 November 1944. A performance flight by WNr. 340 081 (RT+KA) had to be aborted when, after a heavy loss of oil through the vent, the starboard engine failed for lack of oil. As well, the fairing behind the port engine's exhaust had burned through. In the ensuing bad weather landing the aircraft landed long. With the machine approaching the end of the airfield the undercarriage was retracted. There are contradictory statements about the resulting damage. The weekly report of the *E-Stelle* Rechlin speaks of 50% damage, while the FTA detachment's monthly report for November 1944 places the damage at 70%.

Performance measurements continued with WNr. 340 085 (RT+KE). Measuring equipment had to be installed first, however. During installation of the barograph, metal shavings were found in a threaded nipple in the airspeed indicator line (fuselage-wing). To make matters worse, there were leaks in the line. The dynamic pressure measuring system failed repeatedly because of water entering the system lines.

Inspection of the first three production aircraft from ATG and Weser also began at the end of November. A number of deficiencies were discovered, however, the manufacturers were already aware of them.

Between 18 and 25 November a Ju 388 with Jumo 213 E engines also arrived from Merseburg. Before it could begin flying the aircraft's superchargers had to be modified, as all Jumo 213s were to be fitted with superchargers with 7-port internal exhausting. The *E-Stelle* Rechlin also received two (more?) BMW 801 TJ-0 test beds from Junkers.[798]

The Jumo powered aircraft was most likely the Ju 388 L-0 V30, WNr. 300 293 (DW+ZA), which Stabsing. Walter Baist flew from Rechlin to Lärz on 30 November 1944 (1430-1450 hours).

EK 388's aircraft showed marked deviations in oil consumption, from 6.7 to 16 liters per hour, with the average being 11.05

Camera installation in a Rechlin test bed. Vernaleken Collection

liters per hour. Actual fuel consumption was 15% higher than on the engine spec sheet. Probably for this reason, an instrumented engine (WNr. 385 131) was installed in WNr. 300 004 (PG+YD), with which it was possible to select a desired consumption setting.[799]

Engine 329 005, installed in WNr. 340 092, was used to check fuel consumption in delivery condition. According to the available settings chart, this was 4 to 6 percent too high. EK 388 carried out consumption checks whenever the flight schedule allowed.

Between 26 November and 2 December 1944, department E 3 a2 carried out a test climb with external ram air intakes for the first time, using its modified Ju 388 K-0. The aircraft carried a full fuel load of 3100 liters, though the only military equipment on board was the two MG 131s. The service ceiling—the altitude at which climb rate dropped to 0.5 meters per second—at combat power was improved to 11600 meters. As the port turbine was only half engaged during the flight, and the starboard one just one-quarter, this was far short of the engines' maximum boost altitude. In addition to the increase in maximum boost altitude, the external ram air intakes also caused an increase in engine performance, as supercharger air temperature and exhaust pressure were lower. BMW saw this flight as further proof that the airframe's poor behavior at high-altitude must be blamed for the aircraft's poor performance at altitude. This flight is also mentioned in the *E-Stelle* Rechlin's weekly report for the period 26 November to 2 December 1944, which likewise criticizes the airframe's high-altitude performance as "catastrophically bad."[800,801] But the engines also gave cause for complaint. During the test climb the engines smoked heavily from the engine vents, which, not being production standard, were arranged atop the center of the engine with the air intake. When the switch was made from maximum allowable continuous power to combat power at 11200 meters, the cockpit filled with oil fumes. The loss of oil at altitude was so great that oil consumption per engine reached 35 liters and 80 liters per hour at maximum allowable continuous power and combat power, respectively. These measurements were, however, made with the standard electrical contents gauges, which were not particularly accurate. Apart from a measurement error, the most likely cause for such high oil consumption was leaking superchargers, for the modifications to supercharger sealing had not yet been carried out on this machine. After this flight the aircraft went to DLH in Staaken for installation of measuring devices.

The supercharger shift altitudes had to be adjusted for the new ram air intakes, for the shift occurred before the low altitude supercharger's maximum boost altitude was reached—by no means at the same altitude for each engine, however: the starboard super-

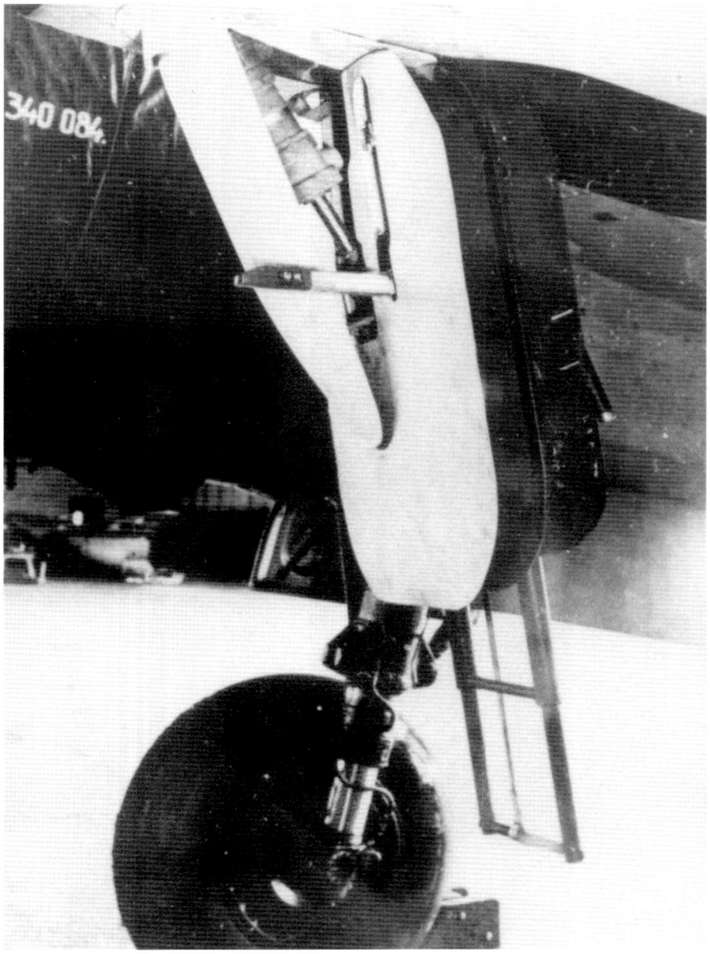

Two views of RT+KD's entry hatch. The ladder telescoped into the hatch. Griehl Collection

charger shifted with the turbine half engaged at 8450 meters, the port not until 9700 meters, with the turbine two-thirds engaged. E3 was, however, satisfied with a night flight at 1000 meters in moonlight. Both systems were absolutely dark from a distance of 30 meters, and plans were made to repeat the test at an altitude of 8-9 km.

Engine starting was significantly improved through the use of aviation starter fuel, with all engines starting after four to five attempts at most. The result, however, was a clear increase in starter wear—on 75% of all starters the coils were charred as a result of frequent run ups. When the cold-start procedure was introduced it was discovered that, in most cases, it was not necessary to add fuel to the oil. Rather, the oil's fuel content was already much too high—up to 20%—and adding more fuel only made the situation worse. The likely reason for this is that many flights were aborted because of suspected or real faults, and most were brief and never got beyond medium altitudes. Many aircraft spent more time in static tests (engine runs) and taxiing than they did in the air.

Testing of the Jumo 213-powered Ju 388 also began at the *E-Stelle* that week. There was also a setback, however, for all Jumo 213 Es produced at Merseburg had to be removed and modified.

In the beginning the tail turret displayed serious shortcomings, which had to be rectified during the course of testing. By 8 November 1944 EKdo 388 had been assigned ten aircraft with the FA 15. There were leaks in the control systems of half of these aircraft, as well as the hydraulic motors of three FA 15s. As the equipment was all brand new, and replacement of the seals on site was impossible, complete replacement of the system was the only solution, and the manufacturer was urged to improve sealing of the units.[802] Askania, the manufacturer, paid too little heed to the proposed modifica-

Reduction of the tail turret's working pressure to 60 ata meant that the hydraulic cooler could be dispensed with. The two openings in WNr. 560 049's bulged bomb bay have already been sealed. Rosch Collection

tions, however, and the problem continued. In January 1945, therefore, the BAL at Junkers contacted Askania's construction inspection and demanded more stringent checks, as the units would otherwise have to maintain an unacceptably large stockpile of spare parts.[803]

The sighting drive's operating pressure was finally reduced from 100 to 60 atm, as the higher pressure made necessary a special cooler to deal with heating of the hydraulic fluid.

The final model tail turret was declared ready for operational use, provided maintenance was effective. Following modification of the belt feeds and changes to the spent belt chutes, operation of the modified mounts had not been verified under g forces and at altitude. High-altitude testing of the periscope and sighting mechanism without the hydraulic fluid cooler was also probably not carried out. Operation at altitudes up to 10 km was trouble free, however. With the exception of the remotely controlled tail turret, the hydraulic system had been proven in the Ju 188; therefore, no further problems were encountered.

On 25 November the trials unit sent telexes to Junkers and ATG, informing them of the Ju 388's shortcomings that were in urgent need of rectification. On WNr. 340 094 a wing attachment nut had not been tightened, although the safety wire was in place. After elevator control had seized on the aileron control rod in WNr. 340 093, a complaint was registered about excessive friction in the pressure tight run-throughs. The control rods were affecting one another, especially at high-altitudes. As well, the hydraulic system was "generally leaky," and it was discovered that the oxygen line from the regulator to the pilot had been bent by the pilot's flying boots while taxiing. A request was therefore made that the line be rerouted. The effectiveness of the rudder trim tab was also criticized. Pedal forces indicated that it had been set incorrectly.[804]

In November 1944 RT+KD was used to test the autopilot. Petrick Collection

In late November 1944, test flights by Ju 388 RT+KD had been used to determine the standard settings for the KS 11 autopilot. These were supposed to be rechecked in a high-altitude flight and another aircraft. In its existing condition the autopilot was assessed as trouble free at all speeds and in all flight states.[805]

Representatives of the Patin company also took part in testing of the autopilot. Several changes were envisaged in the autopilot switches and mounting, however, these were to be resolved directly between Patin and Junkers. The final report said of the autopilot:

> Following the elimination of minor complaints, the autopilot functions acceptably for the aircraft's intended role. The KS 11 control box will be used until the autopilot system is converted to the KS 12. The control box is being changed to meet the requirement for constant turning speed.

Before the performance of the ultimate production aircraft could be measured, the performance of each aircraft had to be checked as it was delivered. The L-0, with three-blade propellers, showed deviations from the projected values in speed and service ceiling. Measurements of production aircraft with four-blade propellers and external ram air intakes, but without the speed improvements resulting from the modification program at Welzow, showed that while rate of climb was up to specification, level speed was off by about 30 kph.

The following table shows the true airspeed measured over a range of altitudes:

Altitude km	Combat Performance	Maximum Duration
0	392	360
2	430	400
4	465	437
6	493	460
7.5	500	–
8	496	–
9.5	503	–
10.5	507	–

Monthly Summary of Ju 388 Trials Unit Aircraft (November 1944)

WNr. 300009	WNr. 340084	WNr. 340085	WNr. 340086	WNr. 340088	WNr. 340090	WNr. 340093	WNr. 340094
1. In Merseburg for modification	1. Starboard fuel pump replaced	1. —	1. Engines missing	1. — 0^{12}	1. Starboard starter replaced	1. Acceptance check	1. “
2. 0^{34}	2. Over land 1^{22}	2. Injection pump replaced	2. Engine change	2. Landing flap “Down” indicator stuck 4^{17}	2. Engine change	2. “	2. “
3. “	3.	3. Hose on starboard engine's injection pump replaced	3. “	3. —	3. Hydraulic problems	3. “	3. “
4. Ram-air intakes switch with WNr. 340085	4. Boost pressure problems, static pressure line frozen 2^{15}	4. Starter replaced	4. “	4. Hydraulic pressure transmitter replaced 3^{29}	4. Landing flaps refused to lock	4. “	4. “
5. 0^{43}	5. Pilot's oxygen line broken at flow meter	5. “	5. “	5. Loose contact on oil temperature indicator	5. Landing flap interlocks replaced	5. “	5. “
6. —	6. Port engine oil temperature indicator replaced	6. Hydraulic interlock for starboard engine oil cooler changed	6. Both generators not working	6. Fuel pressure transmitter replaced	6. “Rich-Poor” indicator port engine replaced	6. “	6. “
7. —	7. Turbine measurements	7. RPM fluctuations both engines 1^{24}	7. “	7. — 0^{08}	7. Port engine hydraulic cylinder for oil cooler ring replaced	7. “	7. Delivered from Merseburg 1^{03}
8. Port engine maximum boost altitude too low 1^{36}	8. Over land 1^{00}	8. Spring on starboard engine's waste gate dump flap broken	8. “	8. Intermittent exhaust flames 2 – 5 meters long from both engines 1^{02}	8. Boost pressure fluctuations 1^{08}	8. Supercharger air line blocked	8. — 1^{23}
9. Turbine measurements	9. 0^{44}	9. No hydraulic pressure at altitude, pitch control problem 1^{48}	9. “	9.	9. Port engine, heavy loss of oil 2^{45}	9. —	9. Acceptance check
10. “	10. Hydraulic trouble 1^{34}	10. Leak in oil cooler	10. Propeller pitch control unit on port engine replaced	10. —	10. Port engine fuel pump replaced	10. Boost pressure fluctuations, long exhaust flame 1^{40}	10. “
11. “	11. Weapons work by Rheinmetall	11. Kärcher heater unserviceable	11.	11. — 0^{09}	11. Starter trouble, starboard engine	11. —	11. “
12. “	12. Undercarriage doors not closing, port oil temp too high	12. Rod on tailwheel door broken, long exhaust flame	12.	12. Intercom interference	12. Starter replaced	12. —	12. “
13. Engine change	13. Weapons work by Rheinmetall	13. Kärcher heater unserviceable	13. Short-circuit	13. —	13. “	13. Starter problems	13. Starboard propeller pitch control unit replaced

WNr. 300009	WNr. 340084	WNr. 340085	WNr. 340086	WNr. 340088	WNr. 340090	WNr. 340093	WNr. 340094
14. Engines removed, aircraft set aside	14. “	14. “	14. Propeller pitch system unserviceable	14. Tire change	14. Fuel pump removed	14. —	14. “
15. “	15. “	15. —	15. “	15. —	15. —	15. —	15. Pressurized cockpit leak
16. “	16. “	16. —	16. “	16. —	16. —	16. Starboard fuel pump replaced	16. Oxygen system leak
17. “	17. Exhaust fairing deformed, short-circuit	17. Weapons work by Rheinmetall	17. Short-circuit	17. —	17. —	17. —	17. —
18. “	18.	18. “	18. “	18. —	18. —	18. —	18. —
19. “	19. Fuel tank pump replaced	19. Work on exhaust	19. Trouble with propeller pitch system	19. Propeller balance weights removed	19. Oil contents indicator transmitter replaced	19. Control lever on rear cockpit bulkhead seized	19. —
20. “	20. Port hydraulic pump not working	20. “	20. “	20. Tailwheel damaged	20. Variable-pitch propeller control unit erratic	20. “	20. —
21. “	21. Hydraulic pump replaced	21. Port oleo collapsed	21. Airspeed indicator not working 0^{07}	21. “	21. 0^{45}	21. “	21. —
22. “	22. Fuel pump replaced	22. Installation of sensors	22. Short-circuit	22. —	22. Problems with propeller pitch control unit	22. —	22. —
23. “	23. Water in static pressure line	23. Port oleo replaced	23. Forward fuselage fuel tank damaged	23. Infrared lighting installed	23. Hydraulic cooler sealed off	23. —	23. —
24. “	24. 0^{20}	24. Hydraulic cooler sealed off	24. Weapons work by Rheinmetall	24. —	24. —	24. —	24. Starboard engine starter replaced
25. “	25. — 0^{59}	25. Installation of temperature gauge in cockpit	25. “	25. —	25. —	25. Short-circuit, switch K 1034 replaced	25. Oil temp gauge replaced because of clear glass window
26. “	26. —	26. Propeller pitch control trouble 1^{47}	26. “	26. —	26. —	26. —	26. Short-circuit
27. “	27. Rudder stop installed 1^{04}	27. Exhaust fairing and pipe cracked, both engines	27. “	27. — 2^{07}	27. Released to Leipzig-Mockau 0^{50}	27. Short-circuit	27. Oil cooler ring hydraulic interlock replaced 2^{14}
28. “	28. Short-circuit	28.	28. “	28. Lever on throttle control rod behind firewall came off because of missing cotter pin	28. “	28. Port engine starter replaced	28. Brake pedal pump replaced 0^{07}
29. “	29.	29. —	29. “	29. “	29. “	29. “	29. —
30. “	30. —	30. —	30. “	30. —	30. “	30. —	30. —
Flying hours October: 2^{53}	9^{05}	4^{59}	0^{07}	11^{14}	5^{26}	1^{40}	4^{47}
Flying hours total: 94^{56}	48^{58}	41^{53}	6^{03}	28^{27}	—	5^{30}	12^{20}

Flight operations by aircraft with standard consumption values resulted in very high rates of fuel consumption, and the calculated flight times could not have been more than 3.5 hours. Newly issued adjustment values from BMW were lower, but their introduction was not assured, because of the appearance of engine problems.

There was less flying in December 1944 than in November (see Appendix 2), and consequently breakdowns were few. The allocation of tasks to individual aircraft remained the same as in November. Performance flights were seriously hampered by bad weather at the beginning of December. It was found that the engines' maximum boost altitude could not be reached with a fully loaded aircraft. As well, because of the higher supercharger air temperatures in flight, at an altitude of 10 km the BMW 801 TJ-0 produced 130 h.p. less than the test bench value. At 11.5 km there was no discernable difference. It was discovered that two more engines had been set 8% too rich.

On 5 December 1944 a flight by RT+KE (WNr. 340 085), which had been converted to external ram air intakes, was aborted because the automatic propeller control device became stuck on takeoff, and led to an overspeed condition. The same problem had been encountered on 26 November. Repeated attempts to manually set the control device proved fruitless. The hydraulic pump on the port engine was defective, and had to be replaced, together with the propeller pitch gearing. The radio system was also not functioning properly.[806] The next week a check flight by WNr. 560 042 (PT+HB) had to be aborted on account of a loose panel, and a planned calibration flight by WNr. 340 093 was called off after the aircraft became stuck—the ground was obviously completely softened. Three BMW 801 TJ-0 engines with about 100 hours showed heavy wear on their supercharger housings.[807]

On 10 December work began to put the V5, which was sitting idle without engines, back into service. WNr. 300 009 was still sidelined without engines, however. The V1 and WNr. 300 292 (DW+YZ) were in Dessau for modifications.

There were personnel changes. Oblt. Ebeling was attached to KG 30 as technical officer, and Hptm. Preis assumed the duties of technical officer with the trials unit. Technical drawings for wiring up the camera equipment arrived at the beginning of December. Photo flights were therefore given high priority in the January flying program, and all were to be made at high-altitude.

A team from ATG added reinforcing rivets to the aircraft in question, and an FTA team completed the tail turret modifications[808] on two aircraft.

When the vertical stabilizer fairing was removed because of modifications to the tail turret, it was discovered that both bearing blocks were cracked. These bearings had to be reinforced immediately, however, the work was delayed, because the necessary documentation was changed twice by the design bureau. According to it, the compressed air lines had frozen twice in the hangar. The requested reinforced bearing blocks were not obtainable, therefore, the existing ones were bored out and strengthened by welding strips of metal onto and into the blocks.

All Ju 388s were briefly grounded by this problem, and in a telex titled "Re: Grounding of Ju 388" dated 23 December, Dr. Stahl requested that aircraft without tail turrets also be checked. To date, at Rechlin the problem had only appeared on aircraft with tail turrets.[809]

Bad weather hampered testing. During a period of cold and frost the fogging of cockpit windows on the inside and frost buildup on the outside became uncomfortably noticeable. The most effective measure against fogging from the inside was blowing warm compressed air onto the windows. Electrically heated wires were supposed to counter icing on the outer surfaces. This would have required converting the two most important windows to silica glass,[810] however, it was not possible to implement and test this change during the course of testing.

Hauptmann Bonow complained "very strenuously" about the fact that the V2 could not be completed at Werneuchen because Junkers had not delivered the long promised wiring diagrams. The wiring diagrams in question were for the oblique armament without round counter, the SN-2 wiring in DR configuration, and the radio range mixer.

Testing in January 1945 was further hindered by cold weather and the associated engine starting difficulties. From the *E-Stelle* Rechlin's weekly report of 27 January 1945:

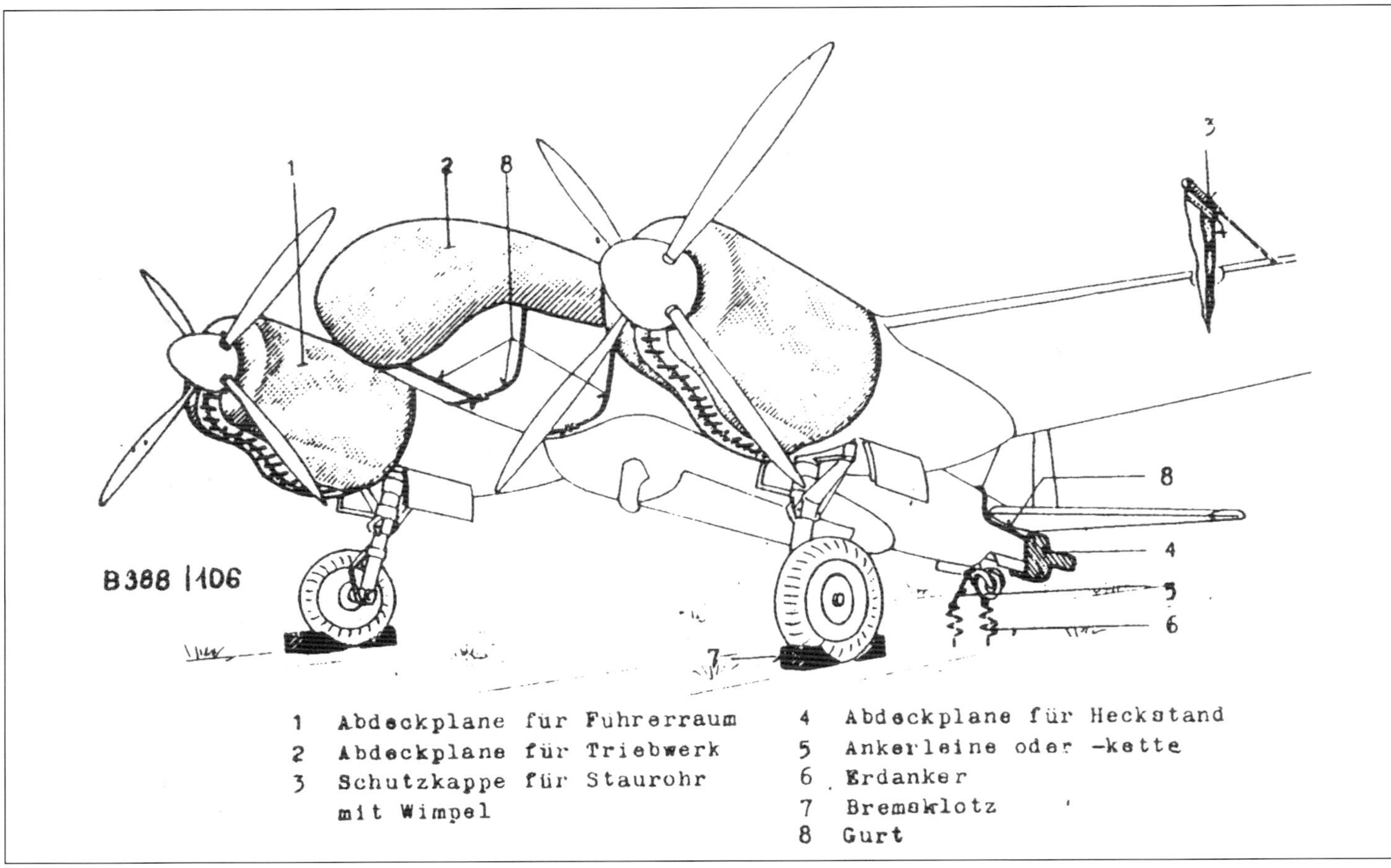

Tarpaulins—seen here in an illustration from the pilot's notes—provided inadequate protection against bad weather in November and December 1944.

"8-388

Aggressive intensification of testing failed to result in any significant increase in flight testing. In most cases serviceability targets could not be met due to starting difficulties in present temperatures. So far flight testing has produced few results. Most importantly, performance measurements have yet to be completed."[811]

Engine starting was poor at temperatures in the -10°C range. This had been anticipated based on results obtained in the Rechlin cold chamber in summer 1944.[812] The problem was made worse by the absence of aviation starter fuel. Regular fuel (B4, 87-octane) was delivered instead. With this, however, the engines could not be started at temperatures around 0°C. Contrary to the general fuel-

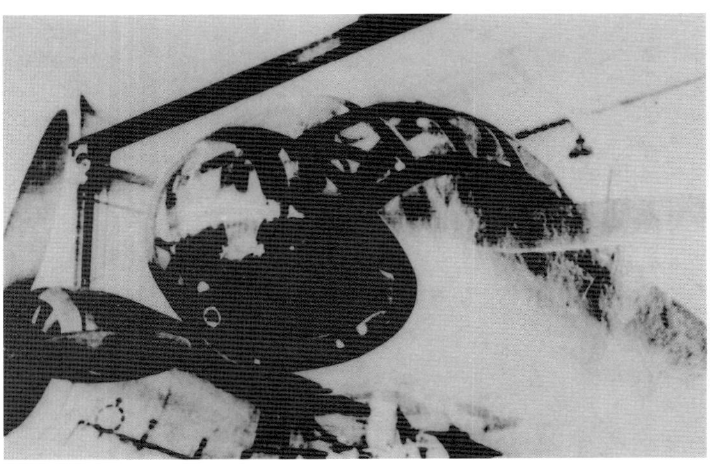

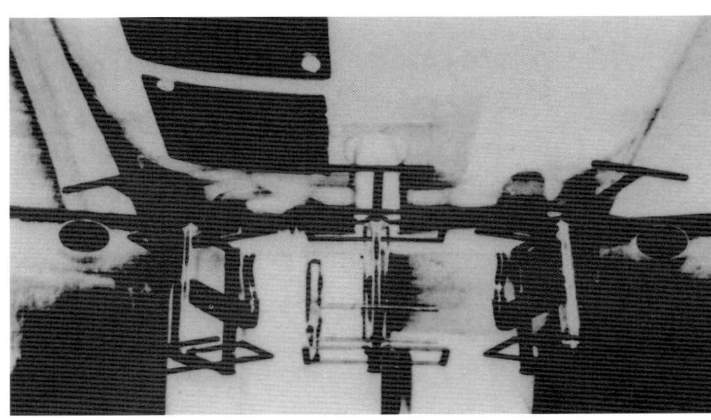

Left: The cockpit of WNr. 340 089 with the movable Hohentwiel antenna beneath it. Right: close up of the antenna installation. Vernaleken Collection

ling manual, starter fuel was required from 0°C, and not -15°C. In tests by BMW, it was found that the engines started easily in temperatures as low as -20°C if the Allweiler HP5 pump was used instead of the SUM primer pump. The HP5 was not in production, however. Nevertheless, use of this pump was considered, but a BMW specialist was later unable to conclusively prove that it was a requirement. Replacement of the SUM pump with the Allweiler pump was ultimately rejected; instead, the SUM pump's starter nozzles were to be moved and fitted with a fine filter, because this seemed the only possible way of improving the TJ engine's starting characteristics.[813]

At least one adjustment flight was carried out in the last weeks of January. In addition to the engine starting problems, testing was also hampered by ice damage to the tail section, and the growing fuel shortage.

The flame damper system was also tested at Rechlin in January 1945. A telex from Dr. Stahl to the BAL at Junkers dated 15 January reveals that the Linde and Junkers versions were about equal up to 10000 meters. Junkers was to provide the results of test flights carried out to determine the speed loss caused by the two systems, so that a decision could be made as to which system would equip production aircraft. The *E-Stelle* had doubts about the estimated difference of 8 to 10 kph. Dr. Stahl also advised that the trials unit urgently required four more sets of flame dampers.[814]

Altogether, by 31 January six zero series (pre-production) aircraft had passed through the trials unit during testing of the reconnaissance version. The bulk of engine testing was carried out with five aircraft. Another aircraft was used to test a search radar, probably the Lorenz FuG 200 "Hohentwiel." This ship detecting radar had an operating frequency of 550 MHz and an output of 30 kW, and was capable of spotting a surfaced submarine at 40 km, and the periscope of a submerged submarine at up to 6 kilometers.[815] There was no intention of installing the FuG 200 in the Ju 388 as standard equipment; however, it was envisaged that the radar might be installed in special aircraft by an appropriate modification facility. On 25 July 1944 Hptm. Stahl announced that a Ju 388 L-0 was to be delivered to Werneuchen for the prototype installation of the FuG 200. WNr. 300 294 was initially envisaged, until it was designated a test bed for the Jumo 213.[816] The aircraft actually used was obviously WNr. 300 292 (DW+YZ), which arrived at Werneuchen on 18 October 1944, and then flew to Dessau on 21 October for modifications.[817]

The FuG 200 is known to have been installed in L-1 production aircraft WNr. 340 089, which went to Werneuchen for this reason on 20 October 1944. A representative of JFM visited the aircraft modification facility there from 25 to 27 October. The Ju 388 was still unknown there, and the inspection and technical personnel had to be briefed on the aircraft. A movable Hohentwiel antenna was to be installed beneath the cockpit. For operation of the antenna the modification facility proposed a hand crank on the right side of the cockpit; however, the DVL rejected this "simple and reliable method of operation." It had developed a hydraulic motor that was to be installed between Frames 2 and 3 beneath the pilot's seat. The JFM representative pointed out that "any installation of hydraulic systems in the pressurized cockpit should be avoided if at all possible, because of the inevitable leakage."[818]

The two L-1 prototypes were used for performance measurements (V5) and tail turret testing (V6). Furthermore, *Erprobungskommando* 388 had a total of 15 L-1 production aircraft at its disposal. These aircraft were used for all testing tasks as serviceability permitted. Of these production aircraft, three each were used by ATG and Weser AG to check their production.

Testing of the Radio System

As stated previously, the Ju 388 L's radio equipment consisted of the FuG 10 K/L, Peil G 6, FuG 16 ZY with supplementary homing, FuG 217, FuG 101 A, FuG 25 a, and Fu Bl 2 F. To a large extent testing of this equipment was carried out during the trials unit's normal flying activities. As well, for a long time WNr. 300 292 was earmarked specifically for testing of radio equipment.

View of the rear cockpit bulkhead, the so-called "Radio Frame," with FuG 10 and the control device for the FA 15 in a Ju 388 L-1 production machine. At the left of the photo is the Bosch buzzer, part of the altitude warning system. Griehl Collection

Function tests revealed that the installation, operation, and accuracy of the FuG 10 K/l, Peil G 6, FuG 101 A, and FuG 25 a systems met existing requirements, and that deviations were within acceptable limits. After the electric interference from the engines had been suppressed, interference voltage from the electrical system with the FuG 217 switched on was within acceptable limits. All of the other electronic devices developed problems or peculiarities, which are described in the following test:

FuG 16 ZY (VHF Voice Radio and Homing)
The antenna system was the same as on all versions of the Ju 88. In terms of emission diagram and range, performance of the FuG 16 ZY did not differ from that of the Ju 88. As determined by E4 in January 1945, however, because of the antenna's location on top of the fuselage (necessary because of the bulged bomb bay), the homing system did not function properly, especially within the last 10 km from the transmitter, in close proximity to the transmitter and directly overhead. For reasons unknown the signals were unclear, making the system unusable. It was therefore decided that the under fuselage installation of the Ju 88 G-6 would be retained for all Ju 388s with no bomb bay.

EiV 10 (Intercom System)
There were initially problems with rumbling noises produced by the ignition system, which reached the system via the antenna. It was suspected that the unshielded pressure tight run-through on the rear cockpit wall was radiating. Use of a shielding box, initially tested on WNr. 340 084, seemed to solve the problem. No significant drop in performance was detected at altitudes up to 10000 meters when high-altitude microphones were used. The use of unsuitable microphones, however, led to a serious reduction in performance.

Fu Bl 2 F (Instrument Landing System receiver)
The sensitivity of the KBL 3 F on the beacon mast was seen as insufficient. The loop antenna excitation device for the EBL 2 was the older model, which the E4 had already rejected in August 1943 as insufficiently sensitive, and difficult to adjust. Instructions were issued to install only the improved antenna excitation device (violin case).

FuG 217 antenna installation on the wing of WNr. 340 089, RT+KL, taken in Merseburg at the end of August 1944. EADS

FuG 217 (Rear-Warning Radar)
The FuG 217 Neptun R2 with two operating frequencies was seen as a replacement for the FuG 216. The Ju 388 obviously played a major role in testing this device.[819] Interference from the FuG 217 proved a serious concern during testing by E4. The problem lay in the close proximity of pulse carrying lines and other elements of the electrical system.

Two close ups of the antenna. Also visible under the wing is one of the antennas for the FuG 101. EADS

In order to eliminate this electrical interference, the wiring in the prototype machine (WNr. 300 292?) was rerouted. Measurements were planned for the second half of December with the participation of the manufacturer, Siemens. The FuG 217 also interfered with the FuG 101 if the S 217's transmitter antenna was operated on the same wing in which the S 101 and E 101 (FuG 101 transmitter and receiver) were installed. No further mutual interference of radar sets could be identified.

Despite extensive efforts, it proved impossible to reduce interference by the FuG 217 through measures at the installation site. The interference voltages were reduced to a bearable level, but they were always above the maximum allowable values. The EZ 6 suffered the most interference, and Siemens was working on eliminating this when testing was halted.

Testing Results

The report on prototype testing of the Ju 388 L-1 began with the following summary:

"As of 31 January 1945 the aircraft type Ju 388 L-1 is conditionally ready for operational use. Improvements identified during testing were retrofitted to the delivered aircraft at a modification facility. The power plants and airframe were continually improved in modification stages.

No fundamental difficulties are to be expected with the airframe or engines (apart from cold weather starting).

Very well-trained maintenance personnel are vital if the aircraft is to achieve its full operational potential.

The aircraft's operating altitude lies between 10 and 12 km. In its present condition, *Werknummer* 340085's maximum speed at

Mounting rack for the FuG 217 in the fuselage. The equipment itself is absent. At the top left of the photo is the Kärcher heater. EADS

combat power at an altitude of 10.5 km is 507 kph. Because of reduced performance compared to measurements with internal intakes (*Werknummer* 081), the aircraft was handed over to Junkers for checks. Further performance measurements are temporarily impossible because of the fuel shortage. The achievable endurance will change again because of resetting of the fuel consumption by BMW, but it should only be slightly in excess of 3 1/2 hours."

The technical-tactical assessment for the long range reconnaissance role reads as follows:

"The Ju 388 L-1 aircraft is solely capable of performing in the high-altitude, long-range reconnaissance role. Its performance is only superior to the Ju 88 and Ju 188 aircraft types at altitudes above 10 km, however at low level the Ju 388 L-1 is inferior to these types. When flown with full tanks, the aircraft climbs to 10 km at maximum allowable continuous power, from there at combat power. Service ceiling is 12 km with a maximum speed of 507 kph at 10.5 km. Endurance about 31/2 hours.

The cameras with M1 or M2 motorized drive operate reliably. Camera heating is not necessary; thus the windows in the access hatch, which are often dirtied by dirt, dust and water during taxiing, can also be deleted.

Defensive capability (2 MG 131 in remotely-controlled tail turret) can be enhanced through the addition of a dorsal position as in the prototype aircraft (field of fire from abeam right almost to the aircraft's longitudinal axis to the rear and above). Since only isolated attacks are to be expected at high-altitude, defensive capability appears adequate. The tail turret is still trouble-prone and must be well-maintained.

Visibility is not good because of distortion and reflection caused by the double glazing and frequent fogging. An improvement in this situation is to be expected through care in manufacture and heating or warm air blowing.

Once the aircraft enters service, it is likely that serviceability rates will be low. Poor cold-weather starting characteristics, poor engine accessibility, frequent failures of engine accessories and exhaust system components, the hydraulic and electrical systems' susceptibility to failure, plus freezing of the static pressure lines demand a high level of maintenance work and good personnel. The principal advantage of the pressurized cockpit is the crew's ability to work unhindered at altitude.

Above-average demands are made of the crew, in flying skill—the pilot—to get the best high-altitude performance out of the aircraft, and physically in terms of long flights wearing oxygen masks and much-increased engine noise at altitude, which reduces the effectiveness of the intercom."

In its final report, the Ju 388 trials unit made the following overall assessment of the type's handling characteristics:

"Although the wings and tail section are taken from the Ju 188, the Ju 388 does not fully achieve the flight characteristics of the Ju 188. The exhaust exits on the upper surface of the wing have such a negative effect on flow that characteristics are at high-altitude as well as during takeoff and landing are degraded. A exhaust system based on Eila (?) was procured to improve the situation.

Stability is very poor at maximum gross weight. Stability worsens with increased altitude, complicating high-altitude flights on instruments or at night. Night takeoffs in this relatively unstable aircraft demand special alertness on the part of the pilot. The landing is simple, but landing speed is higher than in the Ju 188 with an empty aircraft up to 180 kph (influence of disturbed flow).

Control effectiveness and forces are normal except for sensitivity of the rudder to incorrect Flettner tab settings, which were observed on the first production aircraft.

With full wing tanks and maximum allowable continuous power, single-engine flight is just possible at low level. Rudder effectiveness is inadequate at indicated airspeeds below 240 kph. Because of the shallow viewing angle through the plexiglass windows, forward view is largely dependent on the quality of the glass. Reflections caused by one-sided lighting and during night flight are extremely bothersome, hampering low-level flight and especially the landing approach. Interim steps have been taken to improve lighting for night flight. Cockpit lighting is set to infra-red."

Operation of the pressurized cockpit was seen as satisfactory at all heights. Complaints during testing were all insignificant. An overpressure of 0.23 to 0.25 atm was consistently maintained. Only once did the pressure regulation fail, because of a broken valve. The initial problems of high cockpit temperatures and oil fumes at high-altitude were finally alleviated through the installation of a cockpit air cooler in the wings, which was to be made adjustable for winter operations.

The navigation equipment was also found to be trouble free. The window for astral navigation requested originally was to be abandoned, however. The only problem was caused by installation of the emergency compass. As long as demagnetization equipment was unavailable, the OK-42 on the forward tension rod had to be replaced with an FK 38 in the nose cone.

Instrumentation was adequate: the requested second turn-and-bank indicator was to be installed beginning with modification stage 6/45, coincident with the introduction of the L-3. The original engine instruments, the smallest size, were to be replaced with similar instruments with clear view lenses to prevent fogging from inside at high-altitudes.

During periods of rainy weather, failures of the dynamic airspeed measuring system due to freezing were common. Even when covers were employed, after sitting in the open for longer periods the lines had to be blown through. The effect of the built in water separator could not be observed by the end of testing because of a change in the weather.

With two 3000-Watt generators to operate the aircraft's systems, the electric power supply was also seen as completely satisfactory. As part of the refinement process, systems had been grouped together in the circuit breaker panel; for reasons of flight safety, however, the combined systems were reassigned individual circuit breakers. In a number of instances the trials unit criticized the aircraft's inadequate waterproofing, which made itself felt, especially during periods of wet weather. Another "refinement measure" also had to be revoked. In keeping with measures taken on other aircraft types, the oxygen supply lines for the radio operator and observer were replaced with a single line. For reasons of safety, the oxygen supply system had to be returned to "Füll 4" status, as per regulations. As well, an emergency oxygen coupling was carried in the aircraft in case of the loss of a breathing apparatus. The seats were also modified for use with the high-altitude parachute: they were fitted with grommets, for use with the parachute's spring snap hooks.

At the conclusion of testing, some of the engine and power plant related measures taken as a result of experience had been implemented, while some were still being worked on. In the opinion of the trials unit, therefore, the reliability and maturity of the engines would just be acceptable when they entered service. The EKdo therefore made a list of requests for reinforcements and design improvements for service engines:

- Improvement of the exhaust turbine system (mountings and operation).
- Improvement of the exhaust pipes.
- Modification of the engine hydraulics for oil cooler and cooling gill operation.
- New mixture setting to reduce fuel consumption. Design improvement to reduce oil consumption (ventilation and mounting of the engine bearings).
- Gap sealing to improve oil cooling in summer.
- Installation of external ram air intakes to improve maximum boost altitude and engine performance.
- Determination of actual engine performance at altitude (not yet carried out on account of measuring equipment failure).
- Strengthening of engine cowlings for diving speeds equivalent to airframe structural strength.
- Improvement of starting performance (measures taken to date unsatisfactory).

The trials unit was convinced that maintenance would determine the aircraft's operational worth, with the engines accounting for the bulk of the work. Only well trained experts could be used to service the aircraft, as untrained soldiers would hardly be capable of coping with the complexity and interaction of the Ju 388's systems. The amount of maintenance time per flying hour by the trials unit made it clear that serviceability rates in winter would depend on the availability of work spaces and hangars. Even with careful maintenance by well trained personnel, operational readiness would hardly be the same as that of other aircraft. The exhaust turbine and associated systems, such as starter, injection pumps, and fuel pumps were forward of the engine bearer ring, which increased maintenance times tremendously. In September 1944 two engines had to be removed because of failed turbine blades, namely the starboard engine of WNr. 500 005 (V5) beginning on 11 September, and the starboard engine of WNr. 340 081 the following day. Following replacement of the turbines the two engines were reinstalled. The process was used to determine the total time for an engine change, which was 6 1/2 hours. They were optimistic that with well trained personnel this time could be reduced by an hour. A starter change, as carried out on the L-1/V5 on 27 October 1944, required 8 1/2 hours.[820] In addition to the required spare parts, a BMW fitter always had to be present while the work was being done to readjust the engine. In this context, it is amazing that the American evaluation team later acknowledged that, in terms of required maintenance, the Ju 388 was on a level with comparable American types.[821]

Disbandment of the Ju 388 Trials Unit

In February 1945 the *E-Stelle* Rechlin requested that Junkers produce a special lightened version of the Ju 388 L-1 as quickly as possible by modifying WNr. 340 085, including removal of its bulged bomb bay.[822] The 1220-liter fuel tank of the Ju 388 L-0 was to be installed in the forward fuselage bay, while the installation of a 700-liter tank like that of the Ju 388 K-1 or the 680-liter tank of *Rüstsatz* B-2 in the rear bay was envisaged. For use in the Ju 388, however, this tank had to be modified by the addition of a second pump.[823] The bottom of the fuselage would now be covered by the standard bomb doors of the Ju 188 D-2, without actuators, and the entry hatch for the rear fuselage of the Ju 88 G-6 or Ju 388 J. The periscope fairing had to be extended to meet the new fuselage contour.

It is impossible to state with certainty whether this work was carried out. The motivation for the conversion, however, is relatively simple to understand. By reducing weight, it was hoped to achieve an increase in operational altitude and improve maximum speed, and the remotely controlled tail turret would surely be more effective than the fixed MG 131s of the L-0 and K-0 series. Furthermore, it was assumed that elimination of the bulged bomb bay would reduce drag, with positive effects on the aircraft's maximum speed.

The high-altitude Zerstörer was also tested by EKdo 388 at Rechlin under the command of Hptm. Bonow (formerly of NJG 100).[824] He related the following:

"On 2 October 1944 I was transferred from the headquarters of NJGr 10 night fighter testing at Werneuchen to *Erprobungskommando* 388 (high-altitude night fighter) at Rechlin, together with my crew and chief mechanic. In addition to me, the unit also had two Oberleutnants with their crews and servicing personnel. I believe I flew the Ju 388 at altitudes up to 10 230 meters. At that altitude flight characteristics were no longer satisfactory. They were especially unsatisfactory for a night fighter aircraft (bad weather aircraft). As well, the tests were flown without armament and ammunition. Testing of the 388 (high-altitude night fighter) ended in late February 1945. Although Prof. Hertel of Junkers had

This photo, one of the last company photographs of a Ju 388 cockpit, was taken in February 1945. The FuG 217 is installed to the right of the blind flying panel. Griehl Collection

done his best to make the Ju 388 ready for service, he was unsuccessful because of conditions at the end of the war."

PG+YD (WNr. 300 004), KS+TN (WNr. 230 694), and DW+ZA (WNr. 300 293) were still at Rechlin on 7 February 1945.[825]

On 14 February 1945 an order was issued for the immediate disbandment of the Ju 388 trials unit, together with the Do 335 trials unit, Trials Unit 26, and the Instruction and Trials Unit 16. All flying personnel made available as a result were placed at the disposal of the appropriate general for employment within his specialty, as were all "shortage specialists," such as aero-engine mechanics, electro-mechanics, oxygen system mechanics, armorers, and inspectors. Any such "shortage specialists" that became available were to be transferred to the Field Maintenance Replacement Battalion based at Eschwege, in northern Hesse. The trials unit's aircraft were to be redistributed by the quartermaster-general in consultation with the K.d.E. If necessary, the responsibilities of the disbanded trials units were to be assumed by other units.[826]

How was this order implemented? The *E-Stelle* Rechlin's weekly report of 18 February 1945 already reported that no Ju 388 flight testing was being conducted "as a result of its having been dropped from the emergency program," and on 8 March the *Kommando der Erprobungsstellen* (K.d.E., or testing stations command) reported the disbandment of Instruction and trials Unit 16 and *Erprobungskommando* Ju 388. The EKdo. 26 was taken over completely by the *General der Schlachtflieger* (general in command of close support aviation), and EKdo. 335 was initially not disbanded at all, as the aircraft remained in the Führer's emergency program, and testing therefore continued.[828] On this subject the *Luftwaffe* High Command also received a similarly worded report from the *E-Stelle* Rechlin. In the meantime, however, the Do 335 V5 and WNr. 240 102 had been parked in the area of the firing range, because there was initial uncertainty as to whether testing would continue.[827]

At its disbandment, the Ju 388 trials unit consisted of four officers, three civilian officials, and 85 NCOs and enlisted men. The

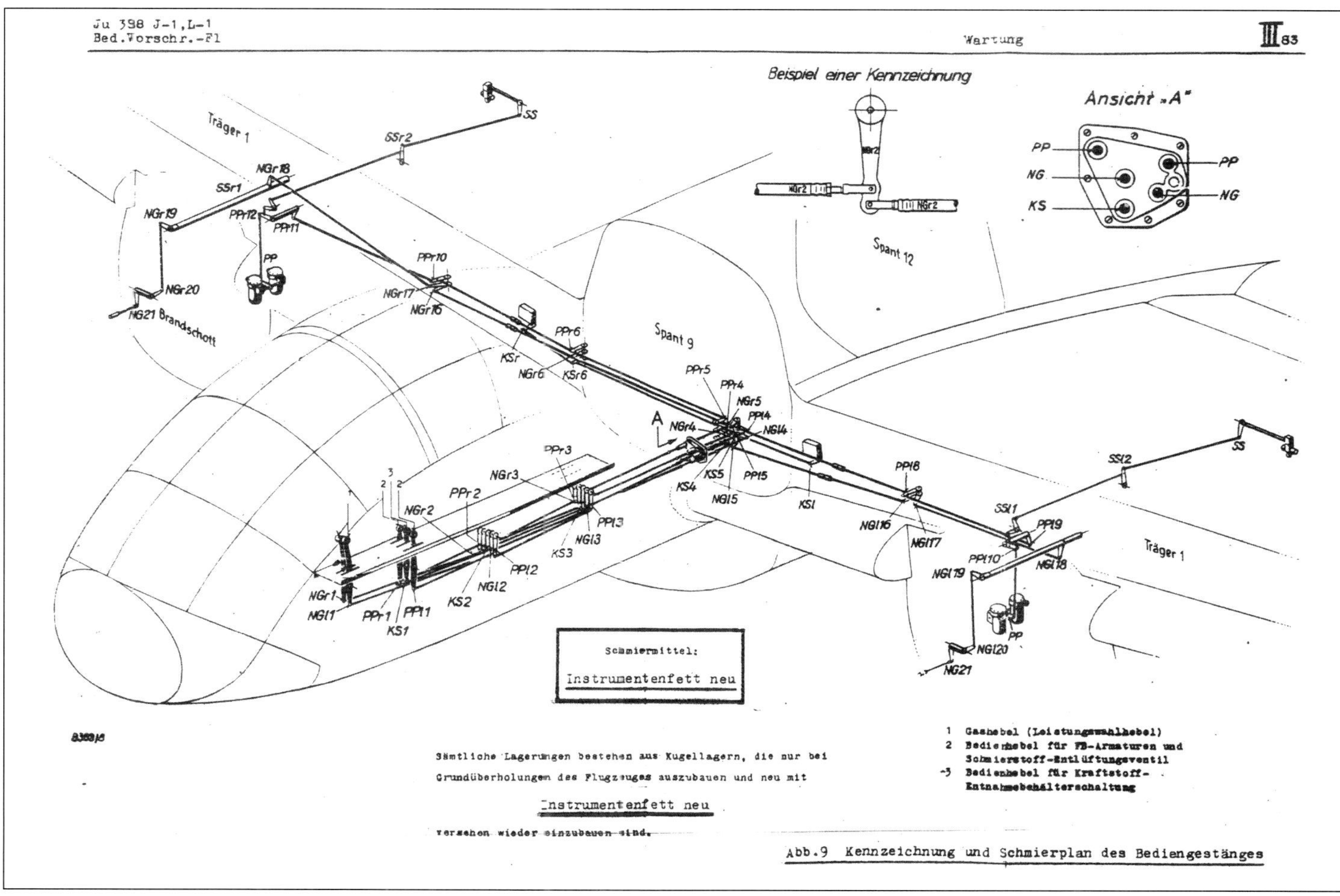

Illustration from the pilot's notes depicting the engine control rods.

OKL Experimental Unit Oranienburg:	1 officer, 8 NCOs
Parachute ACE Gardelegen:	16 NCOs and enlisted men
Army Assembly Camp Munsterlager	1 NCO, 3 enlisted men
Do 335 Trials Unit:	1 NCO, 8 enlisted men
Field Maintenance Replacement Training Battalion Eschwege:	1 Tech. Sergeant, 1 NCO
II Replacement Training Group, Gen. d. A. Pütnitz:	2 officers, 1 NCO
Gen. d. J. Classification Office Berlin-Kladow:	12 NCOs and enlisted men
Gladenbeck Detachment Brandis:	1 officer

majority of these personnel—a government inspector, a maintenance chief, and 34 NCOs and men—were transferred to KG 76. The dispositions of the remaining members were as in the above table.

On 1 March 1945 Knemeyer placed Hptm. Bonow in charge of testing the Ar 234 night fighter at Oranienburg for missions over Berlin.

Even after testing of the Ju 388 was halted the aircraft continued to serve as a test bed for the Jumo 213 E. Two flights were made by Ju 388 WNr. 300 293, DW+ZA, at Rechlin between 23 February and 1 March 1945. It was determined that supercharger surging had been eliminated from the new engines with controllable exhaust opening. Maximum boost altitude was 8 km with gas release at 3,000 rpm, and 9.1 km without gas release at 3,250 rpm.[830]

Tarnewitz

Before Tarnewitz became involved with the production aircraft's remotely controlled tail turrets, it conducted tests with the armament of the pre-production aircraft. At the beginning of July 1944 the Junkers Weapons Design Bureau (Kobü-Waffen) received complaints from the *E-Stelle* Tarnewitz about the Ju 388 K-0's rear firing armament of two MG 131 machine-guns. The angle of the two machine-guns had to be changed from 75° to 60°, as spent casings tended to jam at the 75° cant. There was a suitably modified mounting at Tarnewitz, but the aircraft in which it was to be tested, the Ju 88 V60, was supposedly needed for higher priority testing, such as the FuG 101. *Herr* Steuerlein of the weapons design bureau asked JFM's technical sales and distribution branch to contact the Tech-

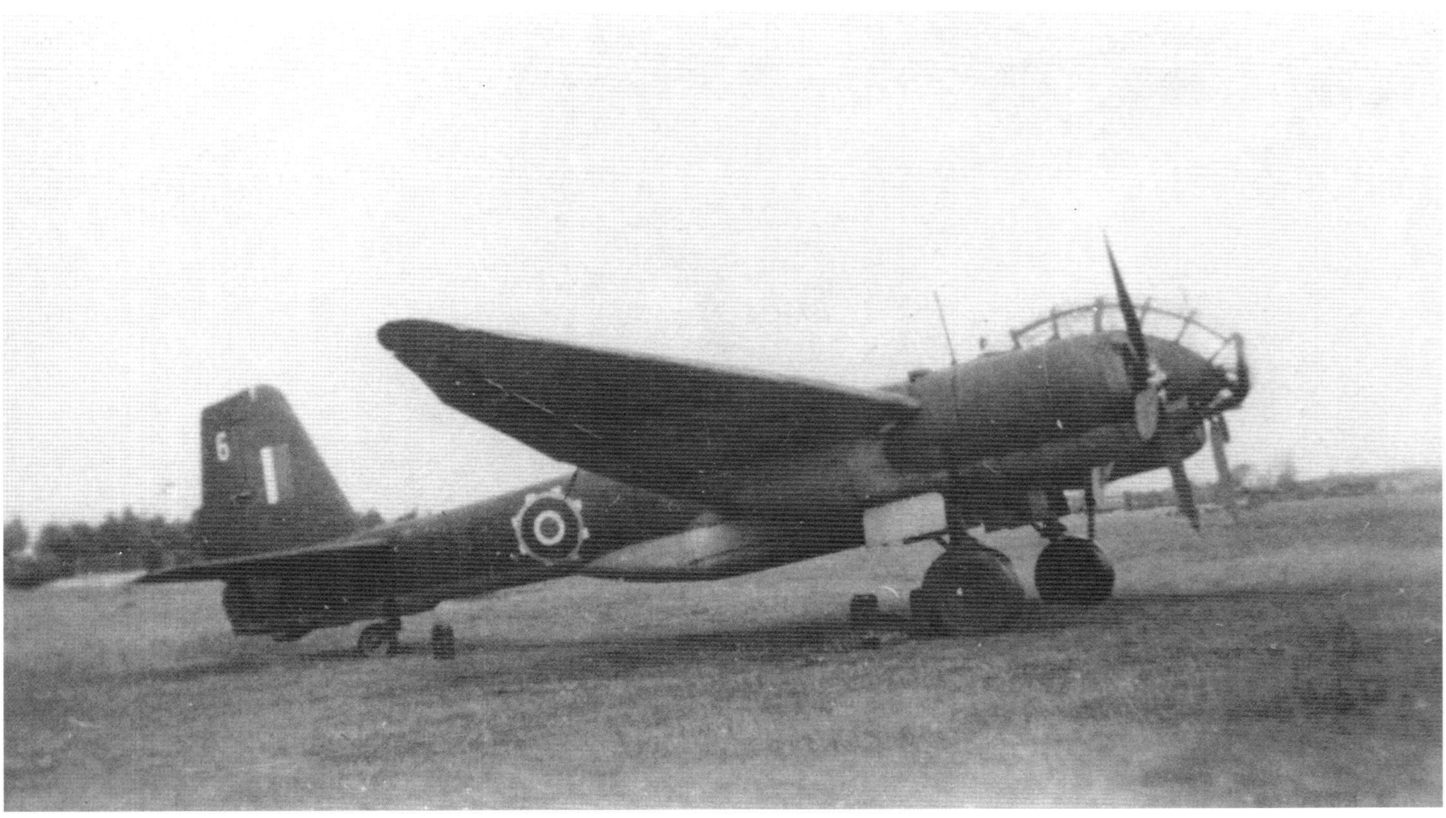

Side view of the Ju 388 V6 after capture by the British, possibly photographed during a stop en route to England. The British identity number "white 6" appears on the tail. Similar numbers were worn by other captured aircraft. Petrick Collection

The exhaust system of the Ju 388 V6 differed from the ultimate production version. It is easy to see why this was called the "Little Boat Solution." Other captured machines are visible in the background, including a Ju 52 and an Me 262, both in British markings. Haubner/Misna Collection

nical Office (GL/C E 2), and request that another test machine be assigned to Tarnewitz. A changed mounting with a modified ammunition feed was to be installed and subsequently tested on the firing range, and in flight at Tarnewitz.[831] The department "Technical Sales and Distribution—Order Processing" (FTV-Au) reacted on 7 July, issuing orders for the gun positions to be modified as soon as possible, and tested before being adopted as standard equipment:

"Besides this weapons installation, the dorsal gun position of the Ju 388 Zero-Series aircraft has still not been cleared and, as soon as the weapon run-through and weapon have been sealed, it must be presented to the *E-Stelle* for testing. It would be advisable to test both in the same machine, and we ask that you investigate to what extent both tasks can be carried out simultaneously without extending the deadline. We ask that we be advised of the likely serial number so that we might submit the necessary allocation application."[832]

Dzulko noted that a serial number was to be selected. Whether or not this intention was carried out is uncertain, however. In any case, no aircraft was fitted with a dorsal gun position by 20 July. Hptm. Stahl, the type supervisor, therefore requested that FSD-M prepare such an aircraft as quickly as possible, even if the pressurized cockpit did not remain sealed during firing.[833]

The tail turret tested at Tarnewitz was obviously removed in England for separate examination. Here the tail turret opening has been covered with a metal fairing. Haubner/Misna Collection

At the urging of Professor Hertel the *E-Stelle* Tarnewitz carried out extensive testing for the purpose of simplifying the tail turret belt feed system. For these experiments the tail turret and belt feeds were mounted on a firing block. The testing station succeeded in reducing the weight of the belt feed system from six to two kilograms by deleting the entire carriage belt feed, significantly improving operating reliability in the process. A total of ten thousand rounds were fired with the new arrangement with no jamming of ammunition belts. The most unfavorable position for the belt was 90° to the side and 45° elevation. There were also casing ejection problems in the 45° position. Overall, the *E-Stelle* judged the static tests so positively that it wanted to test the tail turret in flight as soon as possible, in order to determine whether G-forces would have any effect on its operation. As the *E-Stelle* was still waiting for the V2, it proposed that the V6, which was supposed to be sent to it in the next few days, be modified in accordance with the test results, so as "not to drag out the results of the flight tests unnecessarily."[834] Finally, on 22 August 1944 the Ju 388 L-1 V6 was delivered to the *E-Stelle* Tarnewitz's department E6/IID, which was responsible for testing gun mounts, drives, remote control systems, and hydraulics,[835] for testing of the FHL 131. It was initially used for company testing by Rheinmetall-Borsig. During the course of these the belt feed system was modified; as well, one switch box was replaced. The tail turret subsequently underwent a function check and firing tests to determine its dispersion pattern.[836] Not a single firing flight had been carried out by 2 September 1944, however, as a variety of mechanical and electrical defects were still preventing the weapons system from functioning properly.[837] Testing by the company continued, and one week later there were still problems with the system, specifically short circuits in the gun mount belt feed.[838] Rheinmetall had also failed to eliminate problems during air firing by mid-September.[839] The gun mount belt feed was finally abandoned, and five firing flights were completed without incident by 30 September. By that time the Ju 388 V1, with its fixed forward firing armament of four MG 151 cannon, had arrived at Tarnewitz. Work was also underway to modify its casing ejection system.[840] Static firing tests were completed successfully after the modifications were complete, however, the two rear guns still

Propeller spinners were probably never installed on the V6. Petrick Collection

caused problems during air firing. The modified gun mount without belt feed was tested in the V6, after which it was cleared for production. The ejection of spent belts continued to cause problems, and the aircraft was returned to Rheinmetall-Borsig to have them corrected.[841] On 12 October 1944, after the conclusion of testing at Tarnewitz, the V1 was handed over to EKdo 388 in Rechlin. The *E-Stelle* Tarnewitz accepted the aircraft back, even though Rheinmetall-Borsig had been unable to completely fix the V6's jamming problem. Because of the problems of spent casings and belts jamming, however, those at Tarnewitz reached the conclusion that the FH 131/Z's usability would be severely restricted without redesign. The tests were continued, however.[842] The program for the week of 22-28 October 1944 included the installation and testing by Rheinmetall of a modified spent belt removal arrangement. For its part, the *E-Stelle* experimented with replacing the GF 4 ammunition belt conveyor with the GF 6. On 29 October the V6 carried out a gun firing flight from Tarnewitz crewed by Gatzemeier (pilot) and Kowalewski (1453 to 1526 hours).[843] At the end of October Tarnewitz was also waiting for the Ju 388 V2; a meeting was supposed to take place at ATG in Leipzig on 2-3 November concerning alignment of the tail turret. Concern was also expressed about the "Schräge Musik" (oblique armament) of the Ju 388 night fighter. If the oblique installation from the Ju 88 was adopted, the ammunition boxes for the tail turret would have to be deleted for cg reasons, leaving the tail turret just 300 rounds per weapon in the feed channels.[844] Then, at the beginning of November, it was clear that the most important changes to the FHL 131 Z in the Ju 388 V6 would be retrofitted to production aircraft. Modifications to the spent belt removal system were still being tested, and a simplification of the switch was being examined in the laboratory.[845] One week later static firing tests were carried out with the modified elements of the spent belt removal system. The modified system worked well; sixty rounds were fired with just a single jammed spent casing. Once the aircraft was pronounced ready to fly, air firing tests were to follow. In this context, the *E-Stelle* reported in one of its weekly reports that the AEG company had checked the installation and alignment of the PVE 11 periscopic sight in a production aircraft.[846] On 18 November the V6 was still unserviceable, however. While static testing of the LSK 17 B gun mount switchbox was still going on, performance measurements of the FA 15 remote aiming system had already concluded. Further measurements were to be made with a new system, whose arrival was as yet undecided. The *E-Stelle* also checked the alignment manual for the PVE 11 and FHL 131 Z. Concerning the PVE 11, the weekly report noted:[847]

"As production conditions become more difficult (Ostarbeiter, or workers from eastern territories), a degradation of the PVE 11's operating accuracy must be expected. Tests are under way."

While the upcoming high-altitude flights had initially been delayed because the V6 was unserviceable until 24 November 1944,[848] bad weather caused a further delay up to and including 2 December. As the FA 15's shaft placement had been modified on the production line, the ultimate testing of the FA 15 was to be done with the production version. For production reasons, however (see above), requirements of the FA 15 with respect to looseness, computational accuracy, and altitude resistance had to be made less stringent.[849]

While the long-awaited V2, which was undergoing modifications at Werneuchen, had not yet arrived at Tarnewitz by just before Christmas, the *E-Stelle* nevertheless now had a Ju 388 K-0 at its disposal. Static firing tests with the MG 131 in the pressurized cockpit's dorsal position resulted in the cockpit filling with smoke. As well, earlier experiments had shown that air loss caused by leaks around the gun mount could not be eliminated by leaving a round in the barrel as a sealing measure.[850]

It is doubtful that the Ju 388 V2 ever arrived at Tarnewitz. Testing report 7/45 by the *E-Stelle* Tarnewitz, dated 17 February 1945, says of the Ju 388:

"8-388

In case the type is not cancelled, ferrying of the V2 from Werneuchen to be accelerated."

The corresponding passage in the report has been blanked out by overtyping with "x"—the Ju 388 had been cancelled.

At the end of the war the Ju 388 V6 was still at Tarnewitz. In June 1945 the English flew it to Lübeck before the arrival of the Russians (in July).[852] For more on this see Chapter XI.

Testing by the Experimental Unit of the OKL and VfH

The *Versuchsverband des Ob.d.L* (Experimental Unit of the Commander-in-Chief of the Air Force) was a very special unit, as revealed by a mission plan dated 29 March 1944. Selected prototypes of advanced aircraft were provided by the Technical Office, and these were used to fly combat missions for the *Luftwaffe* operations staff that were beyond the capabilities of conventional units. At the end of March 1944 the unit had on strength one Ar 240 with DB

603 engines (G supercharger) and GM1, and a Ju 88 powered by BMW 801 TJ V engines, which were in the process of being converted to BMW 801 TJ-0s. A Ju 388 L-0 with BMW 801 TJ-0 engines was supposed to arrive in April, and by May two Ju 88 Ts were to be converted to take Jumo 213 E and DB 603 engines with GM1. Had it not been for acts of sabotage in Toulouse, beginning in September (according to the plans), the unit was to have received a Ju 488 with BMW 801 D engines for long range reconnaissance (6300 km), and one with DB 603 engines for high-altitude reconnaissance. As well, by September the unit was supposed to receive a Do 335 with larger wings, with conversion from the DB 603 E to DB 603 L to follow. The document also makes reference to Ju 388 L-0s with Jumo 213 E and Jumo 222 engines, Ju 288 V-series machines with Jumo 222s, and Ju 488s with Jumo 222 engines, although no delivery dates are provided.

The unit also flew combat missions with aircraft modified for "special purposes," through the installation of special fuels, additives, or special equipment.

The OKL.Fl.E. Chief assigned the OKL's experimental unit the task of equipping a Ju 388 L-0 with Jumo 213 E engines for use as a long range reconnaissance aircraft and determining its performance. In mid-August 1944, therefore, the Merseburg-built Ju 388 L-0 V32 (WNr. 300 295, DW+ZC) was taken by rail to Oranienburg for shipment. Assigned to the 3rd Staffel of the Versuchsverband des Ob.d.L, it was fitted with Jumo 213 E engines (power plant 9-8213 D), and in September was flown as T9+DL. The aircraft was equipped with the VS 19 propellers (9-3037.16 V blades, 3.6 m diameter, chord to length ratio 10.5%) envisaged for the Jumo 213, and carried out its first static test on 12 September 1944. It was equipped with cameras for all of its flights, and had an equipped weight of 9475 kg. When all four wing tanks and two fuselage tanks were filled with fuel, the aircraft weighed 12685 kg with a crew of four. During flights with the experimental unit, however, takeoff weight was always 12 metric tons. While the Waffentropfen (ventral weapons housing) with two aft firing MG 131s was absent, one MG 131 was installed in the cockpit dorsal position. The inflatable

The rival Do 335 V3 was experimentally equipped with cameras and flown by the OKL's experimental unit. Griehl Collection

T9+DL's sister machine, the Ju 388 V31, WNr. 300 294 (DW+ZB), was also equipped with Jumo 213 engines. It may be assumed that T9+DL was similar in appearance. At the end of the war the V31 was abandoned at the edge of a wood with smashed cockpit glazing, and damage to the underside of the wing. Ott Collection

raft and emergency transmitter were also absent, and were replaced with ballast. Addressing the mechanical problems encountered during flight operations was severely hampered by difficulties in the procurement of spare parts. As well, bad weather several times brought testing to a halt for days. From its first flight on 16 September 1944 until it crashed on 5 January 1945, the aircraft made just 23 flights with a total of 9 hours, 40 minutes of flying time. As a result, little performance data was gathered. The highest altitude achieved was 10200 m. During almost half of all test flights, which often had to be aborted due to engine trouble, the machine was in the air for less than half an hour. The aircraft flew twice in one day on just two occasions (3 and 21 November 1944). The rest of the time it flew just once per day. By far the longest flight, with a duration of 103 minutes, was made on 21 November 1944.

In view of these facts, a report by the experimental unit describing the aircraft as fit for operations appears extremely optimistic, especially considering its intended role of strategic reconnaissance aircraft. Probably the situation was already so desperate that they had to send into action everything that could stay in the air.

Highly praised with respect to service use was the Jumo 213's good accessibility, which made it easy to service all components. The sparkplugs could easily be changed within an hour.[854] Unfortunately, we have no comparative data for the BMW 801 TJ-0. The only aspect that came in for criticism was the engine access panels, which were found to be overly complicated. According to the experimental unit, servicing could be greatly simplified by modifying this area.

Naturally, testing concentrated on the new engines, whose maximum boost altitude in a climb at maximum allowable continuous power at 300 kph and a dynamic pressure of 450 kg/m^2 was determined to be 9600 (port) and 9550 meters (starboard).

The VS 19 propellers proved quite troublesome in the beginning. During the course of testing, however, it became apparent that the most significant sources of trouble were production related. On 26 September 1944 a directive was issued increasing the basic propeller setting from 22° to 23°. At that setting, engine speed during takeoff was increased from 3,250 to 3,300 rpm. During the second flight on 17 September the port propeller reacted sluggishly to control inputs. Increasing speed governor pressure brought no improvement. The governors on both engines were subsequently replaced. As this failed to eliminate the problem, the source had to lie in the propeller hub itself. Following disassembly by a Junkers propeller specialist, it was found that a screw had sheared off on the crosshead of the port propeller hub. Both propellers were subsequently sent back to Jumo, where serious manufacturing flaws were discovered. Following overhaul by Jumo, the propellers were serviceable.

On 2 October 1944, after just 2 hours, 43 minutes of operation, a crack was discovered in a propeller blade of the starboard VS 9. Close examination revealed that the lengthwise crack, which was located near the blade root, was due to a missing film of glue. As a precaution all of the blades were replaced. Six days later the same propeller was the source of another complaint: the counterbalances separated when the blades were set to high pitch.[855] During the seventh flight on 12 October the port propeller ran continuously on "LOW PITCH." It refused to move to the feathered position, however, it functioned in the opposite direction. A check revealed that the counterbalances had moved up too far, and as a result made contact when "HIGH PITCH" was selected. Few propeller problems were encountered in subsequent testing. During the 17th flight the starboard propeller changed pitch on its own; however, the problem was quickly solved by replacing the speed governor.

During the first static test on 12 September 1944 the coolant became too hot because too little air was passing through the cooler. Additional coolers were therefore installed for the static tests.

Cooling continued to cause problems for reasons that were not completely understood. During the fifth flight on 28 September, coolant was lost through the ventilation outlets while climbing at maximum allowable continuous power, even though the temperature was normal (95°C). The same thing happened on the next flight, however, the loss of coolant was somewhat reduced by fully opening the cooling gills. A subsequent check revealed that the coolant circulation system was in order. The problem in the port engine was solved by moving lines, which it was suspected would reduce leakage, and the replacement of pressure control valves. No more coolant evaporated, even at a coolant temperature of 118°C. On the starboard engine another line had to be moved before the problem was eliminated. After these modifications there was no further loss of coolant, even at high-altitude—with the exception of the 11th flight.

On 29 September 1944 a filling relay on the starboard engine's gearbox began leaking oil after three hours of operation (engine No. 104 161 0102).

The engines of T9+DL also exhibited the supercharger surging previously experienced with the Jumo 213 E; however, it was largely overcome by bleeding air behind the first supercharger stage, and did not hamper testing. During the sixth flight on 2 October the pilot was unable to reduce the speed of the port engine below 2,600 rpm at 8000 meters, thus bleeding over an area of 18.5 cm^2 was not sufficient.

Siemens V 446 E sparkplugs were also tested in combination with the Jumo 213 E. The first problems with the new plugs were encountered after 3 hours, 45 minutes of flying time. When they were removed, three of the 24 were found to be defective. Testing of the Siemens plugs could therefore not continue, and Beru plugs (V 5 F-280 E 43) were reinstalled in the port engine.

The eighth flight on 15 October had to be aborted because of problems with the Beru sparkplugs in the port engine. Finally, acting on instructions from Jumo, the ignition timing was advanced from 23° to 28°. No further ignition problems were encountered.

The very poor efficiency of the engines' air intakes was criticized. This problem had already been identified in flights by T9–EL, a Ju 88 T equipped with Jumo 213 Es. While the air intakes delivered for Ju 388 T9+DL displayed better workmanship, their ram air efficiency was still inadequate, especially at high-altitude, as was determined by the ninth flight on 15 October. As the aircraft climbed, total pressure in the intake was measured, and the difference compared to the total pressure in the pitot tube was recorded. The difference was –60 mm WS on the ground and -30 mm WS at a height of 8000 meters. As a result, the air intake was later replaced with one with a nozzle ring, which led to a clearly higher ram air efficiency of +140 mm WS on the ground and +125 mm WS at an altitude of 10 km. Thus, at 10 km a total difference of 13 mm of mercury was achieved, which was equivalent to a theoretical difference of 400 m. As the nozzle ring could be manufactured on site relatively cheaply, its general adoption was urged because of the resulting improvement in maximum boost altitude.

The flight on 15 October 1944 also resulted in engine problems. While in level flight at 7000 meters the port engine, which had flown 3.5 hours to date, suddenly began losing water from the ventilation, as a result of which the flight was aborted. After landing, coolant flowed from the exhaust port and sparkplugs of cylinder no. 8. The engine was subsequently removed; disassembly at Dessau revealed that the center electrode of the forward sparkplug (Beru V 5 F-280 E 43) was missing. The sparkplug jacket and cooling box had burned through as a result of the defective plug; the hole in the cooling box was 10 mm long and 2 mm wide. A replacement engine (No. 104 161 0110) was installed on 18 October. During a static test on 20 October a slide valve in the gearbox seized, but the problem was fixed quickly. The bleed area of this engine's supercharger had already been increased to 20.1 cm^2, as a result of which it ran completely smoothly from 1500 meters to at an altitude of 8 km. During the aircraft's tenth flight on 25 October a filling regulator on the port engine worked poorly, and was subsequently replaced.

A number of scheduled high-altitude flights by the Ju 388 also had to be aborted because of engine problems. During the ninth flight the starboard engine's supercharger leaked oil, and replacing the gasket on the power control unit failed to fix the problem. Then, during the 9th flight, on 3 November, the port engine's supercharger drive developed problems, not switching back from third to first gear until 1000 meters had been reached. A burr had formed on one of the cones of the snap switch; the actual supercharger drive caused no problems during the entire course of testing. The aircraft was flown to Finow the same day for a supercharger change. The starboard supercharger was replaced twice, for during installation of the first replacement it was found that the hole for the intercooler was not the proper size, causing the intercooler to sit wrong. The oil leakage did not cease, however, until the superchargers of both engines were replaced. The exact cause of the problem was thus never found, as it involved both the engine and supercharger.

On 21 November, after another 19 minutes flying time, the same engine's camshaft was damaged by a broken camshaft follower roller. While climbing through 9000 meters the port engine suddenly began to shake. Suspecting supercharger surging as the cause, the crew increased rpm, which caused the shaking to become worse. The engine behaved exactly the same at 8 and 3 km. The engine was shut down and the propeller feathered; nevertheless, at 270 kph the engine continued to turn at 200-300 rpm.

After landing the injector pump was replaced, as it was suspected of leaking; however, the engine's behavior was unchanged. The cause was finally discovered after the right cylinder head cover was removed: the exhaust camshaft follower roller on cylinder no. 2 had broken and fallen out. The camshaft was sent to Dessau for closer examination.

In the period from 26 November to 2 December, after another hour in the air (3 hr, 19 min), the technicians then had to struggle with leaky hoses on the camshaft lubricator, and a loose screw on the oil filter.[856]

On 20 December, two weeks after installation of the aileron seals, the aircraft's speed was measured in level flight. At an altitude of 10150 m with an outside air temperature of -53°C, the two Jumo 213 Es each consumed approximately 340 liters of fuel per hour at maximum allowable continuous power (2,700 rpm, boost pressure 1.40 atm). This accelerated the aircraft to the thoroughly notable speed of 565 kph.

This flight was also used to investigate the behavior of the engines when power was reduced at 10000 meters and 330 kph. The port engine (bleed area 20.1 cm^2) was throttled back to 2,500 rpm at 1.39 atm of boost, at which point mild supercharger surging began. It proved possible to throttle back the starboard engine, whose bleed area had been increased to 25.7 cm^2, to 2,300 rpm.

The 21st flight on 18 December 1944 had to be aborted shortly after takeoff, as the port engine's boost pressure was just 1.20 atm at combat power. Disassembly of the filling regulator revealed that a cylinder head screw had fallen out and become jammed between the filling regulator's barometric cells. This was a known problem on the Jumo 213 A.

On 29 December the penultimate flight was aborted at an altitude of 6.2 km after the port engine began to vibrate badly, and cylinder No. 3 showed signs of poor combustion. It was found that the valve play had shifted badly.

There were also problems associated with the airframe. The stops for the landing flap cruising position proved too weak, and had to be thickened by 4 mm to prevent the flaps from swinging during flight. Construction was also criticized. A considerable quantity of metal chips, mainly drilling chips, was found when the hydraulic system filters were cleaned. These could only have got into the filters during assembly. As several magnetic valves had to be replaced because of the chips, greater "cleanliness" was urged during assembly of the machines.

The pressurized cockpit's jettisonable canopy was difficult to use, and would not have been easy to jettison in an emergency. This poor fitting canopy was also the main reason for leaks in the pressurized cockpit. Use of the MG 131 resulted in torn rubber seals; consequently, the weapon had to be removed, and the opening closed with the original seal. As a result, during the fifth flight on 28 September 1944, even at 1.4 atm of boost pressure (maximum allowable continuous power) at an altitude of 4700 meters it proved possible to build up a cockpit over-pressure of 0.16 atm, which was equivalent to the air pressure at a height of 2.7 km. During depressurization of the cockpit (sealing check) on 10 October, pressure fell from 0.13 to 0.04 atm in 50 seconds in spite of closed outlet valves. Cleaning of the valves, some of which were dirty, resulted in little improvement. On 23 October Junkers ordered a fitter sent to VfH in Oranienburg to check and repair the cockpit pressurization system.[857]

Also related to the cockpit, criticism was leveled at the recessed switches on the starboard wall. These stuck out too far, making it possible to activate them accidentally. While these switches were better recessed on production aircraft anyway, Oranienburg produced its own temporary fix by covering the switches with a folding plexiglass panel. The modification was obviously carried out by personnel from Merseburg. The Ju 388 type supervisor, Hptm. Stahl, considered the modification unnecessary, however. Oranienburg also complained that the lock on the undercarriage control lever worked poorly. Junkers replied that this was a manufacturing error, because the lock was identical to that of the Ju 188. Because the directive to demagnetize the canopy framework was not issued until early October 1944 after problems with production aircraft, the V32's emergency compass was out by as much as 60°. Junkers proposed moving the compass to a spot next to the navigation panel, as this had produced good results in Merseburg.[858]

After the poor fitting dorsal gun mount had been reworked and the weapon could finally be installed, it was found that the radio operator's seat was so close to the dorsal weapon that aiming was impossible. It was also discovered that the flare cartridges did not fit in the envisaged mounting beside the radio operator's seat. Also criticized was the absence of replacement silicagel cartridges for the moisture traps in the double glazing. Oranienburg was obviously unaware that the silicagel cartridges were regenerative, and after the first set of cartridges was used up they used the moisture traps from WNr. 300 296 (see following text), which as yet was not being flown. A change was subsequently made to the longer lasting, but non-regenerative Kolphosal cartridges (DIN 9087 K).

Sheet steel panels had been installed on the upper wing surfaces to protect them from the hot exhaust gases produced by the BMW 801 TJ-0 engines. With the Jumo 213 these were of course not needed, and it was found that the entirely new wings beneath the panels were already corroded. A recommendation was made that all machines be checked for similar signs of corrosion. In response to this and other criticisms, Junkers advised that in future the Astro-panels, which often leaked, and the steel skinning should be deleted as a rule. Filling the fuselage fuel tanks proved very time consuming. It took about an hour to fill the forward 1220-liter tank, as it lacked a suitable filler opening. According to Junkers, however, improved tanks could not be expected until the end of 1944. On the other hand, no retroactive changes were to be made to the "obsolete" autopilot mounting.

At 1145 hours on 5 January 1945 the Ju 388 V32 took off from Oranienburg on its 23rd flight. On board were pilot *Oberleutnant* Reuter, test engineer K. Heintz of the Dessau flight department, and radio operator Ofw. Runkel. According to *Herr* Paasche, a fitter with the engine division field service, the purpose of the flight was to determine performance at altitude. The runway observer reported that the takeoff appeared normal. About five minutes later,[859] however, disaster struck. According to eyewitnesses, the aircraft gradually went from normal level flight into an ever steeper dive. It finally struck the hard, frozen ground in a shallow dive (<25°), and was completely destroyed. The engine had appeared to be operating normally. The entire crew was killed in this tragic end to testing, which came after 9 hours and 40 minutes of total flying time. The crash site was six kilometers from the airfield in a straight line from the runway.

Preuschen obviously inspected the crash site himself, as he compiled an accident report that included a sketch of the crash site. Destruction of the aircraft was almost total, and much of the wreckage had burned, which complicated the search for the cause. The marks in the ground made by the propeller blades showed that the engines had been running at the same speed, which had been greater than idle. Preuschen reconstructed the accident as follows:

"The aircraft bounced into the air again without any part of the wings touching the ground, although the force of the impact had dislodged some pieces (inner landing flaps, parts of the engine cowlings). Attitude on impact can be determined fairly conclusively by deformation of the test probe and the fact that pieces of glass and parts of the canopy first appear after about 200 meters. There is evidence of fire from the initial fuselage impact point and next to the initial engine impact point, suggesting that a fuselage tank and a port wing tank burst and their contents caught fire. The positions of the most important parts may be seen in the sketch of the site. What remained of the fuselage and wings came to a stop after about 400 meters. Fire from the forward fuselage fuel tank largely destroyed the center part of the fuselage and the wings, and the fire from the starboard outer tank destroyed most of the cockpit. Heavier components such as engines and propeller hubs were all found 100 to 150 meters further away from the crash. The engine cowlings had been completely torn away."

By order of Major Heese, commander of the OKL's experimental unit, the wreckage of the aircraft was inspected on the spot by Stabsingenieur Wiedmann, head of the maintenance section, and Preuschen himself. It was originally suspected that electric elevator trim had mistakenly been set to "Nose Heavy," leading to the crash; however, this was quickly disproved. The trim tab spindle and trim indicators were found to be in the normal position. The surviving components of the elevator controls showed no signs of a

Ernst Seibert (left) and Kurt Heintz after their record setting flight in the Ju 88 V5. Dipl.-Ing. Kurt Heintz, who was copilot for the record flight, was a native of Hamburg, and 30 years old at the time. After completing his studies at technical academies in Munich, Berlin, and Danzig (where he was also head of the Danziger Akaflieg), Heintz became assistant and deputy head of the Aero-Technical Institute of the Charlottenburg Technical Academy. In February 1938 he joined Junkers' fight testing group. He was killed in the crash of T9+DL. Vernaleken Collection

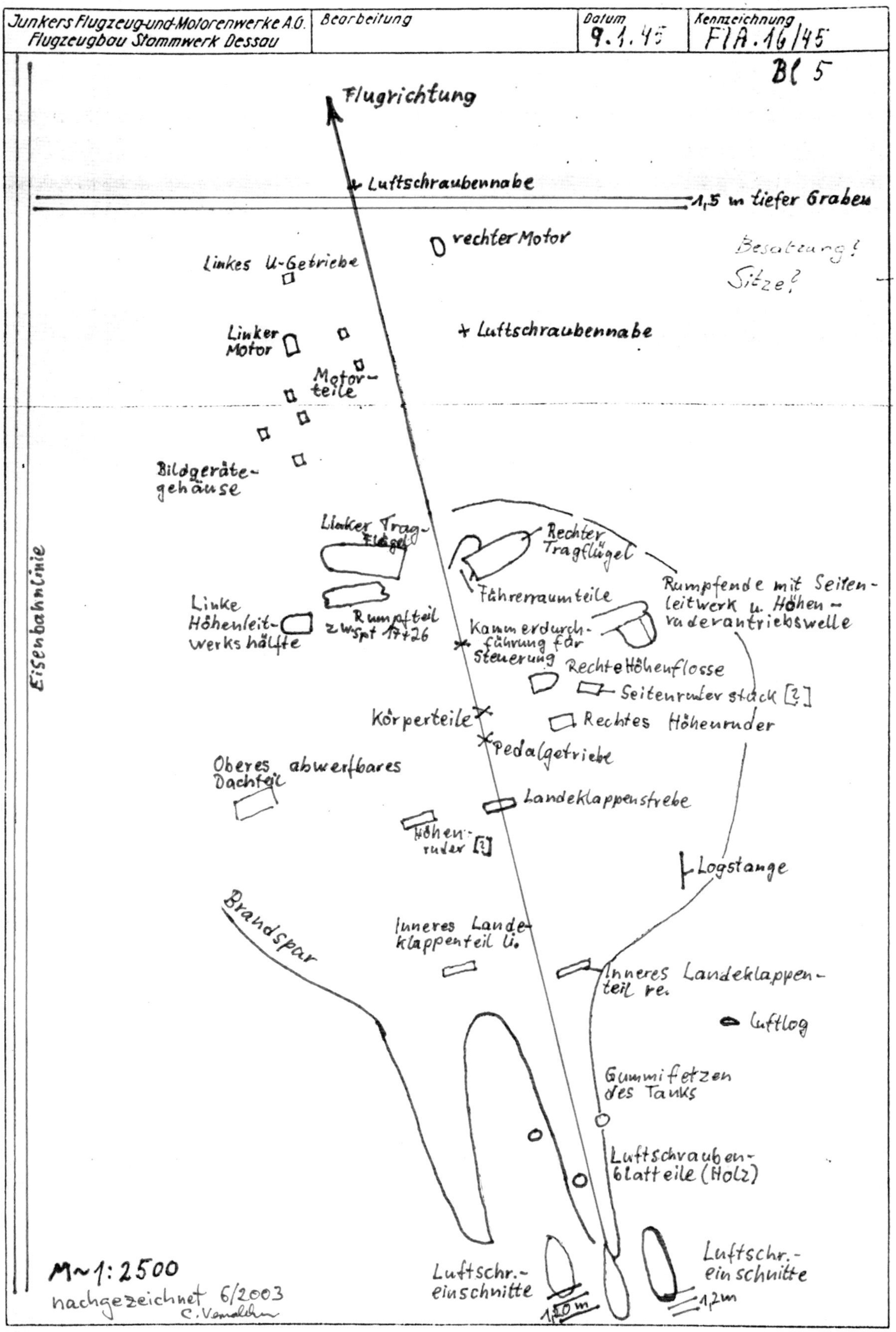

This sketch of the crash site was probably drawn by Preuschen. Vernaleken Collection

break or blockage. When recovered, the hydraulic cylinder for the flaps was in the retracted position. Enemy action, or a reckless "low pass" by the crew, appeared unlikely to have caused the accident, as did the loss of an engine. Reuter's instrument flight experience made it seem unlikely that a sudden fogging of the canopy caused the accident. The simultaneous running away of both propellers in the direction of "Low Pitch" also seemed unlikely, but a detailed examination of the propeller hubs had not yet been carried out. Assuming a defect in the stabilizer incidence and landing flap hydraulics, incorrect setting of the horizontal stabilizer (landing position) on takeoff could possibly have caused the crash. If the trims had been used to compensate for this during takeoff, it could have resulted in retraction of the flaps, which would suddenly make the aircraft nose heavy. Something similar had happened at least once at Dessau, and the pilot, who had little experience on the aircraft, lost about 200 meters of altitude before regaining control. Preuschen could also not exclude a momentary seizing of the controls.[860] In the end, however, the cause of the crash remained unexplained.

Testing of the Ju 388 V32 with its Jumo 213 E engines is one of the few completely documented chapters in the history of the Ju 388. The following is a reprint of the original flight results:

The VfH in Oranienburg also tested the Ju 388 L-0 (WNr. 300 296), which was powered by BMW 801 TJ engines. The aircraft's first flight took place at Oranienburg on 9 November 1944, and by about the end of November both engines had just half an hour flying time. The port engine's starter failed and had to be replaced. Flight tests with the machine were supposed to be stepped up so that it could be used operationally. Fuel consumption measurements were planned first, however.[863] The operational markings applied to the aircraft are unfortunately not known. In February 1945 the VfH equipped a Ju 388 with two Jumo 213 E-1 engines and a water injection system. Two more Jumo 213 E-1s were envisaged for a Ju 88 T. There was no flying in February 1945, however.[864]

No.	Date	Duration	Altitude	Remarks
1	16 Sept. 1944	12 min.	1.2 km	Initial test flight. Aircraft handling in order. Port engine vibrates at M 2. Heavy loss of oil from port propeller.
2	17 Sept. 1944	16 min.		Follow-up test flight. Port propeller pitch mechanism very sluggish.
3	18 Sept. 1944	15 min.		No improvement in port propeller control after increasing governor pressure. Port engine vibrated at M 2. Auxiliary pump switches project too far from the panel.
4	20 Sept. 1944	11 min.		Port propeller still not operating properly even after replacement of rpm governor. Turns slowly in the feathered position.
	Flying halted because of sheared-off screw in port propeller crosshead.			
5	28 Sept. 1944	41 min.	5.0 km	Climb to 5 km. Increase in engine vibration at M 2. Shift altitudes 2400/2650 m at indicated airspeed of 305 kph. At 4.7 km at 0.16 atm cockpit altitude of 2.7 km. Time to fall from 0.13 to 0.02 atm, 24". This short time is due to the poor fit of the canopy.
6	2 Oct. 1944	68 min.	8.5 km	Climb at maximum allowable continuous power to 7.7 km, at combat power to 8.5 km. Shift altitudes 2450/2550 m and 5860/6550 m. Drop-off altitude at 1.43/1.40 atm of boost pressure, 5.8 km. Continuous flow of coolant from coolant vent at temperatures of 95° C/102° C. Coolant loss reduced by fully opening cooling gills. Oil leaked from port vent at combat power. True airspeed in level flight at 5 km at maximum allowable continuous power = 305 kph, boost pressure 1.41/1.40 atm; true airspeed at combat power = 545 kph, boost pressure 1.53/1.52 atm. Supercharger air temperature 88 and 98°. The rearmost window on the starboard side had heavy interior misting, the others relatively little. The hard glass windows had very little ice buildup. Oil fumes in the cockpit air, cockpit over-pressure just 0.16 atm.

				At altitude of 8 km the port engine could only be throttled back to 2,600 rpm. Bleed after the first stage through seven holes is insufficient. There is a lengthwise crack at the root of one blade of the VS 19 propeller. Propeller must be changed.

Flying halted because of crack in starboard propeller blade, problems with supercharger oil leaks and coolant ventilation.

7	12 Oct. 1944	22 min.	4.2 km	After modification of coolant vent, climb to 4000 m at maximum allowable continuous power. The starboard propeller control unit gave problems and refused to go to feathered position. Previously observed coolant loss did not reappear. Cause of the propeller problem: counterbalances striking, as they had been turned up too far.
8	14 Oct. 1944	25 min.	3.0 km	Climb at combat power aborted at 3000 m as port engine vibrated at M 2.
9	15 Oct. 1944	68 min.	8.0 km	Climb to 8.0 km at maximum allowable continuous power. Maximum boost altitude in 2nd gear at boost pressure of 1.37 atm (port engine) 6350 m, 1.40 atm (starboard engine) 5830 m. In level flight at 8 km: Engine speed 2,700 rpm, true airspeed 542 kph, boost pressure 1.43/1.46 atm, supercharger air temperature 110° C; Engine speed 3,000 rpm, true airspeed 565 kph, boost pressure 1.53/1.59 atm, supercharger air temperature 121° C; Level flight at 7 km: Engine speed 2,700 rpm, true airspeed 515 kph, boost pressure 1.41/1.44 atm, supercharger air temperature 107° C. Sudden heavy loss of coolant from port vent after engine speed increased to 3,000 rpm. Coolant temperature 95° C. Flight aborted. Phenomenon reappeared when power applied. After landing coolant ran out of cylinder No. 8 (exhaust pipe and sparkplugs). Starboard supercharger leaked oil badly. Propeller in order. Port cooling flap no longer moves. See Graph 1.

Flying halted because of port engine change and bad weather.

10	25 Oct. 1944	70 min.	2.3 km	Maintenance test and airspeed indicator range flight with new port engine. Level flight at 2.3 km at maximum allowable continuous power, true airspeed = 469 kph, test at combat power broken off due to solid cloud layer. True airspeed at 80 m at maximum allowable continuous power = 436 kph, true airspeed at combat power = 457 kph. Takeoff weight 11.9 tonnes. Port engine's filling regulator very sluggish.
11	3 Nov. 1944	35 min.	5.5 km	Climb at maximum allowable continuous power. Port engine fitted with Siemens V 446 E sparkplugs. Shift altitude 2nd gear 1800/2600 m. 3rd gear port 5400 m. Port had been set on ground at 2.25/6 km. Continuous loss of coolant from port vent at 97° C. When cooling gills were opened it stopped, to begin again in control range. Flight aborted at 5.5 km. In a power-off descent the port supercharger gears refused to shift back, either mechanically or electrically. Not until 1000 m did the 3rd and 2nd gears shift back in rapid succession.

12	3 Nov. 1944	15 min.		Landing at Finow for supercharger change.
	Flying halted because of supercharger change at Finow (needed because of supercharger oil leakage).			
13	8 Nov. 1944	12 min.		Return flight from Finow after supercharger change and exchange of superchargers.
	Flying halted because of supercharger oil leakage, fouling of the oxygen system and bad weather.			
14	20 Nov. 1944		2.0 km	After changing of the 2nd starboard supercharger, exchanging of the port and starboard superchargers and completion of the prescribed modifications by the Junkers team, climb aborted at 2 km as a panel popped open behind the starboard engine and the weather was deteriorating rapidly. Starboard supercharger again leaked oil.
15	21 Nov. 1944	38 min.	4.5 km	Climb at maximum allowable continuous power, aborted at 4.5 km because of MYO.[862]
16	21 Nov. 1944	103 min.	9.0 km	Climb to 9.0 km at maximum allowable continuous power. At 9 km the port engine suddenly began knocking. Supercharger surging was first suspected. Power was increased and at 2,800 rpm the engine began vibrating heavily. Examination of throttled-back behavior at 8 km; port engine completely smooth to 1,500 rpm (bleed area 20.1 cm≤) Starboard engine supercharger surging at 2,100 rpm (bleed area 16.7 cm≤). Power was increased and at 2,800 rpm the port engine also began to vibrate, likewise at 3 km. Switching to M1 or M2 had no effect on this behavior. Problem with injection pump suspected, especially as leaks were found there after landing. The starboard supercharger had again leaked oil. Behavior of VS 19 propeller with 9-30337.16 V blades: At 23° basic setting for takeoff rise to 3,300 rpm. At an indicated airspeed of 270 kph the port engine still turned (about 200-300 rpm).
	Flying halted because of cam roller failure on port engine and bad weather.			
17	6 Dec. 1944	50 min.	7.1 km	Climb to 7.1 km with Beru plugs, as 3 Siemens plugs defective (Siemens V 446 E) after installation of new cooler segments and replacement of broken cam roller on port engine. In level flight at 7.1 km the starboard VS 19 propeller ran on fine pitch. Flight aborted.
18	9 Dec. 1944	24 min.	4.0 km	Port oil temperature climbed rapidly during climb at maximum allowable continuous power. Flight aborted.
19	10 Dec. 1944	49 min.	8.0 km	Climb to 8 km at maximum allowable continuous power. After increasing engine speed to combat power the port engine rpm rose to 3,300. Flight aborted.
	Flying halted because of bad weather.			

20	17 Dec. 1944	75 min.	10.15 km	Following installation of new (larger) cooler segments climb at maximum allowable continuous power to 10.15 km at takeoff weight of 12 tonnes. Climb rate at 9.2 km = 2.5 m/sec. Climb rate at 10 km = 1.38 m/sec. Fall-off altitudes at indicated airspeed of 300 kph (q = 450 kg/m≤) 1st gear 2360/2250 m at boost pressure of 1.35/1.415 atm 2nd gear 5700/5910 m at boost pressure of 1.42/1.44 atm 3rd gear 9600/9950 m at boost pressure of 1.50/1.495 atm Gentle supercharger surging began in the port engine at 1.39 atm at 10.15 km. Investigation of throttled-back behavior at 10 km at indicated airspeed of 330 kph up to 2,500/2,300 rpm good. Bleed area 20.1/25.7 cm≤. Ram-effect at ground level + 140, at 10 km + 125 mm WS. See Graph 2. In level flight at 10.1 km at maximum allowable continuous power, boost pressure = 1.40/1.38 atm, true airspeed according to airspeed recorder 565 kph. Temperature behind supercharger 138° C, behind intercooler 107° C. Cooling flap position 44°. The engines ran very smoothly. Difference between cockpit and external pressure at 10 km = 0.155 atm. Cockpit altitude = 6900 m. The flight had to be broken off due to impending darkness.
21	18 Dec. 1944	9 min.		Climb at combat power aborted soon after takeoff, as port engine boost pressure just 1.20 atm. Cause: cylinder head screw holding Bakelite insert in filling regulator fell out, blocking regulator box. The filling regulator was replaced.
				Flying halted because of bad weather.
22	29 Dec. 1944	26 min.	6.2 km	Climb at combat power aborted at 6.2 km due to increasing vibration in port engine and indications of poor combustion in Cylinder 3. Cause: valve play on the right side, especially in Cylinder 3, had become badly miss-set. Cameras did not work synchronously.
				Flying halted because of bad weather.
23	5 Jan. 1945			Aircraft crashed. Crew (pilot Oblt. Reuter, observer Dipl.-Ing. Heintz, radio operator Ofw. Runkel) killed. Cause of the crash is not known.

Chapter 8: Operational Use of the Ju 388 as a High-Altitude Reconnaissance Aircraft?

The Military Situation in 1943-44

With the strong focus on the production of fighter aircraft, and the cancellation of almost every type that might possibly be used in the strategic reconnaissance role, the situation of the reconnaissance units became increasingly precarious in the summer of 1944. Diesing, head of planning in the RLM, therefore made the following proposal at a Fighter Staff meeting held on 3 July 1944:

"Investigations are to be made as to whether, so to speak, racing aircraft can be made in small numbers for the long-range reconnaissance role, in which the He 177 is a complete failure. The question was raised of building the Ju 488 in small numbers, then the Me 264 and, if other programs exist, small numbers of similar aircraft as prototypes."[865]

At the latest, the material deficit in the field of reconnaissance became apparent when Allied preparations for "Operation Overlord," the invasion of Normandy, caught Germany completely by surprise, followed by the almost total absence of photoreconnaissance results in the following weeks, when the Allies enjoyed local air superiority over the landing area. It was not until about a month after D-Day when a prototype of the Ar 234 reconnaissance aircraft succeeded in photographing the invasion front and the area behind it to the English coast, showing the virtually limitless supply potential of the Allies, and the catastrophic situation facing Germany, that the catastrophic situation in which the reconnaissance units found themselves in mid-1944 became clear. New types were urgently needed as replacements for the standard long range reconnaissance types (the Me 410, Ju 88 T, and Ju 188). New aircraft types had to possess technical superiority factors that would allow them to successfully carry out reconnaissance missions. Two factors came into question: speed and/or altitude. The same applied to bomber aircraft, but in the latter case the units waited in vain for the Ju 388.

There would also have been operational possibilities for the Ju 388 in cooperation between bombers and reconnaissance aircraft. Both used very similar tactical methods to mark targets for the bomber units. Both the *Luftwaffe* and the Allies employed specially trained crews and aircraft equipped with special electronic aids (the German EGON and British Oboe methods and H2S radar[866]) to identify targets, and visually mark them for following bomber squadrons. As the *Luftwaffe* was suffering a significantly higher loss rate over England than the Allies over Germany, the Ju 388, with its ability to operate over long ranges at high-altitude, would have been ideally suited for such pathfinder duties. But offensive operations against England ceased in May 1944,[867,868] when the Ju 388 was still in testing.

The *Luftwaffe*'s use of the Ju 388 was limited mainly to the testing station at Rechlin, and the Vers.Verb. OKL. DW+YY is seen here at Rechlin. EADS

Planned Use of the Ju 388

At a meeting between Milch and Göring on the Obersalzberg in May 1944, the Ju 388 was described as "one of the most important aircraft for employment in the west." They intended to wage the air war against England with the Ju 388 and the Do 335. The Ar 234 and Me 262 were only envisaged in the role of interim bomber, while the future heavy bomber would be a true four-engined version of the He 177. In July 1944 there were plans for the formation of a "new bombing air force."[869] Based on Delivery Plan 226, this was to be formed by December of the same year using the types Me 262, Ar 234, Ju 287, Do 335, and Ju 388. Concrete plans for December 1944 called for two Gruppen of LG 1 to receive the high-altitude bomber version of the Ju 388, all three Gruppen of KG 6 the standard bomber, and the first Gruppe of KG 66 a pathfinder version, while all three Gruppen of KG 26 were to be issued the Ju 388 torpedo-bomber.

In fact, in December these units were still equipped with the Ju 88 S and versions of the Ju 188.[870] Indeed, the K-version of the Ju 388 had been dropped from the production program in autumn 1944.

A similar fate lay in store for the Zerstörer and (high-altitude) night fighter versions. Although production and testing work continued into February 1945, they were cancelled before a single machine could be delivered to a front line unit.

Only for the reconnaissance version are there references to an operational career—albeit a modest one—with the flying units. The operational testing of a handful of aircraft by the OKL's experimental unit has already been described. In June 1944 the Office of the Quartermaster-General (Dept. 6),[871] which was responsible for supply and equipment, wanted to reduce the number of long range reconnaissance Staffeln to receive the Ju 388 from the original 30 to 11, and instead equip six with the Do 335, and six with a variant of the Ju 388 powered by the Jumo 213 E with methanol water injection and the He 177. Furthermore, it had already been decided that the He 177 and Jumo powered Ju 388 would be replaced by the Ju 488, as the Ju 388 "would not last at the front for long anyway." Half a year later, in mid-January 1945,[872] the Ju 388 L-1 was the only conventional aircraft type left in planning for long range reconnaissance alongside the Ar 234 B and C, apart from a minimal number of Do 635s (just two to five aircraft per month) and ten Do 335 A-1s and Hü 211s beginning in 1945.

WNr. 340 101 was built by ATG. Here it is seen parked in front of a hangar, probably at Junkers' "branch office" in Brandis. The tail turret has either been covered with a dark tarpaulin or removed. Petrick Collection

Night Reconnaissance Replacement Training *Staffel* in Brieg

In the second half of 1944 many reconnaissance units were making preparations to convert to the long awaited new types. Among them were a series of units which were to convert to the Ar 234 and Ju 388. Both types required thorough training. With the exception of WNr. 230 156, it is not known how many of the small series of Ju 388 L-0 "Nacht" machines, produced from the K-0, actually reached service units. Aircraft allocations confirm modifications to Ju 188s at Brieg, such as "night wiring." Installation of the remaining camera equipment must have been carried out by a modification team in Jüterbog (Altes Lager) or Kölleda.

Surviving fragmentary 10-day reports by the reconnaissance units[873] between the end of August and the end of October 1944 confirm the existence of a Ju 388 with the night reconnaissance replacement training unit at Brieg, and the conversion of 1.(F)/33 at Jüterbog. By 20 October the latter unit already had seven trained and three partially trained crews. As 4.(F)/11 was located at Welzow air base between February and March 1945, it is possible that this unit, which at the end of 1944 was equipped exclusively with variants of the Ju 188, may have flown the successor type prior to its disbandment in April. So far, however, no proof of this has surfaced. In any case, 4.(F)/11 flew long range reconnaissance missions over Silesia from Welzow until Soviet troops reached the base in mid-April.

Elements of the wiring for the camera installation in WNr. 230 156, originally built as a K-0 at Merseburg. This photo was taken in November 1944. Note the plate bolted over the bomb sight opening in the cockpit floor. Griehl Collection

The air park in Jüterbog (Altes Lager) was used for modification and reequipment measures by reconnaissance units in 1944-45,[874] until Russian tanks also arrived there on 20 April 1945. In the summer of 1944, for example, 3.(F)/123 was preparing for conversion to the Ju 388 L-1.[875] How far these preparations proceeded, however, is not known.

According to strength reports by the night reconnaissance replacement training Staffel in Brieg, the unit had at least one Ju 388 from September to December 1944.[876] It may be assumed that the aircraft was the Ju 388 L-0 Nacht (former K-0) KS+TF, WNr. 230 156. But the unit was supposed to receive additional Ju 388s. On 4 October 1944 Oblt. Denker informed Junkers that twelve Ju 388 K-0s were supposed to be assigned to him for the installation of night photography equipment. The aircraft were to be modified to the same standard as KS+TF, which was then at Dessau. These modifications included installation of external ram air intakes, and the hooking up of the wing deicing. Flame damping systems were to be delivered as soon as they became available.[877]

In between, however, WNr. 230 156 was repeatedly brought back to Dessau, probably for retrofit and repair work. At 1554 hours on 23 September 1944 Flugkapitän Wendel took off from Brieg in KS+TF and ferried it to Dessau, landing there at 1704. The aircraft obviously also spent lengthy periods at Dessau in November and December. The wind deicing system and external ram air intakes were installed there in November.

The logbook of Uffz. Frei, who flew as copilot, reveals that the aircraft was still at Brieg in January 1945. WNr. 230 156 is known to have made the following flights:

Flight No.	Date	Time	km	Pilot	Remarks
413	3.1.1945	12:10 – 12:10	60	Tannapel	circuit
414	3.1.1945	12:15 – 12:22	42	Caesar	circuit
415	3.1.1945	12:25 – 12:32	42	Caesar	circuit
416	3.1.1945	12:34 – 12:40	36	Caesar	circuit
418	13.1.1945	16:17 – 16:25	48	Caesar	dusk circuits
419	13.1.1945	16:30 – 16:37	42	Caesar	dusk circuits
420	13.1.1945	16:40 – 16:51	66	Caesar	night circuits
421	13.1.1945	16:53 – 17:00	42	Caesar	night circuits
422	13.1.1945	17:02 – 17:09	42	Caesar	night circuits
423	13.1.1945	17:12 – 17:20	48	Caesar	night circuits
424	13.1.1945	17:22 – 17:29	42	Caesar	night circuits

A series of photographs of the standard Rb 50/30 camera made by the Zeiss company. The Rb 75/30 differed only in focal length. Handig Collection

On 19 January 1945 the aircraft was flown from Brieg to Damgarten-Pütnitz by the crew of Caesar/Frei (flight no. 427). The flight lasted from 1020 to 1150 hours and covered 540 km.

At the beginning of 1945 there was a series of restructurings among the reconnaissance units still in existence at that time. These included the creation of an Ergänzungs-Aufklärungsgeschwader 1[878] (1st Reconnaissance Replacement-Training Wing) under Obst. Gerhard Kopper on 10 January 1945. Formed from several long range reconnaissance Gruppen, the unit's existence was brief (until 5 March 1945). The unit received a large number of "standard" reconnaissance types, mainly Ju 188 D-2s, even receiving one a day for a while.

Locating a *Luftwaffe* reconnaissance unit close to the air ordnance office responsible for supplying and repairing cameras was an obvious choice. There was also an ATG modification facility on the airfield. It is therefore very likely that the aircraft that "assembled" there from early 1945 were used in a near operational manner. As well, compared to Dessau, Halle, and Leipzig, the airfields in the Thuringia region were largely untouched by air raids.

While the *I. Gruppe* was formed at Bromberg, from 21 January until 4 March 1945 it was at Kölleda. It subsequently became part of "1./EAGr Kölleda," as did the Stab (headquarters) of EAG 1. The II. Gruppe was formed at Pütnitz on 10 January 1945, also from the remnants of other reconnaissance replacement training units. Its 6. Staffel was stationed at Brieg until February. The latter unit is known to have operated the Ju 388. It must therefore at least have had the above described machine, which was used by the night reconnaissance replacement training unit in Brieg.

A photograph of several apparently blown up Ju 388s confirms that there were in fact still Ju 388s at Kölleda at the end of the war. Unfortunately, just one incomplete serial number (WNr. 340 28?) can be made out, but this does indicate an ATG built aircraft. Whether missions were in fact flown from there (over the Eastern Front?), or the unit was just forming and training on the maintenance intensive aircraft, is not known.

Incidentally, the contemporary of the Ju 388 L, the Do 335 V3 reconnaissance aircraft, which was equipped with two Rb 50/30 cameras and a GM 1 tank in its bomb bay,[879] may not have flown any missions from Orly with the Vers. Verb. OKL, as the type was not operationally ready.[880] By that time the future had already arrived in the form of the Ar 234 jet reconnaissance aircraft.

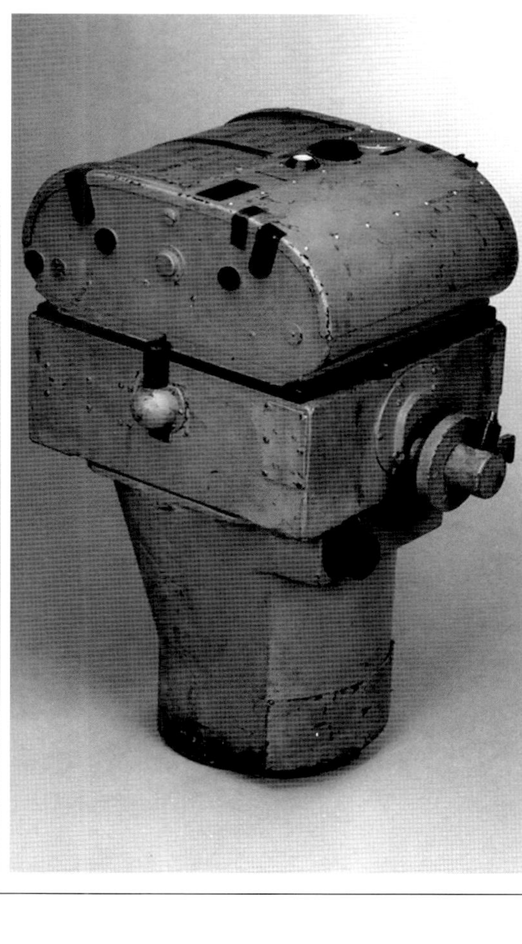

Chapter 9: The Contemporaries of the Ju 388

The assessment of every new aircraft must also always be seen in the context of its contemporaries. Because of the multitude of roles of which the Ju 388 was capable—reconnaissance, high-altitude bomber, heavy fighter, and night fighter—to describe its contemporaries in any detail would far exceed the scope of this work. We have therefore decided to describe three selected piston-engined contemporaries that best embodied the state of technology for their respective operational roles during the period from late 1943 to late 1944. Unlike the rest of the book, in compiling this chapter we have also drawn on the existing literature.

Henschel Hs 128 and Hs 130[881]

One of the few true special purpose aircraft whose development was pursued with a certain determination by the *Luftwaffe* in the years from 1938 to 1944 was the Hs 130 project. Ordered by the *Luftwaffe* as a high-altitude reconnaissance and bomber aircraft, this design, which was derived from the Hs 128, suffered throughout its life from inadequate power plants. The Hs 128 V1 and V2 were ordered by the DVL in 1938 in spite of a complete lack of interest on the part of the military. The aircraft was powered by DB 601 or Jumo 210 engines, but as its development was halted when war broke out, they were never equipped with special mechanical superchargers or exhaust driven turbosuperchargers, which were to have enabled the aircraft to reach altitudes of 17000 meters. The impetus for development of high-altitude power plants was provided by *Oberst* Theodor Rowehl, whose secret unit specialized in (high-altitude) long range reconnaissance missions. One of the developmental focal points was the design of a pressurized cockpit to enable the crew to conduct high-altitude missions without the need for cumbersome special clothing.

Development of the Hs 130 as a three-seat, high-altitude long range reconnaissance aircraft began in September 1938 in response to a requirement issued by the RLM. The first examples were to be equipped with the DB 601 and variants of the DB 605 with two stage mechanical superchargers; however, the Jumo 208 diesel engine was also considered. The Hs 130 A-V1 was first test flown in

This photo of the Hs 128 shows the relatively large size of this high-altitude research aircraft, upon which the later Hs 130 was based. Griehl Collection

The pressurized cockpit of a Hs 130 A or E. Two Hs 130 As, WNr. 054 and 010, were destroyed in a bombing raid on the BMW works in Stuttgart/Echterdingen on 5 September 1944.

April 1940, beginning a lengthy series of trials that ultimately continued until 1945. Already in 1941 the aircraft reached an altitude of 13200 meters with GM-1 power boosting. In June 1942 the aircraft arrived at Dessau for installation of Jumo 208 engines. Once modified, the aircraft, now designated the Hs 130 V1/U1, was transferred to Rechlin. Records show that on 23 December 1943 Rechlin test pilot Borsdorff took the aircraft up on a 22-minute flight, at which time it was probably fitted with the Jumo engines.[882]

After testing at Rechlin, a few aircraft went to the VfH, or to the Vers.Verb. ObdL. The installation of Kärcher heaters for cockpit heating was a source of constant trouble.

In May 1943 Baist flew the Hs 130 V7 at Rechlin. On 12 October 1943 the V10 was flown by pilot Eisermann, who like Borsdorff, and Baist later, also tested the Ju 388. The A-version, which was developed as a reconnaissance aircraft, was followed by the Hs 130 C, intended for the role of high-altitude bomber. The latter was in

This snapshot, obviously taken during a display or an inspection, may be the only authentic photo of a Hs 130 C, and thus an absolute rarity. Griehl Collection

This burned wreckage is probably the remains of a Hs 130 C. Griehl Collection

fact a completely new design for which there was no RLM order. Instead, in April 1941 the RLM issued an order for two pre-production aircraft with BMW 801 MA power plants, which were later to be replaced by J or TJ versions. Once again equipped for the reconnaissance role, the C-version had other similarities to the Ju 388 in addition to its power plants, including a defensive armament consisting of two MG 131 machine-guns in a remotely controlled barbette, which was to be installed at the rear of the pressurized cockpit. The three C-series prototypes were initially fitted with conventional BMW 801 engines, and only one of the three machines was refitted with BMW 801 TJs at Rechlin in November 1943. In June 1944 one of the aircraft was demonstrated to a Japanese delegation. In June 1942 the third C-series aircraft was equipped with DB 605 engines. These were replaced by DB 603 As in June 1943, after which the machine went to the Aeronautical Experimental Institute (AVA) in Prague. Finally, in the autumn of 1943, all three aircraft were grounded.

The Hs 130 D was to have been powered by Jumo diesel engines, but the unavailability of its intended power plants kept it on the drawing board. The next version was the Hs 130 E. All of the type's previously unsolved engine problems were to be solved through the use of a so called

Höhenzentrale (a third engine in the fuselage driving an enormous blower), which was always seen as something of a technical embarrassment. To save development time, components from the earlier variants were used. For cg reasons, the forward fuselage of the A-version was lengthened by 18.7 meters, and the 33-meter wing and the undercarriage of the C-version were used. The first example of the Hs 130 E, an unarmed high-altitude reconnaissance aircraft, flew in September 1942, still without the Höhenzentrale. Testing, which was plagued by numerous technical problems, dragged on until mid-1944, when the aircraft was grounded and then scrapped. The second machine flew in November 1943, for the first time with the Höhenzentrale. Other E-series prototypes followed, but like all the C-series machines they ended up engine test beds.

Although most of the technical problems associated with the development of high-altitude aircraft were solved with time, the dozen or so Hs 130s that were built were ultimately just expensive objects of study with similar performance parameters to the Ju 388. A Henschel crew is said to have reached an altitude of 14450 meters (47,408 feet). *Oberst* von Lossberg was probably right when he declared that the *Luftwaffe* should not and could not afford two such types.

In February 1944 Milch ordered work on the Hs 130 cancelled immediately. Most of the Hs 130 Cs were to be scrapped. The Hs 130 E-1's ten under construction was disassembled, and the material recycled. Just one Hs 130 E is known with certainty to have been parked on a Henschel company airfield. From that point on the field belonged to the Ju 388.

Hütter Hü 211

The Hütter Hü 211, which was based on the Heinkel He 219, which will be described in the subsequent section, was a high-altitude reconnaissance aircraft that never got beyond the project stage. To have been powered by Jumo 222 engines, the Hü 211 had a V-tail and high aspect ratio wings reminiscent of a glider.

The Hütter Hü 211 and the Ju 488 came up several times during discussions on the topic of long-range reconnaissance at Fighter Staff meetings in early June 1944. The reason was the cancellation of the Ju 290 and Ju 390, which left no suitable types for reconnaissance over the Atlantic. Curiously, Diesing stated that the Hütter project "was under way for the long-range reconnaissance role, based on a 388 with lengthened fuselage and wooden wings."[883] Whether he confused it with the He 219, or whether the RLM development experts were considering the possible use of the Ju 388's pressurized cockpit (see following), is not known.

A wooden model of the Hü 211, displaying the avant garde lines selected by the designers. The high aspect ratio wings are reminiscent of sailplanes. Griehl Collection

The Hü 211's great wingspan is apparent in this frontal view. Griehl Collection

Work on the aircraft definitely progressed beyond the drawing board. In Jumo's report for the second quarter of 1944, the delivery of documentation to the Hütter GmbH company is mentioned under the heading "Airframe Industry, Jumo 222 Sector."[884] In the following quarter, provisional documentation for the "Jumo 222 E/F (double-supercharger engine)" was handed over to Heinkel (for the He 219) and Hütter.

At the beginning of August 1944, Jumo's main plant in Dessau confirmed preparation of a Jumo 222 mockup for the Hütter project, with engine mounts for the beginning of September. Two assembled Jumo 222 E/F II power plants, and possibly one replacement engine were to be provided to Hütter from production at Muldenstein for installation in the airframe. Two more were to be delivered in December and January, four more in February, and finally six in March 1945.[885] In fact, according to the last surviving monthly reports, Junkers envisaged that the few completed Jumo 222s would be used for testing in the He 219 and installation studies in the Hü 211.

Inspections of Hü 211 mockups took place in Ravensburg and Kirchheim/Teck on 12 and 14 December 1944. Representatives of EHAG and the RLM took part, along with staff from the *E-Stelle* Rechlin. Preliminary experiments with integral wing tanks were also going on at that time.[887]

The cockpit of the Hü 211 bore a striking similarity to the pressurized cockpit of the Ju 388. It is possible that consideration was being given to using the Junkers components, which were already available, to speed up development.

The design of the Hü 211's engine gondolas for the Jumo 222—seen here in mockup form—had also reached an advanced stage. Griehl Collection

In a comparison of reconnaissance types dated 16 January 1945,[888] which contains performance data and production quantities based on the aircraft delivery program, the Hü 211 is described as having two Jumo 222 A/B-3 engines and two Rb 50/30 cameras. It was to have reached the impressive speed of 715 kph (444 mph) at an altitude of 7300 meters (23,950'). Its performance was thus superior to that of the Do 335 (413 kph at 26,246') and Do 635 (428 mph at 24,606'), both equipped with two Rb 50/18 cameras.

Two Jumo 222 engines were delivered to HFW in January 1945. After modernization of the cooling system and radiators the engines were to be installed in the first Hü 211. Delivery of two more engines was expected soon. One engine (B-2 No. 42) had to be removed after coolant entered the engine during the acceptance run.[889]

The end of the project was already in sight by mid-January 1945, however. Aircraft Procurement and Supply asked Prof. Hertel to review the Hütter project so that the development committee could reach a "decision on the 211 matter" within a week.[890]

The Major Competitors: He 219 and Ju 388 J

The Heinkel design that ultimately led to the He 219 also had its origins in the specification for a high speed bomber aircraft. Not until 25 June 1941 did the RLM specify the roles for the projected new Heinkel design as high-altitude reconnaissance, high-altitude fighter, and night fighter. In August 1941, at the urging of General Kammhuber, the contract was changed to specify the heavy fighter/night fighter roles against the wishes of the RLM's Technical Office. The first flight took place in November 1942. In common with its two later competitors, the aircraft was originally supposed to have a remotely-controlled barbette mounting two MG 131 machine-guns.[892]

After several changes during the early stages of the project, the He 219's operational role gradually crystallized more and more into that of a night fighter. As result, it became what was probably the *Luftwaffe*'s only dedicated night fighter. When comparing the He 219 to the Ju 388, it should be kept in mind that, as a special purpose aircraft, the Heinkel design had an advantage over the Junkers product, which was a multi-role aircraft.

At the urging of Kammhuber and Heinkel, pre-production aircraft were issued to service units at a very early stage, achieving considerable success in combat. Of particular significance, to the *Luftwaffe* leadership as well, was the He 219's confirmed successes against the DH 98 Mosquito. In spite of this the RLM did not push for adoption of the type; instead, there was a lengthy debate about its cancellation in favor of the Ju 388. The He 219 was also designed for the Jumo 222 in the 2,000 to 2,500 hp performance class, which would have bestowed a sparkling performance. Restrictions imposed on the Jumo 222's development by the RLM meant that, like the Ju 388, the type had to be equipped with less powerful engines. Heinkel turned to the DB 603, which was never available in large quantities.[893]

With the He 219's potential recognized, and the RLM's operational requirements constantly changing, the He 219 came under consideration for the high-speed bomber role along with the Ju 88 V56 (S) and the Ar 240/440.

He 219 V33 with full radar equipment. This machine was used by Telefunken for testing antennas and equipment. Griehl Collection

General der Nachtjäger Kammhuber, who was demanding a specialized night fighter for his units, clearly favored the He 219. In 1943 the competition for this role culminated in two comparative trials. As no Ju 388 was available, Junkers entered a Ju 188.

The first trial between what Milch had viewed as the two main competitors since September 194 (the He 219 and the Ju 188) took place in January 1943.[894] In July 1942 Junkers had produced a cockpit mockup[895] of a Ju 188 night fighter, which established itself in the minds of the RLM decision makers as the phantom of a future night fighter. It was all the more important for Heinkel that the He 219 emerge from the comparative trial as victor, and an initial order was issued for 130 machines.

A second competition was held on 25 March 1943[896] between a Ju 188 E (without bomb racks and armament, flown by *Oberst* von Lossberg), Do 217, and He 219 (flown by Major Streib). The He 219 once again demonstrated its superiority. The next day it was flown by a total of ten pilots, confirming the type's speed advantage of 25-40 kph (depending on altitude).

A third comparative trial was held at Rechlin in mid-June with Milch in attendance. This time the participants were the He 219, Ju 188, Ta 154, and Do 335. Also demonstrated was a He 219 with a three-man cockpit. The Ta 154 was poorly received, while the Do 335, ferried directly and demonstrated by Dornier pilots, caused a great deal of excitement on account of its unorthodox configuration. The type that received the most favorable assessment, however, was the Ar 234, which even held its own against the Me 262 in a dogfight.[897]

In spite of these results, at a state secretary's meeting[898] on 2 August 1943 Milch proposed that the spare production space at Henschel resulting from the cancellation of the Me 410 should be used to build the Ju 188 night fighter, which he rated "at least as good as the He 219." He added that, because of the possible shortages of BMW 801 engines that might result, the Ju 188 should also be equipped with the DB 603, a step that clearly would have come at the expense of He 219 production.

This aircraft was used for ejection seat trials. The seats subsequently proved themselves in service. Griehl Collection

Quantity production of the He 219 began at Rostock in August 1943, and later at Vienna-Schwechat.[899] Mielec and Budzyn were also supposed to join the program.[900] Each time the RLM tried to cancel the He 219 in favor of other types it encountered a determined verbal defense from Prof. Heinkel. The *General der Kampfflieger* (at that time Hajo *Herr*man) confirmed to Prof. Heinkel the RLM's tendency "to assess the He 219 unfavorably because of the strong influence of Junkers."[901]

The Ju 388 J was a design based on a multipurpose aircraft with corresponding compromises. It should have been superior at high-altitude, where the He 219 was not as highly rated because of its power plants. Director Francke hoped to overcome this deficiency by using the DB 627 or DB 628 high-altitude engine (both abandoned), or even the Jumo 222 E/F, which was also planned for later versions of the Ju 388. A true competitor for the Ju 388 could have been born at a conference in Vienna-Schwechat on 26 August 1943, the He 219 B high-altitude variant. Equipped with wooden wings and 2,500-h.p. Jumo 222 A/B or E/F engines, it was designed to have a maximum speed of 615 kph at sea level and an impressive 759 kph at 12500 meters with a range of 2850 km (consuming 3400 liters of fuel).[902] Also considered was a combination of the Jumo 222 with TK 11 and TK 15 for a high-altitude fighter and bomber with an increased wingspan of 30 meters.[903]

Heinkel left no stone unturned in its efforts to improve the He 219's performance, and so the company became interested in mixed propulsion. Experiments began in mid-September 1943 in an effort to increase maximum speed. A BMW 003 jet engine was mounted beneath the fuselage of the He 219 A-010. Although an improvement could be demonstrated, the gains did not justify the cost, for in action the extra engine would have been nothing more than "ballast" most of the time.[904] The aircraft was lost on 13 November, and conversion of the V30 began on 20 January.[905]

One of the He 219's greatest drawbacks was its two man cockpit. The *Luftwaffe* began demanding the development of a three man cockpit in early June 1944.[906] At least one experimental aircraft was in fact fitted with such a cockpit, but it apparently was never introduced into production.

Testing of the Jumo 222 in the He 219 V16 and V19 began in October 1944. One of the engines in the V19 (100 204 0026) had to be changed after just 4 1/2 hours when it began emitting steam from the engine vents, and became difficult to turn over. Metal shavings were discovered in the oil filter, and a seized large end bearing and water entry was the probable cause.[907] In November Jumo 222 No. 100 204 0006, mounted on the He 219 V16, developed a similar problem, but by then the He 219 with Jumo 222 engines had already been cancelled. In its place, testing began of the He 219 V41 with Jumo 213 E engines.

Design work was completed at the beginning of September, and installation of the first two engines was planned within 14 days. The first aircraft was to be ready to fly one month later. EHAG intended to use the last three September production airframes and the first two from October production in Vienna for these conversions. In case these were not available in time, Director Schaberger of Heinkel-Vienna asked *Herr* Carlsson, representative of Special Committee F3 on the Armaments Staff, to authorize use of the "He 211 Ns" (cover designation for the He 219) at the Eger repair facility to ensure that the in service target date was met.[908] These five He 219s with Jumo 213 E engines with MW 50 power boosting were to be readied for immediate front line service. During a static test, however, one Jumo 213 E (100 161 0006) in the V41 broke down because of a foreign object in the engine—sabotage?[909]

On 21 December Director Francke rejected the idea of delivering the He 219 with Jumo 213 engines and no methanol water system. He argued that such a machine would have an advantage of just 30 kph and 2 km greater maximum boost altitude than the version with DB 603 E engines, which would have appeared too little to the front line units. They should therefore wait for the promised new supercharger from Jumo, and therefore hand over three machines to the BAL in December and then two more.[910]

The subject of the high-altitude fighter was raised again in late 1944, a few days after testing of the He 219 with Jumo 213 E and Jumo 222 engines had begun. Heinkel proposed the He 219 with Jumo 222 E/F engines and methanol water injection as an immediate solution for an anti-Mosquito aircraft in place of the Ju 388 ("better a few than none").[911]

Heinkel's high-flying production plans were frustrated by repeated raids by the American 15th Air Force on the Vienna Basin, and with it Heinkel-South's second production and testing facility in Schwechat (on 17 March, 23 April, and 26 June 1944). Constant aerial surveillance by the Allies, which went almost unhindered by the *Luftwaffe*, had revealed that the pre-production series of the new *Luftwaffe* night fighter was being produced there. A precision attack by 146 bombers in three waves hit the factory hard. Until the end of the war, production by Heinkel never again approached what it had been before the attack.[912]

As well, at a meeting at Berchtesgaden on 25 May 1944, Göring ordered production of the He 219 to cease after the existing orders had been filled, citing the raw materials situation, as it was to be replaced as a night fighter by the Ju 388. EHAG also received a directive from the RLM to cancel further development of the He 219. Prof. Heinkel subsequently produced a document with arguments as to why the He 219 should continue in production:[913]

- **Flight Characteristics:** exemplary flight characteristics, confirmed by test stations and front line units
- easy to fly, and suitable even for average pilots
- good single-engine characteristics, and excellent stall behavior as per the latest DVL guidelines (also including yaw, turns)
- spin proof because of twin tails
- **Nosewheel:** simplifies takeoff and landing, prevents swing, thus a great advantage in night landings
- **Great Structural Strength:** He 219 has a load factor of 6, compared to 4.2 for the Ju 388
- **Cockpit:** good view, three-man cockpit already tested, soon to be introduced into production
- **Armament:** 4-6 weapons of 20 or 30 mm caliber, two upwards-firing 30-mm weapons in oblique mounting
- **Favorable Design, Ease of Maintenance:** involvement of construction personnel in design, just 7,000 man-hours from the 1,000th machine compared to "9,000 best case" for the Ju 388
- **Maximum Speed:** about 30 kph higher than the Ju 388 with the same engines
- at present only suitable Mosquito hunter
- installation of Jumo 222 will result in "ideal Mosquito hunter," prototype in June 1944
- **Bombs:** installation of bomb racks envisaged, carriage of up to 1 000 kg of bombs
- **Versatility:** use of the He 219 as night fighter, daylight Zerstörer, and fighter-bomber possible
- **Experiences in Front-Line Operation**:
  "Through our technical field service, which has provided skilled maintenance personnel in support of the front-line units, we have been kept well-informed of the assessment of the He 219 in service use.
  The overwhelming opinion is that the He 219 is the best available design for the night fighter role—viewed as a percentage, successes in combat with the He 219 are considerably greater than with other types. The ejection seats have fully proved themselves in emergencies. Equipped with GM superchargers, the He 219 has recently achieved considerable success in the anti-Mosquito role. It is unlikely that any night fighter pilot will be able to understand why in the future he will have to do without 'his' He 219 and go into combat in an aircraft that is significantly inferior to the He 219 in performance, handling and serviceability."
- **Production:** reduced output of aircraft between rundown of He 219 and startup of Ju 388.

Not only was Heinkel confident that it could maintain the planned production rate of 50 aircraft per month as of November 1944, it believed that a significant increase was possible, as the firm had already been promised additional concentration camp inmates. As Prof. Heinkel saw it, monthly production should be increased to 60 aircraft in December 1944, and then rise continuously to 200 machines per month in August 1945. He also thought it possible to further increase production, which was to take place in Rostock and its dispersed sites (Mielec, Barth, etc), to as many as 250 machines per month by December 1945. Any further increase would have required the involvement of licensed factories, as in the Junkers concern.

Prof. Heinkel's intervention, in his typical fashion, was successful. On 12 June 1944 HDL Saur decided to initiate an "immediate action" to push forward the He 219 with the DB 603 A engine, but with the D-version's supercharger. As well, the He 219 was to be apportioned in such a way that an overlapping of the two types was assured. The Ju 388 was still seen as the ultimate solution, however.[914]

At a meeting in Berlin on 13 June 1944, Prof. Heinkel learned that the decision of 25 May had been reversed, that a follow-up order was envisaged, and that development of the He 219 was to be continued. A version of the aircraft powered by DB 603 AR engines was to be employed as a Mosquito interceptor. Prof. Heinkel nevertheless made a point of bringing forward his arguments in the question of the He 219:

"We nevertheless make a point of this, because it is obvious from our representations that we have always been convinced of the superiority of our design and that we therefore seriously regretted the decision—now reversed—of 25 April 1944, not out of company interests, but because we could not comprehend why a type that has proved so successful in action and is considered the best aircraft by the men who fly it against the enemy, should be replaced by another, less good aircraft."

Prof. Heinkel correctly saw the materials situation argument as invalid, for as the material requirements of the He 219 and Ju 388 were about the same, resources could be divided between the two aircraft—this would have had no influence on the absolute number of completed aircraft. He also refused to accept the justification of type rationalization, for after all, an order had been issued for the Ta 254 at the same time that the He 219 was cancelled.

The criticisms that Heinkel raised in a point-by-point statement of case against the Junkers concern and certain RLM offices were serious and numerous:

- My visit to the Hotel Ritz in the spring of 1940: my proposal to build 100 He 111s instead of 45 was accepted. Correct decision. [Author's note: Prof. Heinkel had followed Göring to Paris to prevent a completely sensible reduction to 30 aircraft per month. This decision was in no way correct, however, for as early as 1941 the obsolescent He 111 "could not even be used over England at night"[915]]

Another view of the He 219 V33 from the right side, Munich-Riem 1944. Griehl Collection

- For the past 1 1/2 years the He 219 has been killed every few weeks and then brought back to life again. Now they are letting it start up in Rostock, and next spring it is to run out again. The gentlemen who are in favor of killing it are unaware of its qualities and performance or its successes, and also have no desire to learn of them.
- The office halts every further development of this type, such as installation of the Jumo 222.
- Hertel originally had no intention of installing the Jumo 222 in the Ju 388. Not until he learned that Heinkel was using the engine did he also install it, and although his prototype did not appear until many months later and ours in three weeks, the decision was made in favor of Junkers for quantity production of the Ju 388 with the Jumo 222.
- If Hertel's monopoly efforts continue to receive such a hearing, there will be a repetition of the disaster with Koppenberg.
- The performance figures, characteristics and target dates contained in Junkers' submissions are always far too optimistic. [...] We lose out because, as is generally acknowledged, we offer realistic deadlines and performance figures in our proposals.
- If the He 219 is better than the Ju 388, Junkers should build the 219 under license. There is no room now for company interest and vanity.
- If the 219 were inferior like the 388, I would immediately build under license without a murmur.
- [Handwritten:] We could have had a fighter capable of 750 kph in the [He] 100. [Me] 109 poor undercarriage.

At the Fighter Staff meeting of 9 July 1944 a document was read out and discussed. It described the impressions gained by the minister during a trip through various factories of the aviation industry. According to it, on Wednesday, 5 July 1944, Heinkel still had no direction to continue with the He 219. The factory in Schwechat concluded that production of the He 219 was to run down. Frydag insisted that was impossible. He had been in Schwechat two days after the bombing raid, and had laid down the program there; but conditions in Vienna were obviously somewhat chaotic, and seemed rather bizarre. An anecdotal extract from the Fighter Staff meeting of 9 July 1944 was therefore allowed to be woven in. Frydag's remarks spurred the following brief exchange:

Saur: With whom have you been speaking? With the director of the coffee house? What's the name of the man who used to sell coffee in the Meindl store [sic!] and is now in charge of the technicians?
Frydag: His status is equal to theirs. I stand behind this man. He has succeeded in throwing out 3,000 unproductives in three weeks.
Lucht: By the way, the technical director is seriously ill at present.[916]

Production of the He 219 was obviously also hampered by poor coordination and wrangling over authority, as the following extract confirms:

Saur: A new prototype series is being built for the night fighter role which will have a three-man cockpit.
Frydag: It is the aircraft we saw in Rechlin. It will be built as a three-seater beginning in November. It was never a single-seater, it has always been a two-seater.
Saur: Would you confirm that and prepare a short memo for the minister?
Frydag: I would be interested in knowing with whom the minister spoke. The document is wrong from top to bottom. The situation in Vienna is intolerable. I can't restore order there if everyone has the right to interfere from the outside.

There is also an order for a change from aluminum wings to wooden ones, but obviously no one could remember it being issued.[917] The following example shows clearly the frustrating arguments that sometimes formed part of the discussions about production of certain types, and how decisions in the RLM were obviously made with shockingly limited expert knowledge. At the meeting in Berlin on 13 June 1944 Saur, head of the Fighter Staff, asked why the He 219 was to be cancelled despite its success in combat. Stabsingenieur Walter Baist of the Erprobungsstelle Rechlin replied that the front line units preferred the Ju 388, whereupon Frydag of the EHAG replied that the Ju 388 had not yet reached a front line unit. How Baist, who at that time had never even flown the Ju 388, reached this conclusion remains a mystery, however.[918]

Prof. Heinkel used this opportunity to put together further arguments in favor of the He 219. His main argument was the operational successes of *Oberstltn.* Streib and *Hptm.* Modrow, who had shot down a considerable number of enemy aircraft while flying from Venlo. The pilots at Venlo were extremely enthusiastic about the aircraft, and according to Heinkel Streib, who until then had obviously assumed that the He 219 would be replaced by the Ju 388, which was to be produced in greater numbers, was now calling loudly for the Heinkel to be kept in production. Adolf Galland made statements to the effect that the He 219 was the best night fighter. As production of the Ju 388 was just beginning, Heinkel was convinced that he could produce 400 to 500 more machines than Junkers by the 1st of September. Prof. Heinkel also stressed the increase in serviceability from the initial 20-25% to 70-80%, thanks to the, in his opinion, outstanding work of the technical field service.

	He 219	**Ju 388 J**
Advantages	1. Modern tricycle undercarriage 2. Fast specialized aircraft 3. Very good view from cockpit 4. Ejection seats 5. About one year development lead 4. Three-man crew 5. Better armor 6. Defensive armament	1. Pressurized cockpit 2. High-altitude engines 3. Mainly proven components from existing Ju 88/188 production, broad production base.
Disadvantages	1. Poor high-altitude performance, mainly because of DB 603 engine 2. Two-man crew 3. Limited space for radar equipment 4. Crew endangered under fire by exposed cockpit and little armor 5. No pressurized cockpit 6. No defensive armament 7. Mass production organizationally more difficult, new design 8. Higher wing loading than Ju 388	1. Not a pure night-fighter 2. Poorer maximum and climb rate than He 219

In addition to these points, EHAG stressed the lower production costs of the He 219. According to plans, 10,000 man-hours would be required to build a He 219 beginning with the 800th machine. By the 1,000th machine, however, this figure could be reduced to 7,000 hours. According to EHAG data, the 1,000th Ju 388 would still require 9,000 man-hours.

Despite all the setbacks in small scale production, Heinkel-South's yearend report viewed further development of the type optimistically. Although the RLM had refused to increase production numbers, six machines had already been built as future Mosquito hunters with Jumo 213 E engines and MW 50 power boosting. These aircraft initially sat idle because of supercharger problems, however, for without methanol water injection they offered no significant performance increase over the DB 603. Heinkel was confident, however, that Jumo would solve this problem within a few weeks.[919] The question of whether it was better to produce the Ju 388 or He 219 remained until shortly before the end of the war. On 3 January 1945 Lucht, a member of the "Aircraft Development High Commission" in Speer's Ministry for Armaments and War Production, sent the following letter[920] to Professor Tank in his capacity as member of the "Special Committee Night Fighters":

"Dear Professor Tank,

I would like to direct your attention to the question of the comparison of the 8-388 and 8-219. In the long run it does not seem advisable to build two night fighter types of roughly the same size at the same time. The 219 is about 30 kph faster than the 388 at all heights, but the 219's landing and takeoff distances are somewhat greater. Otherwise the data for the two aircraft are not that different. Aircraft Procurement and Supply has been in favor of canceling the He 219 for some time, because Heinkel would require additional resources to produce the 219 in large numbers, whereas Junkers meets the requirement as is. According to the most recent production plan, monthly output of the 388 has been or will be reduced (about 100 per month). Given this new state of affairs, I think it very important that we decide whether preference should not be given to the aircraft with the superior performance (219).

I would be very grateful if you and your special committee could give this matter particular attention and come up with proposals or decisions.

With warmest greetings and
Heil Hitler,
yours (signed) Lucht."

The response from the special development committee "Bad Weather and Night Fighter" came just three weeks later. On 22 January 1945 Prof. Tank replied:

"Dear *Herr* Lucht,

Thank you for your letter of 3 January 1945. The performance comparison produced the information in the appendix. It is clearly obvious that neither aircraft has the necessary performance for successful employment against fast bomber aircraft (Mosquito). Both types have sufficient performance for use against four-engined bombers (Boeing, Stirling, etc.).

In selecting the type to be produced, it should be considered that the Ju 388 can be used in a much wider variety of roles than the He 219 on account of its more varied armament (including upwards-firing weapons), its crew of three, its greater endurance and lower wing loading, and that the Ju 388 can be used as a trainer for crews of the high-speed night fighter to be procured in the future, which as a rule will have a three-man cockpit.

The disadvantage of the Ju 388's somewhat lower airspeed and climb rate does not hamper its operational effectiveness and is more than offset by the described advantages. The somewhat greater construction cost is cancelled out by better mass production possibilities compared to the He 219.

With fondest wishes and
Heil Hitler,
Your (not signed)"

Tank's statements about the performance of the two types were largely based on calculations. After estimating the drag producing area of both airframes, the most important performance factors for the DB 603 E (medium altitudes) and Jumo 213 E (high-altitudes) were calculated and compared. The results showed that the He 219 enjoyed a speed superiority of between 35 and 45 kph at all heights between ground level and maximum boost altitude with either power plant. The He 219's rate of climb was also superior, by 1 to 2 m/sec, while its service ceiling was as much as 300 meters higher (DB 603 E). No significant difference could be found in takeoff distance between the He 219 and Ju 388. With its greater fuel capacity, however, the Ju 388 was able to remain airborne up to 40 minutes longer. A detailed comparison is provided in the following table:

DB 603 E (Jumo 213 E)	**Ju 388**	**He 219**
Gross weight	14.200 (14.100) kg	12.550 (12.450) kg
Wing area	56.0 m²	44.5 m²
Wing loading	254 kg/m²	282 kg/m²
Wingspan	22.0 m	18.50 m
Aspect ratio	8.65	7.70
Fuel capacity (no external tanks)	3280 l	2620 l
Endurance at maximum allowable continuous power		
Level flight	2.98 h	2.33 h
Climb	0.24 h	0.20 h
Total (altitude = 6 km)	3.22 h	2.53 h
Armament	2 x MG 151 + 2 x MK 108 in ventral tray 2x MG 151 in oblique fuselage mount FHL 131 Z tail turret	2 x MG 151 in wings and 4 x MG 151 or 4 x MK 108 or 4 x MK 103 in fuselage tray
Crew	3 Man	2 Man
Drag-producing area with Morgenstern antenna, no flame dampers	1.51 (1.49) m²	1.20 (1.18) m²
Takeoff distance at takeoff power (2 x 2,000 h.p.)	640 m	620 m
Performance at special emergency power, no flame dampers		
Maximum speeds:		
At ground level	535 (510) km/h	575 (546) km/h
At maximum boost altitude	609 (648) km/h	650 (693) km/h
Rate of Climb		
At ground level	13.3 (11.4) m/sec	15.4 (13.3) m/sec
At maximum boost altitude	9.8 (6.3) m/sec	11.4 (7.4) m/sec
Service Ceiling	9.9 (11.7) km	10.2 (11.8) km

In spite of the indisputable performance advantages of the He 219 over the Ju 388, there were also rational arguments against mass production of the Heinkel design. In a small irony of fate, almost twice as many He 219s (268) were built as Ju 388s. Many of the "Uhus" went directly to the night fighter units, where they were very popular with their crews, and splendidly maintained by EHAG's technical field service. One would have to doubt, however, whether the technical field service would have been capable of providing such intensive maintenance had the aircraft been mass produced. It was instead probably part of Heinkel's strategy, in cooperation with General Kammhuber, to establish the aircraft against its competitors through demonstrable operational successes. The quartermaster-general's daily reports continued reporting acceptances of "He 219 A-7s" until April 1945, even if only in the "OKL reserve."[922]

The sole surviving He 219 is also in the hands of the National Air and Space Museum, and is soon to be completely restored. One will probably be able to make a direct comparison of the He 219 and Ju 388 in a new branch of the museum in Dulles International Airport.

Successor and Contemporary: Ju 488

Contemporary types with four or more piston engines have intentionally been omitted from this examination. The Ju 488, however, is an exception, as it was a direct development of the Ju 388, and was intended to at least partly replace it. At a meeting of department heads on 7 December 1943, *Oberstleutnant* Knemeyer gave a short presentation on the development of high-altitude aircraft. In his opinion the Ju 488 had to be built, because by 1945 the Ju 388 would be able to adequately cover the high-altitude sector.

The timetable for the Ju 488 at that time was as follows: the first prototype was to be completed in June 1944, the first pattern aircraft in November, and production was to commence in May 1945. Junkers believed that it could complete the aircraft even sooner. At the meeting Knemeyer characterized the Ju 488 as follows:

"The Ju 488 has a long fuselage like that of the existing 88 H and twin fins and rudders like the 288. It has the same tail turret as the 388 and will probably also have a remotely-controlled dorsal

An He 219 A undergoes an engine change, revealing the nacelle designed specially for the DB 603 engine. This probably would have been a complicating factor for the installation of the Jumo 213, or the much larger Jumo 222. Griehl Collection

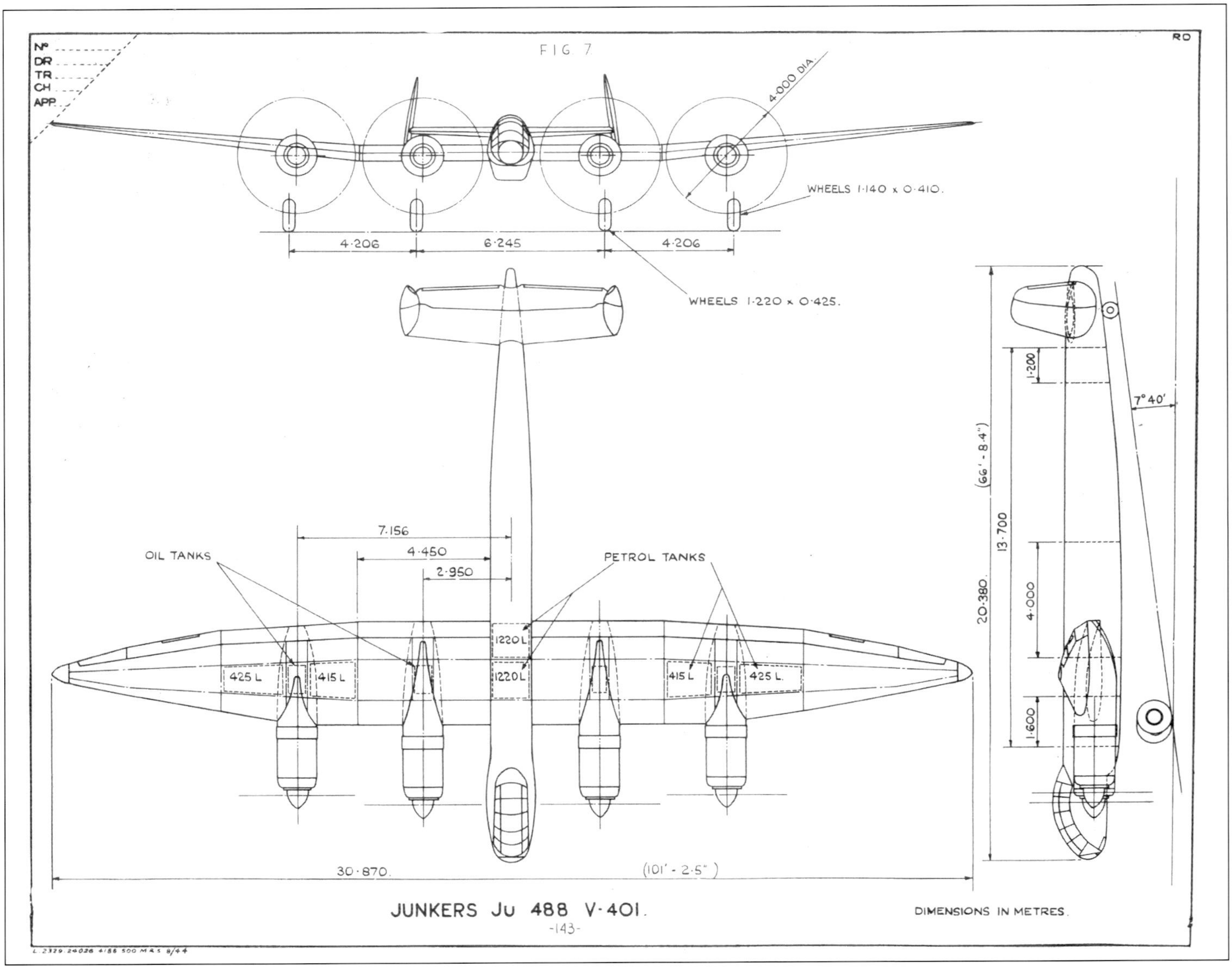

This drawing was produced for U.S. intelligence, and is part of a report on German activities in the French aviation industry. The Ju 488 V401 exhibits significant differences from the V403. IWM

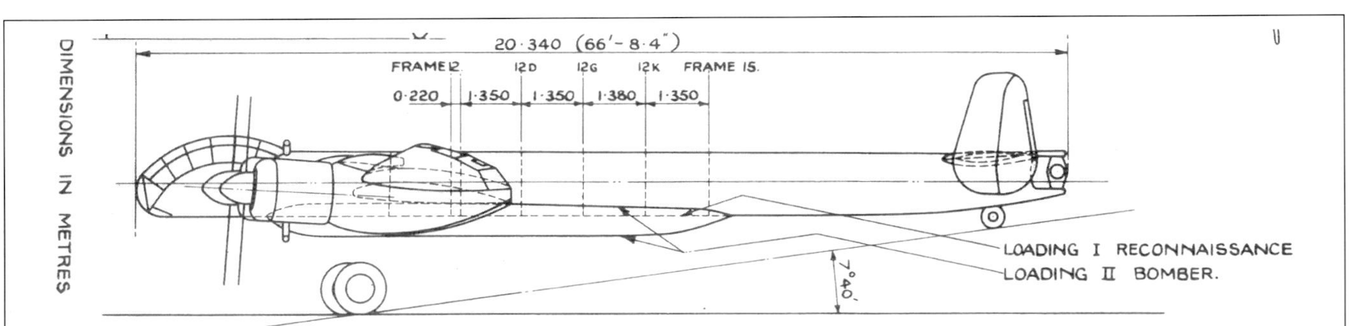

Interestingly, this later version, here identified as the V403, has the original fuselage with long bulged bomb bay. IWM

turret. It differs from the Ju 388 only in its wing center-section. The aircraft has four undercarriage legs and the Ju 388's three-man pressurized cockpit. The bombs are housed in an lengthened bulged bomb bay. Inserted between the cockpit and wing center-section is a section containing a 750-liter fuel tank. A similar section is located between the center-section and aft fuselage."[923]

Junkers received the first advance orders for construction of Ju 488 prototypes in the first quarter of 1944. The orders covered the "Development and Construction of Ju 188 V401 and V402 Flying Mock-Ups," and were valued at 1.9 million Reichsmark. The first four prototypes (the V403 to V406) were to be developed and built for another 3 million Reichsmark [monthly report FV 03/1944]. In the next quarter Junkers responded with a corresponding offer to the RLM. The costs for the first two prototypes were now estimated at 2.2 million RM, and the estimated cost for each of the subsequent machines also rose by 150,000 RM compared to the advance order. Meanwhile, it had also been determined that final assembly of the first two prototypes would take place in Dessau, while a final decision on construction of the V403 to V406 had yet to be made [monthly report FV 06/1944]. In fact, as stated in Chapter III, major components of two Ju 388 K-0s (WNr. 230 152 and 22230 154) were shipped to Toulouse for construction of the Ju 488 V401 and V402.

As previously described in Chapter II, during the development meetings on high-altitude aircraft held on 10 and 20 August 1943, the Ju 488 and Hs 130 had been shelved in favor of the Ju 388, or were "to be pursued as projects only." The Ju 488 later reappeared, however, and as a contemporary of the long lived Me 264 it was one of the great hopes for long range maritime reconnaissance over the Atlantic, until it was realized that this requirement could not be met.

It was quickly agreed that the Ju 488 should enter production as soon as possible. The question now arose, however, as to whether the Hs 130, with what *Oberst* Petersen described as the "extremely cumbersome" Höhenzentrale should be cancelled or, as Knemeyer suggested, be equipped with four engines and "perhaps also find use in the reconnaissance sector." Both the Ju 488 and the four-engined Hs 130 were capable of about 700 kph at combat power at an altitude of 13.5 km. A four-engined Ju 388 could carry a bomb load of 2000 kg, and cover a distance of 4500 km in the maritime reconnaissance role. The range of a four-engined Hs 130 E was more modest—2400 km at a cruising speed of 600 kph at 13000 m. With two Jumo 222 F engines the aircraft's estimated speed at that altitude was 620 kph, and 710 kph with four BMW 801 TJ power plants. Equipped with four Jumo 222 Fs, the Hs 130 would have had an impressive maximum speed of 730 kph at an altitude of 14-15 km.

The participants at the meeting had a hard time in agreeing on anything to do with the Hs 130.

Major Rehm and Generalmajor Kleinrath welcomed additional high-altitude aircraft. General Staff *Oberst* von Lossberg, however, argued that because of shortages of materials and labor, they could not afford two four-engined types, especially as just 30 of the four-engined Hs 130s were to be built. Knemeyer in turn pointed out that most of the design work on the Hs 130 had already been done, and that much of the metal had already been cut.

Pending Göring's approval, it was therefore decided to force the Ju 488. A decision on the fate of the Hs 130 would wait until Frydag had been consulted.

The RLM obviously considered using the Ju 488 prototypes for maritime reconnaissance. A note to this affect appears in the minutes of a meeting with the Reichsmarschall on 1 July 1944.[924]

Chapter 10: License Negotiations with Japan

Since the 1920s, Japan had frequently built foreign aircraft under license, including many German types. Dornier and Rohrbach flying boats, for example, were built under license in Japan. German engineers, including Prof. Alexander Baumann, designer of the Staaken giant aircraft, also provided personal "development assistance." Junkers also established early ties with Japan. Following a demonstration of the Junkers F 13 to a Japanese government delegation, the Japanese military procured their first example of the type in 1922. It served for many years as an air ambulance. Two more float equipped F 13s were procured for the navy, and in 1928 Mitsubishi imported two more, which were then purchased and operated by the Japan Airline Company. In 1928 the Japanese army bought the rights to build the Junkers G 38 under license. A heavy bomber version dubbed the K 51 by Junkers was built by Mitsubishi. But even before the Japanese military finally accepted the plans for the K 51 in 1932, the twin-engined Ki-1 was created from the Junkers S 36. The Ki-2 was a development of this aircraft built by Mitsubishi. By 1935 the latter company had also built a total of six examples of the K 51 as the Ki-20.[925]

Cooperation Becomes Closer

Economic cooperation between Japan and the German Reich was strengthened after the conclusion of the Three Power Pact between Germany, Italy, and Japan on 27 September 1940. Japanese delegations were frequent and welcome guests of the German aviation industry, which was eager to gain Germany's powerful far eastern ally as an additional source of capital. In May 1943 the Junkers Flugzeug und Motorenwerke (JFM) concluded a deal with Kawasaki, in which Junkers agreed to assist Kawasaki in the construction of a large factory in Japan, where the Ju 188 and Jumo 213 were to be built. Thirty specialists were to be made available for a period of three years. Junkers charged 30-million Reichsmarks just for this goodwill effort.[926]

A contemporary German postcard showing a delegation from the Japanese parliament examining a Junkers F13 in Japan. Vernaleken Collection

On 15 August 1944, even while the "Kawasaki Project"—which resulted in a German-Japanese agreement on the relinquishment of copyrights—was still under development, Junkers received a committee of technical officers from the Japanese army and navy air forces led by Major-General Otani. When the technical speeches were over, Otani asked to procure a license for the Junkers Ju 388, and pressed for the soonest possible preparation of the required documentation, apparently for transportation reasons. Otani also expressed interest in acquiring a license for the Junkers Ju 390. The next day, accompanied by Admiral Koshima, the delegation visited various Junkers factories.[927] Although the foreign sales department submitted a written request to the licensing department several days later asking for the preparation of offers and clarification of release questions, the German side's response to the question of issuing a license was slow in coming, despite Japanese reminders of the urgency they attached to the matter. The reasons for this will be addressed later. At least part of the correspondence related to this licensing question survives in the EADS Unternehmensarchiv in Ottobrunn, and a review of these documents suggests that little progress was made in September and October 1944. The first confirmed meeting between Junkers employee Scheibe of the aircraft foreign sales department and Major-General Otani did not take place until 31 October. There were obviously disagreements.[928] Finally, on 7 November 1944 the Mitsubishi subsidiary in Germany, the Mitsubishi Shoji Kaisha GmbH, which was located on Hermann-Göring-Strasse in Berlin, contacted Junkers. In the letter Mitsubishi informed Junkers that it had been tasked by the Imperial Japanese Army to negotiate with Junkers for delivery of technical drawings of the Ju 388 and Ju 390. As Mitsubishi requested delivery of the first set of drawings in mid-November, approval for the release of both types had obviously been given or promised by that time. The drawings were to be made as precisely as if a pattern aircraft were available. The Japanese had the most interest in the Ju 388 K, for they asked for the drawings to be of the bomber version, with supplementary drawings for the night fighter (especially the cockpit) and long range reconnaissance versions. Detailed descriptions, performance tables, and a description of the armament were to be delivered with the drawings. Mitsubishi expected a second set of drawings, together with two sets of descriptions to follow immediately after the first delivery. As well, three more sets of drawings were to be available on demand. The sense of urgency attached to the matter was again stressed in the letter.[929]

In a letter dated 9 November 1944, ministerial director Müller of the RLM apologized to Major-General Otani for the delays:

"The official response to the desire expressed by the Imperial Japanese Army Air Force for the achievement of close technical cooperation in the areas of design and development, especially with the Junkers company, has obviously been delayed by errors in technical processing. I have therefore asked the office of *Oberst*ingenieur Schwencke (TLR-Rü S.J.) to immediately act on your request for an examination of the development-technical conditions. In carrying this out, the Junkers company will be the first to be considered.

An unintended delay has also cropped up in the matter of providing general arrangement drawings of the aircraft types Ju 388, Ju 390, FA 15 and Me 323."[930]

According to Müller, however, the Japanese failure to respond to several requests for more detailed information as to the type and quantity of drawings requested was the reason for the delay in providing the general arrangement drawings. The Japanese wishes would possibly have to be included in the official request list as per the treaty between Germany and Japan. To avoid further delays, however, he said that he had instructed the involved companies to initiate direct negotiations; once the remaining questions had been cleared up he would issue formal approval. The very next day Müller asked Junkers and Messerschmitt to initiate negotiations with the Japanese, and to determine whether they wanted construction or general arrangement drawings. On this occasion he stated the reasons for the delays rather more openly than the day before:

"Through the Armed Forces High Command/Field Economics Office (Foreign) Dept. 3, the Japanese military attaché has asked for the release of drawings for the purpose of studying the above-named types [Ju 388, Ju 390, FA 15, and Me 323, the authors].

It is impossible to determine from the request whether they desire sufficient documentation to enable replication or whether the sole purpose of the drawings is to study the aircraft types. Experience has shown that in requesting the transfer of drawings, what the Japanese desire is usually equivalent to a complete set of replication drawings.

Regrettably, delays have arisen through formal handling by the OKW and the under the conditions necessary inclusion of the request for licenses in the official request list, which has caused a certain resentment on the Japanese side."[931]

On Tuesday, 14 November 1944, the Junkers Flugzeug und Motorenwerke confirmed to Mitsubishi that it had received its two letters of 7 November. One letter, "Delivery of Drawings for the Ju 388 and Ju 390," was the one cited above; the second letter, titled "Copyright and Pattern Delivery for FA 15," is not contained in the holdings viewed by the authors in the Ottobrunn archive. The title suggests that the Japanese were very likely capable of expressing their wishes very clearly. The vague phrasing of the letter concerning the Ju 388 and Ju 390 may therefore indicate that the Japanese had not completely made up their minds, rather than a culturally related problem of understanding.

But back to the letter from Junkers. Mitsubishi was informed that they intended to negotiate directly with Major-General Otani during the course of the week to avoid further delays; Mitsubishi would be informed of the results.[932]

Annoyance with Otani

On the following Thursday, 16 November 1944, a meeting was held between representatives of JFM and Major-General Otani and Lieutenant-Colonel Ishizuga, his future deputy. Junkers was represented by *Herr* Thiergärtner of the licenses and patents department, and *Herr* Scheibe of the aircraft foreign sales department. The meeting began badly, for Otani had misunderstood the first sentence of the RLM letter of 9 November, and taken it as an insult against his person. After this obvious misunderstanding had been cleared up, Thiergärtner and Scheibe, referring to the two letters from Mitsubishi, asked Otani to explain his views on the documentation to be provided for the Ju 388 and Ju 390, as in Junkers' view there had been a change compared to his earlier wishes. The FA 15 was not discussed, as Japan definitely wanted to procure the rights to reproduce this piece of equipment.

"General Otani first declared that he was very angry about the delays caused by the RLM's handling of the matter. He had made the request for approval in August, and had also asked for a statement on the type's development potential. The latter was important in deciding whether or not to build the type under license, and not a single step could be taken without it."[933]

Major-General Otani emphasized that Tokyo needed documentation of the status of both types in order to be able to decide on possible production, as delivery of the aircraft to Japan was out of the question. (By that point in time, submarines were probably the only means left for the exchange of personnel, goods, and documents between Germany and Japan.) The Junkers delegation subsequently informed Otani that pure design drawings were only of value for the study of an aircraft delivered from Germany; if reproduction was the objective, production drawings were at least equally important. They went on to say that by introducing new production equipment and methods and a changed production organization, in the course of development from the Ju 88 A-1 to Ju 388 K-1 Junkers had succeeded in reducing the number of hours required for construction from about 18,000 to just over 9,000—and that in spite of the Ju 388's improved performance and higher weight.

It is not entirely clear whether the figures for the Ju 388 K-1 production hours came from production plans created for the type at that time (and thus possibly not terribly realistic), or whether actual figures for the Ju 388 L-1 then in production were recalculated as required. Based on the fact that the Ju 388 L-1 had only just entered production at ATG and Weserflug, the first option seems more likely. To avoid further lost time, Junkers proposed direct negotiations for licenses for the two types as planned in August, and the conclusion of a supplementary cooperation agreement. It was envisaged that the proposed agreement would be similar to that planned for the Kawasaki project.

Major-General Otani promised to wire Junkers' position to Tokyo. Admittedly, an answer could not be expected for at least 10-14 days, because Junkers could not give any price relation between design drawings and license. Otani complained bitterly about the way the matter was being handled by the Japanese committee, the economics ministry, and various other offices which, contrary to the terms of the treaty, steadfastly refused to give information on the cost of providing construction drawings. His ministry, however, had to know exactly what financial obligations would arise. Both Thiergärtner and Schiebe wrote memoranda to document this meeting. Concerning Otani's remarks about the Japanese committee, Thiergärtner merely wrote:

"He [Otani, the author] said that ambassador Oshima would soon have to take appropriate steps; in any case, the present negotiations were unbearable for the Japanese government."

That is worded very cautiously, for if one believes Schiebe's account there had in fact been a heated argument:

"Several details of the discussion, parts of which were conducted with a sharpness unusual for the Japanese, are to be recorded, as General O. made a number of very pointed remarks. [...] In any case, in his opinion it just won't do for the Japanese committee to accept the German efforts with no offer of a price orientation or without further representations, and then be charged a large sum by way of the relief committee. Japan is, after all, not a German protectorate, and he has the feeling that the German economic offices

Hiroshi Oshima (1886-1975) was Japanese ambassador in Berlin from 1938 to 1939, and from 1941 to 1945. The International Military Tribunal for the Far East declared him a war criminal and sentenced him to death; however, he was pardoned in 1955. Handig Collection

intend to use Japan to compensate for the loss of the European partners, especially in the provision of ideal services. He declared that if the difficulties that he believes were caused by the economic offices increase further, he will have no other choice but to turn directly to the Führer through ambassador Oshima."

Considering the, from a European standpoint, excessive friendliness with which the Japanese had made their requests in earlier letters, the energetic statements by Otani must have caught Thiergärtner and Scheibe completely by surprise and rattled them. Unfortunately, we will probably never know if Otani had acted with clever calculation to achieve the element of surprise, or if he was simply very angry.

For reasons that are obvious from the above quotation, Otani would have preferred to pursue the new projects outside the treaty in direct negotiations. In order to allow for delivery of the drawings within four to six weeks—roughly Christmas of 1944—it was agreed to allow the preliminary work by Junkers to continue regardless, as Tokyo was expected to respond with a binding promise or a request for design drawings for information purposes. Incidentally, the Japanese were particularly interested in the design drawings for the wings and fuselage.

Following the meeting with Major-General Otani, which had ended harmoniously in spite of their disagreements, Scheibe and Thiergärtner met with *Herr* Dinter and Dr. Kambayashi of Mitsubishi in the B.B.,[934] and informed them of the results of the meeting. No further negotiations with Mitsubishi could take place until a response was received from Tokyo, and Mitsubishi's participation in settling of the matter through the Japan committee was questionable anyway. Junkers, on the other hand, had no reservations about involving Mitsubishi, and was even ready in principle to pay a commission.

Meanwhile, the issuing of a license for the FA 15 was being handled through Mitsubishi, as revealed by a letter from Thiergärtner to the Aviation Industry Economic Group, Foreign Department, dated 21 November 1944. Interestingly, Junkers was very interested in processing this contract through the Japan committee. In addition to the complete gun position, the Japanese also wanted to procure copy rights and drawings from Zeiss (periscopic sight) and Rheinmetall (barbette).

Paper Shortage

In Gernrode, in the Harz Region, to where the "JFM Licenses and Patents" department had been relocated, careful preparations were made for the license deal with the Japanese. For example, the patents and patent reports by Junkers in Japan that would be used by the Junkers Ju 388 were researched.[935] The work came to a halt at the end of November, however, when the supply of tracing paper dried up. Dr. Tesch of the Kochstedt Industrial Economics Office[936] was hastily engaged, but even he was unable to help Junkers. According to the Head Technical Office, the following quantities of paper were required for the duplication of construction drawings for the Ju 388 and Ju 390:

500 rolls of BG-transparent, each 90 cm wide 50 m long,
50,000 sheets of DIN A-1 M-paper and
10,000 sheets of DIN A-3 M-paper[937]

Meanwhile, by way of Mitsubishi the Imperial Japanese Army was urging the earliest possible delivery of technical drawings.[938] However, the main suppliers of tracing paper had recently been eliminated through enemy action. On 6 December Thiergärtner turned directly to ministerial director Müller of the RLM, to inform him that because of the difficulties in procuring paper it would be impossible:

"...to even come close to meeting the desired delivery dates expressed by the Japanese. Considering the possibility of further annoying the Japanese, we think it better that General Otani be advised openly about these problems, which initially cannot be overcome, and the likelihood that we will be unable to meet the deadline."[939]

Thiergärtner also wanted to know from Müller whether he wanted to give Otani the bad news, or if he should do it himself. *Herr Herr*mann of the FTH drawings service proposed that they obtain paper from the companies that were to have built the Ju 388, and from Junkers for the Ju 390, as stocks held by them were no longer required due to changes in the procurement program. The usable material could then be sorted out.[940] In a telephone conversation on 9 December, *Herr* Dintner of Mitsubishi again brought up the Japanese impatience. On the same day Junkers assured Mitsubishi in writing that it was working "very intensively on preparation of the requested design drawings," however, it was running into numerous difficulties. Junkers was also reluctant to specify a delivery date for reproduction drawings for the FA 15 because of decentralization of production, and the resulting difficulty in coordination with outside equipment manufacturers and replica companies. As well, Mitsubishi was advised as politely as possible that the reply from Tokyo promised for the end of November had not yet been received.[941]

To overcome the difficulties in the procurement of drawing paper, the RLM's department LF2 II even turned to the Chief of Aircraft Procurement and Supply. Junkers was informed that they

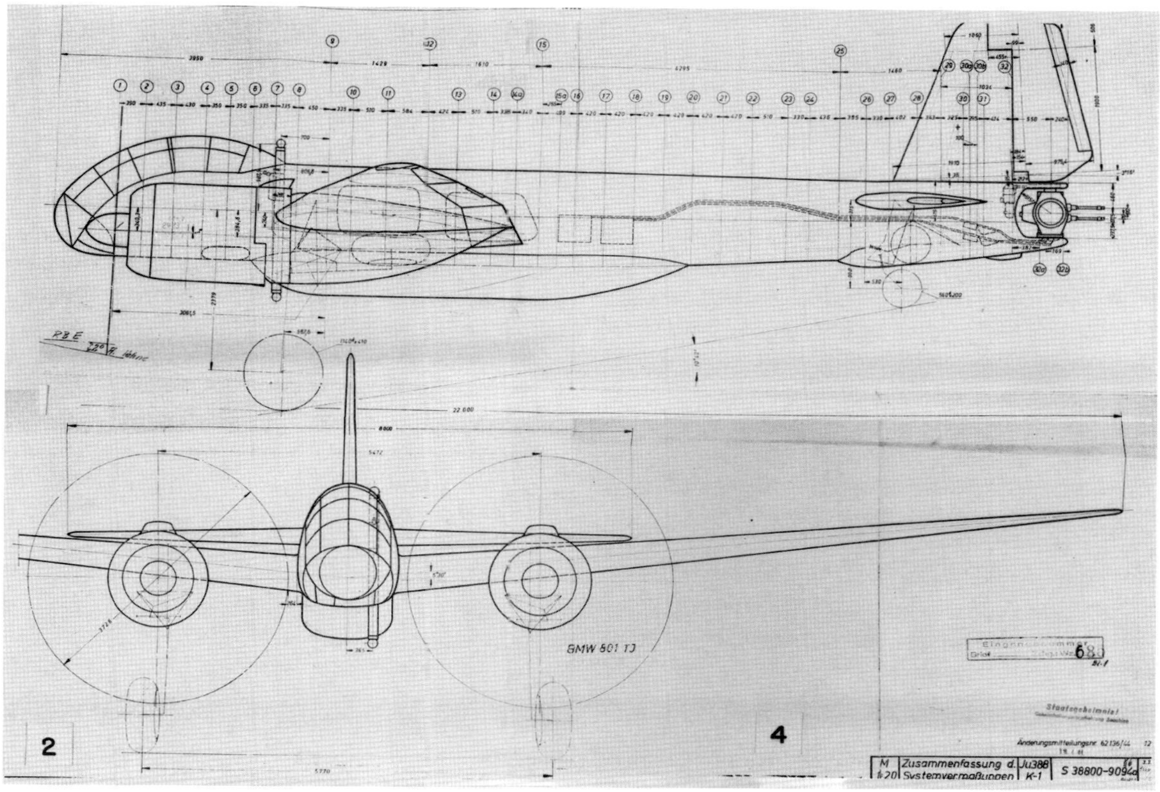

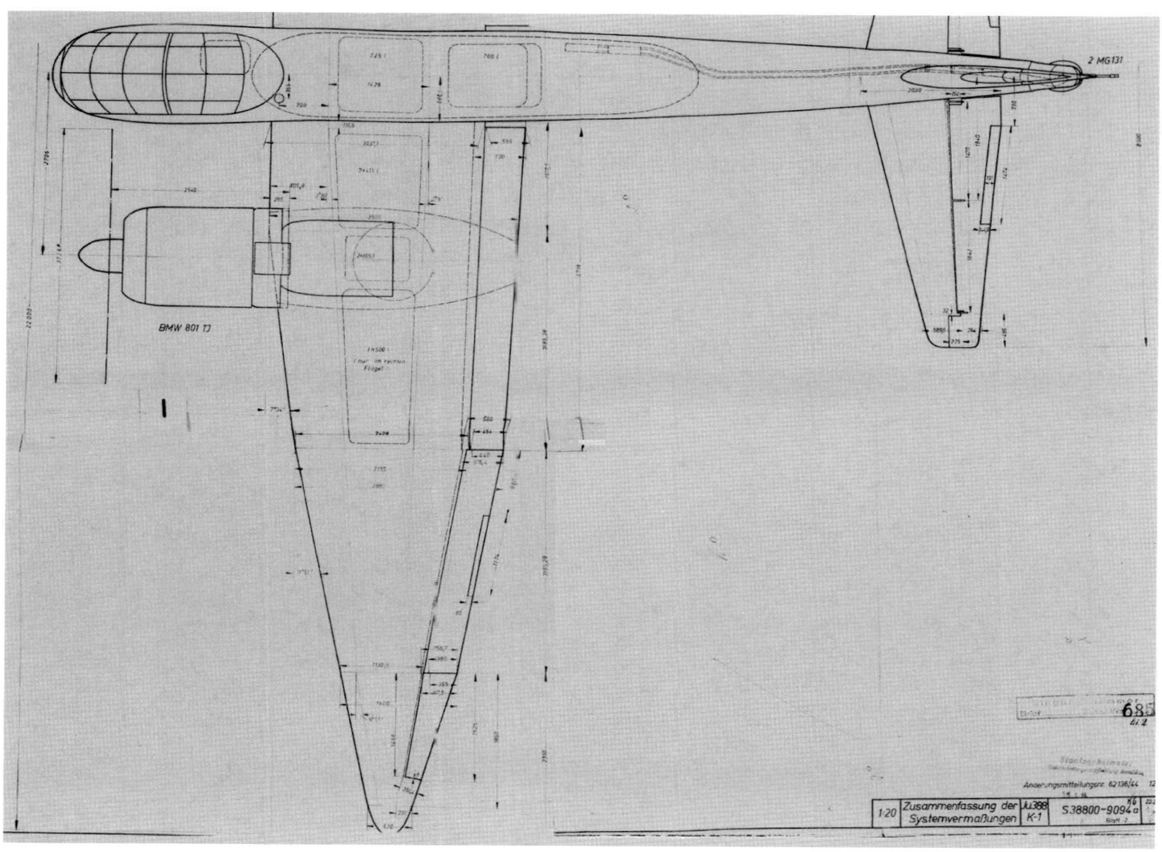

Procurement of sufficient quantities of paper for the copies of drawing sets proved difficult. Here are two general arrangement drawings. Vernaleken Collection felt it appropriate to inform Major-General Otani of these difficulties and "announce the anticipated inability to meet the deadline in a suitable way."[942]

In view of the tremendous quantities of paper required for the reproduction drawings of the Ju 388 and Ju 390, the question naturally also arose as to what mass and weight these would have. According to calculations, a set of drawings for the Ju 388 with an area of 3000m² weighed 330 kg, for the Ju 390 with an area of 5 000 m² it was even higher, 550 kg. With five sets of drawings each, the total weight came to 4.4 metric tons. Folded to DIN A4 and folded, the stacks of DIN A-1 paper would have been 5.4 and 9.1 meters high.[943]

Meeting with Major-General Otani and Major Suematsu about the aircraft and engine types in which the Japanese were interested were held in Raguhn, about 14 kilometers south-southeast of Dessau, and Weiden on 14 and 15 December. Otani informed the Junkers representatives that the Japanese Army had finally decided to procure the licenses for the Ju 388, Ju 390, and Jumo 222. Several details still had to be worked out with Tokyo, however, such as the question of engine accessories, and the desired equipment state. The Japanese Army also intended to procure licenses for the Ju 488, Ju 287, Ju 635, and possibly also for the Ju 248 and Jumo 223. The final decision on this had not yet been made by Tokyo, however. Another meeting with Major-General Otani, Major Suematsu, and Major Kawakita was held in Berlin on 18 December, and Otani promised that the Japanese Army would soon finalize its wishes with regard to the versions, series, and engines in which it was interested. In this context, however, Thiergärtner felt it imperative that Otani be provided with the specifications and performance tables for the above named types that same week, as Japanese notes from previous meetings had proved sketchy and contradictory. Junkers was of course very interested that Otani send his telegraphed report to Tokyo with the correct data, on whose basis the decision on the procurement of licenses would be based.[944] Beyond the license agreements, at the meeting Otani again brought up the cooperation treaty he favored, whose goal was support for Japanese industry in further developing the procured types. Otani also wanted the latest performance data for the Jumo 213, as the Japanese had again become "interested" in the use of this engine in the license produced aircraft. He was also interested in the wooden laminar profile wing developed by Junkers for the Ju 388 and Ju 488. He declared himself in agreement with the plan to initially deliver just two sets of drawings for each type as red prints. He reserved the right, however, to request delivery of another set should the first two sets of drawings be lost on the way to Japan.

While striving to complete the drawing sets as quickly as paper availability would allow, Junkers was also working on a draft agreement. It was based on the license agreement for the Jumo 004 engine.[945]

There was an important step forward in the paper question at the end of December, when Dr. Tesch of the RLM-GL F1 informed Junkers' foreign sales department that the needed paper could be obtained from the Humpert company in Berlin. Of this, 350 rolls with a total area of 10 200 m², enough for at least one set of drawings for the Ju 388 and Ju 390, were in storage at the Hugo Hesch company in Königstein/Elbe.[946]

Numerous details also had to be cleared up, for example, which accessories and components for the Ju 388 and Ju 390 had been developed by subcontractors rather than Junkers, which of the used materials were covered by trade mark rights, and how, for example, to calculate a charge for development of the PVE 11 A-1 (Fl 22 830) periscopic sight, developed jointly by Junkers and Zeiss. Junkers had provided Zeiss with 15 designers, and had only charged Zeiss for the cost of their salaries, while Zeiss billed the RLM for the entire development contract. The periscopic sight had been developed specifically for the FA 15 control system, and was not suitable for use with other drives without modification.[947]

During a subsequent meeting with the Japanese they were given specification sheets for the types 8-488, 8-287, 8-635, 8-248, and Jumo 222.[948]

On 2 January 1945 Junkers transferred responsibility for the actual production of the drawing sets to ATG:

"For the reproduction rights for the type 3-88 within the framework of a secret state agreement with Japan, we require the manufacturing equipment drawings and lists plus the production plans, type plans and other production-related documentation for the fuselage, cockpit, vertical stabilizer and rear fuselage framework in the form of two sets of drawings or copies each. For special reasons we have been given very short deadlines with no additional security guarantees. We require the first set by 9 January 1945, the second set by 26 January 1945; we request that they be delivered to FTH-Zd – Dessau, Franzstrasse, in the Seiler House. The drawing paper needed for this work can be made available to you.

As we have been given the task of producing the remaining reproduction documentation for this job, we must assign to you the task of duplicating the manufacturing equipment and production documents. The codeword "Reproduction Job N 388" are to be used in correspondence and the sending of documentation.

In view of the political importance of this task, and the need for the utmost secrecy, the government agencies are absolutely expecting that we will deliver on time."[950]

This letter clearly shows the position of supremacy that Junkers had over the replica firms. There is no cautious inquiry as to whether ATG can assume the task; rather, Junkers simply orders ATG to deliver the drawings in the shortest time possible. The documentation for the wings was ordered from the Siebel-Flugzeugwerke (SFW), however, the order letter has not survived. All of the documentation was assembled by Junkers and packed in special crates; a special cost center was even set up for the job under the code name "Reproduction Job N 388," K.A.-Nr. 70 99 9720/133/000.[951] Because of the short deadline set by the Japanese, it was decided simply to deliver the actual documentation for the Ju 388 L-1 and J-3, as the necessary drawings for the K-series could not be produced in such a short time. The drawing sets for the Ju 390 represented the A-1 version. On 2 January 1945 Thiergärtner telephoned Dr. Tesch of the RLM and informed him that, because of the short notice delivery deadline, Junkers required the RLM's final release confirmation for the Ju 388 and Ju 390 immediately. He also requested that the release confirmation should explicitly authorize Junkers to deliver the German production drawings without restriction, for a review of the documents for equipment which they might

An in flight photograph of the Ju 390, RC+DA. It was a larger version of the Ju 290 designed according to the building block principle with six BMW 801 engines. Just two prototypes were built. Haubner/Misna Collection

wish to withhold from the Japanese was impossible in the time available. Dr. Tesch assured him that he would immediately send him the requested release confirmation for both types.[952] In fact, the requested release for the Ju 388 L-1 and J-3, and for the Ju 390 and FA 15 with sublicenses, was issued by ministerial director Müller on 5 January. As a precaution, he issued releases for the Ju 488, Ju 287, 8-635, 8-248, and the Jumo 222 E/F and Jumo 213 E engines at the same time.[953]

Junkers was overly cautious in its dealings with the RLM, as a letter to the industrial economics office on 8 January 1945 shows. Although approval had already been given for the delivery of complete sets of drawings for the Ju 388 and Ju 390, Junkers asked for a special letter of approval for delivery of the specification sheets to Major-General Otani! The latter had asked that he be provided documents for his personal use during a license meeting on 5 January; he ultimately received specification sheets, AG lists,[954] and a general arrangement drawing for the Ju 390 dated May 1944.[955]

On 9 January 1945 the Reich Industry Group contacted Junkers with a request that it be sent two draft contracts for the released types. These drafts were to include an estimate and justification of compensation claims (license fees). Negotiations with the Japanese Navy concerning the Jumo 223 were apparently nearing completion, as the Reich group also requested two draft contracts for this engine.[956]

Thiergärtner's letters to the aviation industry economics group (on 12 January 1945) and the Reichsgruppe Industrie (on 16 January) reveal that Junkers was obviously able to meet the delivery target date for the reproduction drawings, probably in part because Siebel-Flugzeugwerke and ATG had been closely involved in producing the documents. Siebel delivered the manufacturing equipment drawings for the production lines for the wings and final assembly, and for this later demanded a license fee of 60,000 Reichsmarks.[957] According to a message dated 31 January, the first set of drawings would be delivered on 10 January, and the second on 27 January 1945. After this happened with the agreement of the RLM, the contractual provisions had to be addressed, for which Junkers needed confirmation from the Japan Commission.

On 29 January Thiergärtner received three draft contracts from the reproduction department of Junkers' head technical office for the FA 15, Ju 388 L-1 and J-3, and the Ju 390 A-1. In their conditions these were similar to the draft for contract No. MK 19 with the Japanese Navy for reproduction of the Jumo 004 jet engine.[959] Thus, the Japanese received the following construction documentation:

a. Technical description of the Type 8-388 J-1/J-3 and 8-388 L-1/L-3 with electrical technical description,

b. all design drawings for the types 8-388 J-3 and 8-388 L-1, including component plans, subgroup drawing lists, and parts lists, plus design drawings and parts lists for dropped weapons and equipment set 8-388 K-1 (bomber),

c. the associated main lists, specifically accessory equipment list, hydraulic equipment list, and electrical equipment list.

d. the associated secondary lists, specifically semi-finished products list, standard part list, cast and pressed parts list, contingency list, and contingency weights list,

e. design standards issues,

f. aircraft handbooks 8-188 E,

g. improvement directive 88 A,

h. loading plan and loading manual,

i. flight report for initial test and acceptance flights with directives,

j. production equipment drawings and production equipment standards issues,

k. production plans with target times and distribution of work processes and materials distribution,

l. inspection equipment list,

m. inspection chart with inspection directives,

n. production directives,

o. surface protection directive.

Forgotten in this list, however, was a publication titled "Service and Operating Manual Fl 8-388 J-1, L-1, January 1945 Issue." No one at Junkers noticed this until 16 March 1945.

It appears that the replication contracts between Junkers and the Japanese Army were in fact negotiated through the Japan Committee. In keeping with its guidelines, they did not extend to materials trade mark rights and experiences. In order to be able to successfully manufacture and operate the license-built machines, the Japanese were supposed to procure the necessary rights from the respective development or manufacturing firms. This applied, for example, to the so-called "Instrument Grease 'New'" (made by Intava, Leipzig), a cold proof grease for gears and bearings, and "Weguper" glue (IG Farben, Leverkusen) for gluing rubber sealing strips onto the cockpit canopy structure.[960]

The research conducted by Junkers for the calculation of compensation claims brought to light some figures that are still of interest today. With the production of 250 aircraft per month, the bare airframe of a Ju 388 J cost 180,000 RM, and the bomber version 200,000. Junkers claimed nine years and 4.1 million design hours in developing the 88 series, beginning with the 88-1 through the 8-188 to the 8-388.[961] Total development costs were 69.1 million.

Assuming an initial license production period of two years to the extent of at least 3,000 airframes, Junkers considered a license fee of 22 million Reichsmarks, or 31.8% of the development costs, to be appropriate. In the case of the Ju 390, the company calculated that it would be satisfied with 27 million RM for production of about 1,000 aircraft, or just 28.8% of the development cost of 93.65 million. For the Jumo 222 Junkers estimated 18 million Reichsmarks, or 42.8% of the 42.1 million RM development cost, based on a total of 6,000 power plants. Percentage wise, the replication rights for the FA 15 were the most expensive of all, on account of the device's almost unlimited potential uses and applications, at three million Reichsmarks, or 57.7% of the development cost of 5.2 million. It was assumed that the Japanese would build at least 4,500 of these devices.

Surviving German records do not allow us to say what became of these ambitious plans in the confusion of war, or whether the delivered drawing sets for the Ju 388 and Ju 390 ever reached Japan. At least documentation and test articles for the Jumo 004 jet engine were delivered at the turn of the year 1944-45 and paid for (20 million RM).[962]

Hostilities in the Far East continued after the end of the war in Europe, and naturally, the Allies were keenly interested in all aspects of German-Japanese cooperation. They learned that drawings for the Junkers G38, two Ju 87s, one Ju 88 A-5, and three Ju 88 A-4s had been delivered to the Japanese. They also discovered that cooperation had extended to drawings of the Ju 388 L-1 and K-3, drawings for the Ju 488, the FA 15, and general data and drawings concerning the 8-635 and 8-163. Their investigations indicated that the transport by submarine envisaged for February 1945 did not take place.[963]

In view of the very real threat posed by the Boeing B-29, which would ultimately drop atomic bombs on Hiroshima and Nagasaki, it is rather surprising that the Japanese leaders were more interested in the bomber version of the Ju 388, and less in the high-altitude heavy fighter. One explanation could be that the Japanese command was increasingly losing touch with reality.

What were the German motives in offering the Japanese their latest and most secret aircraft? Here one must clearly differentiate between the motives of the state offices and the Junkers Flugzeug und Motorenwerke. The failure of its latest product to go into production in Germany, as fallout from the military situation, was a financial disaster for Junkers. It was hoped that by issuing a license to the Japanese the company could achieve at least part of the income expected from the project. For their part, with the end of the war in sight, the state offices wanted to transfer as much technology as possible to Japan, and support its ally in its struggle against the common foe America. Mass production of all the German types offered to Japan would surely have resulted in serious losses to the Americans, however, it is highly unlikely that this could have changed the course of history.

Chapter 11: Captured Aircraft Tested by the Victorious Powers

"Hour Zero"

Immediately after the cessation of hostilities, Allied forces advancing into various parts of the (former) German Reich came upon the remains of the Ju 388 production program, which had been a high priority project until January 1945.

Usually it was advancing combat troops that came upon former German airfields and found the aircraft. Immediately behind them—but sometimes even before them—came special intelligence teams. They drove or flew in small liaison aircraft, some captured from their erstwhile enemy, to locations identified as being of special interest to conduct an initial technical analysis of the equipment and research being carried out there. Documents were secured, especially at production and research sites, and hauled away in trucks. JFM reports for 1945-46 contain long lists of documents that were officially surrendered.

It is notable that a not insignificant number of aircraft were discovered in relatively intact condition, at least to the naked eye. Locations included Altenburg, and possibly also (Bad) Langensalza in Thuringia.

CIOS teams reached the dispersed Heinkel-Hirth operation in Friedrichshall on 13 April 1945, after the area was occupied by the U.S. Army. Since mid-1944 it had been housed in a salt mine in Kochendorf that had been adapted by the Todt Organization.[964] Flight Lieutenant Adderley and Major Saddler, MAP, of the USAAF, and Lieutenant Robinson of the USNR discovered evidence of development and test work being conducted there, and indications of production of exhaust driven turbosuperchargers for the BMW 801 TJ. A ten man British team[965] visited the Bayerischen Motorenwerke plants in Oberwiesenfeld, Munich, and Allach between 24 July and 14 August 1945. In Oberwiesenfeld the team found several BMW

This imposing scene greeted the advancing American 6th Army. At least eight seemingly intact Ju 388 L-1s are visible in this photograph, which was allegedly taken in Langensalza. In the background is a Willys Overland Jeep and an M8 armored car. Rosch Collection

A GI poses beneath the propeller and leaning against the fin of Ju 388 WNr. 340 284, which was in Altenburg at the end of the war. The aircraft, whose jettisonable canopy section lies on the ground in front of the narrow gauge railroad track, may have been one of the last Ju 388s built by ATG. Crow Collection

801 mockups with various exhaust driven turbosuperchargers. It was assumed that production took place in Allach.

CIOS intelligence teams made a very detailed examination of JFM production sites[966] before the areas in which they were situated were turned over to the Soviet Union as per the terms of the Yalta agreement. Among those interrogated in Dessau were managing director Thiedemann, director Zindel, and Prof. Baade. While the intelligence teams were mainly interested in jet engine research, the report also mentioned the condition of the factories and the aircraft found there that were flyable or under construction.

No hints of the Ju 388 were found in Klein or Grofl Schierstedt, Ballenstedt, Egeln, Magdeburg, or Nordhausen. Only in Halberstadt were there clues to participation in the production of Ju 388 night fighters. The Allied interviewer also inquired about cooperation with and transfer of know how concerning jet engines to Japan. The leading Junkers engineers vehemently denied that this had taken place.[967]

Reports about the effectiveness of the bombing campaign also provide clues as to the condition of the factories and production there. Between 4 and 18 May 1945 a report was compiled entitled U.S. Strategic Bombing Survey Aircraft Division Airframe Plant

Report No. 1 Junkersflugzeuge und Motorenwerk Dessau.[968] The production figures offered for the JFM company and its licensees were impressive: Ju 88 (15,183), Ju 87 (5,709), Ju 52 (4,845), and Ju 188 (1,237), for a total of 26,974. The company also produced large numbers of engines: Jumo 211 (66,892), Jumo 213 (9,322), Jumo 004 (5,807), and Jumo 222 (286).

The report contains a partial list of the numerous attacks by the RAF and 8th Air Force on the various factories (tonnage of bombs dropped, numbers of aircraft committed, and casualties). It becomes apparent that the company was a frequent target, especially in the summer of 1944. The heavy raid on Dessau on 7 March 1945, for example, is not included (8th or 15th Air Force), thus the report is incomplete in this regard. Cooperation with Japan is again discussed. Dr. Frydag admitted that significant difficulties had been encountered in production of the Ju 388. The progressive destruction of the transportation system seriously hampered the production effort, as 95% of production was transported by rail to other factories. Another hindrance was the fuel shortage, which made transport by road impossible. Completed aircraft often sat for up to three weeks before they could be test flown. Toward the end of the war test flying came to a complete halt, causing deliveries to the units to come to a standstill.[969]

The report offers the following figures for production of the Ju 388:

Ju 388	**July 44**	**Aug.44**	**Sept. 44**	**Oct. 44**	**Nov. 44**	**Dec. 44**	**Jan. 45**	**Feb. 45**	**March 45**
authorized	1	4	6	14	17	29	41	0	0
actual	1	**4**	**3**	**5**	**20**	**24**	**8**	**8**	**0**

Although photographs confirm the presence of a number of aircraft or their remains in other places—for example, Kaltenkirchen or Kölleda (the ATG modification facility for Ju 388 production aircraft)—it is surprising that so few of the 75 to 100 more or less completed aircraft survived the end of the war. The intelligence teams obviously had trouble finding flyable aircraft that could be shipped home for testing.

In Denmark and Norway, where many long range reconnaissance aircraft continued operating until the end of the war, no Ju 388s were found, while other new aircraft types, such as the He 219 and Ar 234, were not uncommon.[970]

"Trophies" of the Soviet Union

There is contradictory information about Ju 388s captured by the Soviet Union. Although Soviet forces advanced right into central and eastern Germany, where most Ju 388s had been built, it seems that they captured just one or two Ju 388 L-1s. Unfortunately, neither the locations where the machines were captured nor their *Werknummern* are known. The Soviets should have discovered examples of the aircraft when they entered Silesia, in Liegnitz, for example. They would almost had to have found examples of the Ju 388, either complete or under construction, in the central region of Germany (Rechlin, Leipzig, Altenburg, Dessau, Merseburg, etc),

Blown up aircraft at Heidkaten, near Kaltenkirchen. Two Ju 388 vertical stabilizers are clearly visible. The one on the left, finished in the "striped" scheme typically seen on ATG built aircraft, is most likely WNr. 340 119. On the right is the tail of a Weser built machine, WNr. 560 045. The aircraft allegedly belonged to the OKL's experimental unit at the end of the war. Hampel Collection

either in the part they conquered themselves, or the part surrendered to them by the Americans several weeks later. No reliable reports of this have surfaced, however, and we will probably have to wait for further research results from the archives in the east.

There is, however, an observation from early September 1946 of a damaged Ju 388 (30%) sitting at the west end of Rechlin airfield, which had been part of the Soviet occupation zone since July 1945. The engines showed no external signs of damage, and there were also three more BMW 801 TJs on transport wagons.[971]

After the war the Junkers design bureau initially continued work on a number of projects (such as the EF 126, Ju 287, EF 131, etc) in Germany under the Soviet Military Administration (SAD) as the Soviet Stock Corporation (SAG), and later as OKB-1 Dessau.[972] In October 1946 the design bureau was moved to the Soviet Union. The Ju 388 appears to have aroused no great interest as an object of study. Numerous reports were compiled for the SMAD about developments by the German aviation industry, more than 200 about engine developments alone.[973] Among them was "Report No. 337, Interim Project Documents - Ju 388 Jet Engine Test Bed for 004 and 012 Devices." The voluminous report on "German Aviation Research in the Year 1945," written by the scientific director of the German Aviation Research Institute,[974] Prof. Dr. Bock, who surely must have had a good knowledge of the latest *Luftwaffe* developments, makes no mention of the Ju 388.

The following information appeared in the activity reports of the "Technical Plant Management" of JFM Dessau for the Soviet occupation until the summer of 1946:

1. April 1946 "Flight testing Ju 388: 3 pressurized cockpits and one aircraft fuselage transported from Merseburg and Altenburg."
2. In the month of June, in addition to numerous other former JFM projects still being worked on in Dessau under Soviet supervision, there is again mention of the Ju 388: "Delivery of Components". An airfield service was again set up, and work to restore the airfield to operation intensified. In fact the final efforts to restore the airfield were in vain, for in October 1946 the entire work force was about to be deported to continue production and, in particular, flight testing in the USSR.
3. In July 1946 the manufacturing section submitted a report on the state of the remaining Ju 388s. The fact that manufacturing section reported on progress with the Ju 388 suggests that a great deal of work was necessary to make one of the aircraft flyable again.

It is therefore not improbable that an aircraft was sent to the NII VVS test center in March 1946. As well, a Ju 388 is said to have been used in testing related to the EF 126 project. After the surrender Junkers continued working on prototypes of the EF 126 on behalf of SMAD.[975]

Apart from a few published photographs, there is only indirect information that one or more Ju 388s were used in various test programs related to other types whose development continued under Soviet supervision, or which were at least examined by the Soviets.

A Ju 388 is said, however, to have taken part in an impressive demonstration of captured German aircraft held on 18 August 1945. The Ju 388 was part of the flying demonstration, and was (of course) one of the aircraft forced down by Soviet fighters.[976]

In addition, there are photographs[977] showing three aspects of a Ju 388 L-1 marked with red stars. The machine displays all the features of an early reconnaissance aircraft, such as BMW 801 TJ-0 engines without external ram air intakes, fuselage with bulged bomb bay, cameras, FuG 101 beneath, and FuG 217 above the wings. The most noteworthy thing about the photographs is the movable antenna for the FuG 200 that is visible beneath the cockpit. In all likelihood the Soviets captured Ju 388 L-1 WNr. 340 089 (RT+KI), which was equipped with this antenna, at Werneuchen.[978]

Another source from October 1946 that indicates the Soviet occupation forces' interest in the Ju 388[979] is the Junkers report "State of Junkers Research at Time of Takeover by the Soviet Mission in 1945." Like the reports compiled by the western powers, it contains a description of the destruction inflicted on the various plants:

"The main Dessau plant has been very badly damaged by bombing. Attacks on Dessau and other plants came rather late (1944), however, and by then dispersed operations were almost all up and running, allowing design and production work to continue. The wind tunnel at Dessau was completely destroyed. Production of prototypes could not be resumed and was partially continued at Bernburg. In contrast, the destruction in the engine works at Dessau was not quite as bad. Some test stands were back in operation, and the design bureau in particular was able to work again.

The Bernburg plant was largely destroyed, however production in the dispersed facilities reached a high level and was roughly equivalent to what it had been in 1942, with 165 Ju 88s and Ju 188s per month."

The activities under American occupation were described as follows:

"The few completed types that had not been destroyed were taken away by the Americans, and a number of the leading engineers, especially the turbine builder Dr. Franz, were kept under watch, and were later forced to go to the USA."

Based on their report, it seemed that it was particularly worthy of mention to the Russian interrogators that,

"[...] the Dessau plant [has] many years of experience in the building of pressurized cockpits (Ju 49, Ju 86 P, Ju 288). The pressurized cockpit for the Ju 288, designed for a multi-man crew, may be the most highly developed product of this type by the German aircraft industry."

Finally, they made a list of the aircraft types in which the Soviets might be interested, including the Ju 188, Ju 288, and the Ju 388 J-1 and L-1:

"As for piston-engined aircraft, the Ju 188 and Ju 388 may be of interest for study purposes. One or another type of the Ju 88 may have fallen into the hands of the Soviet Union in the east. A Ju 388 could easily be made flyable using salvaged components and completed engine parts."

The following technical data was provided for the Ju 388 with Jumo 213 engines:

"Data: with Jumo 213 E with 2 x 1,740 takeoff power, gross weight 13 700 kg, landing weight 12 000 kg, maximum speed at 0.3 km altitude 375 kph, at altitude of 6 km 480 kph, at 12 km 530 kph, takeoff run 900 m, 1 300 m to height of 20 m, range 2 000 to 2 500 km. Service ceiling 12 800 m. Extensive documentation (operating manuals) and photographs are available for these two types."

In fact, the high-altitude aircraft found a purpose. During the course of testing and development of the DFS 346.1 to 346.3 (D) as the Samolyot 346 P[980] in the Soviet Union there was a requirement for a high-altitude aircraft capable of carrying the aircraft to the necessary altitude before its rocket engine was started. The abovementioned Ju 388 was used for this purpose. A total of 24 flights are believed to have been made in the Soviet Union.[981] The DFS prototypes carried by the Ju 388 were flown by *Herr* Ziese, an experienced Siebel test pilot, and weighed between 1880 and 2180 kg.

One poor quality photograph that has been published many times purports to show the aircraft used to carry the DFS high-altitude machine. The Ju 388 in the photo does not have the bulged bomb bay. A close examination of the photographs reveals that the aircraft is similar to RT+KE in every visible respect—with the exception of the bulged bomb bay, and the red star on the tail. If this photo is authentic, it would prove that the planned conversion of RT+KE to the "lightened condition without bulged bomb bay" was in fact carried out. After its use as a "mother ship" all trace of the aircraft vanishes.

The conclusion of the Ju 388's role of "flying laboratory" in the late 1940s was its use in testing of the Type 150 hydro-mechanical control system. This system, the first development by the interned German experts that was not based on a wartime Junkers project, reached the final assembly stage in 1951. It did not fly until October 1952, however, and was ultimately used in preliminary experiments leading to the Baade 152, a civilian aircraft designed and built in the DDR (East Germany).

Unless there is a surprise waiting in one of the CIS states, we will have to assume that none of the "trophies" survived. Interestingly, however, the operating manual dated December 1944 which the Soviets brought from Dessau in October 1946 has survived.

Great Britain

The British captured the Ju 388 V6 at the Armament Testing Station E6 in Tarnewitz, and assigned it the Air Ministry number AM83.

The captured aircraft was obviously also examined by the USAAF Air Intelligence Team of the 2nd Tactical Air Force. On 21 May it compiled a brief report (No. 143)[982] that described the significant technical characteristics of the type, indicating an interest in the type:

Ju 388 L76 – with BMW 801 TJ-0 engines in very good condition. This aircraft is fitted with a small rear revolving turret, in which are mountings for two MG 131s (one above the other) remotely controlled and fired from the cockpit. This turret is marked FHL Z/1-B3, Fl.204309, Hersteller Rft. In the rear end of the cockpit is the control lever and next to it is a large periscope sight which projects above the fuselage, where it is covered by a glass bowl, and below

Three American soldiers pose with WNr. 340 307 in an ATG hangar in Leipzig-Mockau. Petrick Collection

Two more aircraft abandoned at Altenburg. In the foreground is WNr 340 116; the number "05" is visible on the tail of the aircraft in the background, thus it could be WNr. 340 105. Crow Collection

the fuselage, forward of the bomb doors, where it is faired in sheet metal. This gives an all-round view. This is marked PVE 11 A-1 (Fl52779). The aircraft is fitted with FuG 217 and FuG 214 (three vertical aerials below each mainplane) and a blister bomb bay below the fuselage. No bomb racks are fitted. In the rear end of bomb bays are mountings for two cameras, apparently of low type, as the space between mountings and fuel tank is approximately 15". It has been arranged that this aircraft be guarded and arrangements made to have it flown to an airfield occupied by the R.A.F. for safe custody.

On 30 May 1945 Air Technical Intelligence, Headquarters, 2nd T.A.F. (M) of the Royal Air Force compiled another, more detailed report about the aircraft captured in flyable condition. It includes twelve photographs of the aircraft already wearing British insignia, with a large white "6" on the tail. The remotely controlled tail barbette and periscope system, identified as "FHL 131 Z/1 B-3" in combination with "PVE 11 A-1," obviously sparked much interest. Overall, the report suggests that the ATI technicians had a good knowledge of the Ju 188, with which the Ju 388 was compared in the report. As expected, neither a bombsight nor bomb release equipment was present.[983]

The smoldering remains of a Ju 388 L-1 at Merseburg, apparently set on fire by retreating German forces. Taken by Robert L. Tremblay (small picture), it shows how little time there was between the end of hostilities and the arrival of logistical units of the U.S. Army. Tremblay's unit had the job of clearing away wrecked aircraft with bulldozers so as to allow American units to use German airfields. Tremblay Collection

Another photo of Merseburg airfield at the end of the war. At least four Ju 388s can be seen, though the three in blast pens appear to be no more than wrecks. The destruction of the crew rooms suggests that they were blown up by the Germans themselves. In the foreground covered with netting is WNr. 560 044. It appears to be relatively intact, the only obvious damage being to the undercarriage. Rosch Collection

The Allies were obviously interested in the Ju 388's pressurized cockpit. Here one is seen next to a Me 163 B rocket powered fighter, which has been made ready for transport. The photo was possibly taken in Brandis or Merseburg. Haubner/Misna Collection

Ju 388 V6 as a war prize in British markings. All of the prototypes built at Dessau and the V8 were finished in the old splinter scheme. Petrick Collection

The Ju 388 V6 at the captured aircraft display in Farnborough. In the background is a Fw 200, while the tail of a He 219 is visible on the right. Petrick Collection

As Tarnewitz was in an area that was to be handed over to the Soviets, in June the V6 was flown to Lübeck/Blankensee. First, however, the necessary 100-octane fuel had to be brought in. The German pilot Gatzmeier was brought back from the POW camp near Gravesmühlen for the flight. He and his mechanic were not allowed to wear parachutes during the flight, or exceed certain speeds and altitudes. As well, as a precaution the aircraft was given an escort of Spitfires in spite of its British roundels.[984]

In August a British pilot flew the aircraft from Schleswig to Matching. The next stop was Farnborough, where it arrived on 1 September. It formed part of the big display of captured enemy equipment in October-November 1945, and remained with the General Duties Flight from September to November 1945, and until April 1946. Transferred to the College of Aeronautics in Cranfield, Bedfordshire, it served as an object of study until finally being scrapped in 1948. The starboard engine gondola, undercarriage, and the tailwheel installation were, however, preserved there until 1967.[985]

Curiously, few accounts of testing by the Royal Aircraft Establishment (RAE) have appeared to date, not even in recently published histories of the testing establishment, although the aircraft appears in numerous photos.

USA

Although a number of machines are known to have fallen into Allied hands, it appears that just one aircraft survived the turmoil of the terrible war to the present day. For some time it has been located in the depot of the National Air and Space Museum (NASM) in Silver Hill, Maryland. The story of its capture by the Allies is closely tied to the well known USAAF intelligence team of Colonel Watson. In an undertaking dubbed "Operation Lusty" the unit, which was dubbed "Watson's Whizzers," more or less systematically "collected" advanced German aircraft, even during the final, often chaotic stages of the war. One of the teams moved from Halle/Saale to Merseburg, where it confiscated a Ju 388 L-1 with WNr. 560 049. Members of the 99th Air Service Squadron of the 9th Air Force's 365th Fighter Group were brought in from Fritzlar to inspect the aircraft and, if possible, make it ready to fly.

Crew chief Tommy Haworth and his men found the Ju 388 fully fueled, and on jacks in a hangar. Fearing booby traps, the aircraft was inspected with extreme caution and thoroughness. Every access hatch was opened, and the men looked behind every removable panel for explosives. The aircraft was jacked up to test the undercarriage, and the landing flaps were also checked. The air-

The captured WNr. 560 049 at Merseburg, still in German markings. The tranquility of the scene is especially noteworthy. American soldiers rest in the shadow of the aircraft's wings. Note the black *Werknummer* and black outline swastika on the tail, a typical feature of aircraft built by Weser. Rosch Collection

Front view of the same aircraft—two GIs are still enjoying the shade. Rosch Collection

American technicians thoroughly inspected WNr. 560 049, and restored it to flying condition. A truck serves as a work platform for the men servicing the engine. Bettag Collection

Above: The aircraft was carefully searched for booby traps, then Thomas Haworth started the starboard engine. Here it belches smoke, while the port engine remains stationary. Right: Herbert Bettag, a metal aircraft builder, strikes a victor's pose by the tail. The last number of the *Werknummer*, painted in white, is impossible to identify. As all other photographs of WNr. 560 049 show a back serial number, this is most likely another Weser built machine in front of one of the hangars at Merseburg. Bettag Collection

craft was obviously not yet fully equipped, for no cameras were found in the bomb bay. Before the Ju 388 could be flown, the cockpit had to be given a thorough inspection. Though Haworth had an interpreter, the man had no knowledge of aircraft. Every time he flipped a switch or pressed a button, Haworth feared that the aircraft would explode. Finally he felt safe enough to start the engines. Engine temperatures quickly reached the critical point, and after being run for five minutes they had to be shut down again.

The aircraft was test flown by Allied crews at Merseburg before the decision was made to fly it to Kassel-Waldau on 20 May. As the aircraft still bore German insignia, the aircraft was given an escort of three P-38s. Flown by an American crew, which included a Major, the aircraft arrived at 1530 hours. Watson himself may have been on board.

Original plans by Watson's Whizzers called for the aircraft to be flown via Prestwick to Iceland, after which it would cross the North Atlantic by way of Greenland, ultimately arriving at Wright Field. In fact, it was transported to the USA aboard the aircraft carrier HMS Reaper, together with a number of German aircraft considered interesting (Me 262, Ar 234, Do 335, He 219, Ta 152, etc), in an operation dubbed "Sea Horse." Like most of the other aircraft involved in the operation, the Ju 388 was flown to Cherbourg/Querqueville on 17 June. The sea voyage began on 20 July 1945 and lasted eight days. All of the captured aircraft were unloaded in Newark, New Jersey, from where they were flown or shipped by rail to Freeman Field.

The captured Ju 388 L-1 was initially assigned the code FE-4010, which was soon changed to T-2-4010. Captured equipment

Both engines are running. Note the removed canopy section, and the two men standing in the opening. A Major was also in the cockpit for the photo, but to be on the safe side Haworth carried out the first engine start himself. Bettag Collection

This group photo in front of the maintenance hangar in Merseburg shows the entire technical crew in front of the repaired aircraft (from left to right): Corporal Tilton, Sergeant Houser, Sergeant Stagmeyer, Corporal Mcantee, Master Sergeant Pursley, Sergeant Neiles, Corporal Ralph Mackey and Sergeant Alden (standing), plus Staff Sergeant Williams, Staff Sergeant Bettag, and Tech. Sergeant Thomas Haworth, commanding the team. The hangar is a typical *Luftwaffe* maintenance hangar with convex roof, examples of which may still be seen in Brandis and Grossenhain, for example. Bettag Collection

The aircraft was ferried to Kassel-Waldau in German markings with an escort of P-38 fighters. By the time these photos were taken, the aircraft had been given temporary American markings. At which stage of the transfer the photos were taken is not known. Crow Collection

WNr. 560 049 in inaccurate German markings applied for testing in the USA. Petrick Collection

The aircraft during a tire change at St. Dizier. Le Fana de l'Aviation

in the U.S. armed forces was either given the identifying designation "EB" for Evaluation Branch, "FE" for Foreign Equipment, or "T2" for T-2 Office of Air Force Intelligence.

The latter compiled a detailed testing report on the aircraft, which was overwhelmingly positive in its findings. In terms of handling, the aircraft seemed equal to comparable American reconnaissance aircraft. The BMW 801 TJ high-altitude engines received favorable comment, including with respect to reliability. The only negative aspect was ease of maintenance, as a result of its compactness. This potential disadvantage was offset, however, by the fact that an engine could be replaced by just three men in only 1.5 hours. The report even went so far as to recommend adoption of the typical 801 control system for comparable American engines. The hydraulically operated tail barbette was viewed as a novelty, as comparable Allied systems were electric. The testing on which the report was based was carried out at Freeman Field. The aircraft was even demonstrated to the press there in September 1945. On 30 September of the same year it went to Wright Field, from where it returned to Freeman in July 1946, and was maintained in flyable condition. After the end of its active flying career it was supposed to be stored at Davis-Monthan AFB, but in September 1946 it was in fact placed in storage in Park Ridge, Illinois. 560049 was placed next to its rival (He 219) in a large hangar where C-54 transports had previously been built. This undoubtedly saved it from the fate suffered by many captured *Luftwaffe* aircraft, namely of being left in the open to rot. From there it was taken to the NASM's facility in Silver Hill, where it remains to this day, broken down into major subassemblies, waiting for restoration. As the aircraft is still in relatively good shape, it is to be put on display unrestored in the NASM extension in Dulles International Airport.

This photograph was probably taken at a display—if not, it is difficult to explain the people sitting on the hangar roof in the background. Petrick Collection

Two views of the aircraft at the display of captured equipment in Dayton. Petrick Collection

Above and right: The last stop for the aircraft before storage in Silver Hill was a hangar at O'Hare Airport in Chicago. These photos were taken before the aircraft was placed in storage. When the Korean War broke out the hangar was required for other purposes, and the aircraft was moved to Silver Hill. Crow Collection

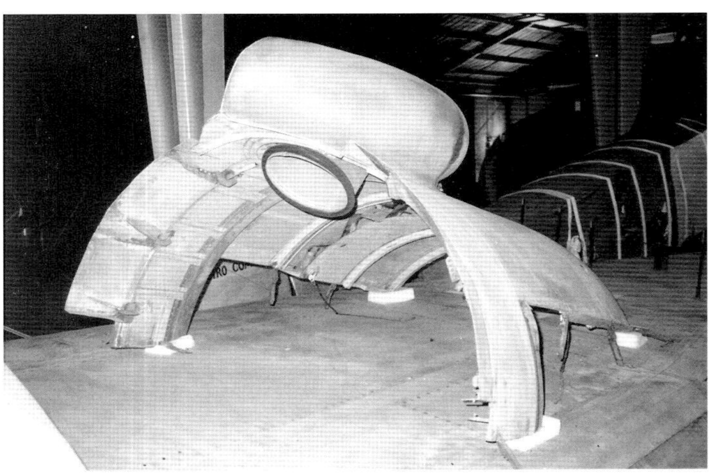

Close up of the lower engine nacelle section with cooling gills, and the external ram air intake of the BMW 801 TJ on the NASM aircraft. Handig Collection

The American owners kept a record on the fin and rudder of Ju 388 560 049. Thus, today at Silver Hill the original black *Werknummer* applied by Weser is visible, as is the small red FE-4010 and the later captured aircraft code T-2-4010. Handig Collection

Chapter 12: Résumé

The Junkers Ju 388 represented both the conclusion and apex of the entire Ju 88 product palette. A process of constant development and improvement of the original Ju 88 resulted in a propeller driven aircraft with conventional engines that represented the state of the art and which, according to American testing reports, was at least equal to comparable American developments. The only reservations about the Ju 388's operational suitability concerned the high maintenance requirements of its BMW 801 TJ-0 engines.

The disastrous consequences of the total absence of aerial reconnaissance were recognized, at the latest, during and after the invasion of Normandy. The idea of countering the enemy's superiority in the air with a superior aircraft that would allow photo reconnaissance missions to be flown was entirely correct, and resulted not just in the Ju 388, but the Ar 234, as well. The Ar 234 even achieved a number of operational successes, something the Ju 388 never had the chance to do, as just two aircraft are known to have reached frontline units. Most Ju 388s were used merely as test beds for the highly complex and finicky BMW 801 TJ power plants.

Apart from a pressurized cockpit and remotely controlled defensive armament, the Ju 388 offered no significant advantages over the Ju 88 G-6 and G-7, which represented a mixture of the Ju 88 and Ju 188. The offensive armament was almost identical, and there was little difference in the airborne radar carried. When the B-29 failed to appear in the skies over Europe, justification for production of the Ju 388 J fell visibly. The catastrophic military and economic situation ultimately led to its cancellation in favor of the Ta 152 and Ju 88 G.

The failure of "Operation Steinbock" made the hopelessness of bomber operations plain for all to see, and so in the autumn of 1944 the K-series was dropped in favor of fighter production. The J-1 and L-1 series followed in February 1945.

The small number of aircraft actually built is striking. Approximately 75 individual aircraft can be identified, while about another two dozen were at least in an advanced state of construction. Development costs were nonetheless enormous. The limited number of completed aircraft can probably be traced to the change in delivery plans in favor of fighter aircraft, and problems with the BMW 801 TJ played a not insignificant role.

The authors are conscious of the fact that, in spite of years of research, we have surely not been able to view all the surviving documents concerning the Ju 388. New, previously unknown facts about the Ju 388 will continue to surface, and these may shed a different light on some aspects of this type. As there is currently not a single book on the Ju 388, in contrast to many other late aircraft developments (which were therefore insignificant to the outcome of the war), we have decided to document the state of our private research in this book. The Ju 388 theme is thus far from closed for us. We are still interested in information of any kind concerning the Ju 388, and we will use our website (http://Ju388.de) to publish new information and corrections. Please use the method that best suits you to contact us if you have additional information for us on the Ju 388, or if you wish to convey your impressions of the book.

Appendix 1: List of *Werknummern* and Manufacturers' Codes

The following is a list of known aircraft, arranged by *Werknummer*. A considerable number of aircraft have been identified from documents, an indication of the aircraft's significance, and the efforts made to push the program from prototypes, to pre-production, to full production.

A number of questions remain, however. Many references are contradictory, and some *Werknummern* found in company albums and photo captions neither appear in any of the documents, nor fit into any of the *Werknummer*[988] or manufacturer's code blocks so far identified.

The gaps in the following list may have been caused by enemy action or other production interruptions. Codes in *italics* indicate that the code has not yet been confirmed by logbook entries or in connection to the *Werknummer*, but rather, that it is the result of extrapolation from verifiable manufacturer's codes. Letters that were easily confused with one another, such as I/J and O/Q, were often avoided in these codes, and this creates further difficulties.

Werknummer Manufacturer's Code	Variant Prototype	Remarks/Use/Fate	Sources
230 151 to 160 KS+TA to KS+TJ	—	Ju 388 K pre-production series, built using major components of the Ju 188	—
230 151	Ju 388 K-0	First Ju 388 K-0 pre-production machine. Maiden flight at Merseburg on 29 May 1944 by pilot Lehmann and flight engineer Lange. Additional flights by Lehmann from Merseburg: 1 June 1944: Safety check flight 10 June 1944: Radio test flight 13 June 1944: Autopilot test flight 14 June 1944: Follow up autopilot test flight 16 June 1944: Follow up radio test flight 23 June 1944: Follow up radio test flight 26 June 1944: Two follow up radio test flights 28 June 1944: Follow up radio test flight 7 July 1944: Autopilot test flight On 8 July 1944 Lehmann ferried the aircraft to Dessau, where it was used as a test bed by the flight department. Several test flights from Prague. Retrofitted with external ram air intakes in November 1944, but no honeycomb filters in front of the intakes. Repeated engine problems in December.	BMW short report No. 40 via field office in Dessau.
		Port engine 385 185, starboard engine 385 172 (September 1944) Port engine 323 044, starboard engine 385 172 (7:37 and 41:17 hr, 24 Nov. 1944) Port engine 323 044, starboard engine 385 172 (9:26 and 43 hr, 9 Dec. 1944)	BMW short reports Nos. 29, 40, and 41
230 152	Ju 388 K-0 Ju 488 V401	The major components of this machine were used to build the first prototypes of the Ju 488 in Toulouse.	Amendment IV to construction directive 7 388 26 10/000 of 12 Aug. 1944

Werknummer Manufacturer's Code	Variant Prototype	Remarks/Use/Fate	Sources
230 153 KS+TC	Ju 388 K-0	Probably completed at Merseburg in June 1944. At 1730 hours on 6 July 1944 radio check flight from Merseburg by pilot Lehmann, accompanied by Reimann. FuG 217 check flight accompanied by Kuhnt on evening of 7 July, also from Merseburg.	Lehmann logbook, flight numbers 4307 and 4310.
		From late July 1944 test bed with EKdo 388 in Rechlin. Flown from Merseburg to Rechlin by deputy type supervisor Alfred Kuhn on 25 July 1944; the flight lasted from 1545 to 1637 hours. Dark camouflage missing during initial inspection (Findings, Test Report 443/44), 4:30 hr flying time. Three-blade VDM propellers (9-12022.21).	Kuhn logbook Flight No. 1643
		Flights by Oblt. Eisermann in this machine at Rechlin: 19 Sept. 1944 from 1002 to 1010 hours (No. 657) 20 Sept. 1944 from 1432 to 1632 hours (No. 662) 21 Sept. from 0906 to 1112 hours (No. 664) 20 Oct. 1944 from 1421 to 1442 hours (No. 700) November-December 1944 to EKdo for radio tests Port engine 385 215, starboard engine 385 226 (July 1944)	Oblt. Eisermann logbook
		Propellers: WNr. 05497 (port) and WNr. 05471 (starboard), Port engine 385 215, starboard engine 385 226 (16 hr, October 1944).	BMW-Königsberg No. 33
230 154	Ju 388 K-0 Ju 488 V402	Like those of WNr. 230 152, the major components of this machine were shipped to Toulouse to be used in creating "two flying mockups of the special version of the Ju 188".	Amendment IV to construction directive 7 388 26 10/000 of 1 Aug. 1944
230 155 KS+TE	Ju 388 K-0	First flight at Merseburg on 24 June 1944, pilot Lehmann. Also served as radio check flight. Two high altitude flights from there, also by Lehmann, on 27 and 28 June 1944.	Lehmann logbook
		Initial inspection in Rechlin on 31 July 1944. Airframe had 7:40 hr, engines 9:40 hr. Three-blade VDM propellers (9-12022.21) On 22 Aug. 1944 at Rechlin flight to 6500 m with four SC 250 by *Fl.Oberstabsingenieur* Walter Ballerstedt from 1658 to 1745 hours.	Rechlin Inspection Report No. 451/44
		Sustained 70% damage in a raid on Rechlin on 25 Aug. 1944.	Preuschen notes
		Port engine 385 263, starboard engine 385 261 (31 July 1944) Propellers: WNr. 05503 (port) and WNr. 05505 (starboard)	Rechlin Inspection Report No. 451/44
230 156 KS+TF	Ju 388 K-0	Originally K-0, later converted into L-0 *Nacht*. Also identified as L-1 in company album.	
		On 22 July 1944 two flights from Merseburg by crew of Lehmann and Lange, first a high altitude flight, then a speed course flight. On 1 August 1944 at Merseburg, Lehmann carried out a familiarization flight for pilot Holtzschke.	Lehmann logbook.
		External ram air intakes retrofitted in Dessau, and was supposed to be the first aircraft issued to users. See Chapter VIII.3.	BMW report via field office in Dessau.
		Engines 385 275 and 385 274 (8 1/2 and 71/2 hours, 9 Dec. 1944)	BMW Brief Report No. 40, 41

Werknummer Manufacturer's Code	Variant Prototype	Remarks/Use/Fate	Sources
230 157 KS+TG	Ju 388 K-0	Envisaged as test bed for Jumo 213 E, and prototype for the Ju 388 J-3 series. Handover to flight operations planned for 12 Aug. 1944.[989] Maiden flight at Merseburg on 17 August 1944 by pilot Lehmann. Radio and autopilot check flight on 18 August (pilot Lehmann). Ferried from Merseburg to Dessau on 21 August 1944, again by Lehmann.	Amendment IV to construction directive 7 388 26 10/000 of 1 Aug. 1944 Lehmann logbook.
230 158 KS+TH	Ju 388 K-0	First flight at Merseburg on 25 July 1944, pilot Lehmann. Also served as radio check flight. Follow up radio test flight and autopilot check flight (pilot Lehmann) on 27 July 1944. Performance check (closed course flight) by crew of Lehmann and Lange on 28 July 1944, maximum speed 405 kph at ground level.[990] Next day a high altitude flight with Lehmann as pilot. In operation memo Fp.-M 83/44 by the Merseburg production inspector the machine is described as undergoing the necessary modification work, on 11 Aug. 1944 it was parked at Merseburg without engines.[989]	Lehmann logbook.
230 159 KS+TI	Ju 388 K-0	Parked at Merseburg without engines on 11 Aug. 1944.[989]	
230 160 KS+TJ	Ju 388 K-0	Parked at Merseburg without engines on 11 Aug. 1944.[989]	
230 691 to 230 700	Ju 388 K-0	Second batch of Ju 388 K pre-production aircraft, no known photos.	
230 691 KS+TK	Ju 388 K-0 V22	According to Amendment IV to Construction Directive 7 388 26 10/000 dated 1 August 1944, aircraft envisaged as test bed for the Jumo 213 E and prototype for the Ju 388 J-3 series. Final assembly was supposed to begin on 12 Aug. 1944.[989]	
230 692 KS+TL	Ju 388 K-0	Construction at an advanced stage in August 1944.[989]	
230 693 KS+TM	Ju 388 K-0	Completed as Ju 388 K-0 at Merseburg, first flight on 18 October 1944 by pilot Lehmann; the flight also served as radio and autopilot check flight. Follow up radio and autopilot test flight on 20 October, further autopilot test flight on same day. High altitude flights from Merseburg by Lehmann on 21 and 29 October and 3 November 1944 (accompanied by Kerksiek). Factory flight at Merseburg on 5 Nov. 1944.	Lehmann logbook. Schreiber logbook.
230 694 KS+TN	Ju 388 K-0	Completed as Ju 388 K-0 at Merseburg, first flight on 29 October 1944 by pilot Lehmann; the flight also served as radio and autopilot check flight. Factory flight at Merseburg on 2 Nov. 1944. Appears in logbook as K 14, meaning 14th K-0. Follow up autopilot test flights by Lehmann on 13, 14, and 16 November 1944.	Lehmann logbook. Schreiber logbook. Lehmann logbook
230 695 KS+TO	Ju 388 K-0 V9	Mentioned in Junkers offer No. 3766 of 28 Nov. 1944 for conversion to Jumo 222 A/B engines. Engine test bed, probably to Ju 388 L-0 standard without bomb bay. Completion was planned for January 1945, depending on engine availability; delivery to KdE.	

Werknummer Manufacturer's Code	Variant Prototype	Remarks/Use/Fate	Sources
		On 25 Nov. 1944, however, the aircraft was heavily damaged in Hangar II during an attack on Merseburg.[991]	
230 696 KS+TP	Ju 388 K-0	Completed as Ju 388 K-0 at Merseburg, first flight on 17 November 1944 by pilot Lehmann; the flight also served as radio and autopilot check flight. Test flight on 18 November, follow up autopilot test flights on 22 and 23 November, high altitude flights on 23 and 24 November (two on each day), in each case with Lehmann at the controls.	
230 697 KS+TQ	Ju 388 K-0 V23	According to Amendment III to construction directive 7 388 00 10/007/003 of 17 Oct. 1944, it was one of five test beds with Jumo 213 E engines, and was to have a Dachlaurin fire extinguishing system installed in the wings and fuselage. Completely destroyed in Hangar II during a raid on Merseburg on 25 Nov. 1944.	Damage report 28 Nov. 1944
230 698 KS+TR	Ju 388 K-0	Transfer to *E-Stelle* Udetfeld for use as experimental aircraft for high altitude ballistics envisaged (Amendment V to construction directive 7 388 26 10/000 of 23 Oct. 1944). Acceptance flight, radio, and autopilot check flight at Merseburg on 28 Nov. 1944	Schreiber logbook
230 699 KS+TS	Ju 388 K-0	Nothing known to date about its construction.	
230 700 KS+TT	Ju 388 L-0/ *Nacht*	Completed as Ju 388 L-0/*Nacht* at Merseburg, and first flown on 4 January 1945 by pilot Lehmann (initially still recorded as K-0). Lehmann recorded the aircraft as an L-0/*Nacht* in all subsequent logbook entries: 9 January 1945: Follow up autopilot test flight 18 January 1945: High altitude flight 20 January 1945: High altitude flight 23 January 1945: High altitude flight	
230 721 - 230 740	Ju 388 K-0	Third block of Ju 388 K pre-production aircraft, some completed as prototypes, some to L-0/*Nacht* standard.	
230 721	Ju 388 J-0 V10	According to Amendment IV to Construction Directive 7 388 2 10/000 of 1 August 1944, initially envisaged as prototype J-2 with Jumo 222 A/B. Subsequently planned for completion as Ju 388 J-1 with BMW 801 TJ. In Merseburg completed as Ju 388 J-0 with two MG 151 and two MK 108, supposed to be sent to Werneuchen as quickly as possible with weapons system unserviceable. BAL report of 15 Dec. 1944. First flight on 3 January 1945 by pilot Lehmann; the flight also served as radio and autopilot check flight. Follow up test flight on the following day.	BAL report dated 15 Dec. 1944. Lehmann logbook
230 723	Ju 388 L-0/ *Nacht*	Completed as Ju 388 L-0/*Nacht* at Merseburg, and first flown on 12 January 1945 by pilot Lehmann; this also served as radio and autopilot check flight. Another high altitude flight, also at Merseburg, is known to have taken place on 17 January 1945.	Lehmann logbook.

Werknummer Manufacturer's Code	Variant Prototype	Remarks/Use/Fate	Sources
230 724	Ju 388 L-0/ *Nacht*	Completed as Ju 388 L-0/*Nacht* at Merseburg, and first flown on 12 January 1945 by pilot Lehmann; this also served as radio and autopilot check flight.	Lehmann logbook.
230 725	Ju 388 K-0 V24	Test bed for Jumo 213 E and prototype for Ju 388 J-3 series, determined by Amendment IV to construction directive 7 388 26 10/000 of 1 Aug. 1944. According to Amendment V of 23 Oct. 1944 the V24 was to be cancelled, therefore, it probably was not built, or was built as a "normal" K-0.	
230 727	Ju 388 K-0 V11	Test bed with Jumo 222 A/B, probably to Ju 388 L-0 standard without bomb bay. Heavily damaged in Hangar II during attack on Merseburg on 25 Nov. 1944, windows of pressurized cockpit damaged, canopy deformed, fuselage and wings pierced in several places.	Damage report 28 Nov. 1944.
230 729	Ju 388 K-0 V25	Test bed for Jumo 213 E, and prototype for Ju 388 J-2 series, determined by Amendment IV to construction directive 7 388 26 10/000 of 1 Aug. 1944. According to Amendment V of 23 Oct. 1944 the V25 was to be cancelled, therefore, it probably was not built, or was built as a "normal" K-0.	
230 733	Ju 388 K-0 V12	As per Amendment IV to Construction Directive 7 388 2 10/000 of 1 August 1944, envisaged as Ju 388 J-2 prototype with Jumo 222 A/B engines. Was finally supposed to be completed as Ju 388 J-1 with BMW 801 TJ engines. Components in Hangar II sustained serious damage in raid on Merseburg on 25 Nov. 1944. Damage report 28 Nov. 1944.	Amendment IV to construction directive 7 388 26 10/000 of 1 Aug. 1944.
230 734	Ju 388 K-0 Ju 388 L-0/ *Nacht*	A photograph taken in May 1945 shows a nearly complete Ju 388 K-0, without engines, with a large "34" on the nose. Given the advanced stage of construction reached by other aircraft from this Werknummer block, as documented by the damage report of 28 Nov. 1944, it seems likely that this uncompleted machine is *Werknummer* 230 734 (the 34th Ju 388 K-0).	
230 737	Ju 388 K-0 V13	This machine was envisaged as an engine test bed with Jumo 222 A/B engines. Probably to Ju 388 L-0 standard without bomb bay. The fuselage, *Werknummer* 5225, sustained 30% damage in Hangar II during the air raid on Merseburg on 25 Nov. 1944.	
230 901-230 910	Ju 388 K-0	Final block of Ju 388 K pre-production aircraft. Most envisaged as prototypes.	
230 901	Ju 388 J V14 V14/2	First envisaged as V14 prototype for Ju 388 J-2 with Jumo 222 engines. Role changed to Ju 388 J-1 prototype with BMW 801 TJ engines, probably renamed V14/2 at same time. Ultimately fitted with Jumo 213 E engines. Still under construction in Feb. 1945.	Construction directive 7 388 0010/012/001
230 902	Ju 388 K-0 V27	The major components were supposed to be delivered to DLH Staaken in final assembly state at the beginning of September 1944 for conversion to DB 603 U test bed. Construction cannot be confirmed.	Amendment IV to construction directive 7 388 26 10/000 of 1 Aug. 1944.

Werknummer Manufacturer's Code	Variant Prototype	Remarks/Use/Fate	Sources
230 903	Ju 388 J-2 V15	Mentioned in Junkers' offer for conversion to Jumo 222 A/B. Completion envisaged for January 1945, delivery to KdE. According to construction directive 7 388 00 10/006 000 of 30 Oct. 1944 was to become sole Ju 388 J-2 operational *Zerstörer* with Jumo 222 A/B engines.	
230 904		Construction cannot be confirmed.	
230 905		Construction cannot be confirmed.	
230 906	Ju 388 K-0 V28	DB 603 U test bed with DLH Staaken, as specified in Amendment IV to construction directive 7 388 26 10/000 of 1 Aug. 1944. Construction cannot be confirmed.	
230 907	Ju 388 K-0	Construction cannot be confirmed.	
230 908	Ju 388 K-0 V18	Engine test bed for Jumo 222 with no special installations, probably to L-0 standard without bomb bay.	
230 909	Ju 388 K-0 V19	Engine test bed for Jumo 222 with no special installations.	
230 910	Ju 388 K-0 V20	Engine test bed for Jumo 222 with no special installations, probably to L-0 standard without bomb bay.	
300 001-300 010	**Ju 388 L-0**	**According to construction directive ordered as L-0.**	
300 001 PG+YA	Ju 388 L-0 V7	This Ju 388 prototype first flew in December 1943 at Dessau, and formed the backbone of the Junkers test program. Test flights in Dessau and Prague. See Chapter V.	
300 002 PG+YB	Ju 388 L-0 V8	The aircraft was completed at Merseburg in February-March 1944, and was equipped with engines 385 182 and 385 178. Maiden flight by Lehmann on 15 March 1944. On 12 Oct. 1944 flown by Oskar Huber for engine measurements at 900 m, and on 13 Oct. to 12000 m. At end of November-beginning of December 1944 was in Dessau for conversion to external ram air intakes, after which it was to go to Rechlin.	LM logbook.
		Engines 385 182 / 385 219 (September 1944) Engines 385 226 / 385 219 (October 1944) Engines 385 225 / 385 274 (1:18 hr, 24 Nov. 1944)	BMW Brief Report Nos. 29, 33, 40
300 003 PG+YC	Ju 388 L-0	Under construction in Merseburg in March 1944. Engines already installed. Maiden flight and radio check flight by Lehmann and flight test engineer Lange at Merseburg on 18 Apr. 1944. Two days later a high altitude flight and an autopilot flight by the same crew. Follow up test flight by Lehmann and BMW engineer Schabert on 24 Apr. 1944. Subsequent flights by Lehmann from Merseburg: 27 Apr. 1944: Radio and autopilot check flight 29 Apr. 1944: Radio and night test flight 4 May 1944: High altitude flight (with Lange)	

Werknummer Manufacturer's Code	Variant Prototype	Remarks/Use/Fate	Sources
		5 May 1944: Test flight from Merseburg to Dessau and back (with Lange) 10 May 1944: Two follow up test flights 11 May 1944: Follow up autopilot test flight	
		On 13 May 1944 ferried by Lehmann from Merseburg to the Fürstenfeldbruck for use as a test vehicle by the BMW flight department.	
300 004 PG+YD	Ju 388 L-0	Completed in Merseburg with engines 385 187 and 385 183 in March 1944. Maiden flight by Lehmann on 22 March 1944. Subsequent flights by Lehmann from Merseburg:	
		25 March 1944: Test flight (with Lange) 25 March 1944: Follow up radio test flight 28 March 1944: Test flight (with Lange) 29 March 1944: Test flight (with Lange) 31 March 1944: Follow up test flight (with Lange) 1 April 1944: Follow up test flight 11 April 1944: Follow up radio test flight (with Lange) 12 April 1944: High altitude flight (with Lange) 12 April 1944: Two follow up radio test flights 13 April 1944: Follow up radio test flight 14 April 1944: Follow up radio test flight 15 April 1944: Two high altitude flights (with Lange) 18 April 1944: Autopilot test flight 19 April 1944: Autopilot test flight 19 April 1944: Autopilot and radio test flight 20 April 1944: Radio test flight 24 April 1944: Follow up and autopilot test flight	
		On 27 April 1944 ferried from Merseburg to Dessau by Lehmann. Subsequently used as a test vehicle by the *E-Stelle* Rechlin.	
		Ferried from Rechlin to Staaken by pilot Alfred Kuhn on 29 Apr. 1944, from 1417 to 1447 hours. At Rechlin in June 1944.	Kuhn logbook, flight no. 1595.
		Test flight to 12000 m by pilot Brandt on 7 July 1944 from 0527 to 0642 hours at Rechlin, same on 8 July from 1118 to 1200 hours (10000 m). Test flight by pilot Hasche (?) to 12000 m on 21 July 1944 at Rechlin, from 1621 to 1837 hours.	Unidentified logbook, flight nos. 672, 674, 697.
		Check flight at Rechlin on 19 Aug. 1944 from 1405 to 1552 by pilot Baist.	Baist logbook, flight no. 3070
		Probably temporarily stationed at Fürstenfeldbruck near Munich.[989] Test flight at Dessau on 19 Nov. 1944 from 1358 to 1426 hours by *Flugkapitän* Wendel. Converted to external ram air intakes and returned to Rechlin on 21 Nov. 1944.	Wendel logbook
		Engines 385 200 / 385 259 (September 1944) Engines 321 221 / 385 259 (October 1944) Engines 321 221 / 385 195 (December 1944)	BMW Brief Report No. 29, 33, 40

Werknummer **Manufacturer's Code**	**Variant Prototype**	**Remarks/Use/Fate**	**Sources**
300 005 PG+YE	Ju 388 L-0	Under construction at Merseburg in March 1944. Engines already installed. In Rechlin in June 1944. Factory test flight at Merseburg on 2 June 1944. Test flight at Rechlin by pilot Hasche on 29 July 1944, from 1517 to 1535 hours. During endurance trials, on 9 Aug. 1944 pilot Kuhn flew the aircraft from Rechlin to Gotenhafen, from 1050 to 1210 hours. Slightly damaged in raid on Rechlin on 25 Aug. 1944. Crash landing because of engine failure on 30 Aug. 1944.	Unidentified logbook, flight no. 707. Kuhn logbook, flight no. 1654
		Port engine 385 170, starboard engine 385 260 (6 July 1944) Propellers (3-blade): WNr. 50121 (port) and WNr. 50135 (starboard)	Rechlin inspection report No. 355/44
		Port engine 385 272, starboard engine 385 260 (30 Aug. 1944) Engines 385 189 / 385 195 (Oct. 1944)	BMW Brief Report No. 29, 33.
300 006 PG+YF	Ju 388 L-0	BAL acceptance flight at Merseburg on 20 June 1944. Flight from Rechlin on 31 July 1944 by pilot Cuno, 1540 to 1653 hours, climb to 8500 m.	
		In Köthen at the end of September 1944, where the BMW engines were removed for installation in the V2.	BMW brief report No. 32
		Several flights by Lehmann in March 1945: 9 March 1945: Cross country flight Merseburg-Delitzsch-Merseburg 14 March 1945: Test flight at Merseburg 16 March 1945: Test flight at Merseburg	Lehmann logbook
		On 31 March 1945 ferried from Merseburg to Köthen	Lehmann logbook
300 007 PG+YG	Ju 388 L-0	BAL acceptance flight on 14 June 1944 at Merseburg, follow up test flight the same day.	Lehmann logbook
		Flight from Rechlin on 8 July 1944 by pilot Brandt, 0527 to 0642 hours, reached 12000 m. On 13 July 1944 flown by Heinz Borsdorff from Rechlin to Lechfeld.	Unidentified logbook, flight no. 673
		Several flights by deputy type supervisor Alfred Kuhn: 8 July 1944: flight from Rechlin to Dessau (1430 to 1505) 8 July 1944: flight from Dessau to Merseburg (1630 to 1655) 8 July 1944: flight from Merseburg to Rechlin (1824 to 1945) 15 July 1944: flight from Rechlin to K**** (1434 to 1506) 15 July 1944: flight from K**** to Rechlin (1716 to 1744) 19 July 1944: flight from Rechlin to Brandenburg (1337 to 1410) 19 July 1944: flight from Brandenburg to Dessau (1521 to 1541) 19 July 1944: flight from Dessau to Rechlin (1630 to 1709) 20 July 1944: flight from Rechlin to Brandenburg (1337 to 1410) 20 July 1944: flight from Brandenburg to Dessau (1521 to 1541) 20 July 1944: flight from Dessau to Rechlin (1620 to 1701) 1 Aug. 1944: flight from Rechlin to Dessau (0950 to 1041), test flight to altitude of 10 km 1 Aug. 1944: flight from Dessau to Rechlin (1650 to 1725), test flight to altitude of 10 km Completely destroyed in attack on Rechlin on 25 Aug. 1944.	Kuhn logbook, flights no. 1628-1638

Werknummer Manufacturer's Code	Variant Prototype	Remarks/Use/Fate	Sources
300 008 PG+YH	Ju 388 L-0	First flight on 7 June 1944 at Merseburg, pilot Lehmann; this also served as the radio check flight. Subsequent flights by Lehmann from Merseburg: 8 June 1944: Autopilot test flight 9 June 1944: Follow up radio test flight 10 June 1944: Two speed course flights (with Lange) 19 June 1944: High altitude flight (with Lange) 21 June 1944: Factory test flight Merseburg-Dessau-Merseburg 28 June 1944: BAL acceptance flight	Lehmann logbook
		Performance measurement flight from Rechlin on 14 July 1944. Pilot Kuhn climbed the aircraft to an altitude of 11 km in a flight that lasted from 1350 to 1552 hours.	Kuhn logbook, flight no. 1648
		Sent to Munich-Riem for icing flights. On 22 Oct. 1944 inspected by JFM employees at the "JFM Field Test Office Mü-Riem," bullet damage repaired, and aircraft returned to flight operations. Installation of external ram air intakes planned.[992] December 1944 belly landing in Plauen following engine failure.	
		Engines 385 214 / 385 257 (40 and 34 hr, December 1944)	BMW Brief Report No. 42
300 009 PG+YI	Ju 388 L-0	Flights by pilot Lehmann from Merseburg: 24 June 1944: Follow up radio test flight 24 June 1944: High altitude flight (with Lange) 27 June 1944: Speed course flight Merseburg-Dessau (with Hotopf) 27 June 1944: Return flight Dessau-Merseburg (with Hotopf) 30 June 1944: BAL acceptance flight	Lehmann logbook
		Experimental aircraft at Rechlin in September and October 1944.	JFM FTA 4/44
		On 2 Oct. 1944 the aircraft was flown from Tarnewitz to Rechlin by Oberstabsing, Pfister, and an unidentified observer as part of the long term testing program. The flight lasted from 1737 to 1800 hours.	Pfister logbook, flight no. 3333
		To Merseburg on 5 Oct. 1944 for modification work and repair of tailwheel damage.	JFM FTA 4/44
		Safety flight and follow up flight at Merseburg on 2 Nov. 1944.	Lehmann logbook
		High altitude flights from Rechlin on 22 and 25 Sept. 1944 by pilot Kuhn, 1:10 and 0.35 hours, respectively. Landing site on 2nd flight illegible.	Kuhn logbook, flight no. 1658/59
		Engines 385 193 / 385 258 (90 hr., October 1944) Engines 385 193 / 385 258 (94 hr., November 1944)	BMW Brief Report No. 33, 37, 44
300 010 PG+YJ	Ju 388 L-0	Used as test vehicle by BMW. BAL acceptance flight by pilot Lehmann at Merseburg on 29 June 1944. On 1 July 1944 ferried from Merseburg via Munich-Oberwiesenfeld to Fürstenfeldbruck by Lehmann.	Lehmann logbook.
300 290 – 300 300	**Ju 388 L-0**	**Merseburg-Built L-0s**	

Werknummer Manufacturer's Code	Variant Prototype	Remarks/Use/Fate	Sources
300 291 DW+YY	Ju 388 L-0	Converted Ju 188 F-1, new cockpit, BMW 801 TJ engines, but old 188 vertical tail, weapons blister with MG 131 Z and rearview periscope, rearward-firing MG 131 in cockpit. Test flight by pilot Brandt on 26 July 1944, from 1218 to 1258 hours, climb to 10000 m. Test flight by pilot Brandt on 31 July 1944, from 1546 to 1646 hours, climb to 12000 m. Completely destroyed in an attack on Rechlin on 25 August 1944. Port engine 386 225, starboard engine 385 272 (6:35 hr., 22 July 1944) Propellers 9-12110.22 (4-blade), WNr. 1589 (port) and 1574 (starboard)	Unidentified logbook, flights no. 703 and 708
300 292 DW+YZ	Ju 388 L-0	Speed course flight from Merseburg on 21 July 1944 by crew of Lehmann and Lange. In Werneuchen in August 1944.[989] From Werneuchen to Dessau on 18 Oct. 1944 for modifications. According to EKdo 388 December report in Dessau for modifications.	Lehmann logbook
300 293 DW+ZA	Ju 388 L-0 V30	Initially equipped with BMW 801 TJ engines, cooling fan failure (385 269) at Merseburg in July 1944. High altitude flight with these engines on 27 July 1944, speed course flight the next day, both by crew of Lehmann and Lange. Engines removed on 11 Aug. 1944, aircraft grounded.[989] From October 1944 test bed with Jumo 213 E engines at Junkers Merseburg. First flight with new engines at Merseburg on 10 October 1944, crewed by Lehmann and flight test engineer Gödecke. Flight also served as radio check flight. Subsequent flights by Lehmann from Merseburg: 12 Oct. 1944: Autopilot and high altitude test flight, followed by high altitude performance flight (with Gödecke). 13 Oct. 1944: High altitude performance flight (with Gödecke) 15 Oct. 1944: High altitude flight (with Gödecke) 20 Oct. 1944: High altitude flight (with Kerksiek) On 30 Nov. 1944 flown from Rechlin to Lärz (1430 to 1450) by Stabsing. Walter Baist accompanied by Uffz. Wagner. Flight with Jumo 213s.	Lehmann logbook Preuschen Jumo monthly report for Oct. and Nov. 1944
300 294 DW+ZB	Ju 388 L-0 V31	Initial test, radio and autopilot check flight, with BMW 801 TJ-0 engines at Merseburg on 25 July 1944 (pilot Lehmann). Second test flight by crew of Lehmann and Lange the next day. On 11 Aug. 1944 parked without engines.[989] Initially refitted with Jumo 213 A-1 engines. Initial flight with new engines, which also served as radio and autopilot check flight, by crew of Lehmann and Gödecke on 13 Oct. 1944. High altitude flight by same crew the next day, follow up autopilot check flight on 15 October. According to Amendment I to construction directive 7 388 26 10/001 of 26 Oct. 1944, it was to be converted to take Type 9-8213 power plants (Jumo 213 E) as soon as the engines were delivered to Rechlin. When the war ended the aircraft was still present, parked at the edge of the forest.	Lehmann logbook

Werknummer **Manufacturer's Code**	**Variant** **Prototype**	**Remarks/Use/Fate**	**Sources**
300 295 DW+ZC T9+DL	Ju 388 L-0 V32	Jumo 213 E engines. Vers.Verb. Ob.d.L. (*Luftwaffe* C in C's Experimental Unit)	Report of 27 Feb. 1945
300 296 DW+ZD	Ju 388 L-0 V33	Aircraft with BMW 801 TJ engines. First flight by VfH Oranienburg on 9 Nov. 1944. Engines 385 285 / 323 046 (November 1944)	BMW Brief Report No. 39
300 297 DW+ZE	Ju 388 L-0	Supposed to be completed at Merseburg on 20 Aug. 1944. Maiden flight in fact took place on 16 Sept. 1944, pilot Lehmann. This flight also served as radio and autopilot check flight. Follow up radio and autopilot check flight on 18 Sept. 1944, high altitude flight two days later. Lehmann flew the aircraft for the last time at Merseburg on 23 Sept. 1944, and its subsequent fate is not known.	Lehmann logbook
300 298 DW+ZF	Ju 388 L-0 V34	Supposed to be completed at Merseburg on 30 Aug. 1944.	
300 299 DW+ZG	Ju 388 L-0	Maiden flight on 22 Oct. 1944, pilot Lehmann. This flight also served as radio and autopilot check flight. On 8 Nov. 19rr ferried from Merseburg to Fürstenfeldbruck by Lehmann. Employed as test bed by BMW. Diving tests at end of November 1944, engines already fitted with external ram air intakes. Aircraft parked at Fürstenfeldbruck at end of war. Port engine 385 279, starboard engine 385 281 (November 1944)	Lehmann logbook Photos, BMW test report 1113/44.
300 300 DW+ZH	Ju 388 L-0	Maiden flight on 2 Nov. 1944, pilot Lehmann (also radio and autopilot check flight); high altitude flights on following two days. Was supposed to be ferried to Munich as soon as possible. At war's end aircraft was obviously parked at Fürstenfeldbruck.	Lehmann logbook Preuschen 6 Nov. 1944
340 081 – 340 120[993] **340 281 – 340 290** **340 301 – 340 310** **340 401 – 340 420**		**First aircraft, 340 081 to 089, from Merseburg; from 340 090 production by ATG in Leipzig-Mockau, Leipzig-Seehausen, and Kölleda**	
340 081 RT+KA	Ju 388 L-1	First L-1 production machine, initial flight by Lehmann at Merseburg on 17 Aug. 1944. High altitude test flight, radio and autopilot check flight, by *Flugkapitän* Baltabol on 21 Aug. 1944 (1730 to 1935) at Merseburg. Subsequent high altitude flight by Baltabol on 27 Aug. from 1440 to 1620 hours. Performance measurements by crew of Lehmann/Gödecke on 31 August (high altitude performance and speed course flight) and 1 September (speed course flight). Accepted by EKdo 388 at Merseburg, poor construction criticized. Flights by pilot Kuhn from Rechlin: 30 Sept. 1944: check flight to altitude of 10 km (1452 to 1626 hr.), flight not recorded in technical field office report 9/44. 2 Oct. 1944: high altitude flight to 10 km (1507 to 1611 hr.) 3 Oct. 1944: high altitude flight to 10 km (0830 to 0933 hr.) 3 Oct. 1944: test flight to 10 km (1340 to 1410 hr.)	Lehmann logbook Baltabol logbook Lehmann logbook Kuhn logbook, flights no. 1662, 1664, 1665, 1671, 1672, 1675, 1681, 1682.

Werknummer Manufacturer's Code	Variant Prototype	Remarks/Use/Fate	Sources
		21 Oct. 1944: test flight to measure high altitude performance to altitude of 11 km (1345 to 1607 hr.) 27 Oct. 1944: flight from Rechlin to Lärz, starboard engine failure (1456 to 1520 hr.) 9 Nov. 1944: performance measurement flight to 10 km (1442 to 1530 hr.) 17 Nov. 1944: performance measurement flight to 11 km (1147 to 1341 hr.) Crash landing on 17 Nov. 1944, 70% damage.	Technical Field Office Nov. 1944
		Port engine 321 225, starboard engine 385 268 (18.4 hr., October 1944)	BMW Brief Report No. 33
340 082 RT+KB	Ju 388 L-1	Nothing known to date about its construction. It may have been the airframe built for structural tests.	
340 083 RT+KC	Ju 388 L-1	First test flight, autopilot and radios, by *Flugkapitän* Baltabol at Merseburg on 25 Aug. 1944. Speed course and high altitude performance flights by crew of Lehmann and Gödecke at Merseburg on 30 Aug. 1944. Further flights by Baltabol are documented in an extract from the FIPI JFM Merseburg takeoff log kept by the head of flight operations: 25 Aug. 1944: test flight—autopilot, radios and climb to high altitude (1710 to 1715 hr.) 29 Aug. 1944: autopilot and radio test flight (1640 to 1745 hr.) 30 Aug. 1944: high altitude flight (1400 to 1535 hr.) Taken on strength by EKdo 388 on 6 Sept. 1944, poor construction criticized, also used for testing by BMW at Rechlin; on 11 Sept. 1944 overturned at Fürth, sustaining 65% damage; does not appear again in EKdo 388 records. Subsequent fate unknown, perhaps scrapped?	Baltabol logbook Lehmann logbook BMW Brief Report No. 40
340 084 RT+KD	Ju 388 L-1	First test flight, autopilot and radios, by *Flugkapitän* Baltabol at Merseburg on 28 Aug. 1944. Speed course flight two days later by crew of Lehmann and Gödecke, high altitude flight on 1 Sept. 1944. Further flights by Baltabol are documented in an extract from the FIPI JFM Merseburg takeoff log kept by the head of flight operations. Poor construction criticized. A fuel injection line broke during the flight. Engine endurance tests. Test flown by pilot Alfred Kuhn in Rechlin on 25 Nov. 1944, reaching altitude of 9 km (1415 to 1510 hr.) Autopilot tests and electrical energy measurements carried out by EKdo 388 in November 44.	Baltabol logbook Lehmann logbook Kuhn logbook, flight no. 1685
		Port engine 385 265, starboard engine 385 269 (24.5 hr., October 1944) Port engine 385 265, starboard engine 385 261 (39.5 and 19.5 hr., November 1944) Port engine 385 265, starboard engine 385 261 (41.75 and 22 hr., November 1944) Port engine 385 265, starboard engine 385 261 (44.5 and 25 hr., December 1944)	BMW Brief Reports No. 33, 37, 39, 42

Werknummer Manufacturer's Code	Variant Prototype	Remarks/Use/Fate	Sources
340 085 RT+KE	Ju 388 L-1	First flight on 18 Sept. 1944 by pilot Lehmann; this also served as radio and autopilot check flight. Two high altitude flights by Lehmann and Gödecke on 19 Sept. 1944. Lehmann also carried out a follow up autopilot check flight and a safety check flight on this day.	Lehmann logbook
		Arrived from Merseburg on 29 Sept. 1944, appears in EKdo 388's monthly report of 2 Oct. 1944, assigned to long-term engine testing, poor construction criticized. External ram air intakes installed in November 1944.	BMW Brief Report No. 40
		Two flights by pilot Kuhn on 26 Nov. 1944, from Rechlin to Lärz and back (1156 to 1206 and 1300 to 1347 hr.), performance measurements during return flight. In November 1944 used for performance measurement and camera trials. In February 1945 aircraft was supposed to be converted as quickly as possible to lightened condition without bomb bay.[994]	Kuhn logbook, flights no 1687, 1688
		Port engine 315 320, starboard engine 321 212 (7.5 hr., October 1944) Engines 387 532 / 321 213 (11.75 and 37.33 hr., November 1944) Engines 387 532 / 321 213 (13.5 and 39 hr., December 1944) Engines 387 532 / 321 213 (15.5 and 41 hr., December 1944)	BMW Brief Report No. 33, 37, 40, 42
340 086 RT+KF	Ju 388 L-1	Merseburg production aircraft, maiden flight by Lehmann on 14 Sept. 1944. Flight also served as radio and autopilot check flight. Two high altitude flights on 15 Sept. 1944, one with Gödecke as observer.	Lehmann logbook
		Taken on strength by EKdo 388 on 23 Sept. 1944, poor construction criticized. In November 1944 used by EKdo 388 for diving trials.	JFM Technical Field Office (FTO) 4/44 and 7/44
		On 31 Dec. 1944 flown by *Flugkapitän* Wendel at Dessau from 1050 to 1120 hr., also on 4 Jan. 1945 from 1536 to 1515 hr.	Wendel logbook
		Port engine 315 315, starboard engine 315 334 (7.5 hr., October 1944)	BMW Brief Report No. 33
340 087 RT+KG	Ju 388 L-1	Maiden flight by pilot Lehmann at Merseburg on 25 Sept. 1944. This flight also served as radio and autopilot check flight. Two more radio and autopilot flights the same day. On 26 September high altitude flight by crew of Lehmann and Gödecke, also from Merseburg. Arrived Rechlin from Merseburg on 2 Oct. 1944, to EKdo 388. Belly-landing at Lärz on 8 November 1944.	Lehmann logbook
		Port engine 321 223, starboard engine 315 321 (5 hr., October 1944) Port engine 321 223, starboard engine 315 321 (24.75 hr., November 1944)	BMW Brief Report No. 33, 37.
340 088 RT+KH	Ju 388 L-1	Maiden flight by pilot Lehmann at Merseburg on 27 Sept. 1944. This flight was also supposed to have served as radio and autopilot check flight, however, the aircraft developed engine trouble that resulted in a forced landing.	Lehmann logbook
		On 28 Sept. 1944, loss of power in port engine (WNr. 323 048) during takeoff at Merseburg, engine seized, metal shavings in supercharger air collector.	BMW Brief Report No. 32

Werknummer **Manufacturer's Code**	**Variant Prototype**	**Remarks/Use/Fate**	**Sources**
		Combined test flight and radio-autopilot check flight on 8 Oct. 1944, again by Lehmann. High altitude flight on 14 Oct. 1944 by crew of Lehmann and Gödecke; subsequent high altitude flight on 17 Oct. 1944. BAL test flight at Merseburg on 21 Oct. 1944.	Lehmann logbook
		On 27 Oct. 1944 from Merseburg to EKdo 388, used by trials unit for flame-damper tests on November 1944.	FTA Report 7/44
		Engines 387 530 / 323 049 (16.37 and 18.75 hr., December 1944)	BMW Brief Report No. 40
340 089 RT+KI (?)	Ju 388 L-1	Maiden flight by pilot Lehmann at Merseburg on 30 Sept. 1944. This flight also served as radio and autopilot check flight. From Merseburg to EKdo on 4 Oct. 1944. Transferred to Werneuchen on 20 Oct. 1944.	Lehmann logbook
		Port engine 385 275, starboard engine 315 345 (5 hr., October 1944)	BMW Brief Report No. 33
340 090 RT+KI (?)		First test flight by *Flugkapitän* Baltabol at Leipzig-Mockau on 26 Sept. 1944. subsequently flown by Baltabol from Leipzig-Mockau to Altenburg on 8 Oct. The aircraft was subsequently transferred from Altenburg to EKdo 388 on 21 Oct. Flown by	Baltabol logbook
		Harry Böttcher in Rechlin on 8 Nov. 1944. Kuhn flew the aircraft there on 9 Nov. from 1121 to 1406 hours, reaching an altitude of 9.6 km. Kuhn again flew the machine on 21 Nov. from 1130 to 1215, reaching an altitude of 9 km. Transferred to Leipzig-Mockau on 27 Nov. 1944.	Kuhn logbook, flights no. 1680 and 1683
		Port engine 317 317, starboard engine 315 325 (13 hr., November 1944) Port engine 315 317, starboard engine 315 325 (13.75 hr., November 1944)	BMW Brief Report No. 37, 39
340 091 RT+KK	Ju 388 L-1	General test flight with autopilot tests at Altenburg on 29 Oct. 1944 by *Flugkapitän* Baltabol.	Baltabol logbook
340 092 RT+KL	Ju 388 L-1	ATG production aircraft. Construction inspection test flight at Altenburg on 19 Oct. 1944, crewed by pilot Karstens and observer August Köpp. Arrived at Rechlin on 1 Nov. According to BMW brief report no. 40, the aircraft had BMW 801 TJ-0 engines while at Rechlin. Fuel consumption tests in November with engine 329 005. Taxiing accident caused by undercarriage failure at the beginning of November 1944. Used by EKdo 388 for hydraulic tests in November 1944.	August Köpp logbook, flight no. 182
		Engines 329 005 / 315 345.[996] (3.5 hr., December 1944)	BMW Brief Report No. 42
340 093 RT+KM	Ju 388 L-1	ATG production aircraft. General test flights at Altenburg on 17 Oct. 1944 from 1654 to 1712, and on 18 Oct. from 0857 to 0917 hours, crewed by Behrend (pilot) and Köpp (observer).	August Köpp logbook, flight no. 180/1.
		On 29 Oct. became the first machine issued to a reconnaissance unit, from Altenburg. In November 1944 used by the trials unit for autopilot tests and electrical energy measurements. Engines 323 047 / 385 278 BMW Brief Report No. 40	JFM FTA Report 7/44
		Port engine 323 047, starboard engine 385 278 (4.5 hr. flying time, December)	BMW Brief Report No. 40, 42

Werknummer Manufacturer's Code	Variant Prototype	Remarks/Use/Fate	Sources
340 094 RT+KN	Ju 388 L-1	First test flight, radios, from Leipzig-Mockau to Altenburg on 22 Oct. 1944 by *Flugkapitän* Baltabol.	Baltabol logbook
		ATG production aircraft. Construction inspection test flight by *Flugkapitän* Baltabol and August Köpp on 31 Oct. 1944, from 1416 to 1436 hours.	August Köpp logbook, flight no. 185.
		Deputy type supervisor Kuhn test flew the aircraft at Rechlin on 8 November, from 1022 to 1145 hours, reaching an altitude of 10 km.	Kuhn logbook, flight no. 1679
		Report dated 2 Dec. 1944 states that aircraft was used by trials unit for testing of flame dampers and observation.	
340 095 RT+KO	Ju 388 L-1	ATG production aircraft. Aircraft evaluation flight by *Flugkapitän* Baltabol and observer August Köpp from Altenburg on 4 Nov. 1944, from 1136 to 1150 hours.	Baltabol logbook
		Production inspection test flight by pilot Darre and observer August Köpp from 1028 to 1040 hours on 9 November.	August Köpp logbook, flights no. 188, 195
		At modification detachment in Welzow on 5 Dec. 1944	
340 096 RT+KP	Ju 388 L-1	ATG production aircraft. Production inspection test flight from Altenburg on 8 Nov. 1944, 1535 to 1546 hours, by pilot Karstens and observer August Köpp.	August Köpp logbook, flight no. 194
		At modification detachment in Welzow on 5 Dec. 1944	
340 097 RT+KQ	Ju 388 L-1	Ferried from Leipzig-Mockau to Merseburg by Lehmann on 29 Oct. 1944. Combined test flight and radio-autopilot check flight there on 2 Nov. Subsequent test flights on same and following day.	Lehmann logbook
		Flight safety inspection at Merseburg in November 1944.	
340 098 RT+KR	Ju 388 L-1	ATG production aircraft. Ferried from Leipzig-Mockau to Merseburg by Heinz Schreiber on 30 Oct. 1944. Flight safety inspection at Merseburg in November 1944.	Schreiber logbook
		Ferried from Merseburg to Dessau by Lehmann on 22 Nov. 1944.	Lehmann logbook
		At the beginning of December was just being fitted with external ram air intakes and was viewed as a replacement for the V3, which had been damaged at Rechlin.	
		Still being flown on 20 Feb. 1945, vibration measurements.	EL Memo 20 Feb. 1945
		Engines 387 537 / 387 616 (no flying time, 24 Nov. 1944) Engines 387 537 / 387 616 (4 hr., 9 Dec. 1944)	BMW Brief Reports No. 40, 41
340 099 RT+KS	Ju 388 L-1	ATG production aircraft. Ferried from Leipzig-Mockau to Merseburg by Heinz Schreiber on 29 Oct. 1944. Flight safety inspection at Merseburg in November 1944.	Schreiber logbook
		Follow up radio and autopilot check flight at Merseburg, pilot Lehmann, on 2 Nov. 1944.	
		A further autopilot test flight, also with Lehmann at the controls, on 4 Nov. 1944.	Lehmann logbook
340 100 RT+KT	Ju 388 L-1	Construction inspection test flight in Altenburg by *Flugkapitän* Baltabol on 17 Nov. 1944. Baltabol flew the aircraft again on 25 November.	Baltabol logbook
		At modification detachment in Welzow on 5 Dec. 1944	
340 101 RT+KU	Ju 388 L-1	ATG production aircraft. Initial test flight on 7 Nov. 1944 from 1429 to 1447 hours and production inspection test flight from 1551 to 1558 at Altenburg, crewed by pilot Darre and observer August Köpp.	August Köpp logbook, flight no. 192/3

Werknummer Manufacturer's Code	Variant Prototype	Remarks/Use/Fate	Sources
		At modification detachment in Welzow on 5 Dec. 1944 Photo confirmation: in front of hangar on Barndis-Waldpohlenz airfield near Leipzig, no manufacturer's code.	
340 102 RT+KV	Ju 388 L-1	On 24 Feb. 1945 ferried from Welzow to Merseburg (230 km) by pilot Lehnhoff, from 1530 to 1610 hours.	Lehnhoff logbook, flight no. 1282
340 103	Ju 388 L-1	Factory test flight with radio and autopilot checks by FL Baltabol at Leipzig-Mockau on 5 Dec. 1944. Additional radio and autopilot checks during second flight that afternoon. Aircraft evaluation flight next day.	Baltabol logbook
340 104	Ju 388 L-1	ATG production aircraft, at modification detachment in Welzow on 5 Dec. 1944.	
340 105	Ju 388 L-1	Flown from Leipzig-Mockau to Altenburg by *Flugkapitän* Baltabol on 11 Nov. 1944. Factory test flight with radio systems check. Autopilot checks and aircraft evaluation flight by same pilot two days later.	Baltabol logbook
		Production inspection test flight from Altenburg by *Flugkapitän* Baltabol and observer August Köpp on 14 Nov. from 0943 to 0955 hr.	August Köpp logbook, flight no. 197
340 106	Ju 388 L-1	Factory test flight from Leipzig-Mockau to Altenburg by *Flugkapitän* Baltabol on 20 Nov. 1944, autopilot, radio and aircraft evaluation flights on 22-23 November.	Baltabol logbook
340 107	Ju 388 L-1	Construction cannot be confirmed.	
340 108	Ju 388 L-1	Construction cannot be confirmed.	
340 109	Ju 388 L-1	ATG production aircraft. Aircraft evaluation flight on 5 Dec. 1944 from 1359 to 1421 hours by pilot Karstens. Production inspection test flight by *Flugkapitän* Baltabol on 7 December from 0945 to 1003. Observer for both flights from Altenburg August Köpp.	August Köpp logbook, flight No. 202/3.
340 110	Ju 388 L-1	Factory test flight, radio and autopilot checks, from Leipzig-Mockau to Altenburg by *Flugkapitän* Baltabol on 3 Dec. 1944. Aircraft evaluation flight and additional autopilot check/adjustment (?) from Altenburg on 7 December, also by *Flugkapitän* Baltabol.	Baltabol logbook
340 111	Ju 388 L-1	ATG production aircraft. Aircraft evaluation flight from Altenburg on 2 Dec. 1944, 0930 to 1002 hours, by *Flugkapitän* Baltabol with observer August Köpp.	August Köpp logbook flight no. 201.
340 112	Ju 388 L-1	Factory test flight, radio and autopilot checks, by *Flugkapitän* Baltabol from Leipzig-Mockau to Altenburg on 28 Nov. 1944; aircraft evaluation flight carried out a day later.	Baltabol logbook
340 113	Ju 388 L-1	To date nothing known about construction.	
340 114	Ju 388 L-1	Expected to join flying program at Altenburg in November 1944, but was not flown due to weather. Ferry flight to Altenburg planned on 14 Dec. 1944.[996,997]	BA-MA RL 3/4315

Werknummer Manufacturer's Code	Variant Prototype	Remarks/Use/Fate	Sources
340 115	Ju 388 L-1	Expected to join flying program at Altenburg in November 1944, but was not flown due to weather. Ferry flight to Altenburg planned on 14 Dec. 1944.[997] Factory test flight, radio and autopilot checks, by *Flugkapitän* Baltabol from Leipzig-Mockau to Altenburg on 16 Dec. 1944. The aircraft evaluation flight was carried out on 17 Dec., however a further autopilot check flight was required on the following day.	BA-MA RL 3/4315 Baltabol logbook
340 116	Ju 388 L-1	ATG production aircraft. Flight to Kölleda planned for 21 Dec. 1944.[997] Parked at Altenburg at end of war. Existence confirmed by photographs. Flown on 25 Jan. 1945. Engines 385 186 / 387 789.	BA-MA RL 3/4315 J. Crow Photo Preuschen
340 117	Ju 388 L-1	ATG production aircraft. Flight to Kölleda planned for 21 Dec. 1944.[997] Aircraft evaluation flight (high altitude) by *Flugkapitän* Baltabol at Altenburg on 27 Jan. 1945 from 0940 to 1036 hours. On 30 Jan. high altitude flight from same location by pilot Behrend and observer August Köpp, 1455 to 1549 hr. On 2 Feb. 1945 high altitude flight from Altenburg by pilot Behrend and observer August Köpp, 1354 to 1520 hr. Port engine 388 323, starboard engine 387 767	August Köpp logbook, flights no. 213, 215, 217.
340 118	Ju 388 L-1	ATG production aircraft. Flight to Kölleda planned for 21 Dec. 1944.[997] Aircraft evaluation flight (high altitude) from Altenburg on 27 Jan. 1945 by pilot Karstens with observer August Köpp, 1531 to 1622 hr. High altitude flight by same crew on 4 February from 1438 to 1530 hr. Aircraft parked in open at Altenburg at end of war.	August Köpp logbook, flight no. 214/8. J. Crow Photo
340 119	Ju 388 L-1	ATG production aircraft. Factory test flight, radios and autopilot, from Altenburg on 21 Dec. 1944 by *Flugkapitän* Baltabol. Flight to Kölleda planned for 23 Dec. 1944.[997] High altitude flight from Altenburg by *Flugkapitän* Baltabol on 20 Jan. 1945 from 1009 to 1039 hr. Aborted due to excessive rpm in unidentified component of starboard engine. Baltabol's last flight before end of war. According to Preuschen reached altitude of 12 km on or about 25 Jan. 1945. Blown up at Heidkaten near Kaltenkirchen at end of war. Possibly on strength with ObdL Experimental Unit at that time. Port engine 385 125, starboard engine 329 016	Baltabol logbook Preuschen
340 281	Ju 388 L-1	ATG production aircraft. Flight to Kölleda planned for 24 Dec. 1944.[997] On 1 Feb. 1945 high altitude flight from Altenburg by pilot Behrend and observer August Köpp, 0912 to 1036 hr. At war's end parked at Altenburg with canopy partly destroyed.	BA-MA RL 3/4315 August Köpp logbook, flight no. 216 J. Crow Photo
340 282	Ju 388 L-1	At present nothing known about construction of this machine.	
340 283	Ju 388 L-1	Status of construction by ATG (Plant 3) on 19 Jan. 1945: Fuselage with tail surfaces and wings present (no engines, tank pumps, etc.).[998]	BA-MA RL 3/3415
340 284	Ju 388 L-1	Flight to Kölleda planned for 27 Dec. 1944.[997] Parked at Altenburg at war's end.	J. Crow Photos

Werknummer Manufacturer's Code	Variant Prototype	Remarks/Use/Fate	Sources
340 285	Ju 388 L-1	At present nothing known about construction of this machine.	
340 286	Ju 388 L-1	Flight to Kölleda planned for 27 Dec. 1944.[997]	BA-MA RL 3/4315
340 287	Ju 388 L-1	Flight to Kölleda planned for 21 Dec. 1944.[997] In fact, pilot Baltabol carried out initial factory test flight with radio and autopilot checks on 16 Jan. 1945.	BA-MA RL 3/4315
340 288	Ju 388 L-1	Flight to Kölleda planned for 28 Dec. 1944.[997] Factory test flight with radio and autopilot checks from Leipzig-Mockau to Kölleda by FL Baltabol on 30 Dec. 1944.	Baltabol logbook
340 289	Ju 388 L-1	Status of construction by ATG (Plant 3) on 19 Jan. 1945: fuselage with tail surfaces and wings in place (no engines, tank pumps, etc.).[998]	BA-MA RL 3/4315
340 290	Ju 388 L-1	Flight to Kölleda planned for 29 Dec. 1944.[997] The aircraft's fuselage was damaged by machine-gun fire in a strafing attack on Kölleda on 22 February 1945.	BA-MA RL 3/4315
340 291	Ju 388 L-1	Factory test flight with radio and autopilot checks from Leipzig-Mockau to Kölleda by *Flugkapitän* Baltabol on 28 Dec. 1944. The aircraft's fuselage was damaged by machine-gun fire in a strafing attack on Kölleda on 22 February 1945.	Baltabol logbook
340 292	Ju 388 L-1	Status of construction by ATG (Plant 3) on 19 Jan. 1945: fuselage with tail surfaces and wings in place (no engines, tank pumps, etc.).[998]	
340 293	Ju 388 L-1	At present nothing known about construction of this machine.	
340 294	Ju 388 L-1	Status of construction by ATG (Plant 3) on 19 Jan. 1945: fuselage with tail surfaces and wings in place (no engines, tank pumps, etc.).[998]	
340 295	Ju 388 L-1	Status of construction by ATG (Plant 3) on 19 Jan. 1945: fuselage with tail surfaces and wings in place (no engines, tank pumps, etc.).[998]	
340 296	Ju 388 L-1	Flight to Kölleda planned for 30 Dec. 1944.[997]	BA-MA RL 3/4315
340 297	Ju 388 L-1	Factory test flight with radio and autopilot checks from Leipzig-Mockau to Kölleda by *Flugkapitän* Baltabol on 16 Jan. 1945.	Baltabol logbook
340 298		At present nothing known about construction of this machine.	
340 299	Ju 388 L-1	Status of construction by ATG (Plant 3) on 19 Jan. 1945: fuselage with cockpit present (wings in plant for 2 of 5 machines).[998]	
340 300	Ju 388 L-1	At present nothing known about construction of this machine.	
340 301	Ju 388 L-1	Status of construction by ATG (Plant 3) on 19 Jan. 1945: fuselage with tail surfaces and wings present (no engines, tank pumps, etc.).[998] At an advanced stage in construction at end of war, but without engines. Parked in damaged hangar in Leipzig-Mockau. (USAF Photo)	

Werknummer Manufacturer's Code	Variant Prototype	Remarks/Use/Fate	Sources
340 302	Ju 388 L-1	Status of construction by ATG (Plant 3) on 19 Jan. 1945: fuselage with tail surfaces and wings present (no engines, tank pumps, etc.).[998]	
340 303	Ju 388 L-1	Status of construction by ATG (Plant 3) on 19 Jan. 1945: fuselage with cockpit present (wings in plant for 2 of 5 machines).[998]	
340 304		Status of construction by ATG (Plant 3) on 19 Jan. 1945: fuselage with tail surfaces and wings present (no engines, tank pumps, etc.).[998]	
340 305	Ju 388 L-1	Status of construction by ATG (Plant 3) on 19 Jan. 1945: fuselage with cockpit present (wings in plant for 2 of 5 machines).[998]	
340 306	Ju 388 L-1	Status of construction by ATG (Plant 3) on 19 Jan. 1945: fuselage with tail surfaces and wings present (no engines, tank pumps, etc.).[998]	
340 307	Ju 388 L-1	Status of construction by ATG (Plant 3) on 19 Jan. 1945: fuselage with tail surfaces and wings present (no engines, tank pumps, etc.).[998] At an advanced stage of construction at war's end, but without engines. Parked in a damaged hangar in Leipzig-Mockau (USAF Photo).	
340 308	Ju 388 L-1	Status of construction by ATG (Plant 3) on 19 Jan. 1945: fuselage and cockpit present (wings in plant for 2 of 5 machines).[998]	
340 309	Ju 388 L-1	At present nothing known about construction of this machine.	
340 310	Ju 388 L-1	Status of construction by ATG (Plant 3) on 19 Jan. 1945: fuselage and cockpit present (wings in plant for 2 of 5 machines).[998]	
340 402	Ju 388 L-1	Status of construction by ATG (Plant 3) on 19 Jan. 1945: fuselage present, no cockpit.[998]	
340 403		At present nothing known about construction of this machine.	
340 404		At present nothing known about construction of this machine.	
340 405	Ju 388 L-1	Status of construction by ATG (Plant 3) on 19 Jan. 1945: fuselage with no cockpit.[998] At advanced stage in construction at war's end, but without engines. Parked in damaged hangar in Leipzig-Mockau. (USAF Photo)	
340 406	Ju 388 L-1	At present nothing known about construction of this machine.	
340 407	Ju 388 L-1	Status of construction by ATG (Plant 3) on 19 Jan. 1945: fuselage with no cockpit.[998]	
340 408	Ju 388 L-1	Status of construction by ATG (Plant 3) on 19 Jan. 1945: fuselage with no cockpit.[998]	
340 409	Ju 388 L-1	At present nothing known about construction of this machine.	
340 410	Ju 388 L-1	At an advanced stage in construction at war's end, but without engines. Parked in damaged hangar in Leipzig-Mockau.	USAF Photo
340 411	Ju 388 L-1	Status of construction by ATG (Plant 3) on 19 Jan. 1945: fuselage with no cockpit.[998] At an advanced stage in construction at war's end, but without engines. Parked in damaged hangar in Leipzig-Mockau.	USAF Photo

Werknummer Manufacturer's Code	Variant Prototype	Remarks/Use/Fate	Sources
340 412	Ju 388 L-1	Status of construction by ATG (Plant 3) on 19 Jan. 1945: fuselage with no cockpit.[998]	
340 413	Ju 388 L-1	At present nothing known about construction of this machine.	
340 414	Ju 388 L-1	Status of construction by ATG (Plant 3) on 19 Jan. 1945: fuselage with no cockpit.[998]	
340 415	Ju 388 L-1	Status of construction by ATG (Plant 3) on 19 Jan. 1945: fuselage with no cockpit.[998]	
340 4xx	Ju 388 L-1	There were two other machines in ATG Plant 3 without *Werknummern*, status of construction on 19 Jan. 1945: fuselage with no cockpit.[998]	
340 xxx ?	Ju 388 J-1	Present in ATG Plant 3 on 19 Jan. 1945 were: 5 fuselages 3 cockpits 2 sets of wings 1 tail section	
340 xxx ?	Ju 388 J-1	Fuselage visible in the background in one of the photographs of the ATG factory buildings in Leipzig.	USAAF Photo
340 xxx ?	Ju 388 J-1	Wrecked J-series airframe in the ATG hangar; FHL 131 and oblique armament clearly visible. The mottled camouflage finish is different than that on L-1 airframes.	USAAF Photo
340 xxx ?	Ju 388 J-1	Two aircraft without engines and tail sections which were apparently photographed by American soldiers outside the ATG plant in Leipzig.	W. Abele Photos
340 xxx ?	Ju 388 J-1	Photo of four apparently discarded fuselages at ATG Leipzig, possibly those mentioned in the document.[998]	
500 001-500 006[999]	**Prototypes**	**True 388 Prototype Series**	
500 001 PE+IA	Ju 388 V1 J-1	Prototype for the heavy fighter/night-fighter. Flew at Dessau in May 1944, later at Tarnewitz. On 12 Oct.1944 EKdo 388 Rechlin took charge of the machine from Tarnewitz. High altitude flights to 8 km by pilot Kuhn on 13-14 Oct. 1944, from 1457 to 1604 and 1038 to 1141 hours. On strength with EKdo 388 as of 13 Jan. 1945. Engines 385 127 / 385 169 (22.17 hr., 24 Nov. 1944)	Kuhn logbook, flight no. 1669/70
500 002 PE+IB	Ju 388 V2 J-1	Second prototype of the Ju 388 J-1 with MG 151/20s and MK 108s in fuselage tray and obliquely-mounted MG 151s, testing by *Luftwaffe* began in December 1944.	
500 003 PE+IC	Ju 388 V3 K-1	First prototype of the planned K-1 high altitude bomber series. First test flight on 2 April 1944. Used mainly by Junkers for engine trials at Dessau. Probably wrecked at Rechlin on 8 Nov. 1944 while landing after flame damper tests. Engines 385 127 / 385 129 (June 1944) Engines 385 194 / 385 264 (September 1944)	BMW Brief Report No. 37 BMW Brief Report No. 29

Werknummer Manufacturer's Code	Variant Prototype	Remarks/Use/Fate	Sources
500 004 PE+ID	Ju 388 V4 K-1	**Second K-1 prototype, used mainly as experimental aircraft by Dessau flight department.** Prototype without tail turret. **Fitted with a new Fimag intercooler in November, was supposed to receive new engines with improved supercharger sealing. The two K-1s were set aside at Dessau on 20 Feb. 1945.**	
		Port engine 385 212, starboard engine 385 129 (June 1944) Engines 385 196 / 321 226 (September 1944) Engines 321 219 / 321 226 (October 1944) Engines 387 837 / 387 776 (no flight time, 9 Dec. 1944)	BMW Brief Report No. 29, 32, 41
500 005 PE+IE	Ju 388 V5 L-1	First prototype of the L-1 series (long-range reconnaissance). Factory testing at Dessau began in summer 1944, then test bed in Rechlin. High altitude flights from Rechlin by pilot Kuhn on 8 Sept. (0947 to 1020 hr.) and 17(?) Sept. 1944. Damaged in strafing attack on 3 Oct. 1944, no further flights that month.	Kuhn logbook, flights. No 1656-57.
		Port engine 385 189, starboard engine 385 195 (July 1944) Port engine 385 189, starboard engine 385 195 (38.5 hr., October 1944)	
500 006 PE+IF	Ju 388 V6 L-1	Second L-1 series prototype. After factory testing at Dessau, transferred to Tarnewitz for testing of tail turret. The aircraft was captured by the English there at the end of the war and later displayed at Farnborough.	
		Port engine 319 058, starboard engine 305 213 (July 1944).	
560 041 – 560 050?	**Weserflug Liegnitz**	**Series Production**	
560 041 PT+HA	Ju 388 L-1	First production aircraft from Weserflug in Liegnitz. On 8 Nov. 1944 delivered from Liegnitz to Rechlin, where it underwent acceptance inspection. Kuhn test flew the aircraft on 26 November from 1000 to 1101 hours, reaching an altitude of 9 km. Released to "Welzow Modification Facility" on 27 November.	Kuhn logbook, flight no. 1686
		Port engine 385 282, starboard engine 385 284 (December 1944)	BMW Brief Report No. 40
560 042 PT+HB	Ju 388 L-1	Delivered from Liegnitz to Rechlin on 19 Nov. 1944. Performance flight on 6 Dec. had to be aborted after 13 minutes, probably because of exhaust fairing deformation. Ferried from Welzow to Merseburg by pilot Lehnhoff on 27-28 Feb. 1945: Welzow-Grossenhain (115 km) on 27 Feb. from 1035 – 1055 hr. Grossenhain to Schkeuditz (180 km) on 27 Feb. from 1530 to 1604 hr. Schkeuditz to Merseburg (60 km) on 28 Feb. from 0850 to 0900 hr.	Kuhn logbook, flight no. 1691 Lehnhoff logbook, flight nos. 1283-85
		Engines 387 528 / 329 006 (2.5 hr., December 1944)	BMW Brief Report
560 043 PT+HC	Ju 388 L-1	Aircraft delivered from Liegnitz to Rechlin on 9 Nov. 1944. After acceptance check on 22 Nov., aircraft was test-flown by Kuhn (1350 to 1425) and reached 9 000 m. At the end of November the aircraft still had internal ram air intakes. Sent to Welzow modification facility for conversion on 29 November.	Kuhn logbook, flight no. 1684

Werknummer **Manufacturer's Code**	**Variant Prototype**	**Remarks/Use/Fate**	**Sources**
		Port engine 329 017, starboard engine 329 010 (December 1944)	BMW Brief Report No. 40
560 044 PT+HD?	Ju 388 L-1	Production aircraft with slightly different camouflage scheme (separation line between top and bottom colors lower on fuselage sides), fairly dense mottling on tail, white *Werknummer*. At war's end at Merseburg, slightly damaged.	
560 045 PT+HE?	Ju 388 L-1	At 388 L-1 modification detachment in Welzow on 5 December 1944. Blown up at Heidkaten near Kaltenkirchen at war's end?	
560 046 PT+HF?	Ju 388 L-1	Construction not verified.	
560 047 PT+HG	Ju 388 l-1	A photo from the Björn Hafsten collection shows a Ju 388 with burnt-out wing tanks, indications of a white *Werknummer* on the tail. The aircraft is painted almost identically to WNr. 560 049. The aircraft has no manufacturer's code.	
560 048 PT+HH?	Ju 388 L-1	Construction not verified.	
560 049 PT+HI?	Ju 388 L-1	Aircraft assembled by Weserflug in Liegnitz, later captured by the Americans in Merseburg. Sole remaining example of the Ju 388, presently in storage at the NASM facility in Silver Hill.	
Aircraft with unidentified or unexplained *Werknummern*.			
Werknummer Not Known TS+YA	Ju 388	This aircraft was flown from Rechlin to Munich-Riem on 15 July 1944 by *Hauptmann* Stahl and observer Auracher. The flight lasted 113 minutes, from 1122 to 1315 hours. Total distance flown was 660 km, for an average speed of 350 kph. The manufacturer's code, which fits no known pattern, is a mystery. Could it possibly have been KS+TA or PG+YA?	Auracher logbook, flight no. 952.
620 305[1000]		The *Werknummer* appears several times as a photo caption (J 58 995) for an interior view of a Ju 388 L-1 cockpit with FA 15 control system. The WNr. actually indicates a Ju 88 G.	Junkers Company Album
WNr. unknown NJ+TS		Entered in pilot Oskar Huber's logbook as a Ju 388. On 24 January 1945, during the evacuation of Gotenhafen/Hexengrund, he flew the aircraft to Lübeck with four other persons on board. The flight covered 310 km and lasted from 1215 to 1351 hours. The code may be a mix-up, as NJ+?? is more appropriate for a Ju 188 E. There is a photo of a Ju 188 E-3 with "normal" BMW 801 engines, equipped with FuG 200 for the torpedo-bomber role. The aircraft has none of the features of the Ju 388, however.	

Appendix 2: Monthly Record for December 1944 (EKdo Ju 388)

Monatsübersicht der Flugzeuge des E-Kommandos Ju 388 Dezember

Dat.	W.Nr. 340084	W.Nr. 340085	W.Nr. 340086	W.Nr. 340088	W.Nr. 340092	W.Nr. 340093	W.Nr. 340094	W.Nr. 540042
1.	-------	-------	Beide Luftschrauben zum Auswuchten abmontiert	-------	Verbrauchsmessung durch BMW	Tw-Wechsel rechts	Seitenruderanschlag angebaut 0^{30}	-------
2.	-------	-------	•	-------	-------	•	-------	-------
3.	------- 0^{25}	-------	•	-------	-------	•	-------	-------
4.	-------	Sichtscheibe vor Beobachtersitz gespr.	-------	Selbstschalter E 1005 defekt	Seitenruderanschlag angebaut	•	-------	-------
5.	------- 0^{24}	lk. Tw. Prop.-Automatik regelt nicht 0^{24}	Prop.Automatk eingeregelt	-------	-------	Tw. rechts eingeregelt	-------	-------
6.	Hydr.Pumpe am recht. Tw. undicht	Hydr.Pumpe fördert nicht.Klimakühler ausgebaut	-------	Kerzenstörung	-------	-------	Arbeiten der ATG-Kolonne	Abgashutze verformt 0^{13}
7.	Kraftstofförderpumpe und Einspritzpumpe ausgebaut	•	Zu Änderungsarbeiten nach Dessau geführt	-------	-------	-------	•	Abgashutze gewechselt
8.	Anlasser ausgebaut	rechtes Tw. springt nicht an		-------	Hydr.Störungen in 7000 m 1^{28}	-------	Ladedruckstörungen. Hydr.Verblockung u. Strebe am Ölkühlr.gew.	FT-Störung
9.	•	Spornrad gewechselt	500 005	Nietarbeiten der ATG-Kolonne	Am rechten Tw.Drucköltg.zur Turbine gerissen	-------	-------	•
10.	Anlasser und Kraftstoffpumpe eingebaut	-------	War bis jetzt ohne Tw. abgestellt.Triebwerkseinbau	-------	Fahrwerksüberprüfung	-------	-------	Nietarbeiten der ATG-Kolonne
11.	Kerzenstörung	Einmotorenlandung in Lärz.R. Tw.starker Ölverlust 1^{40}	•	-------	•	-------	-------	-------
12.	-------	•	•	Anlasserwechsel rechts	Am recht.Tw.Lalukü u.Abgashutze abgebaut	-------	Arbeiten am Auspuff	-------
13.	-------	• 0^{07}	•	Kraftstofförderpumpe gewechselt	Fahrwerksanzeige "aus zeigt nicht an	-------	•	Überprüfung der Höhenflossenbolzen lt.FS von ATG
14.	Preßluftflasche für Einstiegflasche undicht	•	•	•	•	-------	-------	•
15.	•	Kerzenwechsel beider Tw.	Abgashutzen anpassen	-------	-------	Propp.-Automatik r. Tw. unklar	-------	•
16.	Lagerbock an Seitenflosse eingerissen	•	•	Lagerbock an Seitenflosse eingerissen	Ausgleichbehälter für Bremsöl undicht	Kerzen und Düsen gewechselt	-------	-------
17.	Heckwaffenänderung	Lagerbock für Seitenflosse eingerissen	•	•	Störungen an der Propp.-Automatik	•	-------	R. Tw. schüttelt 0^{10}
18.	•	Zylinder 5 abgebaut	Nacharbeiten zum Tw.-Einbau	•	kerzenwechsel	Zylinder 2 wegen Gewindeschaden gew.	-------	Kerzenwechsel
19.	Seitenruder und Flosse abgebaut	•	•	•	Drehzahlgeber gewechselt	Lk.Tw.-Zylinder 2 u. 14 wechseln	-------	Propp.-Verstellgerät gewechselt
20.	Seitenflosse zur Änd.Durchf.i.d.Werft	Seitenflosse und Ruder abgebaut und in der Werft	•	•	-------	•	Schauzeichen "Start" zeigt nicht an	•
21.	•	•	Lagerbock an Seitenflosse eingerissen	Seitenruder und Flosse zur Werft	Zündkabelstörungen	•	-------	-------
22.	•	•	Seitenflosse abmontiert und zur Werft	•	•	•	-------	Anlasserwechsel
23.	•	•	•	•	-------	•	Anlasserwechsel	•
24.	•	•	•	•	Fg.-Anzeige "aus"zeigt nicht an. Beide Tw.-hoher Ölverbrauch 1^{25}	-------	•	Ladedruckstörungen. Fg.-Anzeige unklar 1^{03}
25.	•	•	•	•	•	-------	-------	•
26.	•	•	•	•	•	-------	-------	• 0^{15}
27.	•	•	•	•	-------	-------	Verstärktes Spornrad eingebaut	R.Bremszylinder gew.
28.	•	•	•	•	-------	-------	------- 0^{06}	-------
29.	•	•	•	•	Beide Tw. zu hoher Ölverbrauch	Änderungen am Heckwaffenstand	-------	-------
30.	•	Aufbau d.Seitenflosse	•	Aufbau d.Seitenflosse	•	•	-------	-------
31.	•	•	•	•	•	•	-------	-------
	Flugstunden Dez.: 0^{49} Flugstd. insgesamt: 41^{19}	2^{11} 41^{18}	--- 26^{23}	--- 17^{13}	2^{53} 7^{34}	--- 4^{10}	0^{06} 7^{44}	1^{41} 5^{35}

Werk-Nr. 230153 hatte weiterhin FT-Einbauten.
" " 340089 war zum FuG 200 und 213 C-Einbau in Werneuchen und traf am 29.12. wieder in Rechlin ein
" " 500001 ist zur Durchführung von Änderungen in Dessau
" " 300292 " " " " " " "
" " 300009 ist ohne Triebwerke abgestellt

Notes

1. Stenographic record of the *Reichsmarschall*'s meeting on 9 July 1944, BA-MA RL 3/2579.

2. Extracts from the *Reichsmarschall*'s meetings of 1 and 29 July 1944, BA-MA RL 3/2579.

3. JFM concept "Introduction of the Japanese Military Commission to the Ju 388" of 14 Aug. 1944, NASM MF R 193.

4. *Luftwaffe* Chief-of-Staff letter No. 1150/42 secret command matter of 25 July 1942. Archiv Griehl

5. TAD Report-13/44 by Hirth Motoren G.m.b.H. of 13 Nov. 1944, NASM MF R 3620 F 1209.

6. Unless otherwise advised, the information is based on the homepage of the USAF Museum, http://wpafb.af.mil/museum/research/bombers, as of February 2003.

7. A type very similar to the Lockheed (Super-) Constellation. It was not chosen for development, as the B-29 was preferred.

8. Unlike the Lockheed B-30 and Douglas B-31, the Consolidated B-32 "Dominator" (unofficially "Terminator") at least reached the prototype stage. A small series of 118 bombers was produced. While the pressurized cockpit and remotely controlled gun turrets had to be abandoned during development. the first prototype flew in September 1942. The first production machine was not delivered until late 1944, however. Fifteen aircraft saw action in the Pacific war, partly in the reconnaissance role. Another 100 machines were built prior to the end of the war, but in September 1945 production was cancelled, and some of the just completed aircraft were scrapped. The B-32's maximum speed was 357 mph (575 kph) at 30,000' (9150 m).

9. A four-engined high altitude bomber developed from the Martin B-26 Marauder with a speed of 345 mph (555 kph) at altitudes up to 39,000' (11 890 m). It was cancelled at the prototype stage in favor of the B-29.

10. Aircraft Procurement and Supply war diary, 29 Jan./4 Feb. 1945.

11. For safety reasons, pressure was reduced just before an actual attack to avoid a possibly fatal drop in pressure in the event of battle damage (see account in "Classics of Aviation" by Patrick Hoeveler, Issue 6, Flugrevue Edition, p. 13f).

12. The first operational aircraft, which were accepted by the USAAF in May 1943, in fact flew to India via North Africa. During the transfer one aircraft was sent to England as a diversionary tactic. The aircraft's teething troubles were not ironed out until April 1944. The first high altitude raid on Tokyo took place in November 1944, see Hoeveler, p. 15ff.

13. *Wellington in Action*, Squadron-Signal Publications Aircraft in Action No. 1076, p. 37. According to this source 67 Mk. Vs were originally fitted with Hercules III radial engines, later rebuilt as Mk. VIs or Mk. VIG Oboe target markers. As with the Ju 388, aircraft with the equipment needed for altitudes above 40,000' (12200 m)—pressurized cockpit, extended wingspan, special heating and climate control systems—demanded lengthy testing. These types were the first British aircraft with pressurized cockpits. They flew from November 1941 to mid-1942 at Boscombe Down, one of the most important British experimental stations, with high altitude engines and turbosuperchargers that had an extremely short life of just 10 hours. In fact, the aircraft only reached 31-34,000 ft (9450-10350 m) at about 270-300 mph (434-482 kph); see Tim Mason, *The Secret Years-Flight Testing at Boscombe Down 1939-1945*, Hikoki Publications 1998, p. 84f and 302.

14. See Alfred Price, *Operation Steinbock-The Baby Blitz 1944* in "Aeroplane," September 2002, p. 74ff.

15. This technical parity is confirmed by newer publications; for example Ken Wakefield, *Pathfinder-Luftwaffe Pathfinder Operations over Britain, 1940-44*, Tempus Publishing Inc., 1999, p. 202. Both methods consisted of a combination of radio and long range ground radar navigation.

16. Shorthand minutes of the meeting on 14 May 1943, Milch reference file, IWM, MF Reel 45, Vol. 42.

17. Injection of an oxygen carrier (usually nitrous oxide) to increase performance above the maximum boost altitude.

18. In this context "displaced" means "separated from the engine." More on this in Chapter IV.

19. Shorthand minutes of a meeting in Rechlin on 25 June 1943, Milch reference file, IWM, MF Reel 45, Vol. 42.

20. Erhard Milch (1892-1972), *Generalfeldmarschall*, 1941-1944 Chief of Aircraft Procurement and Supply (successor to Udet), 1938-1945 Inspector-General of the *Luftwaffe*.

21. *Kommando der Erprobungsstellen* (Testing Stations Command), *Oberst* Edgar Petersen was commander of testing stations from 1941 until the end of the war.

22. Nevertheless, "Junkers" was the customary designation for the company and aircraft plant.

23. U.S. Strategic Bombing Survey, Air Frames Plant Report No. 1, Junkers Aircraft and Aero-Engine Works, Dessau, p. 6, DMM LR 02459. This figure is probably quite reliable, as Junkers employees were interrogated for this report, which was compiled from 4 to 18 May 1945.

24. Apart from the Ju 488, which never flew.

25. Wolfgang Wagner, *Hugo Junkers-Pionier der Luftfahrt*, Bernhard & Graefe 1996, p. 280 f.

26. Peter Kneutz, *Einsteigen zum Flug!*, R. Voigtländers Verlag Leipzig 1938, p. 13.

27. *Luftfahrt-Geschichte 3/73, Henschel Hs 130*, Schmidt-Publikation, Stuttgart, p. 1.

28. Junkers-Chronik, p. 72, DMM LR 02462.

29. *Luftfahrt-Geschichte 5/73, Junkers Ju 86 P-1*, Schmidt-Verlag, Stuttgart, p. 2. The Ju 86 P was originally designated the "H" for high altitude aircraft. The "H" also appears temporarily in association with Ju 188 prototypes with BMW 801 engines in the Junkers company albums in the DMM.

30. Dietmar Hermann, *Focke-Wulf Ta 152-The Path to the High Altitude Fighter*, Aviatik Verlag 1998, p. 141f; performance data for the Ta 152 H-0 without and the Ta 152 H-1 with MW 50 and GM 1 injection, indicating that it was at least equal to the late war versions of the Supermarine Spitfire Mk. XIV and North American P-51 D Mustang.

31. "Turbine Air-Jet Engines," or "TL-Engines" for short, was the standard term of the day for what we today call jet engines.

32. Milch himself used a Ju 188 E-1 as his personal transport. The aircraft, with the manufacturer's code ST+GL, was allegedly given to Milch as "Chief of Aircraft Procurement and Supply with the rank of a State Secretary," see Green, "Warplanes of the Third Reich," p. 489.

33. An irony of fate that he was not incorrect. What he could not know, however, was that it would be with the French naval air arm, which would operate the Ju 188 until 1947. Some of the E/F machines were captured, however, after the ceasefire the Aeronavale had S.N.C.A. de Sud-Est in Toulouse build 12 E-series machines, in part using German components; see Green, p. 490.

34. Minutes of a meeting on the Jumo 222 on 14 Jan. 1944, NASM MF R 2825 F 199.

35. Amendment IV to the JFM construction directive 7 388 26 10 / 000 "Construction of 50 Ju 388 K-0 (formerly Ju 188 K-0) bomber aircraft" of 12 Aug. 1944, NASM MF R 193.

36. **C**ombined **I**ntelligence **O**bjectives **S**ubcommittee, jointly operating British and American intelligence units, unlike the "BIOS teams," which were purely British units.

37. "German Activities in the French Aircraft Industry," report by CIOS Team No. 18, Combined Intelligence Objectives Subcommittee G-2 Division, SHAEF (Rear) APO-413, IWM.

38. Apart from the Ju EF 126, which was conceived as a suicide machine.

39. Weekly report of the *E-Stelle* Rechlin for the period 6-12 Aug. 1944 (12 Aug. 1944), BA-MA RL 36/86.

40. Armaments Staff Report of 25 Sept. 1944, point 7/925/1245, Griehl Collection. The Me 264 was cancelled at the same time.

41. Described in detail with original documents in Stephen Ransom, *Zwischen Leipzig und der Mulde-Flugplatz Brandis 1939-1945*, Stedinger Verlag, p. 71 ff.

42. The famous fighter pilot of the First World War (62 victories), who between the wars built aircraft and enjoyed great success as a stunt pilot. He committed suicide in Berlin on 17 Nov. 1941.

43. See Lutz Budrass, "*Flugzeugindustrie und Luftrüstung in Deutschland 1918-1945*," p. 738.

44. Pilot's notes for the Ju 88 A-4 and variants derived from it, October 1942; Hafner Archive.

45. See Griehl, , "*Ju 88 Star of the Luftwaffe*," p. 74. Number of D-6s built based on a reader's letter from Peter Achs in Jet & Prop Foto-Archiv Band 8, p. 94. Amendment Sheet No. 2 to the Ju 88 A-4 Aircraft Handbook dated March 1944 states that the D-6 and D-7 differed, in that the ventral gondola was deleted from the latter variant.

46. JFM construction directive 71880010/025/000 of 21 Oct. 1942, NASM MF R 193.

47. Report for the month of March 1943 by JFM aircraft marketing of 15 Apr. 1943, NARA MF 177 35.

48. EHF file memo No. 171/43 of 28 March 1943 concerning comparative trials at Rechlin. The first comparative trial between the Ar 240 V10, Ju 88 V56 (S prototype), Me 410, and He 219 took place there on 19 Jan. 1943; see *E-Stelle* report in Hans-Jürgen Becker, "*Schwere Jäger und Zerstörer der Luftwaffe 1933-1945*," Motorbuch Verlag 1999, p. 125 f.

49. According to Gebhard Aders ("Geschichte der deutschen Nachtjagd 1917-45"), there were two mockups (p. 371). Hanfried Schliephake ("Flugzeugbewaffnungen," p. 155) claims that three "Ju 188 R-0s" were fitted with two MK 103s. Griehl ("Taktische Militärflugzeuge in Deutschland 1925 bis heute," p. 62) also mentions a Ju 188 R prototype. The possibility of mixup with Ju 88 G or Ju 388 prototypes cannot be excluded.

50. Report for the month of September 1943 by JFM aircraft marketing dated 22 Oct. 1943, NARA MF 177 35.

51. Bukowski/Griehl, "*Junkersflugzeuge 1933-1945-Der Illustrierte Originalbericht des Professor Brunolf Baade an die Sowjetische Militäradministration, Dessau 1946*," Karl Müller Verlag, Erlangen, p. 52.

52. Report for the month of May 1943 by JFM aircraft marketing dated 15 June 1943, NARA MF 177 35.

53. ATG TH-Overview 90541 "Aircraft Fabrication Progress Report," status as of 13 Dec. 1944, dated 15 Dec. 1944, BA-MA RL 3/1277.

54. According to Delivery Plan 227, Issue 1, of 15 Dec. 1944, *Lufthansa* in Staaken was to convert the last 22 of a total of 52 Ju 88 T-3 between Dec. 1944 and Apr. 1945, BA-MA RL 3/1045.

55. Report for the month of June 1943 by JFM aircraft marketing dated 22 July 1943, NARA MF 177 35.

56. Report for the month of September 1943 by JFM aircraft marketing dated 22 Oct. 1943, NARA MF 177 35.

57. Report for the month of December 1943 by JFM aircraft marketing dated 25 Jan. 1944, NARA MF 177 35.

58. Excluding the V55, mathematically one comes up with 57 Ju 88 S-1s. This might be because, according to the December report, production of Ju 88 S-2 aircraft had been changed into S-1s.

59. Griehl, "*Ju 88-Star of the Luftwaffe*," p. 80f.

60. Injection of an oxygen carrier (usually nitrous oxide) to increase performance above the maximum boost altitude.

61. Shorthand minutes of the meeting on 24 May 1943, Milch file, IWM, MF Reel 45, Vol. 42.

62. Construction directive 7 188 26 10/001 of 24 May 1943 NASM MF R 193, also see transcript!

63. It was not unusual to "open up" a new hundred-thousand block for a new type. The Ju 388 used two. By the way, the Ju 188 also began with the WNr. 10 000, and the Ju 88 G with the block 700 000.

64. *Lotfe = Lotfernrohr*, optical bombsight.

65. Amendment I to the JFM construction directive 7 388 26 10/001/000 of 6 Aug. 1943, NASM MF R 193.

66. Report for the month of September 1943 by JFM aircraft marketing dated 22 Oct. 1943, NARA MF T 177 35.

67. Letter from Junkers to the Reich Minister of Aviation and Commander-in-Chief of the *Luftwaffe* of 14 Dec. 1943, EADS Unternehmensarchiv Ottobrunn

68. Amendment I to JFM construction directive 7 188 00 10/037/001 of 26 May 1943, NASM MF R 193.

69. Amendment I to JFM construction directive 7 188 00 10/038/001 of 31 May 1943, NASM MF R 193.

70. Shorthand minutes of the meeting of 26 June 1943, Milch file, IWM, Reel 45, Vol. 2.

71. Report for the month of June 1943 by JFM aircraft marketing dated 22 July 1943, NARA MF T 177 35.

72. See Griehl, Do 217/317/417, p. 199 f.

73. See Petrick/Mankau, p. 52 f.

74. Shorthand minutes of the meeting of 25 June 1943 in Rechlin, Milch reference file, IWM, MF Reel 45 Vol. 42.

75. Quoted from Hentschel, "*Die Geheimen Konferenzen des Generalluftzeugmeisters*," Bernard & Graefe, p. 156.

76. Shorthand minutes of the meetings on 2 and 3 Sept. 1943 in the Reichsjägerhof, participants included Göring, Milch, Gen. d. Flieger Korten, Gen. d. Kampfflieger Peltz, Baumbach, Brauchitsch, Diesing, and Knemeyer, Milch reference file, IWM, MF Reel/Volume 62.

77. Report for the month of September 1943 by JFM aircraft marketing dated 22 Oct. 1943, NARA MF T 177 35.

78. According to the December report, in addition to the Ju 88 V89 and V90, prototypes for the extended fuselage Ju 88 H-1, two Ju 88 H-1s converted in Merseburg were also delivered in the last quarter of 1943. According to the Junkers chronicle, the Ju 88 H-1 first flew on 2 Nov. 1943 with Rupprecht Wendel at the controls.

79. Report for the month of December 1943 by JFM aircraft marketing dated 25 Jan. 1944, NARA MF T 177 35.

80. If one accepts the arrival stamp (21 Feb. 1944) of the Dept. of Training Materials and Film, whose arrival stamp was usually three to ten days later than the report date, the amendment was probably written in February 1944.

81. NARA MF T 177 35, other new additions to the program were 218 Ju 188 H-2s and 638 Ju 188 G-1s.

82. As to the probable reasons, see Footnote 80.

83. Report No. 93 on the GL/Office Head meeting on 12 Oct. 1943, Archiv Griehl.

84. Griehl, Do 217/317/417, p. 66.

85. Shorthand minutes of the meeting of 4 Nov. 1943 from 1100 to 1300 hours in the Junkers works, Milch reference file, IWM, MF Reel/Volume 63.

86. Report No. 100 on the GL/Office Head meeting on 17 Dec. 1943, Archiv Griehl.

87. Milch reference file, IWM, MF Vol. 41 Reel 44.

88. M. Griehl has provided us with an extract from these notes; the original is considered lost.

89. Copy of the Junkers Chronicle, DMM LR 02462; according to this, various Ju 288 prototypes were demonstrated on 5 Nov. 1943, along with the Ju 390 and Ju 352.

90. David Irving, "*Die Tragödie der Deutschen Luftwaffe-Aus den Akten und Erinnerungen von Feldmarschall Milch*," p. 327, p. 459.

91. Milch reference file, IWM, MF Vol. 42 Reel 44.

92. Karl Koller, General der Flieger, 1943-44 Head of the *Luftwaffe* Operations Staff, 1944/45 Chief-of-Staff.

93. Milch reference file, IWM, MF Vol. 42 Reel 44.

94. Attachment to letter from the *Technische Amt* GL/C-B GL/C No. 1563/43 secret command matter, see Georg Hentschel, "Die geheimen Konferenzen des Generalflugzeugmeisters," Bernard & Graefe Verlag, S.208f.

95. Draft of a file memo on the Gen.Qu.6.Abt. meeting on 12 June 1944, Archiv Griehl

96. Points from a meeting with RM Göring on 1 July 1944, dated 2 July 1944, Archiv Griehl

97. Shorthand minutes of the Reichsmarschall meeting on 9 July 1944, BA-MA RL 3/2579.

98. Excerpts from the Reichsmarschall meetings on 1 and 29 July 1944, BA-MA RL 3/2579.

99. JFM test report Fla 528/44 of 4 Aug. 1944, NASM MF R 186.

100. J. Muttray, "*Anlage und konstruktiver Aufbau der Höhenkammern*," speech to the Lilienthal Society for Aviation Research on 17 January 1941 in Dessau, in: "*Jahrbuch 1942 der deutschen Luftfahrtforschung*," p. I 617 f.

101. JFM construction directive 71880010/025/000 of 21 Oct. 1942, NASM MF R 193.

102. Junkers company album on the Ju 188, DMM LR 02751.

103 JFM memo Kobü-My 543 of 3 May 1944, NASM MF R 2196 F 404

104. Description of the climate control system by an unidentified Junkers engineer dated 7 Feb. 1944, NASM MF R 186.

105. JFM memo Kobü-Fla 352/44 of 5 May 1944, NASM MF R 193.

106. JFM memo Kobü-Rh 3156 of 7 March 1944, NASM MF R 186.

107. JFM memo Kobü-Rh 2971 of 15 Feb. 1944, NASM MF R 2196 F 404.

108. JFM test report No. 1/112/221 of 3 March 1944, NASM MF R 199.

109. JFM memo Kobü-Rh 3155 of 6 May 1944, NASM MF R 2196 F 404.

110. JFM memo FTV-Au 568/44 of 10 May 1944, NASM MF R 193.

111. Telegram from the *E-Stelle* Rechlin E2 to the BAL at Junkers Dessau dated 10 Aug. 1944, EADS Unternehmensarchiv Ottobrunn

112. JFM memo Kobü-Rh 3449 of 7 Aug. 1944, NASM MF R 193.

113. JFM preliminary report Ju 388 No. 1/173 of 27 May 1944, NASM MF R

114. Photo album in LR 02734, 02735, 92751, 02840-42, special collections of the Deutsches Museum.

115. The date of the work is verified by a production drawing made by the participating JFM second-year apprentices which is dated "4th to 24th Jan. 1943."

116. Information from interviews with *Herr* Schönfeld and Dietmar Kuropka of the Arbeitskreis Bernburg, written messages from *Herr* Kurt Rusch and *Herr* Wilhelm Muschter by way of *Herr* Kuropka on 28 Oct. 2002, Karl-Heinz Pitschler, Jürgen Bressem, and an original "Work Report" by Junkers apprentices from Bernburg 1941-42.

117. Junkers Chronicle, p. 80, DMM LR 02462.

118. Griehl, "*Taktische Militärflugzeuge in Deutschland 1925 bis heute*," p. 62.

119. Time sequencing of Junkers company photographs using the Stephen Ransom system.

120. Griehl, "*Ju 88-Star of the Luftwaffe*," p. 35; only one three-seat prototype was built (SL+PC, WNr. 140 561).

121. Luftfahrt International 3/1979, p. 124.

122. JFM memo Kobü-Mu/TL 2628 of 15 Nov. 1944, DMM and NASM MF R 2144 F 822.

123. JFM memo Kobü-Mu/TL FA-15 1/57 of 22 Apr. 1944, DMM and NASM MF R 2774 F 25.

124. JFM memo Kobü-Rh 2707 of 8 Nov. 1944, NASM Shorthand minutes of the meeting of R 2820 F 457.

125. Minutes of Reichsmarschall meeting on 9 July 1944, BA-MA RL 3/2579.

126. Minutes of the fifth Working Party Congress W 39 / g at the Rapido-Werken, dated 30 Oct. 1944.

127. Type testing 8-388, test report on the L-1 series, status as of 31 Jan. 1945, NASM MF R 193.

128. JFM memo Kobü-Rh 2967 of 14 Feb. 1944, NASM MF R 2196 F 404.

129. Interim report S 44/62-0 by the Fluid Mechanics Department of 29 Aug. 1944, EADS Unternehmensarchiv Ottobrunn

130. Letter from the Arado Flugzeugwerke GmbH to JFM on 16 Feb. 1945, DMM LR

131. A very common glass made from quartz sand.

132. JFM memo Kobü-Rh 3087 of 6 Apr. 1944, EADS Unternehmensarchiv Ottobrunn

133. Design drawings for this tank were not completed and turned over to the Raspe company until October 1944, see Kobü-Rh memo 3658 of 16 Oct. 1944, EADS Unternehmensarchiv Ottobrunn

134. Could also be "TW."

135. *E-Stelle* Travemünde, December report by Dept. E2/3, Group E3T, B. No. 611/44 secret command matter, NA RA MFT 177 40.

136. Travemünde's responsibility for fuel jettisoning systems, including for the Do 335, may be easily explained. Tests could be carried out over water without worrying about jettisoned fuel affecting fields and forests.

137. Junkers production document F-388.00.02 of 2 Aug. 1944, DMM

138. Fighter Staff 6 June 1944, Milch reference file, IWM archive

139. Amendment V to construction directive 7 388 26 10/000 "Construction of 50 Ju 388 K-0 bomber aircraft (formerly Ju 188 K-0)" of 23 Oct. 1944, NASM MF R 193.

140. This involved carrying out the modifications described in Change Advance Notices 2098/44, 2119/44, and 2126/44. *f*V 2126/44 concerned changes to the "Belt Insertion Symbols."

141. Amendment VI to construction directive 7 388 26 10/000 "Construction of 50 Ju 388 K-0 bomber aircraft (formerly Ju 188 K-0)" of 5 Dec. 1944, EADS Unternehmensarchiv Ottobrunn

142. *E-Stelle* Rechlin weekly report for the period 23-29 July 1944 (29 July 1944), BA-MA RL 36/86.

143. NASM MF R 2734/F 227, part of a report.

144. Stenographic report on the Fighter Staff meeting of 3 July 1944, BA-MA RL 3/9 p. 192 f.

145. Amendment IV to JFM construction directive 7 388 26 10/000 "Construction of 50 Ju 388 K-0 bomber aircraft (formerly Ju 188 K-0)" of 12 Aug. 1944, NASM MF R 193.

146. Amendment II to JFM construction directive 7 388 26 10/000 "Construction of 50 Ju 388 K-0 bomber aircraft (formerly Ju 188 K-0)" of 19 June 1944. EADS Unternehmensarchiv Ottobrunn

147. Amendment I to construction directive 7 388 26 10/000 "Construction of 50 Ju 388 K-0 bomber aircraft (formerly Ju 188 K-0)" of 4 May 1944, NASM MF R 193.

148. JFM memo Kobü-Rh 3577 of 22 Sept. 1944, NASM MF R 193.

149. JFM memo Kobü-Rh 3207 of 17 May 1944, NASM MF R 2196 F 404.

150. Amendment III to JFM construction directive 7 388 26 10/000 of 18 July 1944, EADS Unternehmensarchiv Ottobrunn

151. Amendment I to construction directive 7 188 00 10/038/001 of 31 May 1943, NASM MF R 193

151a. Installation of the FuG 120a in the Ju 388 K-0 and K-1 was ultimately abandoned; see *f*V 2474/44 vom 30.7.1944.

152. "*Die V1-Eine Dokumentation*," Bechtle Verlag, Esslingen-Munich, 2nd Edition 1991, Wilhelm Hellmold, p. 69 ff.

153. Minutes of the Reichsmarschall meeting on 9 July 1944, BA-MA RL 3/2579.

154. Reichsmarschall meeting of 23/25 May 1944, BA-MA RL 3/2580.

155. Minutes of designers' meeting with the Reichsmarschall on 25 May 1944, BA-MA RL 3/2580.

156. JFM aircraft marketing division monthly report of 30 Oct. 1944, NARA Shorthand minutes of the meeting of T 177 35.

157. ATG TH-Overview 90541 "Aircraft Fabrication Progress Report," status as of 13 Dec. 1944, dated 15 Dec. 1944, BA-MA RL 3/1277.

158. Letter from Dzulko to Zindel dated 9 Jan. 1945, NASM MF R 2196 F 404.

159. Minutes of the Fighter Staff meeting on 6 June 1944, Milch reference file, IWM.

160. FTA = *Flugzeugvertrieb Technischer Aussendienst* (Aircraft Marketing Technical Field Service).

161. JFM Kobü-Entwurf 88/392, p. 2 ff. Page 1 is unfortunately missing. NASM MF R 186; other differences lay in the landing flaps to be used, the ailerons, and the wing and horizontal stabilizer deicing. Remarkably, this comprehensive overview of all bomber, torpedo-bomber, (long-range) reconnaissance, and night fighter variants of the Ju 88/188/388 makes no mention of the Ju 188 S or T versions.

162. Torpedo Setting Controls (instruments in cockpit) and Transmitting Connecting Shafts; see "*Der Lufttorpedo-Entwicklung und Technik in Deutschland 1915-1945*," Friedrich Lauck, Bernhard & Graefe Verlag, 1981, p. 251.

163. Construction directive 7 388 00 10/011 "Development and Prototype Installation L 10 in Ju 388 M-1" of 8 Sept. 1944, EADS Unternehmensarchiv Ottobrunn.

164. *E-Stelle* Travemünde, monthly report for Oct. 1944 by Dept. E 4/5, Group E 5, B No. 535/44 secret command matter of 28 Nov. 1944, NARA Shorthand minutes of the meeting of T 177 40.

165. JFM FTH Overall Planning, production plan E 189, page 4 "Torpedo-Bombers, Transports, Dive-Bombers" of 18 March 1944, BA-MA RL 3/1048.

166. GL/C delivery plan 1486/44 of 15 May 1944, 1st Issue, Page 4 "Aircraft Program Junkers Aircraft" (Special Committee F1), secret command matter 1486/44 of 15 May 1944, BA-MA RL 3/1040.

167. Amendment I to JFM construction directive 7 188 00 10/037/001 of 26 May 1943, NASM MF R 193.

167a. Advance Notification of Alteration *f*V 4382/44 of 26 Apr. 1944, NASM MF R3798 F617.

168. JFM test report memo "Ju 388 general" of 7 Sept. 1944, NASM MF R 193.

168a. Advance Notification of Alteration *f*V 5186/44 of 6 Jul. 1944, NASM MF R3798 F584.

169. JFM-FSD technical description NASM R 2615 F 774.

170. JFM alteration message 64044/44 of 30 Sept. 1944, EADS Unternehmensarchiv Ottobrunn.

171. JFM memo Kobü-Rh 3548 of 11 Sept. 1944, EADS Unternehmensarchiv Ottobrunn.

172. JFM file memo Kobü-Rh 3807 of 27 Nov. 1944, NASM MF R 2196 F 404.

173. The Reich Minister for Armaments and War Production, extracts from No. M.T. 99 12320/45 secret, abbreviated copy dated 5 January 1945, NASM R4098 F650.

174. Armaments Staff joint report from 10 Jan. 1945, BA-MA RL 3/3195.

175. JFM construction directive 7 388 00 10/008/000 of 12 Aug. 1944, EADS Unternehmensarchiv Ottobrunn.

176. JFM memo Kobü-Rh 3714 of 31 Oct. 1944, NASM MF R 2196 F 404.

177. JFM memo Kobü-R 144 of 13 Sept. 1944, DMM.

178. JFM memo Kobü-Rh 3787 of 21 Nov. 1944, NASM MF R 2196 F 404.

179. JFM memo Kobü-Rh 3673 of 19 Oct. 1944, NASM MF R 2196 F 404.

180. JFM journal FTH-FN/88/44 of 12 Oct. 1944, NASM MF R 2196 F 404.

181. JFM memo Kobü-Rh 3230 of 25 May 1944, NASM MF R 193.

182. JFM memo Kobü-Rh 3668 of 19 Oct. 1944, NASM MF R 193.

183. JFM memo Fp-M 58/44 of 27 May 1944, NASM MF R 2196 F 404.

184. JFM memo M/Fb-Ltg. 44/186 of 2 June 1944, NASM MF R 2196 F 404.

185. Ju 188 E-1 Handbook, Part 8G, Vol. 3, November 1943 issue.

186. JFM memo Kobü-Rh 3473 of 14 Aug. 1944, NASM MF R 193.

187. See, for example, JFM memo Kobü-Rh 3669 of 10 July 1944, NASM MF R 2144 F 35.

188. For space reasons, the topic of mixed propulsion will not be addressed, however.

189. JFM file memo 1300/43 of 6 Oct. 1943, NASM MF R 2476 F 1031.

190. JFM memo EL 282/43 of 25 Oct. 1943, NASM MF R 2814 F 385.

191. Report No. 93 on the GL/Office Head meeting of 12 Oct. 1943, Archiv Griehl.

192. Type Testing 8-388, test report on the L-1 series, status as of 31 Jan. 1945, NASM MF R 193.

193. JFM memo Kobü-Ew 4365 concerning complaints about the BMW 801 TJ power plant of 1 Nov. 1943, BMW Archive and NASM MF R 2768 F 320.

194. JFM memo about meeting with *Herr* Schilo in the RLM on 15 Nov. 1943, topic DB 603 and Jumo 213 engines with exhaust driven turbosuperchargers, DMM LR 01078.

195. Characteristic values for the BMW 801 J, K (P8001) engines, undated, MTU archive T.1.6.801.9.1.

196. BMW continuous flow machines report No. 5068 of 3 March 1942, MTU Archive.

197. BMW continuous flow machines report No. 4015 of 29 Jan. 1942, MTU Archive T.1.6.801J.5.

198. Letter from Dipl.-Ing. A. Müller to the Heinkel and Hirth company, Zuffenhausen, of 30 Oct. 1944, NASM Shorthand minutes of the meeting.

199. Registered letter from the Hirth-Motoren works of EHAG to the OKL, Chef TLR/F1-E 3 II of 7 Nov. 1944, NASM MF.

200. BMW continuous flow machines report No. 7000 of 18 March 1944, NASM MF R 3448 F 1161 and MTU Archive.

201. ata = absolute pressure in atmospheres. No longer commonly used.

202. Familiarization lecture for pilots on the BMW 801 TJ power plant in the Ju 388, BMW EZA/2d, of 5 Feb. 1945, NASM MF R3212 F1058.

203. Technical description No. 110145 Ju 388 L-1 and J-3 by JFM, undated, Deutsches Museum.

204. BMW continuous flow machines report No. 4016 of 9 March 1942, MTU Archive.

205. Report by BMW's Main Department for Aero-Engine Testing V 323 E/T 640, running No. 542/43 of 13 July 1943, MTU Archive.

206. JFM memo Kobü-Ew 3257 of 20 Oct. 1942, MTU Archive.

207. BMW, Technical Field Service EZA/2c, report of 6 Oct. 1942, BMW Archive.

208. BMW EZA/2 brief report No. 33 on prototype installation and testing by field offices dated 14 Oct. 1944, NASM MF R2411 F835.

209. Letter from *E-Stelle* Rechlin dated 11 Aug. 1944, NASM MF R 193.

210. Shorthand minutes of meeting on 14 May 1943, Milch reference file, IWM, London, Reel 45, Vol. 42.

211. Shorthand minutes of the meeting on 11 June 1944, Milch reference file, IWM, London, Reel 45, Vol. 42.

212. Report No. 100 on the office chief's meeting on 7 Dec. 1943, Griehl Archive.

213. Shorthand minutes, RLM development conference on 14 Apr. 1944, Milch reference file, IWM, London, Reel 45 ,Vol. 42.

214. The two aircraft envisaged for construction of the Ju 488 had probably come from the original 70 aircraft (20 L-0 and 50 K-0).

215. JFM memo Kobü-L 2349 of 3 May 1944, NASM MF R 2196 F 404.

216. Shorthand report on the Fighter Staff meeting of 18 July 1944, BA-MA RL 3/9, p. 414.

217. Valid only with use of the 9-9500.30 mechanical position indicator. JFM memo Kobü-Ew. 2426 of 17 Aug. 1944, NASM MF R 2761 F 827.

218. JFM memo Kobü-Ew 2411 of 20 July 1944, NASM MF R 2733 F 332.

219. BMW brief report No. 10 on prototype installation and testing by field offices of 16 March 1944, NASM MF R 2284 F 844.

220. JFM memo Kobü-Rh 2428 of 17 Aug. 1944, NASM MF R 2768 F 474.

221. Message from head of development of 16 Nov. 1944, NASM MF R 2196 F 404.

222. JFM memo Kobü-Rh 3401 of 25 July 1944, NASM MF R 2196 F 404.

223. Allgemeine Transportanlagengesellschaft GmbH.

224. BMW EZA/2 brief report No. 41 on prototype installation and testing by field offices of 14 Dec. 1944, NASM MF R 2648 F 380.

225. Complaints about BMW 801 TJ-0 power plant dated 11 Dec. 1944, NASM MF R 8079 F 1081.

226. Journal No. K 1611 of 23 Dec. 1944, NASM MF R 2770 F 191.

227. JFM memo Kobü-Rh 3790 of 21 Nov. 1944, DMM and NASM MF R 2782 F 16.

228. JFM memo FlA. 500/44 of 22 July 1944, NASM MF R 2725 F 1096.

229. BMW EZW report 10/744, running no. EZV 765/44, NASM MF R 2381 F 277.

230. JFM memo 718/44 of 19 Sept. 1944, NASM MF R 2762 F 981.

231. BMW EZA/2 brief report No. 32 on prototype installation and testing by field offices of 9 Oct. 1944, NASM MF 2411 F 838.

232. BMW EZA/2 brief report No. 41 on prototype installation and testing by field offices of 14 Dec. 1944, NASM MF R 2648 F 380.

233. JFM file memo Kobü-L 2637 of 20 Oct. 1944, EADS Unternehmensarchiv Ottobrunn.

234. "Schaper" appears in the memo, however, in view of the numerous surviving reports by an Engineer Schapert of the BMW field service with JFM, one may assume that the name that appears in the memo is the result of an onomatopoeic mix-up.

235. BMW trip report No. 1024/44 by Ing. A. Reichel of 24 Nov. 1944, NASM MF R 2411 F 746.

235a. On this day pilot Lehmann and flight test engineer Gödecke carried out a speed course and high altitude performance measurement flight at Merseburg; see Lehmann logbook, Achs Collection.

236. Fp.-Baum.-Mittlg. Nr. 930/44 of 9 Dec. 1944, NASM MF R 2766 F 696.

237. BMW EZV report No. 1124/44 of 18 Dec. 1944, NASM MF R 2141 F 352.

238. Draft of an engine handbook for the Jumo 213 A-1 and C-0, Dessau, December 1943.

239. Appendix: Engine Development (Otto Maderwerk), DMM, NASM MF 3979 F 190.

240. JFM test report No. 2195-AD of 25 May 1944, DMM.

241. Report by JFM Kobü-Vorentwicklung "Work Carried out in Months of November and December 1944" of 28 Jan. 1945, DMM LR 01078.

242. *E-Stelle* Rechlin weekly report for the period 14-20 Jan. 1945 of 27 Jan. 1945, BA-MA RL 3/2573 p. 93.

243. October 1944 report by the Jumo Development Department of 2 Nov. 1944, NASM MF R 2853 F 190.

244. November 1944 report by the Jumo Development Department of 4 Dec. 1944, NASM MF R 2853 F 204.

245. October 1944 report by the Jumo Development Department of 2 Nov. 1944, NASM MF R 2853 F 190.

246. November 1944 report by the Jumo Development Department of 4 Dec. 1944, NASM MF R 2853 F 204.

247. These aircraft are mentioned in the monthly reports as having had engine failures, which of course does exclude the possibility that the Jumo 213 E operated satisfactorily in other machines!

248. "Answers to the Fighter Command Questionnaire by Oberst Petersen KdE" of 18 June 1945, AIR 40/2836, Public Record Office, Kew.

249. JFM interim report S. 44/12-0 of 24 Feb. 1944, NASM MF R 2612 F 988.

250. Amendment 1 to the JFM memo Kobü-Rh 3193 of 31 May 1944, NASM MF R 2196 F 404.

251. JFM memo M-Fb.Bm./44/282 of 5 July 1944, NASM MF R 2196 F 404.

252. Amendment I to construction directive 7 388 00 10/007/001 "Construction of 5 Test Beds from Ju 388 with Jumo 213 E" of 18 July 1944, EADS Unternehmensarchiv Ottobrunn.

253. Amendment IV to construction directive 7 388 26 10/100 of 12 Aug. 1944, NASM MF R 193.

254. Letter from the JFM-FTV to Chef TLR E Stab 2 of 7 Sept. 1944, EADS Unternehmensarchiv Ottobrunn.

255. JFM construction directive 7 388 26 10/001 "Construction of one test bed each from the Ju 388 L-0 series with Jumo 213 A and Jumo 213 E" of 25 Sept. 1944, EADS Unternehmensarchiv Ottobrunn.

256. Amendment I to construction directive 7 388 26 10/001 "Construction of one test-bed each from the Ju 388 L-0 series with Jumo 213 A and Jumo 213 E" of 26 Oct. 1944, EADS Unternehmensarchiv Ottobrunn.

257. The wording "local modification facility" appears in the original text.

258. Brief report Br. B. No. 28/45 secret by the 3./Versuchsverband des O.K.L. of 27 Feb. 1945, NASM MF R 187.

259. JFM construction directive 7 388 00 10/013/00, "Construction of 6 Ju 388 J-3 Prototype Aircraft for Use as Mosquito Hunters" of 11 Nov. 1944, NASM MF R 193.

260. Amendment I to the JFM construction directive 7 388 00 10/013/00 "Construction of 6 Ju 388 J-3 Prototype Aircraft for Use as Mosquito Hunters" of 20 Nov. 1944, NASM MF R 193.

261. JFM short message from FSD FTV-Au to FSD Fevü-Bm of 9 Jan. 1945, NASM R 193.

262. JFM memo Kobü-Ew 2396 of 17 June 1944, NASM R 2812 F 1192.

263. JFM memo Kobü-Ew No. 2401 of 29 June 1944, NASM MF R 2768 F 303.

264. JFM memo Kobü-Ew No. 2434 of 30 Aug. 1944, NASM MF R 2767 F 411.

265. JFM-FSD technical description, NASM R 2615 F 774.

266. Junkers Chronicle p. 75, DMM LR 02462.

267. Minutes of meeting concerning the Jumo 222 of 14 Jan. 1944, NASM MF R 2825 F 199.

268. Minutes of meeting at the *E-Stelle* Rechlin on 19 Nov. 1944, NASM MF R 3285 F 432.

269. OMW-Kobü file memo Piston Aero-Engines II of 30 Nov. 1944 concerning the status of the Jumo 222, NASM MF R 3279 F 1099.

270. JFM Engine Sales activity report for January to March 1944, NARA MF T 177 35.

271. JFM Engine Sales activity report for April to June 1944, NARA MF T 177 35.

272. JFM Engine Sales activity report for July to September 1944, NARA MF T 177 35.

273. Jumo monthly report for October 1944 by Engine Test Stand Bureau of 4 Nov. 1944, DMM LR 01078 and NASM MF R 2700 F 484.

274. Jumo Engine Test Stand Bureau quarterly report October-December 1944 of 8 Jan. 1945, NASM MF 2853 F 166.

275. Jumo monthly report for December 1944 by the Engine Construction/Technical Field Service Department of 2 Jan. 1945, DMM LR 01078 and NASM MF R 2853 F 218.

276. *E-Stelle* Rechlin weekly report for the period 14-20 Jan. 1945 of 27 Jan. 194, BA-MA RL 3/2573 p. 93.

277. Monthly report for January 1945 by OMW Test Stand 2 concerning Jumo 222, dated 5 Feb. 1945, DMM LR 01078.

278. Letter from Dipl.-Ing. A. Müller to the Heinkel and Hirth company, Zuffenhausen, of 30 Oct. 1944, NASM Shorthand minutes of the meeting of.

279. Registered letter from the Hirth-Motoren works of EHAG to Fl.-Oberstabsing. Mann of the OKL, Chef TLR/F1-E 3 II of 4 Nov. 1944, NASM MF.

280. Letter from the *Technische Amt* GL/C-E No. 11 345/44 secret of 17 Apr. 1944, EADS Unternehmensarchiv Ottobrunn.

281. JFM memo Kobü-Rh 3807 of 27 Nov. 1944, NASM MF R 2196 F 404.

282. JFM Engine Sales file memo of 4 may 1944, NASM MF R 2196 F 404.

283. JFM construction directive 7 388 00 10/005/000 of 11 May 1944, NASM MF R 193. According to this, the advance order originated from 14 March 1944.

284. JFM memo Kobü-Rh 3227 of 23 May 1944, NASM MF R 186.

285. 1st Amendment to JFM memo Kobü-Rh 3227 of 21 June 1944, NASM MF R 186.

286. JFM memo Kobü-Rh 3226 of 24 May 1944, NASM MF R 186.

287. JFM memo Kobü-Ew 2407 of 11 July 1944, NASM MF R 2734 F 208.

288. Amendment I to construction directive 7 388 00 10/005/001 of 12 Aug. 1944, NASM MF R 193.

289. Amendment IV to construction directive 7 388 26 10/000 of 12 Aug. 1944, EADS Unternehmensarchiv Ottobrunn.

290. JFM memo Kobü-Rh 3562 of 15 Sept. 1944, NASM MF R 186.

291. Construction directive 7 388 00 10/006/000 of 30 Oct. 1944, NASM MF R 193.

292. Amendment II to construction directive 7 388 00 10/005/002 "Construction of 6 test beds from Ju 388 with 9-8222 A/B (Jumo 222)" of 13 Nov. 1944, NASM MF R 193.

293. Amendment III to construction directive 7 388 00 10/005/003 "Construction of 6 test beds from Ju 388 with 9-8222 A/B (Jumo 222)" of 25 Nov. 1944, NASM MF R 193.

294. Advance alteration notice *f*V 5723/44 of 21 Nov. 1944, NASM MF R 186.

295. Offer No. 3766 of 28 Nov. 1944 to the OKL TLR forced landing/E-2/A, NASM MF R 2639 F 870.

296. Jumo monthly report for December 1944 by the Engine Construction/Technical Field Service Department of 2 Jan. 1945, DMM LR 01078 and NASM MF R 2853 F 218.

297. Jumo monthly report for December 1944 by the OMW Engine Test-Stand Bureau of 3 Jan. 1945, DMM LR 01078 and NASM MF R 2700 F 504.

298. Jumo monthly report for January 1945 by the OMW Engine Test-Stand Bureau of 5 Feb. 1945, DMM LR 01078 and NASM MF R 2700 F 514.

299. JFM memo Kobü-Entwurf No. 2438 of 20 Sept. 1944, NASM MF R 3262 F 821.

300. von Gersdorff et al., *Flugmotoren und Strahltriebwerke*, p. 103 ff.

301. Registered letter from Hirth-Motoren to BMW of 30 March 1944, NASM MF R 3597 F 161.

302. Letter from Hirth-Motoren to Daimler-Benz of 30 Nov. 1944, NASM Shorthand minutes of the meeting of.

303. Internal File Memo No.5525 Daimler-Benz AG, Development Section, of 8 Apr. 1944, NASM R 2479 F 561.

304. Registered letter from Dir. Schilo to Dir. Nallinger, Daimler-Benz AG, of 5 May 1944, NASM MF R 3204 F 176.

305. Registered letter from Daimler-Benz AG to Dir. Schilo of 11 May 1944, NASM MF.

306. Registered letter from Nallinger and Mann to Dipl.-Ing. Lusznat of the Engine Configuration Working Committee (OKL/TLR/Fl-E 3 VI) of 4 Nov. 1944, NASM MF R 2966 F 856.

307. Internal memo Daimler-Benz AG, Development Section, of 30 Oct. 1944, NASM MF R 2684 F 675.

308. Tabular overview of night fighter types, undated, NASM MF R 2787 F 1041.

309. JFM memo Kobü-Rh 3807 of 27 Nov. 1944, NASM MF R 3620 F 1209.

310. TAD Report-13/44 by the Hirth Motoren G.m.b.H. of 13 Nov. 1944, NASM MF R 3620 F 1209.

311. Letter from Hirth-Motoren to Daimler-Benz of 30 Nov. 1944, NASM MF.

312. Registered letter from EHAG's Hirth-Motoren plant to the OKL, Chef TLR/Fl-E 3 II, of 7 Nov. 1944, NASM MF.

313. Registered letter from EHAG's Hirth-Motoren plant to Blohm & Voss of 8 Dec. 1944, NASM MF.

314. Test report by Hirth-Motoren G.m.b.H., Zuffenhausen, of 9 Dec. 1944, NASM MF.

315. Registered letter from EHAG's Hirth-Motoren plant to Blohm & Voss of 9 Jan. 1945, NASM MF.

316. Registered letter from EHAG's Hirth-Motoren plant to Fl.-Haupting. Soboler, TLR/Fl-E3 II, of 23 Jan. 1945, NASM MF.

317. VDM memo Z 23/XV 4404 A (secret) of 18 Apr. 1944, NASM MF R 2935 F 714.

318. JFM memo Kobü-Ew 2430 of 23 Aug. 1944, NASM MF R 2743 F 976.

319. See Heinrich Beauvais in *Flugerprobungsstellen bis 1945*, Bernhard & Graefe Verlag, p. 9.

320. Letter from Junkers to the Reich Minister of Aviation and Commander-in-Chief of the *Luftwaffe* of 14 Dec. 1943, EADS Unternehmensarchiv Ottobrunn.

321. JFM memo FlA 1115 of 1 Dec. 1943, NASM MF R 186.

322. BAL Inspection Report Ju 388 L-0 300 001 of 15 Dec. 1943, NASM MF R 193.

323. JFM Inspection Report-Pamphlet Ju 388 V7 of 16 Dec. 1943 with comments by Kobü/Rh (Dzulko) of 21 Jan. 1944, NASM MF R 193.

324. System for communication between crew members using headset microphone, today called intercom.

325. JFM memo FlA 1166/43 of 23 Dec. 1943, NASM MF R 186, and logbook of Adolf Körner, flight no. 949, Achs Collection.

326. JFM memo FlA 37/44 of 13 Jan. 1944, NASM MF R 186, and logbook of Adolf Körner, flights no. 950 and 951, Achs Collection.

327. JFM memo FlA 50/44 of 18 Jan. 1944, NASM MF R 186.

328. JFM memo FlA 60/44 of 21 Jan. 1944, NASM MF R 186.

329. JFM memo Fb-Bm 44/88/021 of 24 Jan. 1944, NASM R 2781 F 328.

330. JFM memo FlA 150/44 of 28 Feb. 1944, NASM MF R 186, and logbook of Adolf Körner, flight no. 952, Achs Collection.

331. JFM memo FlA 88/44 of 2 Feb. 1944, NASM MF R 186.

332. JFM memo Kobü-Fla 352/44 of 5 May 1944, NASM MF R 193.

333. JFM interim report P.3174/705 of 2 Feb. 1944, NASM MF R 2615 F 756.

334. JFM memo FlA 98/44 of 4 Feb. 1944, NASM MF R 186.

335. Daily report for 6 Feb. 1944, dated 7 Feb. 1944, NASM MF R 186, and logbook of Adolf Körner, flight no. 953, Achs Collection.

336. JFM memo FlA 108/44 of 10 Feb.1944, NASM MF R 186.

337. JFM memo FlA 107/44 of 9 Feb.1944, NASM MF R 186.

338. JFM memo FlA 123/44 of 16 Feb. 1944, NASM MF R 186, and logbook of Adolf Körner, flight no. 955, Achs Collection.

339. JFM memo FlA 132/44 of 18 Feb. 1944, NASM MF R 186, and logbook of Adolf Körner, flight no. 956, Achs Collection.

340. JFM memo FlA 150/44 of 28 Feb. 1944, NASM MF R 186.

341. Memo Fp.-Halle 226 43/44 of 29 March 1944, NASM MF R 2760 F 1088.

342. JFM memo FlA 142/44 of 25 Feb. 1944, NASM MF R 186.

343. JFM memo FlA 148/44 of 28 Feb. 1944, NASM MF R 186.

344. JFM memo Kobü-Fla 352/44 of 5 May 1944, NASM MF R 193.

345. JFM memo FlA 149/44 of 28 Feb. 1944, NASM MF R 186.

346. JFM memo FlA 163/44 of 3 March 1944, NASM MF R 186.

347. JFM memo FlA 173/44 of 7 March 1944, NASM MF R 186.

348. BMW EZA/2 brief report No. 10 on prototype installation and testing by field offices of 16 March 1944, NASM MF R 2284 F 844.

349. Letter Br.B. No. 226041/44 to Junkers company of 30 March 1944, EADS Unternehmensarchiv Ottobrunn.

350. JFM memo FlA 191/44 of 13 March 1944, NASM MF R 186.

351. It should be mentioned that the Jumo 004 jet engines used in Me 262 and Ar 234 service machines were replaced regularly after just 20 to 25 hours of use.

352. Letter from the BMW Technical Field Service in Dessau of 21 Apr. 1944, Deutsches Museum.

353. JFM short memo of 23 March 1944, NASM MF R 186.

354. JFM short memo of 24 March 1944 and JFM memo FlA 226/44 of 26 March 1944, NASM MF R 186.

355. Damage report 1/53/44 for aero-engines and propellers of 19 Apr. 1944, Landesarchiv Oranienbaum No. 1012.

356. Letter from the Mitteldeutschen Versicherungs-Vermittlungs-Gesellschaft (Mivermi) to the Allianz in Berlin of 18 July 1944, Landesarchiv Oranienbaum No. 1012.

357. JFM memo FlA 373/44 of 16 May 1944, NASM MF R 2762 F 812.

358. Hygroscopic = water absorbing.

359. JFM memo Kobü-Fla 352/44 of 5 May 1944, NASM MF R 193.

360. Kobü-Vers test report 632/9 K 1 of 28 Apr. 1944, NASM MF R 3983 F 1127.

361. An illustration of this heater is found in the operating manual, Part 3, p. 114.

362. JFM memo Kobü-Rh 3273 of 11 June 1944, EADS Unternehmensarchiv Ottobrunn.

363. Telex "Report Ju 388 from 16 June 1944-30 June 1944" of 1 July 1944, EADS Unternehmensarchiv Ottobrunn.

364. JFM memo FlA 601/44 of 18 Sept. 1944, NASM MF R 183.

365. JFM test report FlA 548/44 of 30 Aug. 1944, NASM MF R 3983 F 1096.

366. JFM memo Fla 601/44 of 18 Sept. 1944, NASM MF R 2752 F 183.

367. JFM test report FlA 569/44 of 31 Aug. 1944, NASM MF R 3989 F 526.

368. The yaw angle is the angle between the aircraft's longitudinal axis and the actual direction of flight.

369. BMW EZA/2 brief report No. 29 of 15 Sept. 1944, NASM MF R 2419 F 534.

370. JFM memo Fla 601/44 of 18 Sept. 1944, NASM MF R 2752 F 183.

371. A slight difference from the figure stated earlier!

372. BMW EZA/2 brief report No. 32 of 9 Oct. 1944, NASM MF R 2411 F 838.

373. JFM memo Kobü-Fla 627/44 of 6 Oct. 1944, NASM MF R 186.

374. JFM memo FlA 698/44 of 13 Nov. 1944, NASM MF R 186.

375. BMW EZA/2 brief report No. 40 of 15 Dec. 1944, NASM MF R 2640 F 286.

376. BMW EZA/2 brief report No. 41 of 30 Dec. 1944, NASM MF R 2648 F 380.

377. JFM test report FlA 741/44 of 14 Dec. 1944, NASM MF R 8002 F 877.

378. BMW EZA/2 brief report No. 40 of 15 Dec. 1944, NASM MF R 2640 F 286.

379. JFM memo FlA 742/44 of 15 Dec. 1944, NASM MF R 3975 F 914.

380. JFM test report FlA 741/44 of 14 Dec. 1944, NASM MF R 8002 F 877.

381. JFM memo Fla 738/44 of 14 Dec. 1944, NASM MF R 186.

382. JFM memo Feprü-Flugbetrieb 105/44 of 28 Dec. 1944, NASM MF R 2196 F 404.

382a. Memo concerning meeting of development heads on 20 Feb. 1945.

383. Report by the Head of Flight Operations, Merseburg Construction Site of 1 June 1944, NASM MF R 2767 F 663.

384. BMW brief report No. 10 on prototype installation and testing by field offices of 16 March 1944, NASM MF R 2284 F 844.

384a. Logbook of Heinz Lehmann (extract), Achs Collection.

385. JFM memo FlA 373/44 of 16 May 1944, NASM MF R 2762 F 812.

385a. Logbook, flights no. 355, 359 and 360, Regel Collection.

386. BMW brief report No. 10 concerning prototype installation and testing by field offices of 16 March 1944, NASM MF R 2284 F 844.

387. Test Report-Pamphlets Ju 388 K-1 V3, WNr. 500 003 of 17, 23 and 31 March 1944, NASM MF R 193.

388. Aircraft Final Report Ju 388 K-1/V3 of 30 June 1944, EADS Unternehmensarchiv Ottobrunn.

389. Inspection of the V3 WNr. 500 003 on 24 March 1944, inspector BAL 1098, NASM MF R 193.

390. JFM brief memo FlA 229/44 of 28 March 1944, NASM MF R 186.

391. JFM brief memo FlA 251/44 of 4 Apr. 1944, NASM MF R 186.

392. JFM memo FlA 256/44 of 7 Apr. 1944, NASM MF R 186.

393. Tanks for the V3 to V6 were ordered on 25 May, see JFM memo Kobü-Rh 3232 of 25 May 1944, EADS Unternehmensarchiv Ottobrunn.

394. JFM memo FlA 362/44 of 9 May 1944, NASM MF R 186.

395. JFM test report Kobü-Vers. 2/173/15 K of 26 May 1944, NASM MF R 2768 294.

396. JFM file memo FlA 373/44 of 16 May 1944, NASM MF R 2762 F 812.

397. JFM memo Kobü-Rh 3273 of 11 June 1944, EADS Unternehmensarchiv Ottobrunn.

398. JFM memo Kobü-Rh 3278 of 12 June 1944, EADS Unternehmensarchiv Ottobrunn.

399. JFM memo Fla 505/44 of 25 July 1944, NASM MF R 186.

400. JFM memo Kobü-Rh 3358 of 7 July 1944, EADS Unternehmensarchiv Ottobrunn.

401. JFM memo Kobü-Rh 3405 of 22 July 1944, EADS Unternehmensarchiv Ottobrunn.

402. Amendment to JFM memo Fla 503/44 of 4 Aug. 1944, NASM MF R 186.

403. JFM memo Kobü-Rh 3401 of 25 July 1944, NASM MF R 2196 F 404.

404. JFM memo Fla 498/44 of 19 July 1944, NASM MF R 186.

405. JFM memo FlA 550/44 of 24 Aug. 1944 and file memo Kobü-Elt 387/44 and 395/44 of 25 and 31 Aug. 1944, NASM MF R 2196 F 404.

406. BMW EZA/2 short report No. 29 on prototype installation and testing by field offices of 15 Sept. 1944, NASM MF R 2419 F 534.

407. JFM test report Fla 587/44 of 12 Sept. 1944, NASM MF R 186.

408. BMW EZA/2 short report No. 32 of 9 Oct. 1944, NASM MF R 2411 F 838.

409. JFM memo Kobü-Fla 627/44 of 6 Oct. 1944, NASM MF R 186.

410. BMW EZA/2 short report No. 37 on prototype installation and testing by field offices (5-11 Nov. 1944), ca. mid-November 1944, NASM MF R 2823 F 872.

411. BAL file memo of 21 March 1944, EADS Unternehmensarchiv Ottobrunn.

412. JFM memo Fp.-Zuba. 12/44 of 10 Apr. 1944, EADS Unternehmensarchiv Ottobrunn.

413. Overview of pending changes of 12 June 1944 and related memo Fb.-Bm 44/88/191 of 24 June 1944, EADS Unternehmensarchiv Ottobrunn.

414. Advance alteration notice *f*V 6827/44 of 25 May 1944, NASM MF R 3978 F 30.

415. Advance alteration notice *f*V 6812/44 and *f*V 6813/44 of 30 Apr. 1944, NASM MF R 3978 F 30.

416. Aircraft final report Ju 388 K-1/V4 of 15 June 1944, EADS Unternehmensarchiv Ottobrunn.

417. JFM memo FlA. 429/44 of 19 June 1944, NASM MF R 186.

418. Inspection report FSD Production Inspection of 13 July 1944, EADS Unternehmensarchiv Ottobrunn.

419. JFM memo Kobü-Rh 3393 of 19 July 1944, EADS Unternehmensarchiv Ottobrunn.

420. Amendment to aircraft final report of 29 July 1944, EADS Unternehmensarchiv Ottobrunn.

421. *E-Stelle* Rechlin weekly report for the period 13-19 Aug. 1944 (19 Aug. 1944), BA-MA RL 36/86.

422. BMW EZA/2 brief report No. 29 of 15 Sept. 1944, NASM MF R 2419 F 534.

423. BMW EZA/2 brief report No. 32 of 9 Oct. 1944, NASM MF R 2411 F 838.

424. JFM memo Feprü II 69/44 of 25 Aug. 1944, NASM MF R 2196 F 404.

425. JFM memo Kobü-Fla 627/44 of 6 Oct. 1944, NASM MF R 186.

426. BMW EZA/2 brief report No. 40 of 6 Dec. 1944, NASM MF R 2640 F 286.

427. BMW EZA/2 brief report No. 41 of 14 Dec. 1944, NASM MF R 2648 F 380.

428. Aircraft final report Ju 388 L-1/V5 of 25 July 1944, EADS Unternehmensarchiv Ottobrunn.

429. Inspection Ju 388 L-1, V5-electrical system of 22 Apr. 1944, NASM MF R 193.

430. JFM memo Feprü-Halle 203 No. 17/44 of 7 Apr. 1944, NASM MF R 186.

431. File memo JFM FlA. 373/44 of 16 May 1944, NASM MF R 2762 F 812.

432. JFM brief memo FlA. 368/44 of 15 May 1944, NASM MF R 186.

433. Memo Fp.-Halle 203 No. 27/44 of 8 July 1944, EADS Unternehmensarchiv Ottobrunn.

434. JFM memo Fla 354/44 of 6 May 1944, NASM MF R 186.

435. JFM memo FTV-Au 852/44 of 19 July 1944, NASM MF R 2196 F 404.

436. JFM memo FlA. 506/44 of 25 July 1944, NASM MF R 186.

437. JFM memo FlA. 515/44 of 29 July 1944, NASM MF R 186.

438. File memo JFM Fla. 529/44 of 3 Aug. 1944.

439. Aircraft final report Ju 388 L-1/V6 of 28 July 1944, EADS Unternehmensarchiv Ottobrunn.

440. JFM memo EL 494 of 30 June 1944, EADS Unternehmensarchiv Ottobrunn.

441. Aircraft final report Ju 388 L-1/V6 of 28 July 1944, EADS Unternehmensarchiv Ottobrunn.

442. JFM memo Kobü-Rh 3118 of 19 Apr. 1944 and 1st amendment of 26 Apr. 1944, EADS Unternehmensarchiv Ottobrunn.

443. JFM memo Kobü-Vers. 3c/1345 of 21 March 1944, NASM MF R 2196 F 404.

444. JFM test report Kobü-Vers. 3c/1373 of 20 Apr. 1944, NASM MF R 2780 F 343.

445. JFM test report Kobü-Vers. 3c/1401 of 18 May 1944, NASM MF R 2748 F 506.

445a. Advance Alteration Notification *f*V 4331/44 of 14 Apr. 1944, NASM.

446. JFM test report Kobü-Vers. 3c/1457 of 21 Aug. 1944, NASM MF R 2768 F 345.

447. JFM memo FlA. 380/44 of 18 May 1944, NASM MF R 186.

448. JFM memo EL 494 of 30 June 1944, EADS Unternehmensarchiv Ottobrunn.

449. JFM memo FlA. 511/44 of 28 July 1944, NASM MF R 186.

450. JFM memo EL 513/44 of 12 July 1944, NASM MF R 2146 F 404.

451. JFM memo FlA. 486/44 of 14 July 1944, NASM MF R 186.

452. JFM FSD Production Inspection, inspection report Ju 388 V1 of 17 July 1944, EADS Unternehmensarchiv Ottobrunn.

453. JFM memo Kobü-Rh 3420 of 28 July 1944, EADS Unternehmensarchiv Ottobrunn.

454. JFM memo Kobü-Rh 3399 of 21 July 1944, EADS Unternehmensarchiv Ottobrunn.

455. BMW EZA/2 brief report No. 29 on prototype installation and testing by field offices of 15 Sept. 1944, NASM MF R 2419 F 534.

456. JFM Test Report-Pamphlet "Ju 388 General" of 7 Sept. 1944, NASM MF R 193 and referring JFM memo Kobü-Rh 3611 of 30 Sept. 1944, EADS Unternehmensarchiv Ottobrunn.

457. BMW EZA/2 brief report No. 32 on prototype installation and testing by field offices of 9 Oct. 1944, NASM MF R 2411 F 838.

458. Memo from FSD Production Inspection Feprü-Halle 203 of 29 Sept. 1944, NASM MF R 193.

459. JFM memo Kobü-Rh 3655 of 16 Oct. 1944, NASM MF R 193.

460. JFM memo Kobü-Fla 627/44 of 6 Oct. 1944, NASM MF R 186. According to *f*V 5447/44 of 21 Sept. 1944, this switch was supposed to help prevent excessively high fluid temperatures in the FA 15's hydraulic system

461. Telex from Junkers to *Kapitän* Vogel of the *E-Stelle* Rechlin of 31 Oct. 1944, EADS Unternehmensarchiv Ottobrunn.

462. JFM file memo FTV-Au 1240/44 of 31 Oct. 1944, EADS Unternehmensarchiv Ottobrunn, plus letter from the BAL to *Hauptnautiker* Vogel, *E-Stelle* Rechlin E 4, of same day, NASM MF R 2196 F 404.

463. JFM memo Kobü-Rh 3702 of 26 Oct. 1944, NASM MF R 2196 F 404.

464. Memo No. 4/44 of 1 Nov. 1944 of the FTA Office with *Erprobungskommando 388*, NASM MF R 193.

465. JFM memo Kobü-Rh 3659 of 17 Oct. 1944, EADS Unternehmensarchiv Ottobrunn.

466. Initial inspection Ju 388 J-1 V1 of 2 Nov. 1944, NASM MF R 193.

467. JFM memo Kobü-Rh 3739 of 9 Nov. 1944, NASM MF R 193.

468. JFM memo Kobü-Rh 3762 of 16 Nov. 1944, EADS Unternehmensarchiv Ottobrunn.

469. Advance alteration notice *f*V 22 275/44 of 15 Nov. 1944.

470. JFM memo Kobü-Rh 3777 of 17 Nov. 1944, EADS Unternehmensarchiv Ottobrunn.

471. Priority test report No. 49/44 by the *E-Stelle* Tarnewitz for the period 3-9 Dec. 1944, Archiv Jens.

472. JFM memo FTV-Au 1435/44 of 9 Dec. 1944, EADS Unternehmensarchiv Ottobrunn.

473. Priority test report No. 51/44 by the *E-Stelle* Tarnewitz for the period 17-23 Dec. 1944, Archiv Jens.

474. Memo by development management of 16 Nov. 1944, NASM MF R 2196 F 404.

475. JFM memo M/Fb-Ltg. 44/375 of 4 Oct. 1944, NASM MF R 2196 F 404.

476. Advance alteration notice JFM 6835 and 6836/44 of 1944, NASM R 187 F 30.

477. JFM memo Kobü-Rh 3193 of 15 May 1944 and 1st amendment of 31 May 1944, NASM MF R 2196 F 404.

478. 2nd amendment to JFM memo Kobü-Rh 3193 of 15 June 1944, NASM MF R 2196 F 404.

479. JFM memo M-Fb.Bm./44/266 of 22 June 1944, NASM MF R 2196 F 404.

480. 2nd [3rd] amendment to JFM memo Kobü-Rh 3193 of 21 June 1944, NASM MF R 2196 F 404.

481. JFM memo Kobü-Rh 3421 of 28 July 1944, NASM MF R 2196 F 404.

482. JFM memo of 4 July 1944, NASM MF R 2196 F 404.

483. 4th amendment to JFM memo Kobü-Rh 3193 of 13 July 1944, NASM MF R 2196 F 404.

484. JFM memo M-Fb.Bm./44/282 of 5 July 1944, NASM MF R 2196 F 404.

485. Amendment IV to construction directive 7 388 26 10/100 of 12 Aug. 1944, NASM MF R 193.

486. JFM memo Kobü-Rh 3402 of 21 July 1944, NASM MF R 2196 F 404.

487. JFM meeting minutes No. 402 of 11 Aug. 1944, NASM MF R 2196 F 404.

487a. Logbook of Heinz Lehmann (extract), Achs Collection.

488. JFM memo FlA. 602/44 of 18 Sept. 1944, NASM MF R 186.

489. JFM memo FlA. 607/44 of 23 Sept. 1944, NASM MF R 186.

490. Monthly report for Oct. 1944 by Jumo development department of 2 Nov. 1944, NASM MF R 2853 F 190.

491. JFM memo FlA. 670/44 of 2 Nov. 1944, NASM MF R 186.

492. Monthly report for November 1944 by Jumo development department of 4 Dec. 1944, NASM MF R 2853 F 204.

493. JFM memo M/Fb-Ltg. 44/374 of 4 Oct. 1944, NASM MF R 2196 F 404.

494. JFM meeting minutes No. 402 of 11 Aug. 1944, NASM MF R 2196 F 404.

495. Amendment II to construction directive 7 388 00 10/005/002 "Construction of 5 test beds from Ju 388" of 12 Aug. 1944, EADS Unternehmensarchiv Ottobrunn.

496. JFM memo Kobü-Rh 3590 of 23 Sept. 1944, NASM MF R 193.

497. Amendment III to construction directive 7 388 00 10/007/003 of 17 Oct. 1944, NASM MF R 193.

498. JFM memo Kobü-Rh 3632 of 6 Oct. 1944, NASM MF R 2196 F 404.

499. JFM memo Kovermi Merseburg of 26 Oct. 1944, NASM MF R 2196 F 404.

500. Offer No 3766 to the OKL/TLR FL/E-2/A of 28 Nov. 1944, NASM MF R 2639 F 870.

501. JFM memo FlA. 692/44 of 13 Nov. 1944, NASM MF R 186.

502. JFM memo Kobü-Rh 3878 of 22 Dec. 1944, NASM MF R 196.

503. Report on Jumo 222 for month of January 1945 by OMW Test Stand 2 of 5 Feb. 1945, DMM LR 01078.

504. JFM memo Kobü-Rh 3226 of 24 May 1944, NASM MF R 186.

505. JFM memo M-Fb.Bm./44/266 of 22 June 1944, NASM MF R 2196 F 404.

506. Memo concerning emergency compass installation in 8-388 J-0 V10 of 6 Dec. 1944, NASM MF R 2196 F 404.

507. BAL report on inspection of weapons installation in 8-388 J-0, V10 of 15 Dec. 1944, NASM MF R 193.

507a. Logbook of Heinz Lehmann (extract), Achs Collection.

508. JFM construction directive 7 388 00 10/013/00 "Construction of 6 Ju 388 J-3 prototypes for Mosquito hunting" of 11 Nov. 1944, NASM MF R 193.

509. Amendment I to construction directive 7 388 00 10/013/00 "Construction of 6 Ju 388 J-3 prototypes for Mosquito hunting" of 20 Nov. 1944, NASM MF R 193.

510. JFM memo Kovermi Loebnitz/Be. 28 of 23 Jan. 1945, EADS Unternehmensarchiv Ottobrunn.

511. Amendment III to construction directive 7 388 00 10/012/004 of 2 Feb. 1945, EADS Unternehmensarchiv Ottobrunn.

512. Construction directive 7 388 00 10/006/002 of 29 Nov. 1944, EADS Unternehmensarchiv Ottobrunn.

513. Amendment II to construction directive 7 388 00 10/006/002 of 29 Nov. 1944, EADS Unternehmensarchiv Ottobrunn.

514. JFM memo Kobü-Rh 3767 of 16 Nov. 1944, EADS Unternehmensarchiv Ottobrunn.

515. Amendment V to construction directive 7 388 26 10/000 "Construction of 50 Ju 388 K-0 (formerly Ju 188 K-0) bomber aircraft" of 23 Oct. 1944, NASM MF R 193.

516. JFM memo FlA. 614/44 of 28 Sept. 1944, NASM MF R 186.

517. JFM memo of 11 April 1944, NASM MF R 186.

518. JFM memo Fla 491/44 of 16 July 1944, NASM MF R 186.

519. JFM memo Kobü-Rh 3207 of 17 May 1944, NASM MF R 2196 F 404.

520. BMW EZA/2 brief report No. 29 on prototype installation and testing by field offices of 15 Sept. 1944, NASM MF R 2419 F 534.

521. JFM memo Kobü-Fla 627/44 of 6 Oct. 1944, NASM MF R 186.

522. BMW EZA/2 brief report No. 40 on prototype installation and testing by field offices of 6 Dec. 1944, NASM MF R 2640 F 286.

523. JFM memo FlA. 39/44 of 1944, NASM MF R 3996 F 166.

524. JFM-FTV order request of 23 Sept. 1944, EADS Unternehmensarchiv Ottobrunn.

525. JFM memo Fp.-Halle 203 45/44 of 10 Nov. 1944, NASM MF R 2747 F 728.

526. JFM memo Fp.-Halle 226 98/44 of 17 Nov. 1944, NASM MF R 193.

527. Preuschen: "388 156 serviceable, no O_2."

528. JFM memo FlA. 599/44 of 18 Sept. 1944, NASM MF R 186.

529. Except for *f*V 5706, which had no impact on test results.

530. BMW test report 1113/44 of 13 Dec. 1944, NASM MF R 2784 F 391.

531. MZK = *Motoren-Zweigwerk Köthen*. (Köthen Engine Branch Plant). Other branch plants were found in Leipzig, Magdeburg, and Schönebeck. The main engine plant was in Dessau (MSD), and included the development plant named for Otto Mader (OMW). The MSD had a branch plant for jet engines in Muldenstein and a dispersed operation "Papierfabrik Muldenstein A.G. Muldensteiner Werke"; see series *Luftfahrtgeschichte im Land Sachsen-Anhalt, Band. 4,-Karte zur Luftfahrtgeschichte in Sachsen-Anhalt*, Günther Hirschfeld, Artur Meuer, Dieter Pasemann, Hrsg. Interessensgemeinschaft Luftfahrtgeschichte im Luftsportverband Sachsen-Anhalt, Halle 1998.

532. See: *Auf der Suche nach der Wahrheit-Die Fliegerhorst Bernburg und die Junkers Flugzeug- und Motorenwerke AG, Flugzeugbau, Zweigwerk Bernburg 1936-1945*, special issue *Der Bernburger Ikarus-Mitteilungsblatt des Arbeitskreises Junkerswerke und Fliegerhorst Bernburg*, Hrsg Dipl-Hist. Hanjo Todte, March 1997.

533. JFM overview drawing J.10046, DMM.

534. NASM R 2095 F 312-315 and R 3208 F 582.

535. JFM memo Kobü-Rh 3591 of 25 Sept. 1944, EADS Unternehmensarchiv Ottobrunn.

536. Series *Luftfahrtgeschichte im Land Sachsen-Anhalt, Heft 3,- Dokumentation der 90-jährigen Geschichte der Luftfahrt in Sachsen-Anhalt und des Lufttransportes in der Region Halle (Saale)*, p. 38.

537. GL/C Aircraft Program, Delivery Plan 226, Version 1, of 15 May 1944, BA-MA RL 3/ 1041.

538. GL/C Aircraft Program, Delivery Plan 226, Version 2, of 15 July 1944, BA-MA RL 3/ 1041.

539. Steering committee Aircraft Production, Aircraft Program (Conversion), Delivery Plan 227, secret command matter 1547/44 of 15 Dec. 1944, BA-MA RL 3/ 1045.

540. Steering committee Aircraft Production, Aircraft Program, Delivery Plan 228, Version 1, secret command matter 2936/45 of 16 March 1944, BA-MA RL 3/ 1046.

541. See Delivery Plans 226 in BA-MA RL 3/1040f.

542. Within the factory, the output of completed machines was monitored by the daily reports and 10-day reports (ten-day reporting period as to state of delivered major components). The monthly final report was verified by the construction inspector. Factory management meetings were held every 14 days, after which invoices were sent to the RLM for aircraft delivered based on a previously established price; "*Response to the questionnaire discussed on 9 May 1945*" by the FZ Bernburg to the U.S. Military Administration, dated 12 May 1945, and *Auf der Suche nach der Wahrheit*, both from the Archiv Arbeitskreis Bernburg.

543. CIOS Report FR XXX, NASM MF R-99 and Evaluation Reports #69 and #149, Uni Bochum Collection via Lutz Budrass.

544. JFM memo Kobü-Rh 3279 of 12 June 1944, NASM MF R 2196 F 404.

545. Letter from ATG-Maschinenbau GmbH to Prof. Hertel of 22 Jan. 1945, NASM MF R 193.

546. JFM brief memo from Head of Development Prof. Hertel to Schilling of 26 Jan. 1945, EADS Unternehmensarchiv Ottobrunn.

547. Graphic of 25 Jan. 1945, EADS Unternehmensarchiv Ottobrunn.

548. Shorthand report on the Fighter Staff meeting of 3 July 1944, BA-MA RL 3/9 fol. 192 f.

549. Shorthand report on the Ju 388 startup meeting on Wednesday, 5 July 1944, 1100 hours, BA-MA RL 3/9 fol. 610 f. Obviously only the conclusion of the startup meeting, during which the results were reviewed with Saur, was recorded in shorthand. Details, such as prototypes with Jumo 213 E engines, were thus unfortunately lost. Highlights of the entire startup meeting may, however, be found in the brief report on the Fighter Staff meeting of 5 July 1944.

550. Frydag used the expression, "missed the boat."

551. Production of fuselages was planned as part of the move (together with Erla-Antwerp) to an underground facility with an area of 40 000 m^2 located in a sandstone quarry near Maastricht, extending into Belgium. Meeting during the 8th trip of "Operation Hubertus," 1-3 June 1944. Milch reference file, IWM, MF Reel 13, Vol. 7. In October Allied intelligence became aware of plans for Fokker to build 15 Ju 388 fuselages per month until June 1946. This was confirmed one month later: Fokker Amsterdam was to produce up to 10 Ju 388 J, K, L fuselages per week, AIR 40/179, Public Records Office, Kew.

552. U.S. Strategic Bombing Survey, Air Frames Plant Report No. 3, IWM, London.

553. Stenographic record of the Fighter Staff meeting of 9 July 1944, BA-MA RL 3/9, fol. 479 f.

554. Interim report on the Fighter Staff meeting of 12 June 1944, BA-MA.

555. Stenographic record of the Fighter Staff meeting of 18 July 1944, BA-MA RL 3/9 fol. 380 f.

556. Stenographic record of the Fighter Staff meeting of 13 July 1944, BA-MA RL 3/9 fol. 437 f.

557. Stenographic record of the Fighter Staff meeting of 9 July 1944, BA-MA RL 3/9 fol. 479 f.

558. Stenographic record of the Fighter Staff meeting of 8 July 1944, BA-MA RL 3/9 fol. 501 f.

559. Stenographic record of the Fighter Staff meeting of 9 July 1944, BA-MA RL 3/9 fol. 452 f.

560. *Generalmajor* Ulrich Diesing served as Technical Officer attached to the ObdL from 1942 to 1944, then became "Chief of Aircraft Procurement and Supply." On 26 March 1945 he resigned from the position after a dispute with Saur. He was succeeded briefly by Gen.Ltn. Kleinrath; see Aircraft Procurement and Supply War Diary 16 March-4 Apr. 1945.

561. Stenographic record of the Fighter Staff meeting of 6 July 1944, BA-MA RL 3/9, fol. 517 f.

562. Memo from BAL at JFM Dessau to the BAL at Weser Flugzeugbau, Berlin Tempelhof, dated 22 June 1944, EADS Unternehmensarchiv Ottobrunn.

563. Stenographic record of the Armaments Staff meeting of 2 August 1944, BA-MA RL 3/9 fol. 122.

564. Armaments Staff joint report of 5 Dec. 1944, BA-MA RL 3/2689.

565. Armaments Staff joint report of 24 Dec. 1944, BA-MA RL 3/2689.

566. Armaments Staff joint report of 28 Dec. 1944, BA-MA RL 3/2689.

567. Armaments Staff joint report of 10 Jan. 1945, BA-MA RL 3/3195.

568. Armaments Staff joint report of 9 Feb. 1945, BA-MA RL 3/2691.

569. Armaments Staff joint report of 12 Feb. 1945, BA-MA RL 3/2691.

570. Brief report Aircraft New Construction and Repair of 28 Feb. 1945, BA-MA RL 3/2691.

571. Summary of the meeting on 8 Feb. 1945 App 4579, Berlin.

572. Meeting record No. 2776/45 secret command matter of the Steering Committee for Aircraft Construction.

573. See Olaf Groehler, *Anhalt im Luftkrieg-Anflug auf Ida Emil*, Anhaltsche Verlagsgesellschaft mbH Dessau 1993, p. 59 ff.

574. Landmark cancellations in the development field continued. They affected the Ju 287, which was cut from 20 to just 6 prototypes. Final assembly of the first Ju 488s (Ju 188 V401 and V402) was now to be carried out by FSD, while no decision had yet been made for the V403-V406.

575. Because of rationalization measures made necessary by the war, beginning in September 1944 there was a change from monthly to quarterly reports, none of which appear to have survived.

576. Man-hours for the Ju 88 and production breakdown for the Ju 88, Ju 188, Ju 388, and He 162 from an unidentified source, probably Junkers, written in German and English after the end of the war, NASM MF R 2040 F 152. There is no doubt of the document's authenticity; other surviving documents concerning the companies in question agree with the information supplied here.

577. Telegram from the BAL at Junkers Dessau to the BAL at WMF of 4 Jan. 1945, 0845 hours, EADS Unternehmensarchiv Ottobrunn.

578. GL/C-Aircraft Program, Delivery Plan 226, 1st Issue, version of 15 May 1944, BA-MA RL 3/ 1041.

579. Concerning the changeover problem in air armaments, which led to regular "unemployment" and large scale scrapping actions (esp. at HFW and JFM FZ Schönebeck) in order retain the companies' own, for the most part well trained, employees, see Budrass, *Flugzeugindustrie und Luftrüstung in Deutschland 1918-1945*, Bundesarchiv Vol. 50, p. 875 ff.

580. In addition to Heinkel Nord and Süd, the JFM branch plant at Bernburg (FZB) also participated in final assembly of the He 162 (WNr. block 310001); information from the Arbeitskreis Bernburg and from CIOS reports and the U.S. Strategic Bombing Survey Air Frames Plant Report No. 1.

581. GL/C-Aircraft Program, Delivery Plan 226, Issue 2, version of 15 July 1944.

582. Test report pamphlet of 25 Feb. 1945, NASM MF R 193.

583. JFM memo Fb-Bm 44/88/021 of 24 Jan. 1944, NASM R 2781 F 328.

584. JFM memo Fp.-Zuba. 12/44 of 10 Apr. 1944, EADS Unternehmensarchiv Ottobrunn.

585. BMW brief report No. 10 on prototype installation and testing by field offices of 16 March 1944, NASM MF R 2284 F 844.

586. JFM memo Kobü-Ew. 2372 of 13 Apr. 1944, NASM MF R 2725 F 1124.

587. JFM memo of 4 July 1944, NASM MF R 2146 F 404.

588. JFM memo Kobü-Fla 352/44 of 5 May 1944, NASM MF R 193.

589. JFM file memo FTV-Au 568/44 by the JFM FTV-Contract Processing of 10 May 1944, NASM MF R 193.

590. JFM memo Kobü-Rh 3207 of 17 May 1944, NASM MF R 2196 F 404.

591. JFM memo Kobü-Rh 3155 of 6 May 1944, NASM MF R 2196 F 404.

592. JFM memo M/Fb-Ltg. 44/186 of 2 June 1944, NASM MF R 2196 F 404.

593. Telex "Report Ju 388 from 16 June 1944-30 June 1944" of 1 July 1944, EADS Unternehmensarchiv Ottobrunn.

594. JFM memo of 4 July 1944, NASM MF R 2196 F 404.

595. JFM memo FlA. 605/44 of 20 Sept. 1944, NASM MF R 186.

596. Amendment to JFM memo Fla 503/44 of 4 Aug. 1944, NASM MF R 186.

597. JFM memo No. Fp.-M 83/44 "Completion of modifications to the Ju 388 L-0, K-0 (record of meeting, reference no. 398), NASM MF R 193.

598. JFM meeting record, reference no. 401 of 11 Aug. 1944, NASM MF R 2196 F 404.

599. Interesting: a complaint was made about the missing "dark camouflage" on WNr. 230 153, manufacturer's code KS+TC; this corresponds with the photo of KS+TA K-0.

600. Findings and Inspection Report 443/44 by the inspection group of the *E-Stelle* Rechlin of 29 July 1944, NASM MF R 193.

601. *E-Stelle* Rechlin Inspection Group, Findings and Inspection Report No. 451/44 regarding faults with Ju 388 K-0 WNr. 230 155, KS+TE, discovered on 31 July 1944, report date 2 Aug. 1944, NASM MF R 193.

602. Letter from Merseburg production inspection Fp.-M. 76/44 of 19 July 1944, NASM MF R 193.

603. Letter from Merseburg production inspection Fp.-M. 75/44 of 18 July 1944, NASM MF R 193.

604. ATG, Technical Head Office, overviews "Aircraft Fabrication Program Progress Report" and "Delivered Aircraft," cutoff date 31 Dec. 1944 (dated 2 Feb. 1945), BA-MA RL 3/1277.

605. JFM memo FlA. 605/44 of 20 Sept. 1944, NASM MF R 186.

606. JFM memo Kobü-Flugabteilung FlA. 617/44 of 29 Sept. 1944, NASM MF R 186.

607. JFM memo FlA. 647/44 of 19 Oct. 1944, NASM MF R 186.

608. GL/C-Aircraft Program, Delivery Plan 226, Issue 2, secret command matter, of 15 Sept. 1944, BA-MA RL 3/1041.

609. Report by plant management (August Kürner, acting member of the board, and director of aircraft large series production in the FZB) of 11 May 1945 to the Allied military administration (answer to questionnaire)

about the activity of the Bernburg branch plant, archive of the "Junkerswerke und Fliegerhorst Bernburg" working group.

610. According to a report to the Allied military administration based on JFM records, in 1943-44 FZB produced: Ju 52 (162), Ju 88 C-6 (122), R-2 (173), G-1 (1,219), G-6 (351), S-1 (5), Ju 188 E (171), F-1 (180), A-2 (69), and D-2 (42). The Mistel combination conversions 3a, 3b, and 3c (referred to internally as "Nero 1" and "Nero 2") followed in 1944-45 by special conversion contract between January and March 1945; beginning in 1944, major components were placed in storage in Bernburg and Langensalz because of the fuel shortage.

611. For more on Junkers assembly line production see report by a JFM Assembly Line and License Inspector with Construction Inspection Air (BAL) in "*Auf der Suche nach Wahrheit*," Arbeitskreis Junkerswerke und Fliegerhorst Bernburg.

612. The following decisions regarding dispersal were made at a Fighter Staff meeting held during a visit to Bernburg on 15 March 1944: the FZA was to be relocated to "above ground" sites in Plauen and Kreilingen. Final assembly by FZB had already been reduced to 75 aircraft in March 1944, and was to be dispersed to an above ground facility in Langensalza, and underground in the Heimkehle Tunnel, and later Tarthum. FZH was to go to the Klusberge sandstone quarry. Tarthum was to be the dispersed site for FZS; see Milch reference file, IWM, MF Reel 1-3 Vol. 1, final meeting in FZB. Consideration was also given to establishing a production group for Ju 188 fuselages and final assembly in Slovakia (in Trenschin, near the Skoda works in Dubnica). As well, Thiedemann proposed moving Siebel wing production to Letov. Also authorized was the production of 50 Ju 188 rear fuselages, not by Fokker itself, but by three companies "in the Hellmont area," because the skilled workers of these companies took pressure off JFM. On the way back from a visit to the plant in Dessau, the members of the Fighter Staff discussed the possibility of investing the funds for dispersing the various JFM facilities, estimated at 108-million Reichsmarschall, into a large bomb proof concrete structure.

613. Library of Bochum University (CIOS Reports # 1-#3).

614. Special collections, DMM LR 02459 (CIOS Reports #4 and 5).

615. It is to be assumed that the former leadership showed a tendency to cooperate, because with the hopeless situation in Germany, they hoped that by presenting their advanced technology they might secure future employment. On the other hand, facts may have been falsified as a result of misunderstandings between those being questioned, the translators, and the report writers.

616. Questions concerning construction processes provided deep insights into production. 95% of the transport between branch plants was by rail, and a switch to truck transportation was not possible because of the fuel shortage; consequently, the destruction of the rail system had negative consequences for JFM production. According to American experts the 4,000 to 5,000 alteration directives—all authorized by the RLM—for the entire Ju 88 family of aircraft also had to be taken into account.

617. Günther Hirschfeld, Artur Meuer, Dieter Pasemann, series *Luftfahrtgeschichte im Land Sachsen-Anhalt, Band. 4,-Karte zur Luftfahrtgeschichte in Sachsen-Anhalt*, Hrsg. Interessensgemeinschaft Luftfahrtgeschichte im Luftsportverband Sachsen-Anhalt, Halle 1998.

618. The wording of a questionnaire from the Soviet administration (Marshall Zhukov) of 5 Jan. 1946 about development of the works contains information about the relocation to the MALACHIT (system of tunnels in the Theke Mountains) and MAKRELE I & II factories, and to the dispersed assembly operations in Zellmehl, near Thorgau and Kleinwangen an der Unstrut. See Junkers-Werke Collection #1140; Landeshauptarchiv Sachsen-Anhalt, Oranienbaum. "Partial production of Ju 88 and Ju 388 airframes is confirmed in "*Decknamenverzeichnis deutscher unterirdischer Bauten*" by Hans Walter Wichert, Druckerei Joh. Schulte ISBN 3-9803271-4-0.

619. "Response to the questionnaire discussed on 9 May 1945" by the FZ Bernburg to the U.S. Military Administration dated 12 May 1945, and *Auf der Suche nach Wahrheit*, Working Group Bernburg.

620. Overview of aircraft delivered as of 31 Dec. 1944, Technical Head Office Production Organization, BA-MA RL 3 / 1277.

621. Even on 6 June 1944, the day of the invasion of Normandy, there was a meeting of the Fighter Staff; Milch reference file, IWM, MF Reel 13, Vol. 7, IWM London.

622. Description by Dir. Schade from "*United States Strategic Bombing Survey Aircraft Division Airframes Plant Report No. 3-ATG Maschinenbau GmbH, Leipzig Germany*," dated 16-29 May 1945, IWM London.

623. Armaments Staff meeting of 1 Jan. 1945; BA-MA RL 3/2690. Low rate of deliveries to BAL in spite of relatively high output was sharply criticized, especially at the turn of the year 1944-45. They would not accept that the hopeless fuel situation and poor weather were partly responsible.

624. See *Sachsenflug und Messecharter-Aus der Geschichte der Leipziger Luftfahrt*, Peter Kühne and Karsten Stölzel, Connewitzer Verlagsbuchhandlung Peter Hinke Leipzig 1999, p. 48 f. and *Strategic Bombing Survey Report No. 3* about ATG loc. cit.

625. According to Kühne and Stölzel the third works in Leipzig, in addition to ATG and Erla, was the Junkers-Flugzeugbauwerft (FWL), located at the southwest corner of the airfield.

626. Interestingly, some testing of the Do 217 H (DD+LW), one of the contemporaries of the Ju 388, was carried out at Leipzig-Mockau airfield from may 1942 to November 1944; see *Sachsenflug und Messecharter*, p. 55.

627. WeLei memo of 14 Feb. 1945, BA-MA RL 3 / 4315 with a reference to the retrofitting of rearward facing antennas at Kölleda and SFW-Schkeuditz.

628. ATG memo by management of Plant 4 to Dir. Schade of 6 Feb. 1945, BA-MA RL 3 4315. Management of Plant 4 reported the accumulation of up to 20 cockpits, plus Ju 188/388 fins and rudders for 40 of each type. These had been stored in Hangar 21/Plant 6 and another storage facility. At the beginning of February Plant 4 had a work force of 574 persons, according to "Retainer Report" by the management of ATG Plant 4 of 4 Apr. 1945, BA-MA RL 3 / 4315.

629. ATG TH production preparation, memo of 4 Jan. 1945, BA-MA RL 3 / 4315.

630. *Mitteldeutschland im Luftkrieg 1944 und 1945-Eine Dokumentation zur Geschichte des Luftkrieges in Mitteldeutschland im Zweiten Weltkrieg*, by Fred Dittmann, 2001, Graphischer Kunstverlag Kyffhäuser, p. 47 ff.

631. On 19 Jan. 1945 the Jig Bureau, Leipzig, reported to Dir. Schade that construction of the "first batch" of jigs for the J/L (ATG production) was proceeding "according to plan." "Manufacturing equipment for the second batch (which on 30 Dec. 44 was noted as partially "stopped") still under construction will be added to the first batch or sent to JFM." The jigs built were for the fuselage, canopy, cockpit, empennage, undercarriage, controls, wings, radio and electrical equipment, and armament. JFM was offered large jigs, building cradles, and larger tools from the second batch. Production of duplicate sets for Menibum, Weserflug, and Henschel had also reached an advanced stage. Production of jigs for the Fw 190 A-9/D-12 and Ta 152 was also proceeding according to plan.

632. BA-MA RL III 1277, ATG Production Plan 15 339 for "Production of Ta 152 C1/R11 Aircraft and Components for the RLM" of 26 March 1945.

633. The meticulous reports compiled for the Bombing Survey were supposed to document the limitations imposed by the strategic bombing offensive, and assess the actual degree of damage inflicted by individual air raids on production (before and after the attacks) by on the spot inspections and questioning of plant managers.

634. See *Sachsenflug und Messecharter*, p. 56, *United States Strategic Bombing Survey Aircraft Division Airframes Plant Report No. 3-ATG Maschinenbau GmbH, Leipzig Germany*, dated 16-29 May 1945, IWM.

635. See *Sachsenflug und Messecharter*, p. 57.

636. Detailed guidelines were issued for the most efficient possible test flying of new and repaired J-1 and L-1 aircraft with BMW 801 TJ-0 engines, a further indication that production really got under way in the autumn of 1944. The test flights had to be carried out with as little fuel as possible. JFM FSD Kobü Richtlinien zum Einfliegen, NASM MF R 2761 F 1002.

637. BMW trip report 915/44 of 26 Sept. 1944, NASM MF R 2780 F 926.

638. SO could stand for "Süd-Ost" (southeast), which, viewed from company headquarters in Leipzig, would be an apt geographical designation for Altenburg.

639. Memo from ATG WeLei SO, Technical Business Director, to *Herr* Heller of 21 Sept. 1944, BA-MA RL 3 / 3195.

640. JFM memo FlA. 679/44 of 6 Nov. 1944, NASM MF R 186.

641. BMW EZA/2 trip report No. 1084/44 by Ing. A. Reichel of 24 Nov. 1944, NASM MF R 746 and R 3965 F 179.

642. BMW trip report No. 1110/44 by Ing. A. Reichel of 8 Dec. 1944, NASM MF R 3256 F 1390.

643. ATG Quality Control trip report of 6 Dec. 1944, NASM MF R 193.

644. This obviously refers to fuselages only, as deliveries were made to other plants, and not directly to the *Luftwaffe*.

645. ATG Plant 2 management memo of 5 Dec. 1944, "Difficulties in production of the Ju 388 L and Ju 388 J caused by missing parts," BA-MA RL 3/4315.

646. Memo from management of ATG Plant 3 to ATG Dir. Schade of 12 Dec. 1944, BA-MA RL 3 / 4315.

647. ATG file memo, SO Plant Management to Dir. Schade and TH/L of 11 Dec. 1944, BA-MA RL 3 / 4315.

648. ATG Welei 2 memo of 21 Dec. 1944, BA-MA RL 3 / 1277 (Welei = plant management).

649. Pressurized cockpits were built in Eilenburg by the C. Fleischer company, a dispersed operation of FSD.

650. ATG TH, memo of 23 Dec. 1944, BA-MA RL 3 / 4315.

651. ATG TH, memo concerning delivery plan of 2 Jan. 1945, BA-MA RL 3 / 4315.

652. ATG TH Production Preparation, memo of 4 Jan. 1945, BA-MA RL 3 / 4315.

653. ATG TH, memo concerning construction changes of 4 Jan. 1945, BA-MA RL 3 / 4315.

654. ATG Welei 3, memo of 11 Jan. 1945, BA-MA RL 3 / 4315.

655. ATG TH, memo of 14 Jan. 1945 concerning a letter from Dr. Caspari (WFG operations manager) to Dir. Schade (ATG), BA-MA RL 3 / 4315.

656. ATG TH Welei 2, memo of 16 Jan. 1945, BA-MA RL 3 / 4315.

657. ATG memo Welei SO of 16 Jan. 1945, BA-MA RL 3 / 4315.

658. Armaments Staff brief report of 1 Jan. 1945, see TLR War Diary. Saur complained about inadequate deliveries to the users (in part due to weather) in spite of high production and test flight numbers. He therefore demanded that the accumulations at repair and modification facilities, especially of high performance aircraft, be eliminated by the end of January, BA-MA RL 3/4237.

659. ATG memo Welei 4 of Jan. 1944, BA-MA RL 3 / 4315.

660. ATG memo Welei 3 *Herr* Kuckertz to Dir. Schade of 19 Jan. 1944, BA-MA RL 3 / 4315.

661. ATG memo Welei 3 *Herr* Kuckertz to Dir. Schade of 19 Jan. 1944, BA-MA RL 3 / 4315.

662. In spite of the difficulties, ATG tried to secure future production. On 22 Jan. 1945 ATG asked the Elbag to obtain from Production Committee F1 a procurement directive for ATG for the construction of fuselages and vertical stabilizers for the Ju 388 L and J, including tail turret mounts and the test flying of completed aircraft. ATG memo of 22 Jan. 1945, BA-MA RL 3 / 4315.

663. ATG memo about a tour of Plant 3 of 21 Jan. 1944, BA-MA RL 3 / 4315.

664. ATG brief memo of 26 Jan. 1944, BA-MA RL 3 / 4315.

665. ATG brief memo of 29 Jan. 1944, BA-MA RL 3 / 4315.

666. JFM memo FlA. 692/44 of 13 Nov. 1944, NASM MF R 186.

667. ATG Welei SO two short memos of 13 Feb. 1945, BA-MA RL 3 / 4315.

668. ATG Welei 3 memo of 6 Feb. 1945, BA-MA RL 3 / 4315.

669. ATG Welei 2 short memo of 7 Feb. 1945, BA-MA RL 3 / 4315. These were L-1 fuselages 7127, 7129, 7130, 7131, 7133, 7135-71241, and J-1 fuselages 1070, 1072-1085.

670. ATG Welei 4, memo of 14 Feb. 1945, BA-MA RL 3 / 4315; rearward facing antennas (FuG 220 D/R) were installed on 56 wings, with 36 man-hours required for each wing. At Kölleda two workers from Plant 4 installed antennas on 18 wings. Beginning with the 131st to 140th wings, provisions for the antennas were made in ATG production, and beginning with the 150th they were installed on the production line. Four wings were delivered from SFW/Schkeuditz for conversion. The estimated modification expenditure was 75 to 80 man-hours per wing.

671. ATG Welei 2 short memo of 22 Feb. 1945, BA-MA RL 3 / 4315.

672. ATG Welei SO memo of 24 Feb. 1945, BA-MA RL 3 / 4315.

673. ATG Jig Bureau, status report to Dir. Schade of 28 Feb. 1945, BA-MA RL 3 / 4315.

674. See alphabetical listing of construction inspection offices, status 15 Nov. 1944, BA-MA RL 3/1795.

675. Plan collection 009778 ff, Special Collections DMM.

676. Memo from the BAL at Junkers Dessau to the BAL at Weser Flugzeugbau, Berlin Tempelhof, of 22 June 1944, EADS Unternehmensarchiv Ottobrunn.

677. File memo No. 13/44 concerning FTH-Pl 44 216, page 2, dated 11 May 1944, EADS Unternehmensarchiv Ottobrunn.

678. GL/C Aircraft Delivery Plan 226 secret command matter No. 1486/44, p. 4 of 25, BA-MA RL 3/1040. The equivalent "Ju 188 A-3 (LT) cuto" was to be produced by SFW, with 157 examples in the second half of 1944.

679. The "T" stood for Tempelhof, the second final assembly site after Bremen, in which WFG built other types, including the Ju 87, in the airfield hangars.

680. GL/C Aircraft Delivery Plan 226, version of 15 July 1944, BA-MA RL 3/ 1041.

681. Letter from the BAL at Weser to the Luftwaffe High Command of 11 Dec. 1944, letter no. 116/BAL/Gr/12.44, EADS Unternehmensarchiv Ottobrunn.

682. Memo of 21 Dec. 1944, EADS Unternehmensarchiv Ottobrunn.

683. Report No. 43/45 on the state of BMW 801 TJ-0 modification program of 19 Jan. 1945, BMW Unternehmensarchiv Mobile Tradition, Munich.

684. TLR War Diary, weekly report 5-11 Dec. 1944, Charles Collection.

685. See Hartmund Pophanken, *Gründung und Ausbau der 'Weser'-Flugzeugbau GmbH 1933 bis 1939-Unternehmerisches Entscheidungshandeln im Kontext der nationalsozialistischen Luftrüstung*, Verlag Haunschild GmbH Bremen, p 113 and 160.

686. Primarily through the installation of the oxygen tanks in the starboard wing instead of in the fuselage, as in the Ju 88 and 188.

687. Series *Luftfahrtgeschichte im Land Sachsen-Anhalt, Heft 9, "Die Siebel-Flugzeugwerke Halle (1934-1946),"* H.J. Ebert, U. Mahn, H.d. Tack, Interessensgemeinschaft Luftfahrtgeschichte im Luftsportverband Sachsen-Anhalt, p. 39.

688. JFM memo Kobü-Rh 3661 of 17 Oct. 1944, DMM & NASM MF R 2747 F 318.

689. Trip report on "Visit to Welzow Modification Detachment 388 L-1, Welzow, on 5 Dec. 1944" of 6 Dec. 1944, NASM MF R 193.

690. JFM memo Kobü-Rh 6/45 of 8 Jan. 1945, NASM MF R 2196 F 404.

691. Letter from BAL at Siebel to BAL at Junkers, letter no. 15/1.1945, of 12 Jan. 1945, EADS Unternehmensarchiv Ottobrunn.

692. Report No. 43/45 on the state of BMW 801 TJ-0 modification program of 19 Jan. 1945, BMW Unternehmensarchiv Mobile Tradition, Munich.

693. These wings were then to be used in production of the aircraft by ATG or Junkers. This information is confirmed by Target No. 19/79 in CIOS Evaluation Report No. 149 of 27 June 1945, Bochum University Library. Also confirmed is the switch from the Ju 388 to the Ta 152. The report contains the remark that production of the latter achieved a rate of one wing every 90 minutes.

694. See series *Luftfahrtgeschichte im Land Sachsen-Anhalt, Heft 3,-Dokumentation der 90-jährigen Geschichte der Luftfahrt in Sachsen-Anhalt und des Lufttransportes in der Region Halle (Saale)*, p. 39.

695. Definitely 963 Ju 88 S-3, according to GL/C Delivery Plan 226 version of 15 May 1944, BA-MA RL 3/ 1041.

696. Fragments of the monthly report "Chapter 15-Technical Report by the Henschel Flugzeugwerke GmbH to the RLM," Griehl Collection.

697. JFM monthly report-JFM Aero-Engine Operation January-March 1944, DMM Collection.

698. Stenographic report on the Fighter Staff meeting of 3 July 1944, BA-MA RL 3/9 fol. 192 f.

699. Trip report of 28 July 1944, meeting in the RLM GL/C concerning production of the Jumo 213 and 222, and equipping the He 219 with them, Griehl Collection.

700. Record of RLM meeting on "Exhaust turbine for Jumo 222 and distribution of the rescued Jumo 222 engines to the various airframe companies," on 2 Aug. 1944, Aero-Engine Main Plant Dessau, OMW, dated 7 Aug. 1944, NASM MF R 2825 F 173.

701. Statement of development tasks to 1945 of October 1945, Griehl Collection.

702. *Opel Militärfahrzeuge 1906-1956*, Eckhert Bartels, Podzun Pallas 1994, p. 127 ff.

703. See Momsen/Grieger, *Das Volkswagenwerk und seine Arbeiter im Dritten Reich*, ECON-Verlag 1996, p. 383.

704. See Momsen/Grieger, p. 347 f.

705. *Wilhelm-Heinrich Evers-Teil 2* by Tim Plorin, in Jet & Prop 2/03, p. 31. The second "chief designer of the Ju 88" was the Austrian A. A. Gassner, a friend of Evers. Both were hired by Koppenberg in 1935.

706. See Momsen/Grieger, p. 619 f., in respect to FTH Overall Planning Draft 194 of 1 July 1944, BA-MA RL 3/1096.

707. JFM-FTH Overall Planning Program 180, FTH Overall Plan 44 109, dated 19 March 1944, BA-MA RL 3 / 1048.

708. GL/C Aircraft Delivery Plan 226 secret command matter No. 1486/44, p. 4 of 25, BA-MA RL 3/1040. The equivalent "Ju 188 A-3 (LT) auto" was to be produced by SFW, with 157 examples in the second half of 1944.

709. GL/C Aircraft Delivery Plan 226, BA-MA RL 3/1041.

710. See Momsen/Grieger, p. 854 f. Ibidem, quoted from "Armaments Staff brief report Aircraft Construction and Repair of 16 Nov. 1944," p. 1 BA-MA RL 3 3034a, p. 3, 11, 32, 34 and 48 between Sept 44 and Jan. 45.

711. See Momsen/Grieger, p. 857.

712. See Hirschfeld et al., *Karte zur Luftfahtgeschichte*.

713. Report on official trip by the BAL Dessau to the Naether company, Zeitz, on 24 Jan. 1944, EADS Unternehmensarchiv Ottobrunn.

714. JFM file memo 11/44 "Ju 388 Pressurized Cockpit Production by the Naether company of Zeitz," of 22 Feb. 1944, EADS Unternehmensarchiv Ottobrunn.

715. Vernaleken Collection.

716. The "Möbelfabrik C. Fleischer & Sohn" was a dispersed operation of the main plant in Dessau located in Eilenburg.

717. JFM meeting record Belei No. 402 of 11 Aug. 1944, NASM MF R 2196 F 404.

718. Weekly report of 3 July 1944, NASM MF R 186.

719. JFM memo Kobü-Vers. 3c/1500 of 24 Oct. 1944, NASM MF R 2196 F 404.

720. 1900 m^2 underground production site, *Decknamenverzeichnis deutscher unterirdischer Bauten*, Hans Walter Wichert, Druckerei Joh. Schulte ISBN 3-9803271-4-0.

721. CIOS File Vo. XXV-53, *German Aircraft Industry, Bremen-Hamburg Area*, NARA.

722. At least one more test flight was usually required for aircraft built on the assembly line, in order to eliminate problems discovered during the initial test flight. In spite of Allied bombing, Bernburg reached its production peak in 1944 (after introduction of a 3-shift operation), with a peak rate of 6-6 aircraft per day. Beginning in 1943 Bernburg also built major components for the replica firms ATG, Siebel (wings and fuselages), Volkswagen/Wolfsburg (also wings and fuselage), and Henschel/Schönefeld (see *Auf der Suche nach Wahrheit*, p. 32 f and 46.).

723. FT memo No. 38/44 "Trip Report on Meeting at Ifa Dessau on 22 Sept. 1944," Special Collection Deutsches Museum, NASM MF R 2733 F 360.

724. BMW report 1103/44 on BMW 801 TJ-0 modification program in Welzow (modification parts procurement situation) of 6 Dec. 1944, NASM R 2648 F 752.

725. BMW trip report 1117/44 on the state of 8801 TJ modification program in Ju 388 of 13 Dec. 1944, NASM MF R 2786 F 90.

726. Trip report on "Visit to 388 L-1 Modification Detachment, Welzow on 5 Dec. 1944" of 6 Dec. 1944, NASM MF R 193.

727. JFM memo 1517/44 of 23 Dec. 1944, EADS Unternehmensarchiv Ottobrunn.

728. BMW interim report 43/45 on status of modification program dated 19 Jan. 1945, holdings of BMW Historical Archiv "Mobile Tradition," Munich.

729. Aircraft allocation in February 1945, Gen.Qu. (6. Abt III C) of 2 March 1945, Archiv Griehl.

730. Detailed daily records of August to December 1944, BA-MA RL 2 III 619, and January to April 1945, BA-MA RL 2 III 620.

731. Aircraft allocations (monthly tabular overviews) by the *Oberkommando der Luftwaffe, Generalquartiermeister*, BA-MA RL 2 III 621.

732. Kuhn logbook, flight no. 1595.

733. JFM memo FTV-Au 568/44 of 10 May 1944, NASM MF R 193.

734. JFM memo No. FTVZ 1627 on type testing of the Ju 388 K-0 and L-0 of 19 Feb. 1944, EADS Unternehmensarchiv Ottobrunn.

735. JFM memo Kobü-Rh 3155 of 6 May 1944, NASM MF R 2196 F 404.

736. B. Nr. 603/44 g.Kdos., Program 12 and 13 June 1944 (visit to Rechlin), BA-MA RL 36/41.

737. Telex "Report Ju 388 from 16-30 June 1944" of 1 July 1944, EADS Unternehmensarchiv Ottobrunn.

738. Letter from JFM, Department FTV-Contract Processing, to the *E-Stelle* Rechlin E3 of 3 Aug. 1944, EADS Unternehmensarchiv Ottobrunn.

739. Telex from Junkers Rechlin to IFA Dessau of 6 July 1944, NASM MF R 193.

740. Findings and Inspection Report Ze 355/44 (Ju 388 WNr. 300 005) of 6 July 1944, NASM MF R 193.

741. JFM memo Kobü-Rh 3452 of 7 Aug. 1944, NASM MF R 193 (reaction to letter from the *E-Stelle* of 14 July 1944).

742. *E-Stelle* Rechlin letter of 14 July 1944, NASM MF R 193.

743. *E-Stelle* Rechlin E 3 a 2 letter of 11 Aug. 1944, NASM MF R 193.

744. *Flugerprobungsstellen bis 1945*, Bernard & Graefe Verlag, p. 288.

745. Technical field service detachment with *Erprobungskommando 388* memo no. 2/44 of 31 Aug. 1944, NASM MF R 193.

746. JFM memo FlA. 500/44 of 22 July 1944, NASM MF R 2725 F 1096.

747. File memo Merseburg construction site of 20 July 1944, EADS Unternehmensarchiv Ottobrunn.

748. *E-Stelle* Rechlin weekly report for period 23-29 July 1944, BA-MA RL 36/86. According to Kuhn's logbook (flight no. 1643), the ferry flight had in fact taken place on 25 July 1944.

749. Findings and Inspection Report 443/44 of the *E-Stelle* Rechlin inspection group of 29 July 1944, NASM MF R 193.

750. Memo Fp.-Baumuster 673/44 of 14 Sept. 1944, NASM MF 2725 F 1159.

751. *E-Stelle* Rechlin weekly report for period 30 July-8 Aug. 1944 (5 Aug. 1944), BA-MA RL 36/86.

751a. Logbook, flight no. 336, Regel Collection.

752. See Lerche, *Testpilot auf Beuteflugzeugen*, p. 149 and p. 186 (appendix).

753. Findings and Inspection Report No. 2/44 by the technical field service office with *Erprobungskommando 388* of 31 Aug. 1944, NASM MF R 193.

754. Memo no. 2/44

755. See, for example, http://www.feldgrau.com/hkcwo.html.

756. FTA Ju 388 field office memo No. 1/44 from the of 1 Aug. 1944, NASM MF R 193.

757. Findings and Inspection Report Ze 355/44 (Ju 388 WNr. 300 005) of 6 July 1944, NASM MF R 193.

758. Telex from "Junkers Rechlin" to "Ifa Dessau" re. "Ju 388 / 300 009" of 10 July 1944, NASM MF R 193.

759. *E-Stelle* Rechlin (Inspection Group/Eh/Stw) of 29 July 1944.

760. JFM memo Kobü-Rh 3488 of 21 Aug. 1944, NASM MF R 193.

761. See JFM memo Kobü-Rh 3596, NASM MF R 193, beginning with the first production machine the valves were to be installed on the starboard fuselage wall.

762. FTA Office with *Erprobungskommando 388*, memo of 31 Aug. 1944, NASM MF R 193.

763. JFM memo Kobü-Rh 3472 of 14 Aug. 1944, NASM MF R 193.

764. *E-Stelle* Rechlin weekly report for the period 6-12 Aug. 1944 (12 Aug. 1944), BA-MA RL 36/86.

765. *E-Stelle* Rechlin weekly report for the period 13-19 Aug. 1944 (19 Aug. 1944), BA-MA RL 36/86.

766. If one can assume from this that "005" was not the V5, which had the *Werknummer* 500 005, rather aircraft 300 005 (L-0). This appears justified, because Preuschen always wrote of the V5, and never mentioned the *Werknummer*.

767. FTA Office with *Erprobungskommando 388*, memo no 2/44 of 31 Aug. 1944, NASM MF R 193.

768. BMW EZA/2 brief report No. 29 on prototype installation and testing by field offices of 1944, NASM MF R 2419 F 534.

769. *E-Stelle* Rechlin weekly report for the period 3-10 Sept. 1944 (10 Sept. 1944), BA-MA RL 36/86.

770. *E-Stelle* Rechlin weekly report for the period 10-16 Sept. 1944 (16 Sept. 1944), BA-MA RL 36/86.

771. Technical Intelligence Report A-472 of 10 June 1945, NASM MF R 488.

772. WNr. 340 083 was incorrectly identified as a Ju 88, however, see BA-MA RL 2 III 780.

773. Telex from *E-Stelle* Rechlin to Junkers of 9 Sept. 1944, EADS Unternehmensarchiv Ottobrunn.

774. Memo no. 3/44 of 2 Oct. 1944, NASM MF R 193.

775. JFM FT memo No. 38/44 of 22 Sept. 1944, NASM MF R 2733 F 360.

776. Stenographic record of the Armaments Staff meeting on 2 Aug. 1944, BA-MA RL 3/9, fol. 153.

777. BMW EZA/2 brief report No. 33 on prototype installation and testing by field offices of 14 Oct. 1944, NASM MF R 2411 F 835.

778. FTA Office with *Erprobungskommando 388*, memo no 4/44 of 1 Nov. 1944, NASM MF R 193.

779. Letter from *Erprobungskommando 388*, Abt. Ib, to the type supervisor, Hptm. Stahl of the KdE, of 1 Oct. 1944, NASM MF R 2774 F 51.

780. *Erprobungskommando 388*, Interim Testing Report No. 8 of 4 Oct. 1944, EADS Unternehmensarchiv Ottobrunn.

781. *Erprobungskommando 388*, Interim Testing Report No. 15 of 7 Oct. 1944, EADS Unternehmensarchiv Ottobrunn.

782. *Erprobungskommando 388*, Interim Testing Report No. 12 of 6 Oct. 1944, EADS Unternehmensarchiv Ottobrunn.

783. BMW EZA/2 brief report No. 37 on prototype installation and testing by field offices (5-11 Nov. 1944), undated, ca. mid-Nov. 1944, NASM MF R 2823 F 872.

784. JFM memo FlA. 523/44 of 1 Aug. 1944, NASM MF R 186.

785. JFM memo FlA. 605/44 of 20 Sept. 1944, NASM MF R 186.

786. Letter concerning icing flight by Ju 388/008 PG+YH of 14 Oct. 1944, NASM MF R 2762 F 831.

787. The document contains a contradiction, as later it says of the same aircraft: "RT+KA speed at 11 and 12 km at combat power still to be determined."

788. *E-Stelle* Rechlin weekly report for the period 23-29 Oct. 1944 (20 Oct. 1944), BA-MA RL 36/86.

789. *E-Stelle* Rechlin weekly report for the period 29 Oct.-4 Nov. 1944 (4 Nov. 1944), BA-MA RL 36/86.

790. Copy of a telegram of 24 Nov. 1944, NASM R 193.

791. JFM memo Kobü-Rh 3807 of 27 Nov. 1944, NASM MF R 2196 F 404.

792. *E-Stelle* Rechlin weekly report for the period 5-12 Nov. 1944 (12 Nov. 1944), BA-MA RL 36/86.

793. Trip report by ATG quality control of 6 Dec. 1944, NASM MF R 193.

794. BMW EZA/2 brief report No. 37 on prototype installation and testing by field offices (5-11 Nov. 1944), undated, ca. mid-Nov. 1944, NASM MF R 2823 F 872.

795. FTA office with *Erprobungskommando 388*, memo no. 1/45 of 2 Jan. 1945, NASM MF R 193.

796. JFM file memo FTH-FN 104/44 of 22 Nov. 1944, DMM and NASM MF R 2747 F 683.

797. *E-Stelle* Rechlin weekly report for the period 11-18 Nov. 1944 (18 Nov. 1944), BA-MA RL 36/86.

798. *E-Stelle* Rechlin weekly report for the period 18-25 Nov. 1944 (25 Nov. 1944), BA-MA RL 36/86.

799. BMW EZA/2 brief report No. 39 on prototype installation and testing by field offices in the period 19-25 Nov. 1944 of 30 Nov. 1944, NASM MF R 2640 F 286.

800. *E-Stelle* Rechlin weekly report for the period 26 Nov. to 2 Dec. 1944 (2 Dec. 1944), BA-MA RL 36/86.

801. BMW EZA/2 brief report No. 40 on prototype installation and testing by field offices in the period 26 Nov.-2 Dec. 1944 of 6 Dec. 1944, NASM MF R 2640 F 286.

802. JFM memo Kobü-Rh 3771 of 17 Nov. 1944, NASM MF R 2794 F 325.

803. Letter from the BAL in Dessau to the BAL with Askania in Berlin of 19 Jan. 1945, EADS Unternehmensarchiv Ottobrunn.

804. Telex from the *E-Stelle* Rechlin to the BAL at ATG and JFM of 25 Nov. 1944, EADS Unternehmensarchiv Ottobrunn.

805. *E-Stelle* Rechlin weekly report for the period 26 Nov.-2 Dec. 1944 (2 Dec. 1944), BA-MA RL 36/86.

806. *E-Stelle* Rechlin weekly report for the period 2-9 Dec. 1944 (9 Dec. 1944), BA-MA RL 36/86.

807. *E-Stelle* Rechlin weekly report for the period 8-15 Dec. 1944 (17 Dec. 1944), BA-MA RL 36/86.

808. According to *f*V 6936/44.

809. Telex from Dr. Stahl to the BAL at Junkers Dessau of 23 Dec. 1944, EADS Unternehmensarchiv Ottobrunn.

810. Plexiglass would be melted by the resulting heat.

811. *E-Stelle* Rechlin weekly report for the period 14-20 Jan. 1945 (27 Jan. 1945), BA-MA RL 3/2573 p. 93.

812. *E-Stelle* Rechlin weekly report for the period 17-23 Dec. 1944, BA-MA RL 36/86.

813. *E-Stelle* Rechlin weekly report for the period 10-17 Feb. 1945 of 18 Feb. 1945, BA-MA RL 3/2573 p. 67.

814. Dr. Stahl's telex to the BAL at JFA in Dessau of 15 Jan. 1945, letter no. 436/45 secret command matter, EADS Unternehmensarchiv Ottobrunn.

815. See Fritz Trenkle, *Bordfunkgeräte*, Die deutsche Luftfahrt, Band 7, p. 178 ff.

816. JFM file memo FTV-Au 884/44 of 31 July 1944, NASM MF R 2196 F 404.

817. Technical field service detachment with *Erprobungskommando 388* memo no. 4/44 of 1 Nov. 1944, NASM MF R 193.

818. Undated trip report, signed "Ackermann," NASM MF R 2196 F 404.

819. See the *E-Stelle* Rechlin's testing plan in BA-MA RL 36/88.

820. Letter from EK 388 to Testing Stations Command (KdE) of 1 Oct. 1944, NASM MF R 2274 F 51.

821. Technical Report No. F-TR-1138-ND of the Air Material Command, Wright Field.

822. JFM memo Kobü-Ew 3699 of 3 Feb 1945, EADS Unternehmensarchiv Ottobrunn.

823. JFM memo Kobü-Rh 58/45 of 6 Feb 1945, EADS Unternehmensarchiv Ottobrunn.

824. Letter from K. Bonow to G. Aders of 28 Feb. 1974, Archiv Handig via E. Sommer.

825. Huber logbook, which speaks of a "deadline," and that the aircraft "was also" in Rechlin, or was flown by him there.

826. Order by the Genst. Gen. Qu. 2 Abt. (II B) of 14 Feb. 1945, Az. Lib16.24, Nr. 727/45g.Kdos., Archiv Griehl.

827. *E-Stelle* Rechlin weekly report for the period 10-17 Feb. 1945 (18 Feb. 1945), BA-MA RL 3/2573 p. 64.

828. Testing Stations Command, letter no. 535/45 secret command matter of 8 March 1945, Archiv Griehl.

829. Chef der TLR, Adj./IIb Nr. 1970/45 geh. of 8 March 1945, Archiv Griehl.

830. *E-Stelle* Rechlin weekly report of 3 March 1945, BA-MA RL 3/2573 p. 50.

831. JFM memo Kobü-Waffen 501, "Ju 388 K-0 Rearward-Firing Armament 2 x MG 131" of 3 July 1944, EADS Unternehmensarchiv Ottobrunn.

832. JFM memo FTV-A 814/44 "Testing of the Ju 388 K-0/L-0 Defensive Positions" of 7 July 1944, EADS Unternehmensarchiv Ottobrunn.

833. JFM AV Merseburg construction facility management Hr. Brauer "Delivery of the Ju 388 L-0 and K-0 Aircraft," dated 20 July 1944, EADS Unternehmensarchiv Ottobrunn.

834. JFM memo Kobü-Elt 368/44 of 7 Aug. 1944, NASM MF R 2762 F 803.

835. *Flugerprobungsstellen bis 1945*, p. 337.

836. Priority test report No. 34/44 by the *E-Stelle* Tarnewitz for the period 20-26 Aug. 1944, Archiv Jens.

837. Priority test report No. 35/44 by the *E-Stelle* Tarnewitz for the period 27 Aug.-2 Sept. 1944, Archiv Jens.

838. Priority test report No. 36/44 by the *E-Stelle* Tarnewitz for the period 3-9 Sept. 1944, Archiv Jens.

839. Priority test report No. 37/44 by the *E-Stelle* Tarnewitz for the period 10-16 Sept. 1944, Archiv Jens.

840. Priority test report No. 39/44 by the *E-Stelle* Tarnewitz for the period 24-29 Sept. 1944, Archiv Jens.

841. Priority test report No. 40/44 by the *E-Stelle* Tarnewitz for the period 1-7 Oct. 1944, Archiv Jens.

842. Priority test report No. 41/44 by the *E-Stelle* Tarnewitz for the period 8-14 Oct. 1944, Archiv Jens.

843. Gatzemeier logbook, Carlsen Collection.

844. Priority test report No. 43/44 by the *E-Stelle* Tarnewitz for the period 22-28 Oct. 1944, Archiv Jens.

845. Priority test report No. 44/44 by the *E-Stelle* Tarnewitz for the period 29 Oct.-4 Nov. 1944, Archiv Jens.

846. Priority test report No. 45/44 by the *E-Stelle* Tarnewitz for the period 5-11 Nov. 1944, Archiv Jens.

847. Priority test report No. 46/44 by the *E-Stelle* Tarnewitz for the period 12-18 Nov. 1944, Archiv Jens.

848. Priority test report No. 47/44 by the *E-Stelle* Tarnewitz for the period 19-25 Nov. 1944, Archiv Jens.

849. Priority test report No. 48/44 by the *E-Stelle* Tarnewitz for the period 26 Nov.-2 Dec. 1944, Archiv Jens.

850. Priority test report No. 51/44 by the *E-Stelle* Tarnewitz for the period 17-21 Dec. 1944, Archiv Jens.

851. Priority test report No. 7/45 by the *E-Stelle* Tarnewitz for the period 11-14 Feb. 1945, Archiv Jens.

852. *Flugerprobungsstellen bis 1945*, p. 206, 209, 213.

853. JFM meeting minutes Belei No. 402 of 11 Aug. 1944, NASM MF R 2196 F 404.

854. The work was timed without the knowledge of the personnel.

855. Jumo development department monthly report for October 1944, dated 2 Nov. 1944, NASM MF R 2853 F 190. There are some clear differences between the dates and flight times in the monthly report, and those in the report by the OKL Experimental Unit, which forms the basis of the entire chapter, cited in footnote 861. The problems described are the same, however, but the monthly report contains additional information about less serious problems. Where doubt existed, the dates from the meticulously compiled account of flight results by the OKL Experimental Unit were used or quoted.

856. Jumo development department monthly report for November 1944, dated 4 Dec. 1944, NASM MF R 2853 F 204. Concerning the reliability of dates and flight times see footnote 855.

857. JFM memo Kobü-Rh 3690 of 23 Oct. 1944, NASM MF R 193.

858. JFM memo Kobü-Rh 3690, loc. cit.

859. A radio equipped sentry attached to an anti-aircraft battery who witnessed the accident, stated that it occurred at 1150 hours.

860. "Report on the accident involving 8-388 L-0/300 295/V32, test bed for Jumo 213 E engines at Oranienburg," Kobü-Flugabteilung FIA 16/45, NASM MF R 186.

861. Brief report Br. B. No. 28/45 secret by the 3./Versuchsverband des O.K.L. of 27 Feb. 1945, NASM MF R 187.

862. myo = **M**aschine **ü**ber **O**rt (aircraft overhead), radio warning of approaching enemy aircraft. As there is no letter "ü" in Morse code, the letter "y" was used instead.

863. BMW EZA/2 brief report No. 39 on prototype installation and testing by field offices of 19-25 Nov. 1944, NASM MF R2279 F1025.

864. Jumo Department T.A.-Motortriebwerke, monthly report for February 1945, undated, DMM.

865. Stenographic record of the Fighter Staff meeting of 3 July 1944, BA-MA RL 3/9 fol. 192 f.

866. Like Egon, Oboe used crossing beams from two radio stations on the ground, while H2S (for "Home Sweet Home") was a ground mapping radar developed from an air intercept radar. See Walter Boyne, *Clash of Wings-World War II in the Air*, Simon & Schuster 1994, p. 296.

867. In this regard it is noteworthy that, while isolated raids continued to be flown against England into May 1944 as part of "Operation Steinbock," until the increasingly lethal flak and fighter defenses over the British Isles made losses prohibitive, no notice was obviously taken of the preparations for the invasion of Normandy. See *Pathfinder-Luftwaffe Pathfinder Operations over Britain*, 1940-44, Ken Wakefield, Tempus Publishing Inc. 1999, p. 216.

868 See *Pathfinder-Luftwaffe Pathfinder Operations over Britain*, 1940-44, Ken Wakefield, Tempus Publishing Inc. 1999, p. 222.

869. Overview "Formation of the New Bombing Air Force as per Delivery Plan 226/2" of 8 July 1944, Griehl Collection. The Do 335 was to have equipped just four Gruppen of KG 2, the Ar 234 a total of 14 Gruppen of KG 51, 54, and 76. Two Gruppen were also to have been

formed on the Ju 287. The Me 262 was to have been replaced by the Ar 234 with the exception of two Gruppen.

870. *Die deutschen Kampfflugzeuge im Einsatz*, Dressel/Griehl, p. 158.

871. File memo concerning a meeting at Gen.Qu. 6. Abt. dated 10 June 1944, draft 12 June 1944 *General der Aufklärungsflieger*, Griehl Collection.

872. See overview "Reconnaissance Situation -Tactical/Strategic Reconnaissance Aircraft" dated 15 Jan. 1945, with performance figures, and the quantities to be delivered from the aircraft delivery program for the types (tactical reconnaissance aircraft) 8-262 A-4, 8-109 G-10/R2, 152 E-1 and H-10, 8-228, (strategic reconnaissance aircraft) 8-34 B-1 and C-4, 8-88 T-3, 8-188 D-2, 8-388 L-1, 8-335 A-1, 8-635, 8-211, BA-MA RL 3 1277 Page 33.

873. 10-day reports (fragments) by the reconnaissance units for autumn 1944, Archiv Griehl.

874. Including, among others, 2.(F)/22, 4.(F)/121, 1.-3.(F)/123, 1.(F)/100, 1.(F)/33, see Jürgen Zapf, *Flugplätze der Luftwaffe 1934-1945-und was davon übrig blieb*, Vol. 1, Berlin & Brandenburg, VDM 2001, p. 188, p. 332 f.

875. Griehl, "Ju 88-Star of the Luftwaffe," p. 73.

876. *Luftwaffe* SIG homepage www.luftwaffe.no/SIG/OOB/Nav44-1.html (as of October 2003).

877. JFM memo Kobü-Rh 3624 of 5 Oct. 1944, NASM MF R 2196 F 404.

878. www.ww2.dk/air/recon/regaufklg.html, as of October 2003, based on material in BA-MA RL 2 III.

879. Photo in J.R. Smith/E.J. Creek, *Dornier Do 335 Arrow*, Monogram Aviation Publications, p. 100.

880. Letter from Olt. Erich Sommer, pilot with VersVerb OKL.

881. The description is based on a four part article in FLUGZEUG 2/1999 to 5/1999 by Manfred Griehl, the first detailed representation of this type.

882. Heinz Borsdorff logbook, flight no. 3171.

883. Fighter Staff meetings between 6 and 10 June 1944, Milch reference file, IWM, London.

884. Jumo (JFM engine operation) monthly reports for April-June 1944, Landeshauptarchiv Sachsen-Anhalt, Oranienbaum.

885. JFM Motorenbau main plant Dessau, OMW dated 7 Aug. 1944, minutes of a meeting in the RLM on 2 Aug. 1944 about exhaust turbine for Jumo 222, and distribution of the saved Jumo 222 engines to the various airframe companies, NASM MF R 2825 F 173.

886. Jumo monthly report for January 1945 concerning OMW Test Stand 2's Jumo 222, of 5 Feb. 1945, DMM LR 01078.

887. *E-Stelle* Rechlin weekly report for the period 17-23 Dec. 1944, BA-MA RL 36/86.

888. See "Reconnaissance Aircraft Situation-Tactical/Strategic Reconnaissance Aircraft," dated 15 Jan. 1945, with performance figures and delivery quantities from the aircraft delivery program, BA-MA RL 3 1277 Page 33.

889. Jumo monthly report for January 1945 concerning OMW Test Stand 2's Jumo 222, of 5 Feb. 1945, DMM LR 01078.

890. Aircraft Procurement and Supply War Diary, 15-21 Jan. 1945.

891. Ernst Heinkel Flugzeugwerke AG (EHAG) as of 1 March 1943.

892. In the few known photos of the V1, the only indication of the planned turret positions are gentle steps in the top and bottom fuselage profile.

893. Mainly because the Me 410 failed to live up to expectations. Production was scaled back, making engines available for the "second class" designs Do 217 M/n and He 219.

894. See Hans Jürgen Becker, *Schwere Jäger und Zerstörer der Luftwaffe 1933-1945*, Motorbuch Verlag 1999, p. 151 ff. The first fly-off took place on 19 Jan. 1943, and involved the Ju 88 V56 (S-0), He 219 V1, and Ar 240 V10, ibid. p. 125 f.

895. Confirmed by mockup photos.

896. See Becker, p. 153. The Do 217 was much heavier than the other two types and, as expected, it performed poorly. The Ju 188 E-1, without bomb carriers and defensive armament, was flown by *Oberst* von Lossberg (RLM), while the He 219 was piloted by Major Streib. After speed trials and a mock combat, the Heinkel fighter emerged the clear winner. The He 219 was about 25 kph faster than the Ju 188 and could outturn it, while its rate of climb was almost equal to that of the Junkers aircraft.

897. Report on a demonstration of the Ar 234 in Rechlin on 12 and 13 June 1944 in *Arado Ar 234-der erste Strahlbomber der Welt*, Luftfahrtdokumente LD 21, Karl R. Pawlas 1976, p. 156 ff. Nothing more is known about the Ju 188, except that it flew the same program as the He 219, which was also demonstrated: dive, pullout, and climb with roll.

898. Roland Remp, *Der Heinkel He 219*, Aviatic Verlag 1999, p. 74.

899. Entwicklungsverlauf He 219 *Nachtjäger*, dated 21 Apr. 1944, EHAG, Griehl Collection. The order for 120 pre-production machines (14 per month) went to EHAG Vienna, because it was a "*dedicated development works*," and to avoid disturbing production in Rostock. The first aircraft went to the units in June 1943. Rostock did not receive a "major order" for 50 aircraft per month until mid-1943

900. See Remp, p. 71. During a conference on 3 December 1944 the RLM ordered the Heinkel works in Rostock to switch production from the He 219 to Ju 88 G, while Heinkel South (Vienna/Schwechat) was to produce the Do 335. Neither of these plans was realized, however.

901. Meeting of directors in Jenbach (Tyrol), Prof. Dr. Heinkel, *Generaldirektor* Frydag, *Direktor* Francke et al, 9 Aug. 1943, Griehl Collection.

902. EHAG minutes by Dir. Francke, on 28 Aug. 1943 in Vienna/Schwechat, Griehl Collection.

903. EHAG minutes by Dir. Francke, on 28 Feb. 1943 in Vienna/Schwechat, Griehl Collection.

904. See Becker p. 156. The A-5 version cockpit with "stepped glazing" was envisaged for 50% of the types to be delivered; memo concerning a meeting of 28 July 1944 in the RLM, EHAG/Vienna, Griehl Collection.

905. Fragment of a meeting record of 4 Apr. 1944 at EHAG/Vienna, Griehl Collection.

906. Telex to the K.d.E. Rechlin of 1 June 1944, OKL Gen.Qu. No. 4415/44, BA-MA RL 36/44.

907. Jumo development department monthly report for October 1944, of 2 Nov. 1944, NASM MF R 2853 F 190.

908. Telex from the management of Heinkel-Vienna/Director Schaberger EHW to *Herr* Carlsson, the F3 Special Committee's representative on the Armaments Staff, concerning the He 211 N, dated 4 Sept. 1944, Griehl Collection.

909. Jumo development department monthly report for November 1944, of 4 Dec. 1944, NASM MF R 2853 F 204. Planning already envisaged this in August 1944. 20 production aircraft were to be modified, beginning with three from November and December production. The increased performance achieved with the Jumo 222 E/F with MW 50 was seen as necessary in order to keep pace with the competing Do 435, which was expected to achieve 700 kph at 10000 m. Power plants were to be delivered by Henschel beginning in August. Five prototypes of the He 219 B-1 were also envisaged, and the V19 and V23 were to be equipped with the Jumo 222 B and E/F, and simultaneously "*equipped for operational use*." Record of a meeting on 28 July 1944 in the RLM, EHAG/Vienna, Griehl Collection.

910. Telex from Francke, EHAG technical management, to F3 Armaments Staff dated 21 Dec. 1944, Heinkel Company Archive 1225 (F), Deutsches Museum Munich.

911. Remp, p. 100.

912. Renato Schirer, unpublished manuscript entitled *Chronik der Garnison Zwölfaxing*, Part 2, 1977.

913. Heinkel draft with arguments in favor of the He 219, ca. June 1944, Griehl Collection.

914. Interim report on the Fighter Staff meeting of 12 June 1944, BA-MA RL 3.

915. See Budrass, p. 685 and p. 739.

916. Stenographic record of the Fighter Staff meeting of 9 July 1944, BA-MA RL 3/9, fol. 479 f.

917. Stenographic record of the Fighter Staff meeting of 9 July 1944, BA-MA RL 3/9, fol. 479 f.

918. Another Heinkel memorandum "For possible use with the Reichsmarschall or *Feldmarschall* Milch" with arguments for the He 219, ca. June 1944.

919. *Jahresabschlussgedanken* (Thoughts on the Annual Statement of Accounts-draft) dated 29 Dec. 1944, Heinkel South 56/44, unsigned, Griehl Collection. Also mentioned is Heinkel South's participation in the design of the "Me 264" strategic reconnaissance aircraft, which had not yet been cancelled, and the "Heinkel Twin 335 Project." The latter was to be transferred to JFM because of the urgency attached to Heinkel's "own" He 162 work. Finally, design work for conversion of the Do 335 into a two-seat night-fighter was also being carried out at Vienna-Schwechat; however, construction of a mockup was lagging far behind schedule, casting a poor light on the plant. This competing design was later to be built by EHAG Oranienburg.

920. Letter from Lucht to Tank of 3 Jan. 1945, letter no. 502/45 geh., NASM MF R 8045 F 875.

921. Letter from Tank to Lucht of 22 Jan. 1945, SOE/379/45 geh., NASM MF R 8045 F 875.

922. Daily Aircraft Allocations to the Air Fleets by the Gen.Qu. 6.Abt. III C, BA-MA RL III 620.

923. Report No. 100 on the office chief's meeting on 7 Dec. 1943, Archiv Griehl.

924. BA-MA RL 3/2579.

925. See Fred Gütschow, *Junkers G 38*, p. 140 ff.

926. See Budrass, loc. cit., p. 764.

927. Letter from the Aircraft Sales-Foreign of 21 Aug. 1944, EADS Unternehmensarchiv Ottobrunn.

928. Aircraft Sales-Foreign file memo of 18 Nov. 1944, EADS Unternehmensarchiv Ottobrunn.

929. Letter from the Mitsubishi Shoji Kaisha GmbH of 7 Nov. 1944.

930. Letter from the RLM Wi.-Amt LF 1 V B, Az 66.d.13.-Nr. 6454/44 geh.II.Ang. of 9 Nov. 1944, loc. cit.

931. Letter from the RLM Wi.-Amt LF 1 V B, Az 66.d.13.-Nr. 6454/44 geh.II.Ang. of 9 Nov. 1944, loc. cit.

932. Letter from Junkers to Mitsubishi of 14 Nov. 1944, loc. cit.

933. JFM-Licenses and Patents, file memo 2011/44 of 18 Nov. 1944, EADS Unternehmensarchiv Ottobrunn.

934. Junkers' Berlin office.

935. Memo for Herr Thiergärtner of 27 Nov. 1944, loc. cit.

936. A small town a few kilometers south of Dessau.

937. Letter from the Aircraft Construction Technical Head Office to Herr Thiergärtner of 29 Nov. 1944, loc. cit.

938. Letter from Mitsubishi Shoji Kaisha GmbH to Junkers of 2 Dec. 1944, loc. cit.

939. Letter from Thiergärtner to the RLM, LF 1 VB, z. Hd. of Ministerial Director Müller of 6 Dec. 1944, loc. cit.

940. Letter to FTH-FN of 6 Dec. 1944, loc. cit.

941. Letter from Scheibe to Mitsubishi Shoji Kaisha GmbH of 9 Dec. 1944, loc. cit.

942. Letter from the RLM Economic Office LF 2 II B, Az.66.e.19.11.Nr.8217/44g. of 13 Dec. 1944, loc. cit.

943. Weight and cubic measure of tracing paper for copying task "N," 14 Dec. 1944, loc. cit.

944 JFM-Licenses and Patents file memo of 19 Dec. 1944, loc. cit.

945. Letter from Thiergärtner to FTH-Nachbau of 20 Dec. 1944, loc. cit.

946. Letter from JFM Aircraft Sales-Foreign to Thiergärtner of 27 Dec. 1944, loc. cit.

947. JFM Aircraft Sales-Foreign file memo of 29 Dec. 1944.

948. Letter to the RLM Industry Office, of 8 Jan. 1945, loc. cit.

949. Bemi = Betriebsmittel (working material).

950. Letter from JFM to ATG Maschinenbau GmbH of 2 Jan. 1945 loc. cit.

951. JFN FTH-FN file memo of 2 Jan. 1945, loc. cit.

952. Thiergärtner file memo of 3 Jan. 1945, loc. cit.

953. Letter from the RLM Economic Office RLM LF I 2 II B. Az. 66.d.13.-Nr. 12/45 geh. of 4 Jan. 1945, loc. cit.

954. Items of equipment lists.

955. Letter from Major-General Otani of 10 Jan. 1945.

956. Letter from Reich Group Industry Ztr.15/45g Lu/Qu of 9 Jan. 1945, loc. cit.

957. Letter from the Siebel-Flugzeugwerke to JFM Licenses and Patents of 15 Feb. 1945, loc. cit.

958. FTH-FN file memo 11/45 of 31 Jan. 1945, loc. cit.

959. Letter from JFM Licenses and Patents to Reich Group Industry, Abt. VI-Zentralstelle of 26 Feb. 1945, loc. cit.

960. Letter from Thiergärtner of 9 Feb. 1945, loc. cit.

961. Thiergärtner file memo of 30 Jan. 1945, loc. cit.

962. Budrass, loc. cit., p. 764.

963. *United States Strategic Bombing Survey Aircraft Division Airframes Plant Report No. 1, Junkersflugzeuge und Motorenwerk Dessau*, DMM LR 02459 and 02464.

964. DMM CIOS File No. XXI-5.

965. DMM CIOS Final Report 141.

966. DMM CIOS File No. JFM memo FlA. /44 of 1944, NASM MF R I-36.

967. Concerning this, see reference in *United States Strategic Bombing Survey Aircraft Division Airframes Plant Report No. 1 Junkers Aircraft and Aero-Engine Works Dessau Germany*, 4-18 May 1945, DMM LR 02459.

968. DMM LR 02459 and 02464.

969. Ibid. p. 14.

970. A list of aircraft found in Norway (Griehl Collection) confirms this also, as does the fact that there were no finds in Sweden or Denmark.

971. Horst Götz/Theodor Rosarius file memo of May 1991.

972. See Dimitri Alexeyevich Sobolev, *Deutsche Spuren in der Sowjetischen Luftfahrtgeschichte*, Verlag E.S. Mittler.

973. Jürgen Michels, Jochen Werner, *Die deutsche Luftfahrt, Luftfahrt Ost 1945 -1990*, Bernhard & Graefe Verlag, 1994, p. 23. Originally OTB-1 for Technical Office, later OKB-1 for "Special Design Bureau."

974. Published privately by Dr. Peter Korrell in 1998; a collection of statements by Prof. Dr. Ing. Günther Bock before a committee of Soviet officers in September 1945.

975. FLUGZEUG 5/2000, p. 22, *Junkers Ju EF 126 'Elli"-Eines der letzten Geheimprojekte der Junkers-Werke.*

976. Geust, *Under the Red Star*, Airlife, p. 31.

977. See Geust/Petrov, German Aircraft in the Soviet Union Red Stars # 2, Apali Oy, p. 101 f.

978. Photos and details of this installation are found in Chapter VII, as well as in Alexandrov/Petrov, *Die deutschen Flugzeuge in russischen und sowjetischen Diensten*, Vol. 2, p. 218-221.

979. *First Supplementary Report on Junkers Research*, DMM LR 004934.

980. *BDV Bedienvorschrift Ju 388*, foreword/accompanying text by Dr. Peter Korrell.

981. *Luftwaffe under the Red Star*, Geust, Airlife, p. 126 f.

982. Report from Air 40/179, Public Record Office, Kew, London.

983. Airfield Report No B-143/RAF, probably 30 May 1945, with photo appendix, of 15 June 1945.

984. *Flugerprobungsstellen bis 1945*, Bernhard & Graefe Verlag, (comments regarding Tarnewitz) p. 336.

985. Phil Butler, War Prizes, Midland Counties Publications, 1994, p. 74 f and 94. Founded in 1946, in part to evaluate captured equipment, and with numerous training models for future British designers, the college had an impressive collection. With the destruction of the hangar in which the collection was housed, the last traces of AM 83 disappeared in the late 1960s.

986. Kenneth S. West, *The Captive Luftwaffe*, Putnam. P. 81.

987. Personal communication from Tommy Haworth, 7 June 2000.

988. Obviously *Werknummer* blocks were not always assigned to one aircraft type or production site. Take, for example, the case of the Ju 388 and Me 262, both of which were produced in the 500 0xx block (possibly intentionally, to confuse the enemy?). Furthermore, the Werknummer blocks 340 001-085 and 340 210- 340 360, which were assigned to production of the Focke-Wulf Fw 190 A-7 at Cottbus between December 1943 and March 1944, were allocated twice (see Prien/Rodeike, "Focke-Wulf Jagdflugzeug Fw 190 A, Fw 190 ÇDora' und Ta 152 H," p.238).

989. JFM Belei meeting minutes No. 402 of 11 Aug. 1944, NASM MF R 2196 F 404.

990. JFM memo FlA. 503/44 (amendment) of 4 Aug. 1944, NASM MF R 186.

991. Damage report from Merseburg of 28 Nov. 1944, EADS Unternehmensarchiv Ottobrunn.

992. JFM memo FTAB 95/44 "Tear-Down Inspection on Ju 388 L-0 WNr. 3000008 present location Munich," Archiv Arbeitskreis Bernburg.

993. Fw 190 A-7s were produced in the same block (confirmed, for example, by the loss of WNr. 340 043).

994. JFM memo Kobü-Ew 3699 of 3 Feb. 1945, EADS Unternehmensarchiv Ottobrunn.

995. In contradiction to BMW Brief Report No. 33, which reported 5 hr flying time and installation in WNr. 340 089.

996. ATG SO Plant Management, memo of 20 Dec. 1944, BA-MA RL 3/4315.

997. ATG Overview Plan "Aircraft Pass-Through Ju 388 L-1," ca. 10 Dec. 1944, BA-MA RL 3/4315.

998. ATG Management Plant 3, Kuckertz to Dir. Schade, BA-MA RL 3/4315.

999. The *Werknummer* block was then used for the Me 262.

1000. A photo of a Ju 88 G-6 tail section confirms the WNr. 620 308 (Under the Red Star). III/NJG 2 reported WNr. 620 320 lost in a strafing attack on 24 Feb. 1945, as well as a WNr. 620 320 (Ju 88 G-6) on 3 March 1945. This makes a Ju 388 Werknummer in this serial block highly unlikely.